An Introduction to Circular Economy

Lerwen Liu · Seeram Ramakrishna
Editors

An Introduction to Circular Economy

Springer

Editors
Lerwen Liu
Research, Innovation and Partnership Office
King Mongkut's University of Technology
Thonburi
Bangkok, Thailand

Seeram Ramakrishna 🄳
Mechanical Engineering
National University of Singapore
Singapore, Singapore

ISBN 978-981-15-8512-8   ISBN 978-981-15-8510-4   (eBook)
https://doi.org/10.1007/978-981-15-8510-4

This Springer imprint is published by the registered company Springer Nature Singapore Pte Ltd.
The registered company address is: 152 Beach Road, #21-01/04 Gateway East, Singapore 189721, Singapore

Foreword

The Covid-19 outbreak has exposed the fragility of the current economic model. It has severed various global and complex supply and production chains, contributing to their vulnerability. This crisis has shown again that the current linear take-use-dispose system is failing and that we need to dramatically transform our economic model to progress decisively in the transition towards a truly circular economy, where resources are maintained in the economy for longer, and waste is minimised.

This new economic model needs to be underpinned by the idea of resilience. The circular economy, by changing the way we produce and consume, can create a more performant economic model that prioritises reliability and sustainability. Europe has been working very hard in this direction over the past 5 years, leading the path towards a climate neutral, resilient circular economy.

In this context, the EU has adopted significant measures for a coordinated response to ensure the recovery and at the same put in place its European Green Deal, the flagship of President von der Leyen's Commission. Such measures can also be replicated on a global scale.

The implementation of the European Green Deal and the Circular Economy Action Plan—will be two important pillars of the recovery process. The Circular Economy Action Plan (adopted in March 2020) announces initiatives for the entire life cycle of products, from design and manufacturing to consumption, repair, reuse, recycling in order to bring resources back into the economy. The aim of the action plan is to reduce the EU's consumption footprint and double the circular material use rate in the coming decade while boosting economic growth and job creation.

All sectors can benefit from this approach, investing in large-scale renovation, renewables, clean transport, sustainable products, sustainable food and nature restoration will be even more important than before. This is not only good for our environment, it is also good for our economies because it reduces dependencies by shortening and diversifying supply chains. Finally, there is also a very strong social dimension, with innovation playing a key role in enabling this historic transformation.

I look to the future now with a renewed confidence in human beings and in our capacity to craft a better world. We should use this crisis to reinvent ourselves. We should not waste it. If we want to preserve life in this planet, while at the same time, develop a sustainable and competitive economy which creates jobs, there is no other alternative than going circular.

This amazing book brilliantly covers all these areas arguing convincingly how to promote the circular economy across the board. It illustrates well the challenges we are facing, while at the same time, points out the significant societal, economic and environmental benefits which circularity entails. We have no time to lose, business as usual is not an option, and time is of the essence.

Professors Seeram Ramakrishna and Doctor Lerwen LIU should be praised for their personal and professional dedications leading the reflections on how to develop a more sustainable future for all.

July 2020   Daniel Calleja Crespo
                                          Director-General DG Environment (ENV)
                                                           European Commission
                                                              Brussels, Belgium

Acknowledgement

We are deeply grateful to all the authors of this book for their dedication in sharing their amazing work related to circular economy.

We would like to thank Ms. Sawaros Thongkaew and Ms. Duangkamol Buaban from the STEAM Platform for their valuable infographic work for a number of chapters, especially Chaps. 1, 29 and 30.

Last but not the least, we express our special thanks to Dr. Giulio Manzoni (Microspace) for giving the original sketch of the cover page artwork and Miss Waricha Henchobdee (STEAM Platform) for cover page graphic design.

October 2020
Lerwen Liu
Seeram Ramakrishna

Contents

Introduction and Overview

Lerwen Liu and Seeram Ramakrishna

Abstract This chapter gives an overview of the entire book summarizing all 29 chapters, laying out its structure and linkage of different chapters. This book is purposefully styled as an introductory textbook on circular economy (CE) for the benefit of educators and students of universities. It provides comprehensive knowledge exemplified by practices from policy, education, R&D, innovation, design, production, waste management, business, and financing around the world. The book covers sectors such as agriculture/food, packaging materials, build environment, textile, energy, and mobility to inspire the growth of circular business transformation. It aims to stimulate action among different stakeholders to drive CE transformation. It elaborates critical driving forces of CE including digital technologies; restorative innovations; business opportunities & sustainable business model; financing instruments, regulation & assessment and experiential education programs. It connects a CE transformation for reaching the SDGs2030 and highlights youth leadership and entrepreneurship at all levels in driving the sustainability transformation.

1 Background

This book is written during unprecedented times of recent human history. Three pertinent observations could be drawn from the COVID-19 pandemic. First, clean air, water, food and energy, and hygienic living environment anchored by general healthcare, wellness, mental health, and family support are essential for the survival and sustainability of the human race on planet Earth. In other words, we can only lead healthy lives in a healthy and safe environment. Second, digital technologies including the internet and artificial intelligence (AI) enabled big data analytics, mobile communication devices, cloud-based services enabling point of care, learning

L. Liu (✉)
KMUTT, Bangkok, Thailand
e-mail: lerwen67@gmail.com

S. Ramakrishna
National University of Singapore, Singapore, Singapore

© Springer Nature Singapore Pte Ltd. 2021
L. Liu and S. Ramakrishna (eds.), *An Introduction to Circular Economy*,
https://doi.org/10.1007/978-981-15-8510-4_1

from anywhere and anytime at own pace, and telecommuting have become integral to human society. In other words, the modern society entrenched with digital technologies found them to be necessary in unprecedented times as well as normal times. Third, the modern society is inundated with non-essentials such as travel for leisure, clubbing and entertainment, and window shopping. In other words, the **modern society** consumes far more resources per person when compared to the **pre-modern society**. Modern society is thriving on the abundant supply of materials, energy, and water, and accessible to billions of people around the world. Waste generation commensurate with consumption. Waste is often not adequately recycled, and hence ends up in soil, water, and air environment of planet Earth. In other words, depletion of natural resources and increased pollution of the Earth ecosystem, which in turn affects the health and well-being of human beings. Current ways of modern society are not conducive to ensure sustainability of resources of Earth for the future generations. Hence, the primary objective of circular economy and sustainability efforts is to deliver a **new-modern society** in which the ways of the current society least compromise the needs of future generations. Desired characteristics of the new-modern society encompass the visions of circular economy and sustainability development aimed at protecting the Earth while ensuring improved quality of living and growth. Simply put it is an economic system aimed at eliminating waste and the continual use of resources (https://en.wikipedia.org/wiki/Circular_economy).

Circular Economy is emerging and is evolving rapidly, specially today, when humanity is facing various challenges including climate change, pandemics and environmental devastation, and widening social inequalities. Policymakers, manufacturers and service providers, and consumers are developing Earth-friendly policies, innovating business practices, and changing consumption behavior toward sustainability, respectively. The adoption of emerging technologies and innovative business models are enabling the transformation of a circular and more sustainable economy. A circular and sustainable economy is driven by sustainable consumption and production. Sustainability mindset, action, and behavior of stakeholders in the ecosystem of production and consumption leads to sustainable practices. The most critical driver of an economic transformation is education that shapes the mindset, action, and behavior of all stakeholders including policymakers, investors, researchers, educators, producers, service providers, consumers, and media.

University curriculum are beginning to embrace circular economy and sustainability knowledge in educating future generation of graduates who become the stakeholders of the economic ecosystem. This book is purposefully styled as an introductory textbook on circular economy (CE) for the benefit of educators and students of universities. It provides comprehensive knowledge exemplified by practices from policy, education, R&D, innovation, design, production, waste management, business, and financing around the world. The book covers sectors such as agriculture/food, packaging materials, build environment, textile, energy, and mobility to inspire the growth of circular business transformation. It aims to stimulate action among different stakeholders to drive CE transformation.

It elaborates critical driving forces of CE including digital technologies; restorative innovations; business opportunities and sustainable business model; financing instruments, regulation and assessment, and experiential education programs. It connects a CE transformation for reaching the SDGs2030 and highlights youth leadership and entrepreneurship at all levels in driving the sustainability transformation.

Each chapter of the book (except the first and the last chapter) follows the format of Abstract; Keywords; Learning Objectives; Introduction; detailed coverage of the topic including Concepts/Mechanisms/Methodologies exemplified by Case Studies; Questions and Further Readings as a homework or exercises for students to expand and deepen their learning; and References.

Below capture the key features of the book:

- Addresses Circularity along product value chain and business supply chain with case studies.
- Provides Circularity guidelines including framework, examples, and case studies for policymakers, educators, business leaders, and investors.
- Contains Comprehensive contribution with inclusivity in terms of age (1/3 below 35 years old) and gender (over 40% female) with multidisciplinary background from all five continents.
- Comprises substantial coverage of updated policy, research and innovation, education programs, and business practices on circular economy in the Asian region.
- Includes Life Cycle Assessment and Costing methodology for circular economy practices.
- Presents interconnectivity along the circular value chain and roles of different stakeholders for a circular economy transformation.
- Highlights different driving factors for a circular economy transition including digital technologies, business opportunities and consumer service models, financing, circularity indicators and assessment, policy and regulations, and education.

2 Overview

See Fig. 1.

We design this book to ensure the circularity of the content, starting from the most critical Life Cycle Thinking mindset in Chapter GHEEWALA, zooming in an overview of the macro world of circular city in Chapter KISSER, to mesoworld of industry circular manufacturing ecosystem (Chapter SHI) where industrial symbiosis is practiced (Chapter LA ROSA), and circular supply chain management (Chapter KHOMPATRAPORN) enabling innovative circular business model based on services (Chapter ITKIN). Further zooming in to microworld of products and consumption circularity from Food (Chapter KISSER, Chapter KHOR, Chapter GODOY-FAUDEZ, CHAPTER EMF, Chapter KHOMPATRAPORN, and Chapter CHEN); Materials including plastics (Chapter BALAJI, Chapter MODAK), buildings (Chapter KISSER, Chapter EMF, Chapter HOOSAIN), and textiles (Chapter KEH); Energy (Chapter SEETHARAM on Community Microgrid, Chapter PATIL

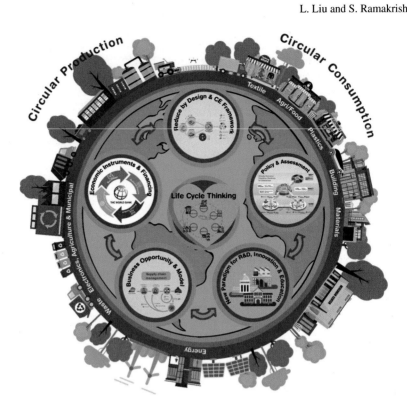

Fig. 1 Infographic of Book Overview Illustrated by Sawaros Thongkaew (STEAM Platform)

on Building Integrated Photovoltaic, and Chapter KHOMPATRAPORN on Virtual Power Plant); Water (Chapter KISSER, Chapter GODOY-FAUDEZ) to Mobility (Chapter EMF). Product design for circularity for building and packaging is also featured in Chapter KISSER, Chapter SHI, Chapter MODAK, Chapter BALAJI, and Chapter TAN.

Moving downstream, in sustainable waste management, extensive coverage includes food waste (Chapter KHOR), waste electric and electronic products (Chapter YU), agriculture and municipal waste (Chapter PETERS, Chapter FUKUDA, Chapter MODAD). An example of upcycling practice through creating eco-art from electronics waste (E-waste) is elaborated in Chapter MALLABADI where the entire artwork creation and its strategic practice for scaling up through education on E-waste are shared.

Circular economy needs a workforce equipped with life cycle thinking mindset, STEM knowledge, and entrepreneurship skills, Chapter SIDDIQUE provides a dedicated overview on circular economy education worldwide and highlighted the STEAM Platform practices on Circular Economy general education module and its youth leadership program. Chapter EMF emphasizes that importance of embedding circular economy principles into teaching across all ages of learning. This supports

Table 1 Summary of book structure

Circular Economy Structure	Topics	Chapter Title	1st Author
Key Concepts and Terminology	**Key Concepts and Terminology**	Key Concepts and Terminology	Mengmeng CUI
Life Cycle Thinking Mindset	Life Cycle Thinking & Assessment	Life Cycle Thinking in a Circular Economy	Shabbir GHEEWALA
The Fabrics of A Circular City	Holistic Picture of Circular City	The Fabrics of a Circular City	Johannes KISSER
Mining	Impact on Water, Energy and Food	Circular Economy in a Water-Energy-Food Security Nexus in a SDGs Framework: Understanding Complexities	Alex GODOY-FAUNDEZ
Production	Industry Circular Manufacturing	Industry Circular Manufacturing	Lei SHI
	Industrial Symbiosis	Industry Symbiosis for Circular Economy: A Possible Scenario in Norway	Daniela LA ROSA
Consumption	Agriculture & Food	Agriculture & Food Circularity	Hung Teik KHOR
	Plastics	Plastics in Circular Economy: A sustainable Progression	Anand BELLAM BALAJI
	Built Environment	Materials Passports and Circular Economy	Mohamed SAMEER HOOSAIN
	Textile	New Paradigm for R&D and Business Model of Textile Circularity	Edwin KEH
	Water	Circular Economy in a Water-Energy-Food Security Nexus in a SDGs Framework: Understanding Complexities	Alex GODOY-FAUDEZ
	Mobility	The Business Opportunity of a Circular Economy	Ellen MacArthur Foundation (EMF)
	Energy: Community Microgrid	Circular Economy Enabled by Community Microgrids	Deva P SEETHARAM

(continued)

Table 1 (continued)

Circular Economy Structure	Topics	Chapter Title	1st Author
	Energy: Building Integrated Photovoltaic	Renewable Energy for Circular Economy: Application of Life Cycle Costing for the Building Integrated Solar PV Systems	Rashmi ANOOP PATIL
Waste Management	Electronic and Electrical Equipment Waste	Recycling of Waste Electric and Electronic Products in China	Kelin YU
	Agricultural Waste	Agricultural & Municipal Waste Management in Thailand	Suneerat FUKUDA
	Municipal Waste	Waste Management Practices: Innovation, Waste to Energy & e-EPR	Stephen PETERS
	Upcycling for Eco-Art	Transforming e-waste to Eco-Art by Upcycling	Vishwanath MALLABADI
R&D	New Paradigm Shift in R&D	New Paradigm for R&D and Business Model of Textile Circularity	Edwin KEH
Technology & Innovation	Digital Technologies	Industry 4.0 & Circular Economy -Deep Dive on Applied Data Analytics	Parvathy KRISHNAKUMARI
	Restorative Innovation	Innovation for Circular Economy	Jovan TAN
Circular Business	Business Opportunities	The Business Opportunity of a Circular Economy	Ellen MacArthur Foundation
	Circular Supply Chain	Circular Supply Chain Management	Charoenchai KHOMPATRAPORN
	Business Model	Circular Economy Business Models and Practices	Anna ITKIN
Financing	Economic Instruments & Financing	Economic Instruments and Financial Mechanisms for the Adoption of a Circular Economy	Santiago ENRIQUEZ
Assessment	Life Cycle Assessment	Life Cycle Thinking in a Circular Economy	Shabbir GHEEWALA

(continued)

Table 1 (continued)

Circular Economy Structure	Topics	Chapter Title	1st Author
	Greenhouse Gas Emissions Life Cycle Assessment	Life Cycle Greenhouse Gas Emissions for Circular Economy	Thumrongrut MUNGCHAROEN
	Life Cycle Costing	The Life Cycle Costing: Methodology and Applications in a Circular Economy	Piya KERDLAP
	Environment, Social and Governance (ESG)	Towards Sustainable Business Strategies for a Circular Economy: Environmental, Social and Governance (ESG) Performance and Evaluation	Rashmi ANOOP PATIL
Education	Sustainability Education	Youth Leadership in a Circular Economy: Education Enabled by the STEAM Platform	Arslan SIDDIQUE
	Education on E-waste and Art Creation	Transforming e-waste to Eco-Art by Upcycling	Vishwanath MALLABADI
Policy Case Studies	India	Circular Economy Practices in India	Prasad MODAK
	Taiwan	Taiwan Circular Economy: Transition Roadmap and Food, Textile & Construction Industries	Shadow CHEN

a mindset shift that will enable future leaders and young professionals to acquire circular economy knowledge, skills, and capabilities which they can take forward within their careers.

Table 1 summarizes the structure of the book and Fig. 1 captures the infographic interpretation of the book content.

The driving force of circularity involves emerging technologies such as digital technologies; research, development and innovation; business opportunities and sustainable business model; economic instruments and financing mechanisms, and assessment and regulation.

2.1   The Role of Digital Technologies

We are entering the era of the 4th Industrial Revolution- Industry 4.0 where digital technologies such as Internet of Things (IoT), Artificial Intelligence (AI), Cloud Computing and Blockchain are enabling transparency and efficiency in our economy.

Digital technologies enable digitization across product lifecycle, from resource extraction, production processes (design, materials, component, module, system), distribution (logistics and retails), consumption to waste management. Chapter HOOSAIN elaborates Materials Passport (MP) for the Built Environment with a highlight of the enabling digital technologies. MP provides information (composition/specification, spatial and life cycle) across the entire value chain of a product and its supply chain from sources to producers, distributors, and consumers/users. This enables re-use, remanufacture, recycle, and recover of materials, components, and systems.

Chapter KRISHNAKUMARI connects Industry 4.0 and the circular economy. Through a digitalization framework, Industry 4.0 proposes the creation of 'digital twins,' embeds interconnected IoT networks, and utilizes Machine Learning, and Big Data Analytics to derive understanding and predictive metrics from manufacturing and industrial data. Industry 4.0 drives the digital transformation toward smart and resilient economy. It focuses how Data Analytics could accelerate a circular economy transition through case studies.

Digital technologies also enable efficient and effective circular supply chain management described in Chapter KHOMPATRAPORN. It also provides circular supply chain transformation strategy through digitalization, collaborative platform, and reverse loop.

2.2   Research and Development and Innovations

Circularity solutions require research and innovation to reach sustainability. Recognizing the urgency of developing a resilient society when humanity is facing unprecedented crisis such as climate change and pandemics, all stakeholders need to act coherently in developing and implementing solutions to secure human survival sustainably. Chapter KISSER, Chapter SHI, Chapter PETERS, Chapter EMF, Chapter MODAK, and CHAPTER CHEN have all discussed the role of the government and multi-stakeholder partnerships in driving circular economy transformation. In particular, Chapter KEH demonstrated public-private partnership (PPP) in developing innovative and scalable circularity solutions enabled by accelerated research and development. The chapter focuses on a case study of a successful implementation of circularity in textile and apparel sector. The case study represents common problems and solutions development methodology applicable to industry sectors. The chapter highlights a new paradigm R&D where all stakeholders (government, research institutes, and industry) have the urgency mindset in solving environmental problems caused by the waste of production of textile and consumption of apparels. A short-term focused target was set, strong partnership and open innovation R&D platform was set up to include supply chain of the entire textile industry. This is to ensure a scalable working solution implementable on both the production and consumptions sites. This PPP is practiced in both co-financing and R&D enabling industry partners to implement a scalable solution in both manufacturing and business.

This case study also demonstrates technology innovation drives circular business innovation, allowing decentralization of product end of life (waste) management improving economic and environmental performance of business.

Deep diving into innovation, Chapter TAN introduces *Restorative Innovation*—an innovation economic model that explains a pattern of innovation-driven growth for innovative solutions designed to restore our health, humanity, and environment. The chapter showcases a number innovative business practices including a cradle to cradle circular business enabled by restorative innovation where the featured company produces materials from bio-based resources, designs customized packaging, services, collects and composts waste, and returns back to earth for regenerating bio-resources.

2.3 Business Opportunities

Chapter EMF demonstrates that circular economy provides a value creation opportunity and solutions framework to address global challenges. This translates to enormous new business opportunities in terms of saving materials cost, avoiding waste management cost, cost saving for improving business efficiency, and new revenue from new business. It also presents enormous opportunities for innovation enabled by emerging technologies such as digital technologies. Circular Economy is viewed as a delivery mechanism for achieving climate change targets, sustainable development goals, and ultimately reaching sustainable development. The chapter shares the outcome of analysis by EMF including a) the circular economy transformation could yield annual benefits for Europe of up to EUR 1.8 trillion in 2030, b) For China, activating broader circular economy solutions in cities could significantly lower the cost of access to goods and services and could save businesses and households approximately USD 11.2 trillion in 2040, and c) For India, the annual benefits could amount to USD 624 billion in 2050 compared with the current development path.

The chapter focuses on business opportunities in three key sectors: the food system in India; the built environment in China's cities; and mobility in Europe. It further quantifies the economic, environmental, and social benefits of these opportunities and explores what are the levers to bring them to scale.

2.4 Business Model

Sustainability of a company is driven by its business model. In a circular economy, business can no longer focus on the pure growth of profit, it has to sustain its operation through taking care of the planet, people, and profit holistically. Chapter ITKIN stresses a sustainable business model (SBM) drives circular economy toward sustainable development. The chapter highlights that circular economy is a functional service economy leading to economic competitiveness. Selling a service enables to create sustainable profits without an externalization of the costs of risk and costs of waste. Case studies of circular business model practices are also elaborated in Chapter TAN, Chapter KHOMPATRAPORN, Chapter MODAK, and Chapter CHEN.

2.5 Economic Instruments and Financing Mechanisms

To implement R&D, business innovation and drive the economic transformation, economic instruments, and financing mechanisms are crucial. Chapter ENRIQUEZ elaborates the importance of incentives that aim to incorporate environmental costs into the budgets of households and enterprises and encourage environmentally sound and efficient production and consumption through full-cost pricing. The chapter recommends incentives to free up and reallocate resources that are currently used in the linear model, as well as to mobilize new funding (sustainable bonds, ESG investment, equity capital) to support a circular economy transition. It stresses that the environmental policy instruments and financing enable investments in eco-design and the adoption and scaling up of new technologies and business models.

2.6 Assessments and Regulations

To drive a circular economy transition locally and globally toward sustainability, monitoring, and assessment is necessary. Although there are not yet standardized sets of circular economy indicators, the European Union, other European countries, and the Ellen MacArthur Foundation (EMF) have developed indicators to measure resource efficiency and raw materials management, materials circularity at the product and corporate levels, such as Buildings As Material Banks (BAMB) Circular Building Assessment and the industry-based circularity dataset initiative.

Circularity assessment tools primarily developed by European organizations include Cradle to Cradle Certified (The Cradle to Cradle Products Innovation Institute), The Circularity Check (Ecopreneur.eu), Circularity Gap Report (Platform for Accelerating the Circular Economy (PACE)), and Circular Business Solutions (alchemia-nova GmbH) are summarized in Chapter KISSER. Other circular economy progress measurement tools including Circulytics by the Ellen MacArthur Foundation, the Circular Transition Indicators by World Business Council for Sustainable Development, Global Reporting Initiative's upcoming circular economy reporting guidelines in the context of waste are summarized in Chapter EMF.

Circular economy-specific regulation, the extended product responsibility (EPR) policy is discussed in a number of chapters including Chapter SHI, Chapter PETERS, Chapter ENRIQUEZ, and Chapter KISSER.

EPR needs to be enforced to all producers globally. Shifting the waste management cost to producers incentivized circular product design, closing the materials loop, and driving service-based business model. In particular, Chapter PETERS stresses that through public–private partnership models the EPR policy enabled by digital technologies offers more efficient and effective route to implementation and accelerates a circular economy transition.

2.7 Sustainability and SDGs

The outcome of circular economy transition needs to be aligned with sustainable development goals and sustainability (social, environmental and economic) as a whole. Chapter GODOY-FAUNDEZ elaborates, for an extractive economy focusing on agriculture and mining, that practicing a circular economy will help to solve problems of the Food-Water-Energy insecurities and achieving sustainable development goals in countries such as Chile.

Circularity does not necessarily lead to sustainability. Life Cycle Assessment (LCA) and Environmental, Social & Governance (ESG) assessments are necessary tools to guide industry and business to achieve sustainability. Chapter GHEEWALA adopts LCA framework to assess the environmental sustainability for sugar cane production and packaging materials. To ensure that circular business is actually environmentally beneficial, it is essential to demonstrate the reduction of life cycle Greenhouse Gas (GHG) emissions. Chapter MUNGCHAROEN demonstrates the calculation of GHG emissions of circular business using methodology of the Intergovernmental Panel on Climate Change (IPCC) guidelines and life cycle assessment standards. In order to guide the economic decision-making for consumers and businesses, Chapter KERDLAP elaborates on Life Cycle Costing (LCC) methodology with case studies on different circular economy business practices. Chapter PATIL further provides a basic overview of the current circularity assessment methodologies and highlighted with case studies that ESG performance can be enhanced through circularity business practices.

Circular economy does not only address environmental sustainability but it also creates social and economic benefits. Chapter KHOMPATRAPORN, Chapter ITKIN, and Chapter EMF discussed both the societal and economic aspect of Circular Economy.

2.8 Policy Case Studies

Without an extensive coverage on circular economy development around world, we feature circular economy in India and Taiwan. In addition to policy coverage of the region, Chapter MODAK provided an overview on India circular economy highlighted the adoption of digital technology in managing waste flow, integrating informal waste recycler into the CE supply chain creating social benefits to the waste pickers, and building multi-stakeholder partnership platform in CE transition. Chapter CHEN shares the Taiwan's transition roadmap that focuses on guiding industry and business toward the circularity transition. The chapter focuses on a few flagships on emerging business in the food, textile, and construction sectors.

Circular economy is evolving rapidly and expanding globally. Some of the case studies shown in this book have not been scaled yet. We hope this book is able to provide guidance for policymakers, investors, corporations, entrepreneurs, researchers, educators, and general public to take action as a consumer and stakeholder to accelerate the transformation through scaling up those practices or innovating more effective practices.

Some of the topics that are emerging such as circularity by design, circularity assessment and regulations, and digital technologies applications will have more extensive coverage with case studies in the second edition to appear in 2021. We will also continue with regional coverage, especially in North America, North Asia, and other part of the world in the next edition.

3 Authors Analytics

We are fortunate to have a total of 66 authors from around the world across 5 continents with multidisciplinary background contributed covering 30 chapters in this book at this inaugural edition to share their learning, knowledge, practices and solutions for acceleration of a circular economy transition. Figure 2a shows the authors demographics based on survey from over two third of total authors participated in the survey. Figure 2b shows the diversity of authors in terms of background, gender, and age group, in particular, authors who are under 35 are in the majority.

A circular economy transition needs each and every one of us to take action both personally and professionally. Effective communication and making a circular economy framework and practices understood by all stakeholder is critical. This book uses Infographics, figures, and tables to make the book content more visual and easy to read and understand. It aims to be inclusive to all. The inclusivity of the book includes diverse age group, geographical location, disciplines, sectors, and stakeholders.

(a)

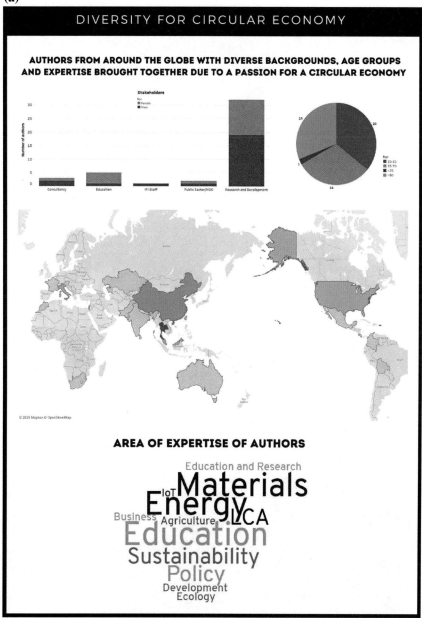

Fig. 2a Author analytics on demographics (Generated Parvathy Krishnan from DAV Data Solutions)

(b)

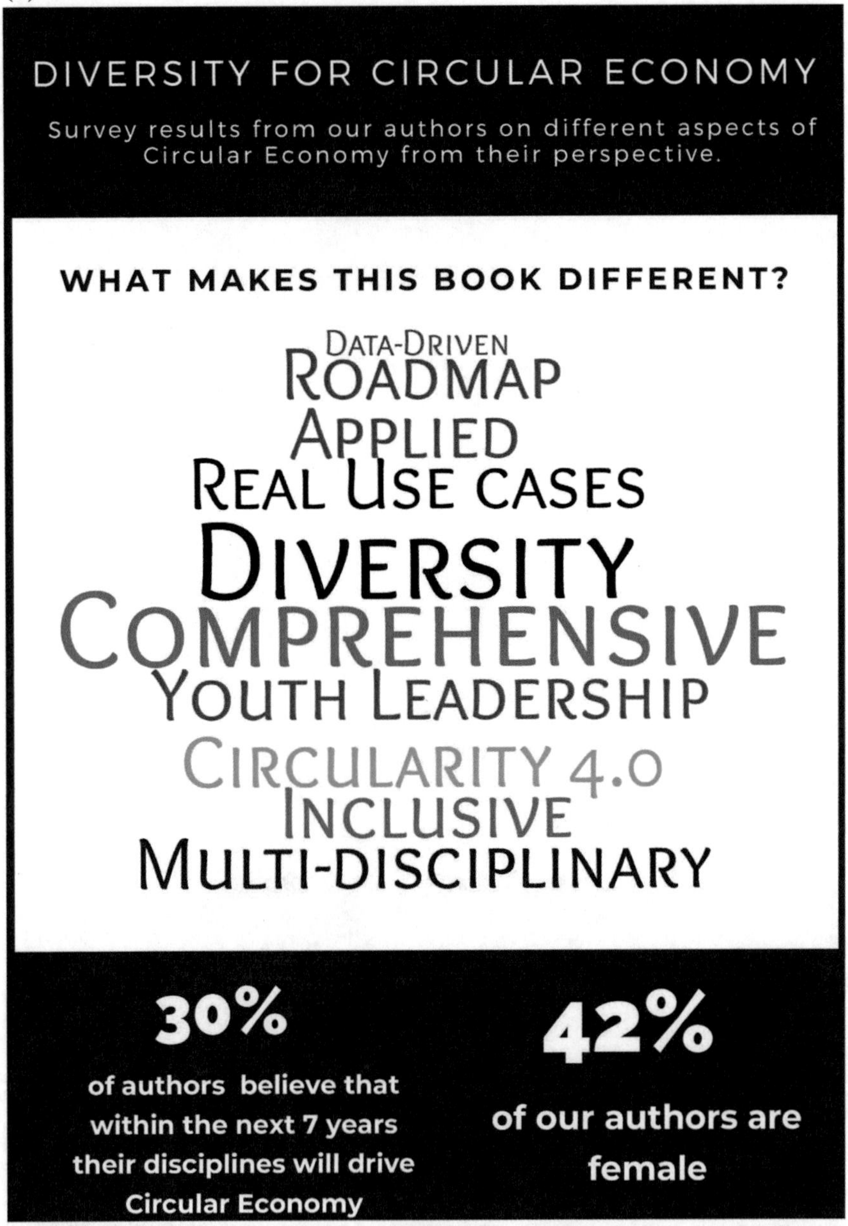

Fig. 2b Author analytics on diversity (Generated Parvathy Krishnan from DAV Data Solutions)

Lerwen Liu is a strong advocate of sustainability through innovation and entrepreneurship education and training focusing on the Asian region. She is a senior advisor in King Mongkut's University of Technology Thonburi (KMUTT) in Bangkok playing a leadership role in (1) streamlining strategy of education, research and innovation, and social impact toward sustainable development goals, industry 4.0, and circular economy, (2) building strategic partnerships with the United Nations, Asia Development Bank, leading universities, government funding agencies and industries worldwide in developing solutions for reaching SDGs and circular economy, and (3) conducting strategic training to KMUTT students and staff preparing the change agents/leaders for the SDGs 2030 and circular economy through her STEAM Platform.

She pioneered education programs of innovation and entrepreneurship toward sustainability, SDGs and circular economy in the National University of Singapore (NUS), Yale-NUS College, and KMUTT. She has organized various platform events in partnership with the UN, ADB, World Bank, and other entrepreneurship-related organizations in driving youth leadership and entrepreneurship towards SDGs and Circular Economy through the adoption of STEM knowledge, and Industry 4.0.

Professor Seeram Ramakrishna *FREng, Everest Chair* (https://www.eng.nus.edu.sg/me/staff/ramakrishna-seeram/), is among the top three impactful authors at the National University of Singapore, NUS (https://academic.microsoft.com/institution/165932596). NUS is ranked among the top five best global universities for engineering in the world (https://www.usnews.com/education/best-global-universities/engineering). He is the Chair of Circular Economy Taskforce. He is a member of Enterprise Singapore's and ISO's Committees on ISO/TC323 Circular Economy and WG3 on Circularity. He also the Chair of Sustainable Manufacturing TC at the Institution of Engineers Singapore and a member of standards committee of Singapore Manufacturing Federation (http://www.smfederation.org.sg). He is an advisor to the Ministry of Sustainability & Environment—National Environmental Agency's CESS events, (https://www.cleanenvirosummit.sg/programme/speakers/professor-seeram-ramakrishna; https://bit.ly/catalyst2019video; https://youtube.com/watch?v=ptSh_1Bgl1g). European Commission Director-General for Environment, Excellency *Daniel Calleja Crespo,* said, *"Professor Seeram Ramakrishna should be praised for his personal engagement leading the reflections on how to develop a more sustainable future for all",* in his foreword for the Springer Nature book on Circular Economy (ISBN: 978-981-15-8509-8). He is a member of UNESCO's Global Independent Expert Group on Universities and the 2030 Agenda (EGU2030). He is the Editor-in-Chief of the Springer NATURE Journal Materials Circular Economy—Sustainability (https://www.springer.com/journal/42824). He is an Associate Editor of eScience journal (http://www.keaipublishing.com/en/journals/escience/editorial-board/). He is an opinion contributor to the Springer Nature Sustainability Community (https://sustainabilitycommunity.springernature.com/users/98825-seeram-ramakrishna/posts/looking-through-covid-19-lens-for-a-sustainable-new-modern-society). He teaches ME6501 Materials and Sustainability course (https://www.europeanbusinessreview.com/circular-economy-sustainability-and-business-opportunities/). He also mentors Integrated Sustainable Design ISD5102 project students. Microsoft Academic ranked him among the top 25 authors out of three million materials researchers worldwide based on H-index (https://academic.microsoft.com/authors/192562407). He is named among the World's Most Influential Minds (Thomson Reuters) and World's Highly Cited Researchers (Clarivate Analytics). Listed among the top three scientists of the world as per the Stanford University researcher study on career-long impact of researchers or c-score (https://drive.google.com/file/d/1bUJrvurVVBbxSl9eFZRSHFi f7tt30-5U/view). He is an Impact Speaker at the University of Toronto, Canada Low Carbon Renewable Materials Center (https://www.lcrmc.com/). He is a judge for the Mohammed Bin Rashid Initiative for the Global Prosperity (https://www.facebook.com/Make4Prosperity/videos/innovation-inclusive-trade/479503539339143/). He advises technology companies with sustainability vision such as TRIA (www.triabio24.com), CeEntek (https://ceentek.com/), Green Li-Ion (www.Greenli-ion.com) and InfraPrime (https://www.infra-prime.com/vision-leadership). He is a Vice-President of Asian Polymer Association (https://www.asianpolymer.org/committee.html). He is a Founding Member of Plastics Recycling Association of Singapore (PRAS). His senior academic leadership roles include University Vice-President (Research Strategy), Dean of Faculty of Engineering; Director of NUS Enterprise and Founding Chairman of Solar Energy Institute of Singapore (http://www.seris.nus.edu.sg/). He is an elected Fellow of UK Royal Academy of Engineering (FREng), Singapore Academy of Engineering and Indian National Academy of Engineering. He received PhD from the University of Cambridge, UK, and The TGMP from the Harvard University, USA.

Key Concepts and Terminology

Mengmeng Cui

Abstract Many of us have heard the phrases "circular economy" and "linear economy". The notion of "circular economy" has been around for at least a few decades, starting with the "open economy" versus "closed economy" articulated by Kenneth Boulding in 1966 in his essay "The Economics of the Coming Spaceship Earth" (To download the essay, please go to: http://www.ub.edu/prometheus21/articulos/obspro metheus/BOULDING.pdf.). Since then, the concepts of **feedback systems**, **cradle-to-cradle**, **closed-loop** and many more essentially circular economy equivalent concepts have flourished and further developed into different branches in resource management, environmental policy, sustainable development and other subjects we are familiar with today from many university curriculums. It is, however, only in recent years, that the circular economy concept as an all-encompassing concept of future economic development model, gained global and cross-sector traction.

This chapter outlines the important sustainability, economics and business concepts that have been developed in the topic of circular economy. Since circular economy is an interdisciplinary concept that cuts across natural science, social science, economics and business domains, it is necessary to explain these concepts in a logical way so to make it easier, not only to understand what it means but also to start thinking about what needs to be done to enable the circular transition. Instead of explaining the concepts one by one, this chapter will use a few cases and embed the concepts in these cases. The chapter will explain these concepts in the following three contexts (Fig. 1):

1. Concepts related to natural cycles of matters
2. Concepts related to symbiosis, both ecological and industrial
3. Concepts related to circular businesses and business transformation.

M. Cui (✉)
PhD candidate in Climate Change and Sustainable Development Policies, University of Lisboa, Lisbon, Portugal
e-mail: mengmeng.cui@edu.ulisboa.pt

© Springer Nature Singapore Pte Ltd. 2021
L. Liu and S. Ramakrishna (eds.), *An Introduction to Circular Economy*,
https://doi.org/10.1007/978-981-15-8510-4_2

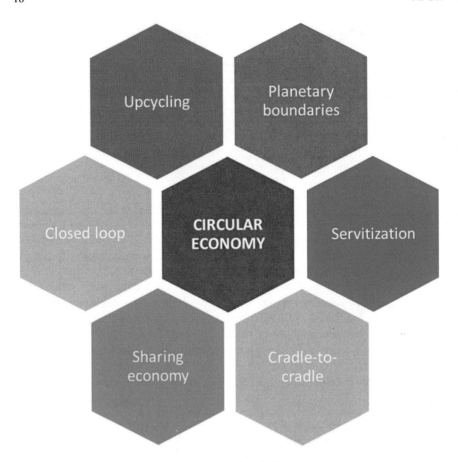

Fig. 1 Concepts around circular economy (non-exhaustive)

1 Objectives

Understand important concepts and terminologies related to the circular economy—in businesses, academic research, industrial and urban planning, agriculture, material, design and many more relevant domains.

2 Overview

The over-simplified understanding of the circular economy could be understood with three main ideas: renewable energy as input into the system; clean water and good water treatment practice; circular use of materials. These concepts may sound dry, but the circular economy is in fact full of stories. This is because the circular economy is

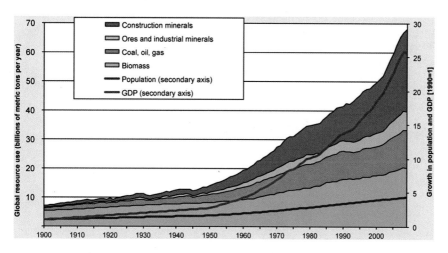

Fig. 2 Global resource consumption (The left axis shows global resource use between 1900 and 2009 measured in billions (109) of metric tons per year. The right axis (1900 = 1) shows the growth in population and Gross Domestic Product (GDP) during the same interval. GDP is measured in constant 1990 Geary-Khamis Dollars. Data source: Krausmann et al. 2009, updated using data available at http://www.uni-klu.ac.at/socec/inhalt/3133.html. *Source* http://www.igbp.net/news/features/features/addictedtoresources.5.705e080613685f74edb800059.html)

about dynamic systems, interactions, human behaviours, cross-industry collaboration and a lot more. Therefore, this chapter will not have a long list of terminologies that one can read online in a few minutes. Instead, the chapter will embed concepts and terminologies in stories, with keywords highlighted and footnotes to provide sources for more information. By the time you finish the stories, these concepts and terminologies should appear familiar and you will be able to proceed to the following chapters for a deeper understanding of the circular economy.

You will be reading a few stories and cases, each covering one important aspect of the circular economy. Before getting into the details, let's first have a look at the **linear model** we have today.

The linear model is often referred to as the **"take-make-throw" model**, which became possible after the first industrial revolution and greatly accelerated in the post-war era, accompanied by a global population boom as illustrated in Fig. 2.

The linear model is wasteful and has brought severe consequences including climate change, biodiversity loss, soil erosion, air and water pollution, etc. In 2009, the Stockholm Resilience Centre published the first **Planetary Boundaries**[1]

---

[1]In 2009, former centre director Johan Rockström led a group of 28 internationally renowned scientists to identify the nine processes that regulate the stability and resilience of the earth system. The scientists proposed quantitative planetary boundaries within which humanity can continue to develop and thrive for generations to come. Crossing these boundaries increases the risk of generating large-scale abrupt or irreversible environmental changes. Since then, the planetary boundaries framework has generated enormous interest within science, policy and practice. https://www.stockholmresilience.org/research/planetary-boundaries.html.

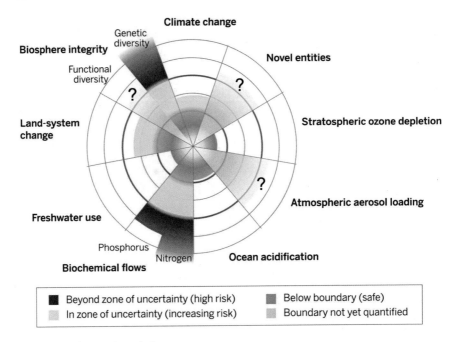

Fig. 3 The planetary boundaries

(watch the TED talk here: https://www.ted.com/talks/johan_rockstrom_let_the_env ironment_guide_our_development) report that described nine crucial earth systems that support human developments on the planet.

The Planetary Boundaries help us better understand where priorities need to be given in order to stay within these boundaries that support our lives on earth. The current state of the nine boundaries clearly calls for action and these actions need to be transformational. This is why the linear model that has brought unprecedented prosperity to mankind has to transition to a different model—a model that can help maintain the integrity of the earth systems that our prosperity relies on. We call this new economic model **the Circular Economy** (Fig. 3).

There are many different definitions of the circular economy. The Ellen Macarthur Foundation defines it as "a systemic approach to economic development designed to benefit businesses, society and the environment" (as illustrated in Fig. 4).[2] Metabolic (a circular economy practitioner) defines it as "a new economic model for addressing human needs and fairly distributing resources without undermining the functioning of the biosphere or crossing any planetary boundaries".[3] Wikipedia defines it more simply as "an economic system aimed at eliminating waste and the continual use of resources".[4]

---

[2]https://www.ellenmacarthurfoundation.org/explore/the-circular-economy-in-detail.

[3]https://www.metabolic.nl/about/our-mission/.

[4]https://en.wikipedia.org/wiki/Circular_economy.

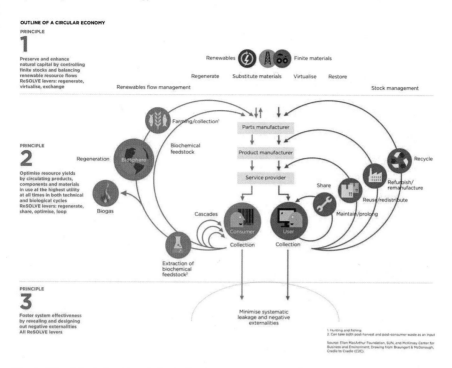

Fig. 4 Ellen MacArthur Foundation's circular economy system diagram

The definition aside, what's important about the circular economy and the new economic model we want to develop and adopt, is the principles and the end goal. Key principles of the circular economy include (but not limited to): design out waste; retain the highest value for the longest time; and maximize renewables. These key principles will be discussed later in the book. The goal of the circular economy is to fundamentally decouple our economic growth and prosperity from the use of resources and environmental impacts. In order to do this, we must rethink our global economy as a system to understand **material flow**,[5] **lifecycle impact**, trade-offs, etc., to implement interventions that would drive the circular transition.

---

[5]The study of material flow accounting (MFA) focuses on the natural resource requirements of national economies, specific economic activities (such as construction and housing, transport and mobility), or geographical units such as cities. MFA accounts for the input of primary materials—biomass, fossil fuels, metal ores and minerals—and semimanufactures and final goods into economic activities. MFA also accounts for the outputs of economic systems including final goods for export, waste and emissions. MFA often conceptualizes the economic system as a 'black box'. There are, however, accounting strategies for material flows within economic systems available as well. https://www.sciencedirect.com/topics/economics-econometrics-and-finance/material-flow.

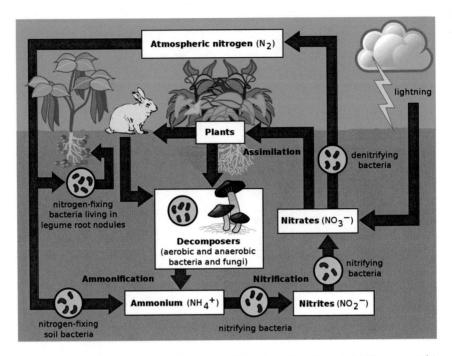

Fig. 5 Demonstration of nitrogen cycle in nature (https://en.wikipedia.org/wiki/Nitrogen_cycle)

3 Story One: The Biological Cycles

It is not difficult to understand that many elements and organic matters in nature go in cycles. Take nitrogen as an example (Fig. 5). Atmospheric nitrogen turns into nitrates that is absorbed by plants, plants are eaten by animals, nitrates are released by the animals into the soil, either fixed by bacteria and returned to plants in the form of ammonia or decomposed by fungus and turns into ammonia, which is digested by other bacteria that return it back to nitrates. This cycle happens with many biochemicals and there is often a balance to it. This is where the **Cradle to Cradle**[6] theory came from—to mimic the biological metabolism of nature, with man-made materials and designs (Fig. 6).

The cradle-to-cradle concept is often used for a single material, or in a single product design process, as shown in the diagram (Fig. 6).

In the circular transition, we need to expand the biological cycles to much bigger systems—the most important one being the food system.

Creating a **closed-loop** food system, in which the output of one process can become the input of another, is urgently needed to feed the growing population. Traditionally, people in many parts of the world practised such closed-loop systems.

---

[6]cradle-to-cradle design: https://en.wikipedia.org/wiki/Cradle-to-cradle_design. Please watch: https://www.youtube.com/watch?v=HM20zk8WvoM.

Fig. 6 The biological cycle
for products for consumption
(https://epea-hamburg.com/
cradle-to-cradle/)

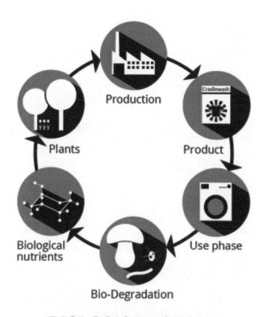

One that has been re-popularized is the "rice-fish-duck" system (as shown in Fig. 7).
It is commonly used in China, Vietnam, Philippines and a few other Asian countries.
It works as depicted in the diagram below, where symbiotic relationships are created
among rice, fish and duck to create nutrient cycles that mimic a natural ecosystem,
in which nothing becomes waste.

This brings us to the next topic on creating **symbiosis**.[7] Symbiosis is a term
first used in ecology, as in our ecosystem, animals, plants and biochemicals form a
relationship where one relies on another. This could be as simple as a relationship
between a suckerfish and a shark, or as complex as a forest among trees, birds, insects,
fungus and nutrient flows. Today, the term "symbiosis" is used far more widely
outside ecological meanings. It is used in industry where "**industrial ecology**[8]" is

---

[7]Symbiosis is originally used to describe any of several living arrangements between members of
two different species, including mutualism, commensalism and parasitism. Both positive (beneficial)
and negative (unfavourable to harmful) associations are therefore included, and the members are
called symbionts. https://www.britannica.com/science/symbiosis.

[8]Industrial ecology is the study of material and energy flows through industrial systems. Industrial
ecology conceptualises industry as a man-made ecosystem that operates in a similar way to natural
ecosystems, where the waste or by product of one process is used as an input into another process.
Industrial ecology interacts with natural ecosystems and attempts to move from a linear to cyclical

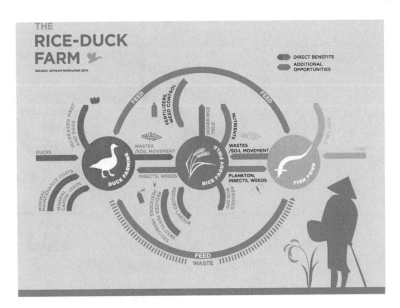

Fig. 7 The rice-duck-fish farming system (https://www.ellenmacarthurfoundation.org/case-stu dies/ecosystem-inspired-farm-yields-large-profits)

studied and its subset "**industrial symbiosis**[9]" is the process by which wastes or by-products of an industry or industrial process become the raw materials for another. Application of this concept allows materials to be used in a more sustainable way and contributes to the creation of a circular economy.

4 Story Two: Creating Symbiosis

The Netherlands, a country that occupies less than 0.1% of global landmass, is the second largest food exporter (by value), next to the United States. The Netherlands has created a highly efficient and symbiotic agriculture practice that allows the greenhouses to use the carbon dioxide emissions from power plants to stimulate plant growth, reuse the waste heat from greenhouses to heat up swimming pools and

---

or closed loop system. Like natural ecosystems, industrial ecology is in a continual state of flux. http://www.gdrc.org/sustdev/concepts/16-l-eco.html.

[9]Industrial symbiosis is the process by which wastes or by-products of an industry or industrial process become the raw materials for another. Application of this concept allows materials to be used in a more sustainable way and contributes to the creation of a circular economy. https://ec.eur opa.eu/environment/europeangreencapital/wp-content/uploads/2018/05/Industrial_Symbiosis.pdf.

schools and reduce water use by 90%.[10] It is a perfect example of creating large scale, cross-sector symbiosis.

In many industrial symbiotic systems, one party's waste is being used as input for another. This process, depending on the nature of how the waste is used, can either be "**upcycling**[11]" or "**downcycling**", sometimes simply referred to as "recycling". I found the below diagram that illustrates the difference between upcycling and recycling. When it comes to the different entry point of the supply chain—upcycling returns the material further upstream of the supply chain, adding more value and often longer time span. Theoretically speaking, it should ensure that the material can be upcycled infinitely, whereas "downcycling", or "recycling" doesn't often concern what happens to the materials after the first recycling cycle. In the example in Fig. 8 with PET bottles, even though when PET is made into textile, the value may go up and lifespan may increase, textile recycling is still difficult and not practised at a large scale. Therefore, if PET is recycled into textile, it would usually end up with only one recycling cycle and become waste again in a short time. Therefore, to distinguish if a material is upcycled or downcycled, three key factors are to be considered. The first factor is value—does the value of material (or the product it goes into) increase or decrease. The second factor is lifespan—does it lead to longer lifespan or shorter lifespan. The third factor is future recyclability—can the material, or the product be recycled again and again.

5 Story Three: Circular Business Transformation

Businesses are constantly reinventing themselves. To transition from a linear model to a circular model, some companies may find themselves crossing over to completely different industries and offering a very different set of products and services. In the first story, we talked about the biological cycle of circular economy. In the cradle-to-cradle design methodology, there is another cycle—**the technical cycle**.[12] The technical cycle is often more difficult to achieve and this is where business innovation is most needed.

According to Accenture Strategy's research, there are five models for businesses to close the technical cycle. There are a number of variations of similar models developed by other companies and organizations, for simplicity purpose, we will stay with the five described in the following diagram.

It is noticeable that the technical flow in Fig. 9 and the business model value chain in Fig. 10 are similar in nature. The technical cycle guides product design to ensure

---

[10]https://www.nationalgeographic.com/magazine/2017/09/holland-agriculture-sustainable-farming/.

[11]Upcycling refers to a process that can be *repeated in perpetuity of returning materials back to a pliable, usable form without degradation to their latent value—moving resources back up the supply chain..* https://intercongreen.com/2010/02/17/recycling-vs-upcycling-what-is-the-difference/.

[12]In the technical cycle, materials that are not used up during use in the product can be reprocessed to allow them to be used in a new product.

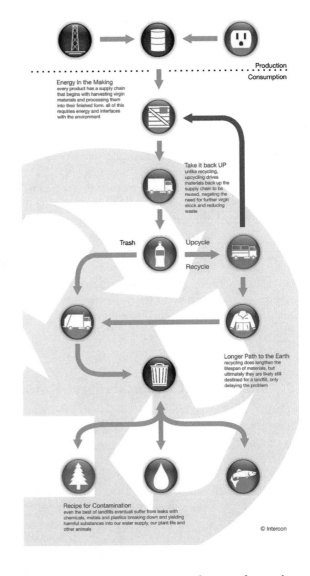

repairability, ease of dissembling, recyclability and other key factors of a product that would make the circular business models possible (Fig. 10).

Let's have a closer look at the models and how business can transition to a circular economy.

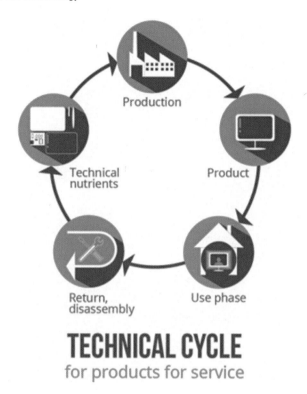

Fig. 9 Technical cycle of Cradle-to-Cradle design (https://epea-hamburg.com/cradle-to-cradle/)

5.1 Model 1. Circular Inputs

It refers to using renewable, bio-based, regenerative or recycled input. This applies to both energy, water and materials. This model used to be seen as an independent step in business value chain and therefore only concerns procurement function in a company. However, it is more and more seen as an integral part of closing the product loop—supplying materials or parts from the same products that have been recycled. One example is the aluminium used by Apple in its computers and phones. Apple sources recycled aluminium and is planning to close the aluminium loop for all their products.

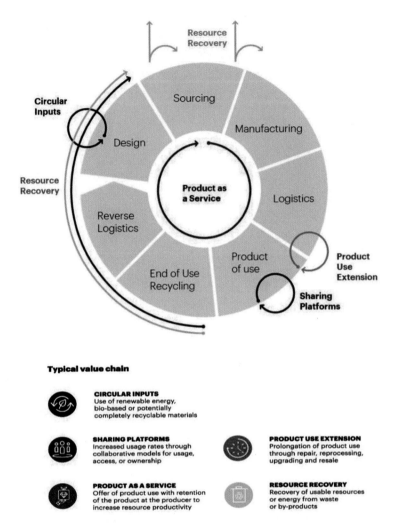

Fig. 10 Accenture's 5 circular business models (https://www.weforum.org/agenda/2020/01/how-can-we-accelerate-the-transition-to-a-circular-economy/)

5.2 Model 2. Product Use Extension

This refers to repair, refurbishing and **re-manufacturing**[13] to extend product life. It is most commonly found among electric and electronic producers, particularly in high-value equipment such as medical equipment. Some consumer goods companies are

---

[13]Remanufacturing is a process of returning a used product to at least original performance specification from the customers' perspective and giving the resultant product a warranty that is at least equal to that of a newly manufactured equivalent (Ijomah, 2002; Ijomah et al., 2004). https://www.sciencedirect.com/topics/engineering/remanufacturing.

also practising product life extension model. The most well-known one is Patagonia's lifetime warranty for return and repair, leveraging a team of 45 full-time employees to provide repair and re-design services.

5.3   Model 3. Product as a Service

This is perhaps one of the most important models for the circular transition. It is also referred to as "**servitization**" in many places. The shift from selling a product to selling a service requires three key changes: mindset, design and ownership. First, it requires a total mindset shift from the linear production-sales (end of responsibility) business model, to a full life cycle service model. Second, in many cases, it also requires a different design of the product. Many of our products are designed for short lifespan—so it's cheaper to produce and we need to replace them more often, which works perfectly in the linear model. But the product-as-a-service model requires robust products that don't break, can easily be repaired, or parts can easily be replaced, and eventually, be taken back to be remanufactured or recycled. The more robust the product, the less cost to the company to provide lifetime care. The easier to remanufacture, refurbish or recycle, the higher value the company can extract from the product at the end of its life. Lastly, ownership shifts. If the company that produces the product ultimately owns the product, even when it does not work any longer and therefore becomes waste, it means the company is the one responsible to take care of it at the end of its life, hence internalize the cost to eventually dispose of the product or parts of the product.

Some of the product-as-a-service examples overlap with our next model—the sharing platform, which may cause some confusion. Therefore, we will use the Michelin tyre example to just see how servitisation works. Michelin now offers "tyre as a service" where they provide tyre management service. Customers are charged by the distance driven. This enables the company to retain ownership of the tyre, ensure their customers return their tyres at the end of life, help customers drive more safely by detecting when the tyres need to be replaced, resulting in reduced carbon emissions and improve recycling.[14]

5.4   Model 4. Sharing Platform

We will not spend much time explaining this model as in recent years, it has penetrated many domains of our life, most noticeable ones are in mobility, holiday rental and fashion. Some of the big companies include AirBnB, Uber, Grab, Mobike and so forth. The sharing platform model has always existed in the past among small

---

[14]Please read the case here: https://www.michelin.com/en/activities/related-services/services-and-solutions/.

networks of people. New digital technologies enabled large-scale, secure and traceable sharing schemes that can flourish into global businesses. The key concept to understand with regard to the sharing platform business model, is the concept of utility. The sharing platform takes advantage of the commonly under-utilized resources, especially the high value ones such as real estate.

5.5 Model 5. Resource Recovery

I put this model last not because it is the most important model, but because it is the last resort in a circular economy. All products at some point will reach their end of life. When all the other models don't work anymore, it is important to recover valuable materials and resources that can be used in other functions through material and energy recovery.

6 The Way Forward

Lastly, what will we do to drive the circular transition? First, let's look at three crucial policy instruments for circular economy.

6.1 Extended Producer Responsibility (EPR)[15]

EPR has been around for more than 30 years. You can think of it as a form of waste management tax put on the producers. It often requires producers/manufacturers to establish collection and treatment systems, including setting up separate collection points and forming recycling partnerships. EPR is most commonly used for electric and electronic products that have higher post-consumer value and bigger environmental impact. Recently, more countries are piloting EPR in the packaging industry, sometimes in the form of a **deposit-return system**.[16]

---

[15]OECD defines Extended Producer Responsibility (EPR) as an environmental policy approach in which a producer's responsibility for a product is extended to the post-consumer stage of a product's life cycle. An EPR policy is characterized by 1. the shifting of responsibility (physically and/or economically; fully or partially) upstream towards the producer and away from municipalities; and 2. the provision of incentives to producers to take into account the environmental considerations when designing their products. http://www.oecd.org/environment/extended-producer-responsibility.html.

[16]https://en.wikipedia.org/wiki/Deposit-refund_system.

6.2 Standardization

The second policy is standardization. According to the Circular Economy Guide for Practitioners, "Standardization is the process of establishing uniformity across manufacturing materials and processes. Potential benefits of standardization include lower production and procurement costs through economies of scale, easier and less expensive repair and replacement, and faster and more efficient processes, for example"[17]. Standardization can reduce collection and sorting cost and improve quality of recycling, **inter-changeability** (e.g. electronic devices chargers are slowly being standardized, so more and more devices now use universal USB type-C connecter) and enable **modular design**[18] which are essential to the circular economy.

6.3 Public Procurement

The last policy concept to discuss is not exactly a policy instrument, but an important means to initiate and boost circular economy development. **Public procurement**[19] or **government procurement** leverages the public sector's purchase power to drive demand for more sustainable products and services. The public sector also bears additional responsibilities for its people and society. Therefore, sustainable public procurement is commonly used as a means to stimulate the market. More governments are adopting **circular procurement**[20] measures, which are a set of rules and criteria for any company or their products to be used by the government.[21] For example, a requirement in recycled content and **FSC certificate**[22] in paper product is often used.

---

[17]https://www.ceguide.org/Strategies-and-examples/Design/Standardization.

[18]Modular design, or modularity in design, is an approach (design theory and practice) that subdivides a system into smaller parts called *modules* (such as modular process skids), which can be independently created, modified, replaced or exchanged between different systems.

[19]Public procurement refers to the process by which public authorities, such as government departments or local authorities, purchase work, goods or services from companies. https://ec.europa.eu/growth/single-market/public-procurement_en.

[20]Circular procurement sets out an approach to green public procurement which pays special attention to "the purchase of works, goods or services that seek to contribute to the closed energy and material loops within supply chains, whilst minimising, and in the best case avoiding, negative environmental impacts and waste creation across the whole life-cycle". https://ec.europa.eu/environment/gpp/circular_procurement_en.htm.

[21]Please read: https://ec.europa.eu/environment/gpp/pdf/CP_European_Commission_Brochure_webversion_small.pdf.

[22]https://www.fsc.org/en/page/forest-management-certification.

6.4 Science-Based Targets[23]

Aside from the policy instruments, the non-profit organizations have also been developing tools and methodologies to facilitate the circular transition. Among them, the Science Based Targets Initiative is one of the most important concepts developed by a group of top scientists. The Science Based Targets Initiative is a collaboration between CDP, the United Nations Global Compact (UNGC), World Resources Institute (WRI) and the World Wide Fund for Nature (WWF) and one of the We Mean Business Coalition commitments. **It champions science-based target setting as a powerful way of boosting companies' competitive advantage in the transition to the low-carbon economy.** Science-based targets provide companies with a clearly defined pathway to future-proof growth by specifying how much and how quickly they need to reduce their greenhouse gas emissions.

7 Conclusion

This chapter is an introductory chapter to get familiar with some of the key concepts in circular economy. We started with introducing the planetary boundaries, which show us which of the earth systems are most in need of interventions. The chapter closes with science-based targets, which is an initiative that helps us focus on what matters. The circular economy is a means to an end. We need to transition to a circular model because our planetary boundaries are crossed and the earth systems are destabilized. Many of the concepts discussed in this chapter are related to how the circular economy could be created, by both the private sector and the public sector.

8 Key Take-Away

Understanding of the circular economy at a high level;
    Understanding of planetary boundaries and science-based targets;
    Understanding measures for both public and private sectors to transition to a circular economy.

---

[23]https://sciencebasedtargets.org/about-the-science-based-targets-initiative/.

9   Table of Useful Resources

No.	Name of organization	URL	Category
1	Accenture	https://www.accenture.com/us-en/about/events/the-circular-economy-handbook	Consulting
2	BSI	https://www.bsigroup.com/en-GB/standards/benefits-of-using-standards/becoming-more-sustainable-with-standards/BS8001-Circular-Economy/	Government
3	Circulate News	https://medium.com/circulatenews	Media
4	Circular Economy Club	https://www.circulareconomyclub.com/	NGO
5	Circular Design Guide	https://www.circulardesignguide.com/	Initiative
6	Cradle-to-cradle Product Innovation Institute	https://www.c2ccertified.org/	NGO
7	Ellen Macarthur Foundation	https://www.ellenmacarthurfoundation.org/	Thinktank
8	European Commission	https://ec.europa.eu/environment/circular-economy/index_en.html	Government
9	ISO	https://www.iso.org/committee/7203984.html	INGO
10	Metabolic	https://www.metabolic.nl/	Consulting
11	OECD	https://www.oecd.org/environment/waste/recircle.html	Government
12	Raconteur	https://www.raconteur.net/sustainability	Media
13	Science Direct	https://www.sciencedirect.com/search/advanced?qs=circular%20economy	Media
14	Science Based Targets	https://sciencebasedtargets.org/	Initiative
15	Stockholm Resilience Center	https://www.stockholmresilience.org/research/planetary-boundaries/planetary-boundaries/about-the-research/the-nine-planetary-boundaries.html	Thinktank
16	UNEP	https://www.unenvironment.org/circularity	Government
17	World Economic Forum	https://www.weforum.org/projects/circular-economy	Thinktank
18	WBCSD	https://www.wbcsd.org/Programs/Circular-Economy	Association

Mengmeng Cui Ms. Cui is a senior sustainability and business strategy consultant. She has more than 10 years of experience in multiple sustainability areas including climate change, circular economy and sustainable cities. Originally from Beijing, China, she holds a bachelor's degree in Asia Pacific Studies, specializing in environmental sociology, from Ritrsumeikan Asia Pacific University in Japan. She also holds a master's degree in Environmental Science, Policy and Management from the European Commission's Erasmus + program with Central European University, Lund University, University of Manchester and University of Aegean.

For the past 9 years, Mengmeng lived in Singapore with her family. She's a member of the Singapore National Mirror Committee for the ISO Circular Economy Standard TC323 and is often consulted by the government on e-waste, packaging and other circular economy policies. During her time with the Accenture Strategy-Sustainability team, Mengmeng advised leading businesses and government agencies including Panasonic, IKEA, major electronic brand, pharmaceutical company and multiple ministries and agencies in Japan and Singapore in the past 9 years. Her work in many countries has provided her with a deep knowledge of sustainability in Asia.

She recently set up her own consultancy Asia Pathway to provide circular economy consulting services. She is also the Asia sales representative for the Dutch circular economy consulting firm Metabolic. She is currently residing in Lisbon, Portugal, to complete her PhD study in Climate Change and Sustainable Development Policies.

Life Cycle Thinking in a Circular Economy

Shabbir H. Gheewala and Thapat Silalertruksa

Abstract In the millions of years of evolution, nature has developed very efficient systems that move all elements and substances in cycles so that there is no waste. Humans, on the other hand, have recently developed industrial systems in the last few centuries that have a linear flow, extracting resources from nature and discarding them as waste after a brief period of use. Solutions to handle pollution have moved from end-of-pipe treatment to cleaner production and now towards a circular economy. A circular economy tries to move away from this linear model in trying to extend the life of products and services while minimizing burdens to the environment. To ensure that there are actually environmental benefits, a life cycle thinking approach is essential. This philosophy is developed in the chapter and life cycle assessment is introduced as an essential tool for environmental evaluation. Case studies on sugarcane biorefinery and packaging materials are provided to illustrate the utility of life cycle assessment in ensuring environmental benefits when approaching circularity.

Keywords Cleaner production · Circular economy · Life cycle assessment · Linear economy · Recycling

Learning Objectives

- Understand the development of environmental management system since the traditional end-of-pipe treatment concept to the life cycle concept
- Distinguish between the linear and circular environmental management system
- Be able to explain the key steps of life cycle assessment (LCA)
- Having ideas from case studies of LCA for assessing systems to ensure optimal solutions for the system and avoiding problem shifting

S. H. Gheewala (✉)
The Joint Graduate School of Energy and Environment, Center for Energy Technology and Environment, King Mongkut's University of Technology Thonburi, 126 Prachauthit Road, Bangkok 10140, Thailand
e-mail: shabbir_g@jgsee.kmutt.ac.th

T. Silalertruksa
Department of Environmental Engineering, Faculty of Engineering, King Mongkut's University of Technology Thonburi, 126 Prachauthit Road, Bangkok 10140, Thailand

© Springer Nature Singapore Pte Ltd. 2021
L. Liu and S. Ramakrishna (eds.), *An Introduction to Circular Economy*,
https://doi.org/10.1007/978-981-15-8510-4_3

- Be able to apply life cycle concept for evaluating the circular economy system or comparative assessment of products/processes.

1 Introduction: Circularity in Nature

Nature has an inherent capability to deal with various kinds of waste which is often referred to as its *carrying capacity*. In fact, there is no waste per se in natural cycles as the output from one process becomes an input for another. One example is the food chain where the primary producers (plants) take carbon dioxide from the atmosphere, and nutrients and water from the soil to produce organic matter via photosynthesis. The primary producers are consumed by the primary consumers (e.g. herbivores) that are in turn consumed by the secondary consumers (e.g. carnivores) that may be consumed further by the tertiary consumers. However, finally, all the producers and consumers are mineralized by decomposers (worms, bacteria, etc.) and the nutrients released back to the environment. This *circularity* in nature, fuelled only by the energy from the sun, is an important example for us to emulate. Waste (solid, liquid and gaseous) produced by human activities can also to some extent be handled by nature through its carrying capacity; the organic fraction of solid waste, for example, quickly decomposes in warm and humid climates under the action of microorganisms. However, when the carrying capacity of nature is exceeded because of the huge amount of waste being generated and also because of the presence of xenobiotics (e.g. pesticides, chemicals, etc.) with which nature is not familiar, there is an accumulation of waste leading to adverse effects on the ecosystems as well as human beings. To deal with this, end-of-life waste treatment technologies have been developed so that the intensity of waste from production processes and other anthropogenic activities could be reduced before discharging to the environment.

2 Shifting the Paradigm from End-of-Pipe Treatment to Cleaner Production

End-of-pipe waste treatment, as it is often called, is a feature of the so-called linear economy Fig. 1. Here, the resources are taken from nature, transformed through manufacturing processes into products for satisfying certain functions, used and then managed by waste treatment processes before being discharged back to the environment. This "take-make-use-discard" model has been used for many decades and is also currently being used to a large extent. However, from the point of view of the product manufacturers, waste treatment is a "non-productive" activity as money is invested into it without any economic returns. In other words, it is almost literally "money down the drain". In the 1990s, this traditional model was challenged by the introduction of waste minimization, pollution prevention or cleaner production which focused on reducing the amount of wastes being produced rather than on

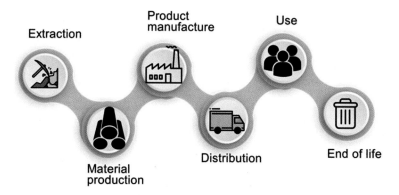

Fig. 1 Generic linear economy model

treating the waste once it has already been produced—a proactive (preventative) rather than reactive approach. Conceptually, this is far superior to the traditional end-of-pipe waste treatment because reducing the waste not only reduces the need for waste treatment but also conserves valuable raw materials and energy inputs that would otherwise be diverted into (unintentionally) "producing" the waste Fig. 2. The full value of resources extracted is not yet included in the current accounting systems as the cost of the natural capital is still an externality. Also, the environmental burdens of mining and transportation/transformation are not included in the cost of the resources. Cleaner production would result in more product being obtained from the same amount of raw materials and energy inputs and at the same time, less waste would need to be managed. Thus, it is certainly a step in the direction towards sustainability. The concept of cleaner production initially focused on the factory (production or process unit) itself, the idea being to minimize the use of resources, energy and reduce waste emissions Fig. 3. With a view towards continual improvement, cleaner production was integrated into the management system of the company via the environmental management system (ISO 14001). More recently, the scope of the assessment and cleaner production efforts have been extended beyond the company boundaries to include the entire supply chain (ISO 14001: 2015) [7].

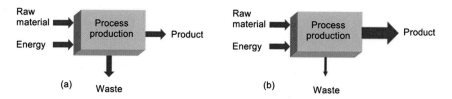

Fig. 2 Generic production process, **a** conventional, **b** cleaner production

Fig. 3 Cleaner production in factories

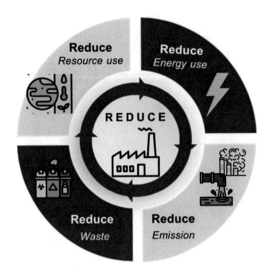

3   Beyond Cleaner Production to Life Cycle Thinking

The inclusion of the entire supply chain or life cycle in the analysis of a product (or service) is very important for several reasons. One of the major reasons is that when we focus only on individual parts of the life cycle and try to optimize each one, we may not end up optimizing the entire system; the combination of a series of sub-optimal solutions may not always lead to the optimal solution for the entire system. This is a powerful message that is observable when we try to apply it in, for example, product design. For instance, a large part of the impacts from cars is produced by the combustion of fuels in the engine to generate motive power; the fuel usage is quite significantly influenced by the weight of the vehicle. The (dead) weight of the vehicle itself usually tends to even exceed that of the passenger(s). Shifting from a steel chassis to a composite carbon fibre material would significantly reduce the weight of the car and consequently the fuel usage and emissions thereof while maintaining the strength and security. However, this will create issues at the end-of-life where steel is virtually fully recyclable, whereas composite materials are relatively very difficult to recycle. Thus, the environmental problem has, in effect, been shifted from the use phase of the car to the end-of-life. Focusing only on the use phase of the car will lead to such problems in the future. As another example, replacing the conventional internal combustion engine cars with electric vehicles shifts the environmental problems from the tailpipe of the car in case of the conventional cars to the power plant which produces electricity for the electric vehicles. Focusing only on the car use, particular tailpipe emissions, will not reveal this shifting of burdens. In both examples, this shifting of environmental burdens from one part of the life cycle to another can easily be observed when we consider the entire life cycle during the environmental assessment preferably at the design stage itself.

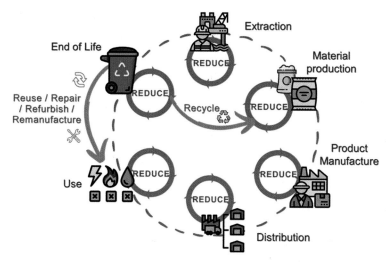

Fig. 4 Circularity and life cycle of products

Thus, the entire life cycle of the product needs to be considered as shown in Fig. 4. Cleaner production principles reducing the use of resources and waste emissions are then applied at every life cycle stage. However, the most important thing is to consider the environmental burdens of all life cycle stages to avoid problem shifting as described in the examples above. The outer circle in Fig. 4, though looking more "circular" than Fig. 1, is still an extension of the linear economy model. When it is combined with the (green coloured) arrows inside, the shift to the circular economy model becomes apparent. This is further explained in the next section.

4 Linear to Circular Economy

When we compare Figs. 1 and 4, we can make several important observations on how the linear economy can be transformed into the circular economy. The "take-make-use-discard" model of the linear economy gets transformed into the circular model at several levels—micro, meso and macro. At the micro level, one particular process or factory can reduce resource use and emissions (as in Fig. 3 which is also represented as the smaller circles in Fig. 4. At the meso level, a group of factories or an entire industrial sector could be included where the unusable output (we try not to call it "waste") from one could feed into another in an industrial park or industrial ecology-type system. This has led to the inclusion of industrial symbiosis practices in industrial parks. The macro level would have this exchange at the level of the entire economy.

The circular economy, though, goes beyond the reduction of resource use and environmental emissions. It focuses more on extending the life of the products that have already been produced through reuse, repair, refurbishment and remanufacture. Recycling of materials, being preferable to the discard of products at the end of life, is probably the last in the hierarchy of circularity because prolongation of product life through the means mentioned above retains the value of the resources and energy invested in making the product. These would be lost when recycling materials, though virgin resources could still be preserved.

Linear economy adds value at every step of the product life cycle, but then the value drops at the end of life. At sale, ownership and liability of risks and waste pass from the manufacturer to the buyer. So in this approach (which is currently the norm), a product once sold ceases to be the responsibility of the producer/manufacturer and becomes the responsibility of the user/buyer who is then the product owner. On the other hand, circular economy attempts to maximize the value at each point in a product's life (even beyond use). Different approaches may be required to make this work—extended producer responsibility (EPR) requiring the producer to share/take responsibility of the product even after sale (e.g. at the end of life), the producer selling the service provided by the product to the user rather than the product itself and shared ownership are some examples of these approaches which have been addressed in more details in other chapters and are thus not repeated here.

5   Life Cycle Assessment

As mentioned before, life cycle thinking is essential in assessing systems to ensure optimal solutions for the system and avoiding problem shifting from one stage of the supply chain to another. This is relevant even when assessing the options related to circularity. As we will see through some examples later on, options such as recycling which intuitively make so much sense may not always be the better option as compared to the use of virgin materials. Thus, life cycle assessment or LCA has been developed as a tool to facilitate the environmental evaluation of products, services and systems throughout the life cycle. It has been standardized in ISO 14040:2006 and ISO 14044:2006 and will be presented here briefly [5, 6].

Life cycle assessment often referred to as "cradle-to-grave" assessment, is a tool for the evaluation of environmental impacts of a product (or service) throughout its entire life starting with the extraction of raw materials from which it is made, through manufacturing of materials, manufacture of products, use of the products, reuse and final disposal at the end of life Fig. 5. Transportation in all the intermediate stages is also included (represented by arrows in Fig. 5. LCA is typically focused on environmental assessment, however, there are related tools such as life cycle costing (LCC) and social life cycle assessment (S-LCA) which use the framework of LCA for addressing economic and social impacts. The phases of an LCA as outlined in ISO 14040:2006 are presented in Fig. 6.

Goal and scope definition: An LCA begins with goal definition where the intended application of the study is defined along with the reasons for carrying out the study and the intended audience for whom the study is being performed. It is also to be

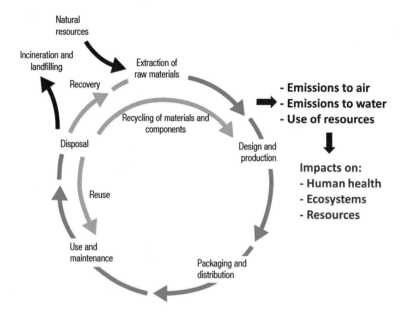

Fig. 5 Product life cycle assessment framework

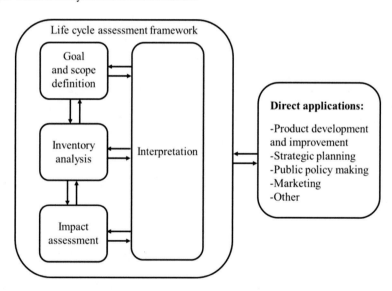

Fig. 6 Stages of an LCA [5]

decided right since the inception stage itself whether the study results will be used for making comparative assertions that will be made public. Once the goal of the study is defined, the scope is formulated which essentially includes the system boundary of the study; what is to be included in or excluded from the study is transparently defined with justification, where necessary. Data requirements, allocation procedures for co-products, assumptions and limitations of the study are laid out at this stage.

One important thing that needs to be defined here is the *"functional unit"* which is the quantification of the function/service provided by the product. This serves as a fair basis of comparison; LCA studies are usually comparative by nature, either comparing competing products providing the same service or comparison of a product with itself when improvements are made. So, for example, if a reusable ceramic mug is to be compared with a single-use polystyrene cup, obviously a comparison between one mug/cup of each type would not be fair. A single-use polystyrene cup can, by definition, only be used once whereas a ceramic mug can be used many times. In such a case, defining a certain amount of function/service (i.e. a functional unit) is the way out. So, for example, a function could be defined as "Serve as drink container for 200 mL of hot beverages three times a day for one year". To satisfy this function, 3×365 or 1,095 single-use polystyrene cups would be required, whereas only one ceramic mug would be sufficient. However, it is very likely that the ceramic could be used longer than a year. So if we were to assume the life of the ceramic mug to be 4 years, we would assign only 1/4th of the production and disposal impacts of the ceramic mug to the functional unit as it has been defined for only 1 year, whereas the mug can be used for 4 years. Washing of the ceramic mug after each use should also be included; on the other hand, there is virtually no impact from the use phase of the single-use polystyrene cups. So fair comparison of the environmental impacts of a single-use polystyrene mug with a reusable ceramic mug would be as illustrated in Fig. 7.

Inventory analysis: Once the goal and scope have been defined, the next step is to collect the life cycle inventory data comprising inputs and outputs from each process in every life cycle stage. Input data comprises the resources, materials and energy used and output data products/co-products as well as emissions to air, water and soil,

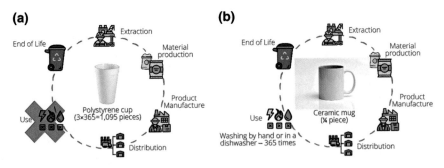

Fig. 7 Comparison of **a** polystyrene cup and **b** ceramic mug on a functionally equivalent basis

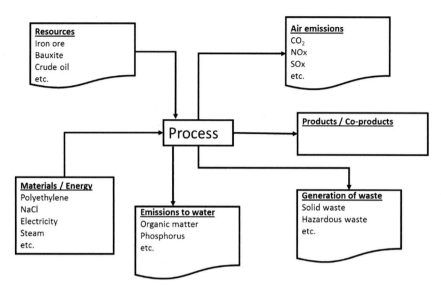

Fig. 8 Generic unit process showing examples of common inventory items

and solid waste Fig. 8. Data on transportation (particularly fuel use) must also be collected.

A lot of the basic data (referred to as background data) such as extraction of minerals and fossil fuels, production of materials (e.g. iron, polyethylene, etc.), grid mix electricity production, etc., are obtained from national/international databases or other literature. It is not practicable or even meaningful to obtain primary data for such items each time a new LCA is conducted. On the other hand, process data (e.g. amount of materials used, amount of electricity and fuels used, etc.) are usually collected directly from the production facility (primary activity data). Once all the data for the individual processes are collected, the processes are linked to each other via the functional unit to complete the life cycle. All inputs are traced back until nature and all products/emissions traced forward until they are released to the environment. Every *exchange with the environment* whether it be resource extraction or release of emissions to air, water, soil, etc., contributes to environmental burdens that must be quantified.

Impact assessment: Once all the inventory data are collected and verified (via triangulation, mass and energy balances, etc.), these must be translated in potential impacts; this is done via life cycle impact assessment. Several methods have been developed by various research groups that could possibly be used for making the impact assessment calculations. Broadly speaking, life cycle impact assessment methods are divided into two groups—midpoint impacts and endpoint impacts (or so-called areas of protection). The midpoint impacts are connected to the endpoint impacts (or final damages) through damage pathways. The general structure of life cycle impact assessment is shown in Fig. 9 [8].

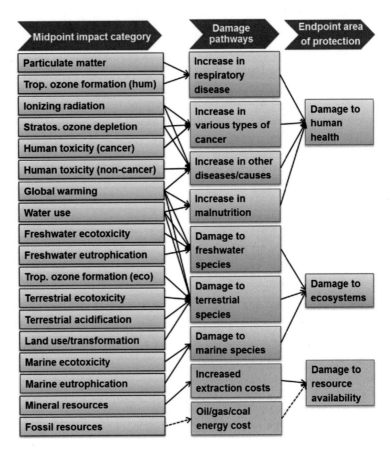

Fig. 9 Structure of a typical life cycle impact assessment method (ReCiPe 2016 v1.1 methodology)

Interpretation: The final step in an LCA entails translating all the scientific calculations conducted in the previous steps (particularly inventory analysis and impact assessment) into results that would be useful to the intended audience in line with the goal and scope definition. Environmental hot spots are identified and suggestions made for possible improvements of the environmental profile.

The application of LCA to real-life problems particularly related to circularity is illustrated via two case studies in the sections below.

6   Case Study I: Sugarcane Biorefinery

There is a perception that the bioeconomy is inherently circular because biological resources are renewable. However, in practice, it could be quite linear because it is possible for resources to be used faster than they are replenished and might not be returned safely and appropriately to the biosphere to rebuild natural capital. In Thailand, the sugarcane industry is recognized as a key industry for the national policy on bioeconomy and circular economy. There is an increasing attraction that the co-products obtained from the sugarcane industry, e.g. cane trash, bagasse, molasses, filter cake, etc., can be used for many applications by processing in a biorefinery to produce a wide range of products, e.g. bioethanol, electricity as well as chemicals including a variety of biopolymers. The biorefinery concept is therefore gaining interest as a promising approach for enhancing the competitiveness of the sugarcane industry as waste can be utilized and the high-value products can be created over the whole sugarcane production system Fig. 10. This is especially for the cane trash, filter cake and vinasse which is abundant and remains as waste and is a burden to the industry requiring proper treatment and management. The poor management system can cause drawbacks to the environment such as the open burning of cane trash in traditional harvesting practice results in air pollution (particularly fine particulate matter) and the leakage of vinasse to the river can cause eutrophication impact as well as other wastewater pollution.

An LCA, therefore, has been used to assesses the environmental sustainability of sugarcane biorefinery systems including sugarcane cultivation and harvesting, sugarcane milling and by-product utilization, i.e. bagasse for steam and electricity, molasses for ethanol and vinasse for fertilizer and soil conditioner. The comparative assessment has been done between the base case and the new biorefinery options as shown in Fig. 10. The base case represents the conventional farming practices with

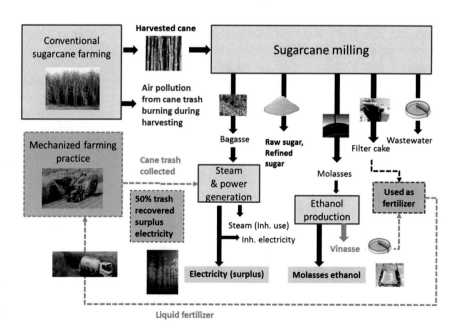

Fig. 10 Sugarcane biorefinery system (green-dotted lines indicate improvement options)

cane trash burning before harvesting, sugar milling, molasses ethanol production and steam and power generation from bagasse. The new biorefinery option represents the implementation of mechanized farming with the utilization of cane trash for electricity generation and use of vinasse as fertilizer and soil conditioner by mixing together with the filter cake, a by-product from sugar mills and returning to cane growers to reduce the use of chemical fertilizers [9]. Since the circular use of sugarcane biomass needs the additional processes as well as input resources which in turn might not be able to overcome the environmental benefits obtained from waste utilization in the biorefinery, LCA is therefore applied to confirm whether or not the improvement of sugarcane cultivation and harvesting practice, e.g. green cane production along with integrated utilization of biomass residues through the entire chain would help reduce the environmental impacts of the main products derived from sugarcane, e.g. sugar and ethanol. The LCA results showed that the potential impacts on climate change, acidification, photo-oxidant formation, particulate matter formation and fossil depletion could be reduced by around 38%, 60%, 90%, 63% and 21%, respectively. This already includes the trade-off with the increased energy and material use for mechanized farming, transport and fertilizer production processes.

7 Case Study II: Packaging Wastes and Its Circulation

Packaging provides several functions, e.g. storage, protection and preservation of products as well as information to consumers including marketing. After the use of the products, packaging is often discarded without further use and is known as a source of solid waste problem in society, especially plastic packaging which is currently of utmost concern. Currently, awareness is being raised on reducing packaging waste, especially the promoting of 3Rs, i.e. Reduce, Reuse and Recycle which directly link to the circular economy concept. Reduction of packaging is generally encouraged to people in daily life, e.g. reducing the use of plastic bags when shopping in a supermarket is the most common campaign. Recycling of packaging waste is often encouraged as one of the circular approaches to divert the packaging wastes away from the landfill by converting those wastes into reusable materials. Nevertheless, too often the way packaging is produced, used and discarded fails to capture the economic benefits of a more "circular" approach and harms the environment [1]. In addition, although recycling is often promoted, it is not always that the material recovery from packaging will be sustainable when compared to the energy and resources used for the material recovery process itself.

Hence, there is a need to identify the appropriate "circular" approach for packaging waste that can really benefit the environment as well as the ecosystem. Different types of packaging materials also require different management approaches. An LCA-type case study by Gujba and Azapagic [2] showed the comparative carbon footprint (life cycle greenhouse gas emissions) of different packaging types i.e. carton, glass, polyethylene terephthalate (PET) and high-density polyethylene (HDPE) bottles and aluminium and steel cans used for four beverage product categories, i.e. milk, fruit

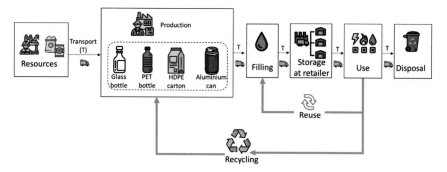

Fig. 11 Life cycle of packaging and circular approaches

juice, water, beer and wine in the UK. The end-of-life options were considered in different proportions including the re-use of packaging, virgin and recycled materials, incineration with energy recovery, incineration without energy recovery and landfilling. Figure 11 shows the simplified life cycle of packaging and its circular approaches.

The results showed that the carton packaging has the lowest carbon footprint in the range of 90–111 kg CO_2e/1000 L of beverage and glass bottle the highest, between 150 and 761 kg CO_2e/1,000 L. Carton and plastic (HDPE and PET) drink packaging had lower carbon footprints than aluminium, steel and glass. However, with higher recycling rates, aluminium and steel cans were comparable to PET. Generally, reuse of glass bottles is preferred to recycling and landfill. This is because recycling of glass containers can save some energy but not a significant quantity compared to reuse. The primary energy saved is around 2,321 MJ/t glass or only about 13% of the energy required to make glass containers from virgin raw materials [3]. If collection and transportation distances for the glass containers are quite high, even these modest savings from recycling could be totally lost. The variation of carbon footprint is found for even the same type of packaging material, mainly influenced by the size and weight of the containers and the recycling rates. It must be noted that different countries having different primary energy sources for electricity generation will result in the varying carbon footprints of recycling. Hence, LCA is required to ensure that the recycling process is beneficial in terms of environmental impacts as compared to the use of virgin materials. The findings also pointed out that for the same type of material, larger containers have a lower carbon footprint than the smaller ones due to the lower amounts of materials required per functional unit. The main hot spots for all types of packaging are the manufacture of raw materials and packaging.

Calculation Example: A Calculation of GHG Emissions from Recycling of Packaging Systems

Task: A beverage company would like to compare the "cradle to grave" GHG emissions of different beverage packaging systems for selecting the one with the lowest

GHG emissions. The functional unit is defined as the amount of packaging required to deliver 1,000 L of beverage. Five scenarios of packaging systems considered for delivering the beverage and the end-of-life scenarios are as follows:

Scenario 1: Glass bottle (100% virgin input) and landfill (after use).
Scenario 2: Glass bottle (100% recycled input) without landfill.
Scenario 3: Reuse of Glass bottle (100% recycled input) without landfill.
Scenario 4: Aluminium cans (100% virgin input) and landfill (after use).
Scenario 5: Aluminium cans (100% recycled input) without landfill.

The specific information of beverage container weight and materials used for top, body and label of the container is shown in Table 1 and the simplified GHG (CO_2eq) emission factors of materials and processes relevant to those five packaging systems are listed in Table 2. Assuming that truck (16-tonne capacity) is used for transportation. The transport distance of containers from factory to container is 200 km, the transport distance of glass and aluminium scrap (after use) to the factory for recycling is 300 km and the transport distance for landfill is 50 km. It must be noted that only the body part of the container is assumed to be landfilled and recycled.

For Scenario 3, the glass bottles can be reused for 20 times, it means that the total number of glass bottles required to deliver 1000 L of beverage would be decreased from 2000 to be only 100. However, there will be an additional requirement for bottle cleaning after each use resulting in the use of about 0.1 L hot water and about 0.2 g detergent per bottle per time of reuse. The extra transport distance for collecting the bottle to reuse is assumed to be 100 km.

Calculation: The comparative results of "cradle-to-grave" GHG emissions for the five scenarios can be calculated by multiplying the material data or activity data to, e.g. the material requirements to fulfil the functional unit Table 1 and their associated GHG emission factors Table 2.

Results and Discussion

The "cradle-to-grave" GHG emissions of the five scenarios showed that Scenario 3, i.e. Reuse of Glass bottle (100% recycled input) without landfill has resulted in the lowest GHG emissions, followed by Scenario 5, i.e. aluminium cans (100% recycled input) and Scenario 2: glass bottle (100% recycled input). Meanwhile, Scenario 4: aluminium cans (100% virgin input) has the highest GHG emissions. It can be seen that the extension of lifetime of packaging via the reuse of glass bottles can significantly reduce the GHG emissions due to the reduction of material use and waste for recycling (Table 3).

Table 1 The container details

Container type	Glass bottle	Glass bottle (Reuse)	Aluminium can
Capacity (L)	0.5	0.5	0.5
Weight	Total = 0.2016 kg/bottle (Top = 0.00085 kg, Body = 0.2 kg, label = 0.00075 g)	Total = 0.2016 kg/bottle (Top = 0.00085 kg, Body = 0.2 kg, label = 0.00075 g)	Total = 0.02 kg/can (Top = 0.003 kg, Body = 0.015 kg)
Material for top (cap)	Aluminium alloy $(AlMg_3)$	Aluminium alloy $(AlMg_3)$	Aluminium alloy $(AlMg_3)$
Material for body	Glass	Glass	Aluminium
Material for label	Kraft paper	Kraft paper	–
Reference flow			
Number of containers required to meet the functional unit	2,000	100	2,000
Weight of material for the containers (kg per 1000 L beverage)			
Body weight	400	20	30
Top weight	1.7	0.085	6
Label weight	1.5	0.075	-
Transport (tonne.km)			
Transport of containers from factory to retailer	80.6	80.6	7.2
Transport of glass bottle for reuse	–	40	–
Transport of glass bottle and aluminium can scrap for recycling	120	6	9
Transport of glass bottle and aluminium can scrap for landfilling	20	–	1.5
Reuse of glass bottle (unit per 1000 L beverage)			
Hot water (m^3)		0.2	
Cleaning agent (kg)		0.4	

The assessment also showed that although the GHG emission factors of glass are significantly lower than aluminium can per kilogram of material due to the lower energy required for mineral extraction and processing, the production emissions per functional unit become comparable due to the higher weight of the glass bottles as compared to the aluminium cans. On the other hand, the recycling of aluminium will reduce the high amount of energy required for the production of virgin aluminium which in turn results in the significantly low GHG emissions of

Table 2 Given GHG
emission factors

Material/Process	Emission factor	Unit
Glass (100% virgin input)	0.41	kg CO_2eq/kg
Glass (100% recycled input)	0.25	kg CO_2eq/kg
Aluminium cans (100% virgin input)	8.01	kg CO_2eq/kg
Aluminium cans (100% recycled input)	2.12	kg CO_2eq/kg
Aluminium alloy (AlMg$_3$)	4.22	kg CO_2eq/kg
Kraft paper	1.08	kg CO_2eq/kg
Landfilling	0.02	kg CO_2eq/kg landfilled material
Transport by 16-tonne truck	0.05	kg CO_2eq/tonne.km
Hot water	1.10	kg CO_2eq/m^3
Cleaning agent	2.30	kg CO_2eq/kg

*GHG Emission factors are modified from ICF [4] and Thailand
Greenhouse Gas Management Organization (TGO) [10]

the recycled aluminium as compared to the virgin aluminium. The recycling of glass
also helps reduce the energy required for the virgin glass production but the credit
is not that dominant as compared to the aluminium recycling because high energy is
also required for the glass cullet melting process. From this example, the recycling
of aluminium cans and glass bottle can reduce GHG emissions by 66% and 35%,
respectively. Interestingly, the combination of reuse of glass bottle and recycling of
the glass after the end of life in Scenario 3 can reduce the GHG emissions of the
glass bottles by around 93%. Hence, when assessing the circular use of material, the
life cycle concept is necessary for implementation. In addition, based on the circular
economy concept, the other options such as the replacement of glass and aluminium
with other materials can be further investigated .

Table 3 GHG emissions for different packaging system scenarios

Life cycle GHG emission	GHG emissions (kg CO_2eq/functional unit)				
	Scenario 1	Scenario 2	Scenario 3	Scenario 4	Scenario 5
Beverage container production					
• Top (cap)	7.17	7.17	0.36	25.32	25.32
• Body	164.00	101.41	5.07	240.30	63.49
• Label	1.62	1.62	0.08	–	–
Beverage container cleaning (Reuse case)					
• Hot water			0.22		
• Cleansing agent			0.92		
• Transport of bottle for reuse			2.16		
Total "Cradle to gate" GHG emissions	*172.79*	*110.21*	*8.81*	*265.62*	*88.81*
Transport of container					
• Factory to Retailer	4.35	4.35	4.35	0.39	0.39
End-of-life scenario					
Transport of glass bottles and aluminium can scrap for landfill	1.08	–	–	0.08	
Landfilling	8.00	–	–	0.60	-
Transport of glass bottles and aluminium can scrap for recycling	–	6.48	0.32		0.49
Total "Cradle to grave" GHG emissions	*186.23*	*121.04*	*13.48*	*266.69*	*89.69*

8   Questions

1. Explain why an LCA is required for assessing the circular economy system.
2. Discuss the pros and cons of industrial waste recycling for energy production, material production and landfill in terms of the potential environmental impacts that may occur from each life cycle stage.
3. Develop the appropriate functional unit and system boundary for comparing the conventional food container made from Polypropylene (PP), Polyethylene (PE) with the bioplastic food container (Polylactic acid: PLA)

a. What is the appropriate functional unit for a fair comparison?
b. Identify the life cycle stages of each packaging production
c. Identify the potential end-of-life scenarios of each packaging
d. List the key environmental interventions that should be considered in each life cycle stage and end-of-life scenarios
e. Explain the pros and cons of each packaging production system including the potential environmental benefits/burdens from different end-of-life scenarios.

Suggested Reading
Life Cycle Assessment

Hauschild, M. Z., Rosenbaum, R. K., & Olsen, S. I. (Eds.). (2018). *Life cycle assessment: Theory and practice.* Springer.

Jolliet, O., Saadé-Sbeih, M., Shaked, S., Jolliet, A., & Crettaz, P. (2016). *Environmental life cycle assessment.* CRC Press.

Klöpffer, W., & Grahl, B. (2014). *Life cycle assessment: A guide to best practice.* Weinheim: Wiley-VCH.

Circular Economy

Stahel, W. R. (2019). *The circular economy: A user's guide.* Routledge.

References

1. EC. (2015). *A European strategy for plastics in a circular economy* (pp. 1–22). European Commission.
2. Gujba, H., & Azapagic, A. (2011). Carbon footprint of beverage packaging in the United Kingdom. In M. Finkbeiner (Ed.), *Towards life cycle sustainability management* (pp. 381–389). https://doi.org/10.1007/978-94-007-1899-9_37.
3. Gaines, L. L., & Mintz, M. M. (1994). *Energy implications of glass-container recycling.* Argonne National Laboratory.
4. ICF. (2006). *Documentation for greenhouse gas emission and energy factors used in the Waste Reduction Model (WARM): Containers, packaging, and non-durable good materials chapters.* ICF International, U.S. Environmental Protection Agency.
5. ISO. (2006a). *ISO 14040:2006 Environmental management—Life cycle assessment—Principles and framework.* International Organization for Standardization.
6. ISO. (2006b). *ISO 14044:2006 Environmental management—Life cycle assessment—Requirements and guidelines.* International Organization for Standardization.
7. ISO. (2015). *ISO 14001:2015 Environmental management systems—Requirements with guidance for use.* International Organization for Standardization.
8. RIVM. (2017). *ReCiPe 2016 v1.1 A harmonized life cycle impact assessment method at midpoint and endpoint level, Report I: Characterization, RIVM Report 2016-0104a.* National Institute for Public Health and the Environment.
9. Silalertruksa, T., Pongpat, P., & Gheewala, S. H. (2017). Life cycle assessment for enhancing environmental sustainability of sugarcane biorefinery in Thailand. *Journal of Cleaner Production, 140*(Part 2), 906–913.

10. TGO. (2020). *GHG Emission Factors for Carbon Footprint of Product (CFP), Thailand Greenhouse Gas Management Organization (Public Organization).* See also https://thaicarbonlabel.tgo.or.th/products_emission/products_emission.pnc.

Shabbir H. Gheewala is a professor at the Joint Graduate School of Energy and Environment (JGSEE), Thailand where he teaches Life Cycle Assessment and has led the Life Cycle Sustainability Assessment Lab for almost 20 years. He also holds an adjunct professorship at the University of North Carolina Chapel Hill, USA. His research focuses on sustainability assessment of energy systems; sustainability indicators including circularity; and certification issues in biofuels and the agro-industry. He is a national expert on product carbon and water footprinting in Thailand. Shabbir mentors a national research network on sustainability assessment and policy for food, fuel and climate change in Thailand.

Dr. Thapat Silalertruksa is a lecturer at the Department of Environmental Engineering, King Mongkut's University of Technology Thonburi (KMUTT). His research works involve with land-water-energy-food nexus assessment; sustainability assessment of food and energy systems; carbon and water footprinting; life cycle assessment (LCA) of products and development of sustainability indicators for bioeconomy and circular economy. He has involved with more than 25 research and consultancy projects on sustainability assessment and implementation of environmental management system and tools for Thai industries and Thai society over the past 20 years especially the work on cleaner technology promotions and the life cycle sustainability assessment of sugarcane and palm oil supply chain.

The Fabrics of a Circular City

Johannes Kisser and Maria Wirth

Abstract Cities are centers of human and economic activity but also of resource use and waste. This gives cities both a critical and a promising role to support the transition to a circular economy by keeping incoming products and resources in the loop. A circular city is a place, where people share the resources they have, facilitated through business models that avoid losses and build on maximizing resource productivity. This requires a redesign of biological and technical material cycles in a way that their value can be maintained at the highest possible level for as long as possible. This chapter addresses questions such as: What could a circular city look like? What does this mean in practice? Where can we already see a transition? And what can urban policymakers and other actors do to realize circular cities? In this chapter, we will explore the resource streams flowing into cities, their main challenges and opportunities, solutions to close biological and technical cycles, the ways to measure progress, the actors and their roles within the circular city fabrics, and finally, a case comparison of circular city practices.

Keywords Circular economy · Cities · Food · Bio-based · Nature-based solutions · Nutrients · Resource loops · Urban planning · Wastewater

Learning Objectives

- Cities are currently major resource sinks within the open-ended linear economy.
- Cities can use their human and economic force to become resource turntables.
- By recirculating resources within cities and city-regions.
- Cities can drive forward circular economy as a whole.
- Circular cities can be inspired by natural processes.

J. Kisser (✉) · M. Wirth
Alchemia-nova, Vienna, Austria
e-mail: jk@alchemia-nova.net
URL: https://www.alchemia-nova.net

M. Wirth
e-mail: maria.wirth@alchemia-nova.net

Alchemia-nova, Athens, Greece

- Transformation towards circular cities is underway.
- Technologies are available but institutional barriers remain.

1 Introduction

Cities are centers of human and economic activity. More than 50% of the global population lives in cities and they raise over 80% of global GDP [1]. They also consume 75% of natural resources, produce 50% of global waste, and 60–80% of greenhouse gas emissions [2]. Currently, major flows of material, energy, and water are taken from the environment, processed to value-added products, and delivered to cities, where their value is consumed and ultimately lost or discarded. In Europe, a region that is poor in primary resources, this open-ended linear flow has created a strong import dependency. The worldwide mega-trends of urbanization, over-exploitation of resources, environmental degradation, and climate change call for an urgent transformation of cities from import-dependent, bottomless resource sinks into turntables capable of keeping valuable resources in continued reuse. In the next decades, 90% of urbanization will take place in the developing world [3], where an increase in living standards, in the current linear system, will result in a surge of resource consumption. By closing material, water and energy loops, cities can become resilient and succeed in providing decent, affordable living conditions in the long term while operating within the planetary boundaries; they should be decoupled from resource use and bad environmental impact [4].

In fact, their specific attributes of cities make them inherently well equipped to model such a transformation. Urban centers have always been breeding grounds for new forms of societal and lifestyle concepts and offer ever-increasing room for innovation. Their immense human, economic, innovative, and resource capacity puts cities in the perfect position to drive forward the transformation toward a circular economy. Instead of resource sinks, cities could function as resource turntables, where products and materials stay in use.

Definition

Cities are settlements with an urban center above 50,000 inhabitants [5].

A **circular city** is one that mitigates waste by keeping products, materials, and resources in use and regenerates natural systems.

Nutrients are necessary for biological metabolism and are usually distinguished between macro- and micronutrients and human or plant nutrients.

Nature-based solutions are solutions that are 'inspired and supported by nature, which are cost-effective, simultaneously provide environmental, social and economic benefits and help build resilience. Such solutions bring more, and more diverse, nature and natural features and processes into cities, landscapes and seascapes, through locally adapted, resource-efficient and systemic interventions' [6].

Resource recovery means the recovery of materials from by-products or waste. In our understanding it does not include energy recovery.

Reuse in our context means using a certain material again either for its original purpose or for a different function without significantly decreasing its value.

2 The Dichotomy of Cities as Circular Economy Hotspots

Large volumes of energy, materials, water, and food enter urban households, businesses, and public spaces. In 2007, 7 billion people lived in urban areas, consuming 75% of global energy and materials [4]. Cities are also hotspots of direct and indirect water and energy consumption, meaning the water embedded in food and materials, as well as water-for-energy and energy-for-water. As mentioned, cities produce half of the global waste. Within the linear economic system, the scale of resource inflows and waste has created cities as significant open-ended resource sinks, relying on a consistent inflow of primary resources.

Not only the volume but also the economic value of resources reaches its peak and diminishes in cities. From production in rural areas and downstream along the value chain, food and materials increase in economic value. They reach the urban consumer at their peak value. Once purchased, their value decreases and is finally destroyed when disposed to waste bins and sewage pipes. Where products are designed for short-term or single use, a larger resource inflow is necessary to produce new products, replacing the losses to waste. As waste, the value of resources becomes negative, i.e., resulting in costs of waste management and/or for human health. Within a circular economy, the economic value of products is maintained and exploited again and again through reuse, refurbishment, repair, and, in worst case, recycling.

Urban waste management struggles with the combination of huge volumes of waste produced by citizens each day and the spatial limitation for the effective handling and final disposal of waste. These factors add pressure to hastily remove waste from human contact in order to protect public health. While landfilling remains a dominant practice in Southern and Eastern Europe, many European cities incinerate their residual waste. While this may solve the spatial problem and recover energy through waste to energy processes, the economic potential of resources is still terminated, so the one-way train continues.

However, precisely these attributes of cities that challenge waste management, provide multiple windows of opportunity for the development of a circular economy often faster than in other sectors.

Commodities

In urban apartments, people cannot store as many products and tools at home. Short distances between users of tools, repair shops, collection, and re-distribution centers make new business models based on sharing, repairing, and reuse much more practical, manageable, and user-friendly. Door-to-door delivery or picking up purchases in person is much more feasible in densely populated areas. Traffic congestions make cars a burden for urban inhabitants but also enable the efficient operation of public transportation. The abundance of resources and goods, as well as the economic force in cities, enable internal diversification of processing techniques needed to re-circulate "spent" resources and goods of all kinds. From nutrient, water and biomethane recovery from wastewater for urban farming and agriculture at the outskirts, via reassembly of building components all the way to specialized electronics refurbishment stores, tool libraries and 3D-printing at maker-spaces—cities can have it all due to their size. This suggests that cities could emerge as leaders in the overall transition to a circular economy.

Building Materials

Cities are characterized by their built environment. This includes commercial buildings, streets, bridges, and open spaces for pedestrians, parking, and other uses. The construction, renovation, repair, and demolition of these structures requires vast material input and currently also generates the largest waste stream in the European Union (EU), with 850 million tons each year, representing 25–30% of total waste [7]. Construction and demolition waste (CDW) often contain bulky and heavy materials, including concrete, wood, asphalt, gypsum, metals, bricks, glass, plastics, etc., and low in hazardous substances. The high resource value bears a high potential for recycling and re-use, with suitable technology for separation already well established and economical [8]. CDW can be prevented through on-site management measures and re-use strategies, which include proper purchasing, handling of materials on-site as well as re-using existing building structures, using standardized components and educating onsite workers [9]. Some EU countries like the Netherlands, Malta or Luxembourg show high CDW recycling rates of up to 70%, which corresponds to the 70% recycling goal for non-hazardous CDW set out in the EU Waste Framework Directive (2008/98/EC). Still, most of CDW in across the rest of EU is landfilled [9].

Example

In 2019, the City of Vienna issued new goals within its Smart City Wien Framework Strategy 2019–2050 [10], committing to the reuse of 80% of the components and materials extracted when dismantling and renovating buildings. The city aims to achieve this urban mining goal by drawing on analytical tools, such as "digital twinning" of buildings representing material demand and supply sides (i.e., buildings as material banks) based on Building Information Modeling (BIM). The five lines of action reach across municipal administration, education, training and research, the economy, governance and digitalization.

Biological Nutrients

Nutrients are taken from the topsoil of agricultural lands, enter cities as food and are then removed as waste. Every year, 2.8 billion tons of organic waste occur in cities and less than 2% of nutrients contained in this organic waste are looped back to food production [11]. Urban inhabitants continuously discard large volumes of nutrients into wastewater streams via human excreta, biodegradable kitchen waste, and hygiene products. So far, the economy has lost around USD 23.3 billion of agricultural nutrients in human excrements alone [12] and additionally loses 100 million tons of bio-waste through the disposal of organic waste every year [13]. These nutrients are largely lost to the atmosphere (via nitrification, incineration), landfilling and the environment (residual nutrient loads in effluents). At the same time, Europe imports 30% of nitrogen, 71% of phosphorus, and 73% of all potassium fertilizers [14] used. Instead of being treated as a burden, organic residues could be transformed into fertilizer or nutrient-rich irrigation water that is safe for reuse, thereby returning "spent" nutrients to soils, where they enter a new lifecycle for food production. By applying currently available technologies, cities could capture nutrients from wastewater and metabolize them to produce fertilizer or soil amendment. This way, secondary nutrients can replace imported synthetic fertilizers produced from primary mineral reserves, avoid their high-carbon footprints while also reducing Europe's external resource dependency.

In addition to fertilizer, many other valuable resources can be recovered from urban wastewater using biological technologies, including nutrient-rich irrigation (fertigation) water, biopolymers, alginates, cellulose, construction material, and fundamental ingredients for energy production (biogas, biofuel, electricity, heat). Wastewater treatment has the potential to shift from a burden to a profit-generating resource factory [15].

Example

Innovative urban farming systems are emerging all over the world, e.g., "Farm.One," a multistory underground farm in New York [16]. Instead of relying on freshwater and fertilizer imports, they could easily draw on secondary nutrients and water already abundantly available in the city. Some urban-farming plots have already started using pre-treated wastewater for irrigation and fertigation, mostly in pilot scale, e.g., ROOF WATER-FARM [17] in Berlin and HOUSEFUL [18] in cities in Austria and Spain. These local, decentralized resource cycles can maximize process efficiencies for recovery and reuse, as well as reduce pressure from urbanization and dependence on long, distant food value chains [19]. Nutrient loads in urban wastewater are so high that nutrients recovered in cities could not only supply urban farming but excess nutrients could also be returned to peri-urban and rural fields.

Important

Cities bear opportunities to develop circularity internally but also to recover and return secondary resources back to production lines located beyond the city's borders. Considering limited space in cities, agricultural and industrial facilities beyond city borders and the benefits of rural development, it is not possible or desirable for a circular economy to develop exclusively internal to a city. Not all secondary resources could be efficiently utilized, and all needs unlikely covered. Large streams of resources flow into cities from the outside. But in the linear economic model, little is given back to where they came from.

Therefore, a circular city can be understood as the turntable of incoming resources. It can loop and cascade the resources as efficiently as possible and return excess resources back up the supply chain. This requires a redesign of the urban system and the interactions with its peri-urban and rural surroundings by city-regional planning.

The layout and design of cities change the way materials and products move around them and their inhabitants.

3   A Vision for Circular Cities

So, what can a circular city look like? (Fig. 1)

In a circular city, buildings, infrastructure, and products are designed to last. They are durable and easy to maintain and repurpose. Modular configuration makes it possible to replace, dismantle, and reassemble parts of a product. Objects can be cleaned, repaired, and shared, and are rarely *owned*. Instead, they are *used* and transferred among users to where they are needed. Tools are rented, washing machines are paid for by number of washing cycles, public transportation vehicles are used as a service, and privately owned automobiles are no longer part of the cityscape, having become obsolete.

Materials are sourced locally, non-hazardous, and made from renewable feedstock. Some objects or parts of them may have to be discarded after final use, but their chemical composition allows for separation and recycling with minimal energy input. Materials are largely bio-based, meaning that they can be cascaded from biomass. After final use, they can either be composted and applied as fertilizer to produce new biomass or processed directly to secondary bio-composite materials and industrial chemicals. Reverse logistics extend backwards all the way up the supply chain. Take-back models organize the collection and sorting of packaging materials for

Fig. 1   Circular city
©alchemia-nova

direct reuse or recycling. Residues from construction, industrial production, and agriculture provide input materials for other value chains.

The city's **infrastructure** is designed for decentralized production and services. Inhabitants have access to nearby repair shops, collection points, and centers for resource recovery and decentralized manufacturing, e.g., with 3D-printers. As neighborhoods become hubs within the urban system, people increasingly live, work, and play in close proximity, permitting access by foot, bicycle, and public transportation. Materials are circulated locally with additive manufacturing as a key technology for repair and upgrading. Reverse logistics link cities with their peri-urban and rural surroundings, making return flows possible, and closing cycles.

Nature is integrated into the city, delivering ecosystem functions for resource recovery, enhancing the energy balance, air quality, and wellbeing in the city. Nature inspires the design of functional biological systems to keep water and nutrient flows in cycles. Nutrients are extracted from wastewater and reused for food production. Residual biomass and organic waste are processed to materials for commodities and building materials in connected biorefineries. Residues of these refineries are used for biogas production and conversion to heat and electricity. The city runs on clean, renewable energy while residual heat is reused in buildings. Instead of leaving and relying on individual transportation in search of nature, they can meet and find recreation within local green spaces.

Important

A **circular city functions like an ecosystem**, where waste is always a resource for something else. It thrives by maintaining value and generating value from the same material resources again and again. Within a circular city, material productivity attains its highest potential while eliminating waste and reducing costs. New business opportunities emerge in decentralized production and service hubs, bringing jobs to neighborhoods and incentivizing the development of new skills. Green infrastructures and restored natural ecosystems in cities improve urban air quality, carbon balances, human health, and wellbeing. The harnessing of local production capacity makes the city more resilient.

By keeping products and materials in use and maintaining their value, we design out waste and pollution from cities. If resource cycles are internal to a city, production must comply with urban regulations. This means that negative externalities are contained within the city and become more visible and the true cost of products is revealed.

These principles can be applied to all urban resource flows, including construction materials, commodities, and food:

Built Environment

Buildings are used as material stockpiles, or "material banks." Digital twins of these material banks exist through 3D BIM, which can be used in the construction or refurbishment process instead of 2D models. BIM can provide information on the composition and location of materials stored in the built environment. Available materials could be redistributed over resource platforms, and planners design a building based on existing materials. These building materials are designed to be modular, repairable, deconstructable, and reusable, with their lifetime extending as long as possible, recycled only as the last resort. Buildings may integrate additional functions, which needs to be taken into account during a construction or refurbishment process.

Case Study Venlo city hall

The city hall of Venlo, Netherlands, has successfully realized the vision of a building inspired by Cradle to Cradle®. The compatibility with a technical cycle was one of the four main criteria during the design phase of the project alongside air and climate quality, integrated renewable energy, and enhanced water quality. Materials were chosen according to their applicability for biological and technical cycles, where residues are raw materials for new products, Materials have an added value for users and the environment. Separability and recyclability were core to the selection of building materials and suppliers, based on comprehensive content lists. As such, surfaces inside the building remain uncoated.

The interior of the building is subject to a take-back regime which foresees its return to the producer after 10 years with a predefined residual value [20] (Fig. 2).

Case Study circular building materials

Businesses have recognized the value in taking back used items. Reverse logistics is slowly penetrating into all layers of supply chains on the path to a circular economy [21].

Fig. 2 Powered by Mostert De Winter bv, The Netherlands, photographer Hans van Cooten

Fig. 3 Daas ClickBrick® system ©Daas Baksteen, published by MaterialDistrict

Desso, a carpet producer, offers a carpet leasing service option that includes installation, cleaning, maintenance, and eventually removal. Customers no longer own the carpet. Instead, the carpet is provided as a service by the manufacturer, who will take the carpet back for recycling at the end of service. The manufacturer can then produce new carpets from this material [22].

Another best practice in the field of reverse logistics is Daas ClickBrick®, a product by the Dutch company Daas Baksteen. This brick renounces the use of glue and relies on a dry stack system, allowing for the bricks to be installed and taken apart easily. Consequently, ClickBrick® can be removed from a building site and reused without a compromise in quality [23] (Fig. 3).

Commodities

In a circular city, commodities are no longer used just once. They are repaired and refurbished. Single-use items, including packaging, vanish from the circular city. There are suppliers for parts if things break. Circular product design minimizes the complexity of material composition for more efficient recycling. The system of commodities will be governed by take-back regimes, extended producer responsibility, and supplied largely by bio-based virgin materials. People gain access to what they need in new ways. Instead of being owned, commodities are used—through sharing or renting, or through product-as-a-service contracts. Functions are provided, but not by selling material goods. Companies don't sell light bulbs, they provide lighting, not radiators but heating, not a washing machine but washing cycles, not cars but transportation. Companies employing product-as-a-service models deliver the functional service but maintain ownership over material goods. This way, they are incentivized to provide products with longer lifetimes and minimal maintenance instead of selling sheer quantity. With commodities like cars, tools, furniture having become a burden in densely populated places, the sharing economy has become mainstream via eBay, NeighborGoods, willhaben, and Poshmark for items, tools and clothes, liquid and sharenow for vehicles, coworking spaces, and many more.

Food Supply

Inspired by nature, a circular urban food system can work in endless nutrient loops. Nature does not know the concept of waste. In nature, one residue is always food for something else. Besides inspiration, nature-based solutions already enable nutrients to be recovered from urban wastewater and safely reused on-site for urban farming while also providing co-benefits such as converting CO_2 to edible or combustible biomass and improving air, water, and soil quality [24]. This way, urban farming can make the use of secondary resources directly where they are available. The projects HOUSEFUL [18] and HYDROUSA [25] (see below) are examples of on-site nutrient and water loops. Urban greens, parks, and underutilized infrastructures can be used for urban farming, and the bounty distributed via soft urban infrastructures such as regular markets with local food suppliers. Excess nutrients can be returned to green belts around cities and rural areas.

Case Study

HOUSEFUL (www.houseful.eu)

This project demonstrates the holistic view of buildings and embeds a series of additional functions into them, co-creating and testing circular business opportunities, and fully embracing the capabilities of nature-based solutions. These technologies include nutrient and water capture through green walls, green roofs, vertical ecosystems, localized energy generation, urban farming, and automation of the systems to minimize maintenance. This project addresses both single buildings and a community at neighborhood scale as entities within the urban system (Fig. 4).

Case Study

HYDROUSA (www.hydrousa.org)

This project demonstrates how to keep water and nutrients in loops. HYDROUSA works with passive systems like the sun to desalinate saltwater, capture rainwater and humidity through surface structures and process wastewater using nature-based solutions. The nutrients in the treated wastewater are applied to a one-hectare biodiverse agroforestry field. The establishment of restorative agriculture is central to the project, restoring ecosystem functions to integrate wastewater treatment with resource reuse and food production. HYDROUSA also showcases the cyclic exchange of nutrients and water between communities and agricultural production in a peri-urban context. Nutrients are harvested as food, consumed, transported by wastewater, metabolized in biological systems, and reapplied to fields. Nutrients are converted back to food with the sun as the energy source (Fig. 5).

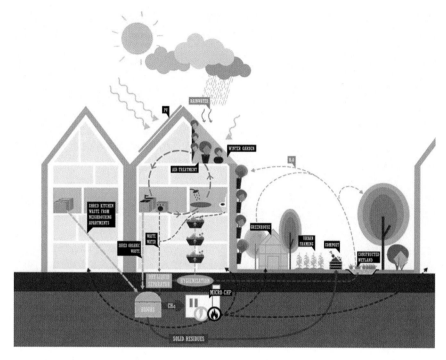

Fig. 4 HOUSEFUL biological cycle overview ©alchemia-nova

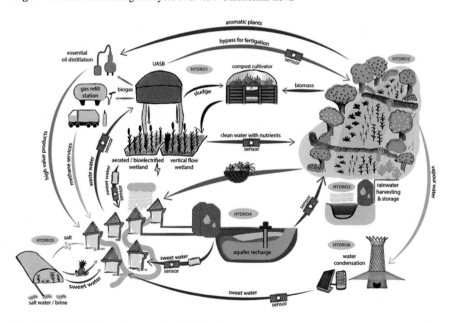

Fig. 5 HYDROUSA cycles overview ©alchemia-nova

Key Points

- Design products in a way, that they can keep their value at the highest possible level for as long as possible.
- Make all products repairable, refurbishable, and recyclable.
- Set quality as the ultimate goal.
- Circular supply chains and reverse logistics are key.
- Make use of things that are already available and design processes to make them accessible.
- Make the use of underutilized infrastructure and integrate additional functions.
- Think in systems and be inspired by nature.

4 Actors in the Circular City Fabrics

Cities are socio-technical systems [4]. They are driven by people who live and work in proximity. More and more planning processes in cities are conducted in a transparent and often even co-creative way. They are driven by urban planners and city authorities, while policymakers set the policy and legal framework. Academia provides innovative technology and solutions, while businesses try to get the innovation to market while developing and operating business models that realize the circular economy. These actors must work together to achieve circular cities and a wider circular economy. The following actions illustrate how these actors rely on each other and drive forward the transition from different angles of the circular city fabric.

Build Partnerships

- City governments can enable, lead, and involve key stakeholders from across the public and private sectors.
- Partnerships between circular grassroots initiatives, experts, financing bodies, and municipal authorities need to be sought.
- Top-down measures can follow sustainable bottom-up initiatives and should also have a big systemic picture in mind.
- We can learn from strategies of other pioneering cities [26] and existing methodologies to define the way forward (e.g., from the Ellen MacArthur Foundation [27] or from the European Investment Bank [28]).

Enable Sustainable Financing

- Financing infrastructure needs to include also quality goals and long perspectives, rather than short-term profits including cheapest materials possible, i.e., enabling impact investment.
- Circular public procurement can also play an important role in pioneering business cases.

- To achieve an approximation of a true-cost accounting, reflecting the real value for our ecosystem, something like a bonus-malus policy or ecotaxation for rewarding sustainable products or processes and penalizing unsustainable or linear ones.

Design Circular

- Enabling and incentivizing better production through business support and advisory services that focus on design and business development.
- Extended Producer Responsibility (EPR) can also be a strong tool for enhancing eco- or circular design at the producer level [29].
- Several concepts have been developed to assess and promote life-centered design on macro- and microeconomic levels, e.g., happiness indexing, or economy for the common good.
- Multifunctional design of the built environment, commodities, and even food systems require the cooperation of many disciplines in research and administration.

Manage Resources Wisely

- Public actors can provide resource management infrastructure that facilitates services such as collection, sorting, reuse, and recycling.
- Digitalization can enable tracing and sharing resource flows in cities.
- Policymakers can benchmark and steer progress through measuring circular economy at the entry, process and exit points of resources and materials—in the best case following widely accepted indicators [30].

Co-create Strong Reforms

- Fiscal reforms incentivize the way materials are handled, they can steer the adoption of service versus sales-based business models and the cost of labor for all services, especially labor-intensive specialized repair.
- Direct regulation or banning of certain materials and products, such as single-use packaging, certain building materials, construction practices, and even food production can steer business and consumer choices.
- Entrepreneurs and innovators play the role of pioneers. Policymakers can establish innovation niches, where they can test sustainable innovations that are not yet in line with the current policy frameworks (see Green Deal in the Netherlands [31]).

Measure Progress

- Municipality management also includes gathering information about the current standing of the city at different levels. This information needs to be measurable to compare different practices in cities.
- Finding this common language also means standardizing community management as such. These protocols are currently being developed at different levels and still need to be harmonized (see also Subchapter 5).
- The best or most valuable measurable factors will form the key performance indicators (KPIs) of a city and can include public procurement, materials stocks, water, nutrient, and food management among others.

Circular economy needs to be translated to all levels of understanding. Information materials should be available for policymakers, planners, research, and funding agencies, and the wider public in general. Educators must be brought on board to convey awareness and knowledge of circular economy principles to the next generation of consumers and producers.

Tip

Multifunctional, life-centered, and participatory design can help to create products and systems that are fit for a fully circular city and generate positive outcomes for the environment, society, and economy. To summarize, the following steps can provide the foundation:

- Engage with other circular initiatives and learn together.
- Enable participation and give space to circular initiatives to grow.
- Educate at all levels.
- Develop strategies with milestones and measure progress.
- Cooperate with the surrounding area and facilitate the exchange of nutrients and secondary materials.
- Learn about circular assessment (e.g., Cradle to Cradle Certified® [32], The Circularity Check [33], the Circularity Gap Report [34], Circular Business Solutions [35], and Circulytics (EMF, 2019).

5   Circular Cities Best Practice

Circular City Indicators

Indicators can help to inform policies and to streamline the implementation of circular economy strategies. Cities can also use indicators to report their efforts and achievements. However, there is no shared view on circular economy indicators for cities.

Many indicators have been developed to assess and compare levels of circularity, such as the EU Monitoring Framework for the Circular Economy complementing the

European Commission's Resource Efficiency Scoreboard and Raw Materials Scoreboard [36]. The ten indicators are listed and described in COM (2018) 29 final [37]. They cover aspects of production and consumption, waste management, secondary raw materials, and competitiveness and innovation. Other sets of indicators include the European Union Eco-innovation Action Plan including indicators for sustainable resource management, societal behavior, and business operations [38], the Eurostat collection of indicators to measure circular economy at the national level [39], the Ellen MacArthur Foundation's Material Circularity Indicator to measure the product or company-level circularity [40], the Cradle to Cradle product standard [32], as well as sector-specific indicators such as the BAMB Circular Building Assessment [41] or the German Sustainable Building Council's recommendations [42]. The government of Luxembourg leads the Circularity Dataset Initiative, which aims to develop an industry standard to report circular practices [43]. This approach of harmonization is also fully in line with the new European Green Deal, which also suggests to develop standards for green reporting among many more supportive policy measures like renovation waves or enhancing nature-based solutions [44].

However, these frameworks are not fully applicable to cities and the corresponding data at city level is also not easily accessible or even publicly available in some cases, as opposed to much more widely available national-level data. Some cities have developed their own progress measurement frameworks, such as the City of Peterborough (UK) [41], or through a research paper, the City of Porto [45].

Amid the diversity of scopes of available case descriptions and data limitations, the definition of circular city indicators is a necessary first step to facilitate cities to collect and provide data to enable assessments, comparison, and progress measurement. Further, the indicators must be applicable to cities at various stages of socio-economic development so that cities can be compared globally. Capturing informal circularity, such as informal separate waste collection and recycling, or illegal dumping of waste, will remain a challenge. Mass balances (material flow analyses) can establish a solid baseline for the identification of formalized and informalized resource flows. This approach is also suggested by Eurostat, the European statistical agency [39].

Sustainable Cities

Meanwhile, there are many frameworks for Sustainable Cities [46], some of which include indicators relevant to circular economy, e.g., volume of solid waste generated, recycling rate, etc. While circular cities are geared toward sustainability, rankings might be different from sustainability rankings alone as sustainability actions might not necessarily contribute to circularity or vice versa. The Urban Agenda, a partnership among EU-level institutions and member states, started to respond to the need for city-level indicators by consolidating CE indicators and issues highlighted by stakeholders [47], but no indicator framework for circularity of cities has been institutionalized or widely accepted so far, which also entails a lack of comparable data.

Case Studies

There are several attempts to collect and share case studies of circular cities worldwide, e.g., by C40 Cities and the EIT Climate-KIC (European Institute of Innovation and Technology's Knowledge and Innovation Community targeted at a zero-carbon economy) [48] and the Ellen MacArthur Foundation [49]. The available municipality-led case studies are very diverse in their scope and approaches. Only little data is available on city-wide achievements (and no city-wide cross-sectoral index). As such, **Brussels** (Belgium) is a frontrunner in the construction and demolition sector, recovering 91% of waste and thus exceeding the European Waste Directive's construction and demolition recovery target of 70% by 2020 [48]. **Samsø** (Sweden) is 100% self-sufficient in energy [40]. A remarkable average of 40% of **Singapore's** water demand is supplied by reclaimed water [50].

Most case studies describe municipality-driven actions that are taking place. They are summarized in Table 1.

6 Conclusion

Cities play a special role within the transition to a circular economy. Currently, they are major resource sinks, but their specific attributes enable them to emerge as drivers of the circular transition. Significant human and economic activity and resource use, population density, and limited space bear pressures and opportunities for the transformation of how the built environment, commodities, and food supply are managed. Cities have the human and economic force to recirculate resources internally and in exchange between city-regions. Business models based on reuse, repair, refurbishment, and recycling are highly efficient in cities, hence a host of circular innovations and initiatives have already successfully established in cities. As pioneers and resource turntables, cities play a vital role in driving forward the transformation toward a circular economy as a whole.

Cities can be inspired by natural ecosystems, where waste is always a resource for something else. In addition, nature-based solutions can be integrated into cities to reap multiple benefits for circularity, resilience, and sustainability. Multifunctional, life-centered, and participatory design among diverse yet closely interrelated stakeholders, based on the foundation of education, can enable to tackle remaining institutional barriers. This way, cities can foster an economic system that serves humanity while operating within the planetary boundaries.

Questions

1. What role can cities play within a circular economy?
2. What is life-centered design?
3. What can nature-based solutions achieve?
4. How can we achieve sustainable systemic solutions?

Table 1 Cities for comparison of case studies

Scope	Action	City
City-wide, cross-sectoral strategic level	Circularity strategies and financing commitments	Maribor (Slovenia) [47] London (UK) [47, 48] Glasgow (UK) [47, 48] Brussels (Belgium) [47] Copenhagen (Denmark) [47] Gothenburg (Sweden) [47] Paris (France) [47] Samsø (Sweden) [47] Seoul (South Korea) [47] Peterborough (UK) [48]
City-wide sector-specific strategies	Sharing economy action plan	Amsterdam (Netherlands) [48]
City-wide public information and engagement	Public information and engagement campaign	Brussels (Belgium) [47, 48] New York (USA) (#WearNext campaign promoting New Yorkers to keep clothes in use) [48]
	Citizen and business collaboration center	Kristiansand (Norway) [47]
	Repair network	Vienna (Austria) [47]
City-wide, sector-specific action programs	Circular buildings and urban mining	Amsterdam (Netherlands) [47]
	Building refurbishment	Houston (USA) [47] Paris (France) [47] Sydney (Australia) [47] Vienna (Austria) [47] Berlin (Germany) [47] Helsinki (Finland) [47] Tokyo (Japan) [47]
	Municipal public procurement	Toronto (Canada) [47, 48]
	Water	Aguascalientes (Mexico) [47]
	Low-energy city	Basel (Switzerland) [47]
	Online marketplace for re-using materials	Austin (USA) [47, 48]
	Regulations on the use of plastic bags	Quezon (Philippines) [47]
	Large-scale municipal biowaste to biochar production	Stockholm (Sweden) [47]

(continued)

Table 1 (continued)

Scope	Action	City
	Studies preparing city-wide projects in the fields of textiles, construction & demolition, innovation & acceleration, water, and food	Tel Aviv (Israel) [47]
Local actions	Industrial symbiosis programs	Cape Town (South Africa) [47] Malmö (Sweden) [47]
	Heat recovery	Arras (France) [47]
	World's first circular shopping center	Eskilstuna (Sweden) [47]
	Second-hand stores led by municipal waste companies	Kristiansand (Norway) [47] Vienna (Austria) [51] New York (USA) [47]
	Electronics repair and service centers	Vienna (Austria) [47] Belo Horizonte (Brazil) (computers) [48]
	Cradle to Cradle carpets for city buildings	San Francisco (USA) [48]
	Cradle to Cradle City Hall	Venlo (Netherlands) [48]
Recycling-focus actions	Resource Innovation Campus at the city's landfill and waste processing facilities	Phoenix (USA) [47]

Answers

1. They can use their human and economic force to function as resource turntables, where they re-circulate products and materials, recover and return excess resources back up the supply chain to production lines in peri-urban and rural surroundings.
2. Life-centered design puts the needs of life-supporting functions as design objective.
3. Multi-functional benefits such as resource recovery, wastewater treatment, ecosystem services, improvement of air and water quality, conversion of CO_2 to biomass.
4. By changing from more to better, replacing quantity with quality.

Suggested Reading

Ellen MacArthur Foundation & ARUP—Circular Economy in Cities https://www.ellenmacarthurfoundation.org/our-work/activities/circular-economy-in-cities.

European Commission Circular Economy Package https://ec.europa.eu/environment/circular-economy/index_en.htm.

New European Green Deal https://ec.europa.eu/info/strategy/priorities-2019-2024/european-green-deal_en.

European Circular Economy Stakeholder Platform https://circulareconomy.eur opa.eu/platform/en.

alchemia-nova circular economy https://www.alchemia-nova.net/circularecon omy/.

References

1. World Bank. (2017). *Urban Development Overview.* http://www.worldbank.org/en/topic/urb andevelopment/overview.
2. Ellen MacArthur Foundation & ARUP. (2019). *Circular Economy in Cities.* http://mava-founda tion.org/wp-content/uploads/2019/01/EMF-Circular-economy-in-cities-preview-paper-1.pdf.
3. United Nations Development Programme. (2019). Goal 11: Sustainable cities and communi- ties. *Goal 11: Sustainable cities and communities* https://www.undp.org/content/oslo-govern ance-centre/en/home/sustainable-development-goals/goal-11-sustainable-cities-and-commun ities.html.
4. Swilling, M., Robinson, B., Marvin, S., & Hodson, M. (2013). *CITY-LEVEL DECOUPLING.* Summary for Policy Makers: Urban resource flows and the governance of Infrastructure transitions.
5. Dijkstra, L. & Poelman, H. (2012). *Cities in Europe the new OECD-EC definition.* 16 https:// ec.europa.eu/regional_policy/sources/docgener/focus/2012_01_city.pdf.
6. European Commission. (2015). Nature-Based Solutions. https://ec.europa.eu/research/enviro nment/index.cfm?pg=nbs.
7. Fischer, C. & Werge, M. (2009). *EU as a recycling society.*
8. European Commission. (2019). Construction and demolition waste. https://ec.europa.eu/env ironment/waste/construction_demolition.htm.
9. Ganguly, P. (2012). Construction and demolition waste handling in the EU. *Littera Scr.* 13.
10. Smart City Wien Rahmenstrategie 2019–2050 (2019). https://smartcity.wien.gv.at/site/files/ 2019/06/SmartCityWienRahmenstrategie2019-2050_Beschlussfassung190626.pdf.
11. Ellen MacArthur Foundation. (2018). *Cities and the circular economy for food.* https://www. ellenmacarthurfoundation.org/our-work/activities/cities-and-the-circular-economy-for-food.
12. Jenkins, J. C. (2011). *The humanure handbook: A guide to composting human manure* (Joseph Jenkins Inc.).
13. European Compost Network. (2019). Bio-waste in Europe. *European compost network.* https:// www.compostnetwork.info/policy/biowaste-in-europe/.
14. Fertilizers Europe. (2017). *Industry facts and figures 2018.* http://fertilizerseurope.com/.
15. Ellen MacArthur Foundation. (2017). *Urban biocycles.* https://www.ellenmacarthurfoundat ion.org/publications/urban-biocycles.
16. Krueger, A. (2017). Herbs from the underground. *The New York Times.*
17. Roof Water-Farm. (n.a.). Roof water-Farm. *Roof Water-Farm* http://www.roofwaterfarm. com/en/.
18. Fondazione iCons. (2019). Houseful. http://houseful.eu/.
19. Tull, K. (2018). *Urban food systems and nutrition.* https://assets.publishing.service.gov.uk/ media/5bae42ffed915d259eaa7769/383_Urban_Food_Systems_and_Nutrition.pdf.
20. Detail, I. (2019). C2C inspired: Venlo City Hall. https://inspiration.detail.de/venlo-city-hall- 113981.html.
21. Dekker, R., Fleischmann, M., Inderfurth, K., & van Wassenhove, L. N. (2013). *Reverse logistics: Quantitative models for closed-loop supply chains* (Springer Science & Business Media).
22. Desso. Carpet leasing. *Desso* http://www.desso-businesscarpets.co.uk/services/carpet-leasing/.
23. KDB Baukeramik Vertriebs GmbH. (2019). Daas ClickBrick. *KDB Baukeramik Vertriebs GmbH* https://www.kdb.de/clickbrick.

24. Kisser, J. et al. (2020). A review of nature-based solutions for resource recovery in cities.
25. Impact Hub Athens. (2019). HYDROUSA. *HYDROUSA* https://www.hydrousa.org/.
26. European Union. (2019). *European circular economy stakeholder platform—strategies.* https://circulareconomy.europa.eu/platform/en/strategies.
27. Ellen MacArthur Foundation. (2019). *Circular economy in cities.* https://www.ellenmacarthurfoundation.org/our-work/activities/circular-economy-in-cities.
28. Byström, J. & Continenza, C. (2018). *The 15 circular steps for cities.* 14.
29. OECD. (2019). Extended producer responsibility. https://www.oecd.org/env/tools-evaluation/extendedproducerresponsibility.htm.
30. European Commission, Directorate-General for the Environment, University of the West of England (UWE) & Science Communication Unit. (2015). *Indicators for sustainable cities* (Publications Office).
31. Rijksdienst voor Ondernemend Nederland. (2019). *Green deal.* https://www.greendeals.nl/english.
32. Cradle to Cradle Products Innovation Institute. (2019). Resources. *Resources* https://www.c2ccertified.org/resources/collection-page/cradle-to-cradle-certified-resources-public.
33. European Sustainable Business Federation, WeSustain & Circular Future. (2019). Circularity Check. *ecopreneur.eu* https://ecopreneur.eu/circularity-check-landing-page/.
34. Circle Economy. (2019). Circularity Gap 2019. *circularity-gap.* https://www.circularity-gap.world.
35. Alchemia-nova. (2019). *Circular business solutions.* https://www.alchemia-nova.net/services/circular-business/.
36. European Commission. (2019). Raw materials information system. *EU science hub. raw materials information system (RMIS)* https://rmis.jrc.ec.europa.eu/?page=scoreboard2016.
37. European Commission. (2018). *Measuring progress towards circular economy in the European Union: Key indicators for a monitoring framework.* https://eur-lex.europa.eu/legal-content/EN/TXT/PDF/?uri=CELEX:52018SC0017&from=en.
38. European Commission. (2019). Circular economy indicators. *Eco-innovation action plan.* https://ec.europa.eu/environment/ecoap/indicators/circular-economy-indicators_en.
39. European Commission—Eurostat. (2019). Monitoring framework. https://ec.europa.eu/eurostat/web/circular-economy/indicators/monitoring-framework.
40. Ellen MacArthur Foundation. (2017). Circularity indicators. *Measuring company circularity.* https://www.ellenmacarthurfoundation.org/resources/apply/measuring-circularity.
41. Paez, C. (2018). Environmental footprint and circularity in the construction sector. *Build up. The european portal for energy efficiency in buildings.* https://www.buildup.eu/en/news/environmental-footprint-and-circularity-construction-sector.
42. German Sustainable Building Council. (2019). Circular economy| DGNB. https://www.dgnb.de/en/topics/circular-economy/index.php.
43. Ministère de l'Économie/Luxembourg. (2019). *Circularity dataset initiative.* https://meco.gouvernement.lu/fr/le-ministere/domaines-activite/ecotechnologies/circularity-dataset-initiative.html.
44. European Commission. A European green deal. https://ec.europa.eu/info/strategy/priorities-2019-2024/european-green-deal_en.
45. Cavaleiro de Ferreira, A., & Fuso-Nerini, F. A framework for implementing and tracking circular economy in cities: The case of porto. *Sustainability.*
46. European Commission. (2018). *Indicators for sustainable cities. In-depth Report.* http://ec.europa.eu/environment/integration/research/newsalert/pdf/indicators_for_sustainable_cities_IR12_en.pdf.
47. Urban Agenda for the EU. (2019). *Indicators for circular economy (CE) transition in cities—Issues and mapping paper.* https://ec.europa.eu/futurium/en/system/files/ged/urban_agenda_partnership_on_circular_economy_-_indicators_for_ce_transition_-_issupaper_0.pdf.
48. C40 Cities & EIT Climate-KIC. (2018). *Municipality-led circular economy case studies.* https://www.c40.org/researches/municipality-led-circular-economy.

49. Ellen MacArthur Foundation. (2017). Case studies. *Circular economy in cities* https://www.ell enmacarthurfoundation.org/our-work/activities/circular-economy-in-cities/case-studies.
50. PUB. Singapore Water Story. (2019). *PUB, Singapore's national water agency.* https://www.pub.gov.sg.
51. Magistrat der Stadt Wien. 48er-Tandler. *48er-Tandler* https://48ertandler.wien.gv.at/site/.

Johannes Kisser is the Technical Director of *alchemia-nova.* As a Chemical Engineer, he has conducted research on nature-based solutions for resource recovery, circular bioeconomy, and ecosystem restoration for over 15 years. He has given numerous speeches at forums, conferences and workshops, and holds lectureships on sustainable product design, supply chain management, circular construction, and CSR management at Austrian universities. He has initiated and developed several Austrian and EU-funded innovation projects on circular economy and nature-based solutions. He currently leads the interdisciplinary Working Group on Resource Recovery within the European research coordination project on Circular Cities. Johannes takes a systems approach combining technology innovation inspired by nature, and driving social transformation.

Maria Wirth is a researcher at *alchemia-nova research & innovation gemeinnützige GmbH.* During her studies of Environmental Technology & International Affairs, she specialized on the interactions between societies and the natural environment. This brought her to the United Nations Industrial Development Organization (UNIDO) and Deutsche Gesellschaft für Internationale Zusammenarbeit (GIZ) GmbH, where she supported strategic partnerships as well as the implementation of international projects for sustainable development of agri-food value chains and industries in Asia and Africa. Since 2018 at alchemia-nova, she supports and develops research and innovation projects on topics related to nature-based solutions and circular economy, with a focus on regenerative resource loops in cities.

Industrial Circular Manufacturing

Lei Shi

Abstract Industry is the propeller for the establishment and development of modern society by creating more human-dominated material and energy flows to transform resources into products or provide services. With the expansion of scale and variety, the industrial system has also laid its many negative marks in the natural environment systems. Thus, we need to reshape traditional industrial systems into more green, low-carbon, and circular ones according to circular economy principles. This chapter covers the following three aspects: (1) the hierarchical and circular structure of industrial ecosystems; (2) circular transformation strategies and practices, including eco-design at the product level, cleaner production at the process level, eco-industrial parks at the park level, sustainable industrial transformation at the regional level; (3) policy instruments of extended producer's responsibility.

Keywords Circular economy · Industrial ecology · Industrial ecosystems · Cleaner production · Eco-design · Eco-industrial parks · Extended producer's responsibility

Learning Objectives

- Circular economy aims to develop economies on the basis of material recycling, while the industry is the propeller to make it happen.
- Industry is a multi-scale complex system, each scale hosts numerous opportunities and means for circular transformation.
- Industrial circular economy measures include eco-design, cleaner production, eco-industrial parks, and sustainable industrial transition, etc.
- A complex system engineering is needed to promote and implement industrial circular economy by integrating and coordinating technology, market, and policy.

L. Shi (✉)
School of Environment, Tsinghua University, Beijing, China
e-mail: slone@tsinghua.edu.cn

1 Introduction

Industry system is one of the important components of social and economic system, taking responsibility for transforming raw materials into products. Generally speaking, industry refers to the material production sector engaged in the exploitation of natural resources, the processing and reprocessing of extractive products and agricultural products. In the categories of industrial division, industry belongs to the secondary industrial sector. This definition elaborates on industry from the perspective of industrial form but does not reveal that industry is actually the material carrier of industrial civilization in essence.

To a large extent, industry is the embodiment of technology. It is a system in which people use technology to create products through various industrial organizations. Generally, industries did not appear at the same time but were triggered by a series of technological changes since the birth of the industrial revolution, which profoundly changed the face and connotation of world development [1]. The power of industry lies in the snowballing growth of its own prosperity and material wealth through a series of self-reinforcing and accumulating technological advances. Increasing wealth, in turn, helps push forward technological, financial, institutional, social, and cultural changes, thus turning the whole economy into an intertwined complex system, which is the essence of industrialization. In fact, every developed industrial economy has already undergone a process of industrialization.

The industry has brought about a huge amount of material wealth to mankind. However, it also left a mottled mark on the nature [2]. The first half of the twentieth century witnessed the outbreak of environmental problems in industrialized countries. With more and more emerging industrial sectors created, more and more unpredictable environmental problems emerged including resource scarcity and environmental degradation. More seriously, the scope of influence of these problems extended from local to regional and eventually to global scale. Nowadays, the environmental issues we talk about the most are climate change, POPs, and the more recent plastic wastes [3].

Obviously, the traditional industrial development pattern is not what we hope for. We need to transform it into a more sustainable one. Since the publication of The Silent Spring in 1962, our mankind has started the process of environmental governance. In just half a century, environmental governance has gone through four paradigms, from the initial end-of-pipes management to cleaner production, to product and service system, and finally, to sustainable system transformation. The goal of sustainable system transformation is to establish a green, low-carbon, and circular industrial ecosystem. Among them, circular economy aims to build industrial development on the basis of material circulation, adoption of renewable energy, the improvement resource productivity and material efficiency as far as possible.

Circular economy requires sustainable transformation at every scale of industry which is a complex and multi-scale system. At the product level, circular economy advocates eco-design which involves recycling, remanufacture, easy disassembly, biodegrading, etc. At the production level, circular economy advocates cleaner production, waste minimization, zero-waste discharge, etc. At industrial park level, circular economy advocates eco-industrial park, green industrial park, circular reform, etc. At the regional level, circular economy advocates closed circulation systems. At the global level, circular economy advocates sustainable transformation of global production networks.

In order to promote industrial circular transformation, the field of industrial ecology has developed a systematic pedigree of industrial metabolism methods [4], including life cycle analysis, material flow analysis, substance flow analysis, and environmental input–output analysis. At the same time, from a complex system engineering perspective, industrial circular transformation also needs the coordination of technology, market and policy. Some innovative policies or business models have already emerged, such as extended producer responsibility, product service system, etc.

Industry is a sector that produces goods or related services within an economy.

Industrialization is the period of social and economic change that transforms a human group from an agrarian society into an industrial society, involving the extensive re-organization of an economy for the purpose of manufacturing.

Industrial ecology is the study of material and energy flows through industrial systems. Industrial ecology seeks to quantify the material flows and document the industrial processes that make modern society function [5].

Eco-design is both a principle and an approach to designing products with special consideration for the environmental impacts of the product during its whole life cycle [6].

Cleaner production was defined by UNEP in 1990 as: "The continuous application of an integrated environmental strategy to processes, products and services to increase efficiency and reduce risks to humans and the environment" [7].

Eco-industrial park is an industrial park in which co-located businesses and infrastructures cooperate with each other to seek enhanced environmental, economic, and social performance through exchanging wastes and sharing physical or non-physical resources, such as materials, water, energy, and information [8].

Extended producer's responsibility is a policy approach under which producers are given a significant responsibility—financial and/or physical—for the entire life cycle of the product and especially for the take-back, recycling and final disposal phases [9].

2 The Hierarchy and Circles of Industry

2.1 The Hierarchy of Industry

Industry is one typical multi-scale complex system, from the micro-level of molecules, molecular clusters, and multiphase systems, to medium level of the production processes, factories, and industrial parks, all the way to the macro level of regional industrial clusters and global production networks. Both spatial-scale and temporal-scale spans more than ten orders of magnitudes, as shown in Fig. 1.

The hierarchical feature determines that industry is a nested system which is influenced by both global and local factors. In the aspect of globalization, the current industrial system is already a global production network in which almost all industrial enterprises are embedded and dependent on each other. The vastness of this network has led us to be more or less indifferent to the ecological impacts of industrial production, which means that we are not aware of their environment and human health impacts when we buy products from other places in the world.

In terms of localization, industrial systems can be turned into industrial ecosystems by taking certain localized measures including waste exchanging, infrastructure sharing, and material/energy cascading usages. The classical industrial ecosystem is the Kalundborg industrial symbiosis system in Denmark (see example) [5]. The so-called localization is to gain a competitive advantage and win the competition in the global production network.

"Example Starts"

The industrial symbiosis system of Kalundborg in Denmark is a typical industrial ecosystem (Fig. 2) [10]. This system hosts dozens of industrial enterprises, including coal-fired power plants, oil refineries, enzyme formulation plants, and gypsum board plants, etc. In this system, different commodities have different spatial properties. For example, enzymes and refined products are globally traded commodities, while electricity, building materials, and thermal networks are regional commodities. Over almost half a century, this system has formed a dense and efficient material and energy exchange network between enterprises through waste exchange and infrastructure sharing. By industrial symbiosis, this system achieves a win–win situation

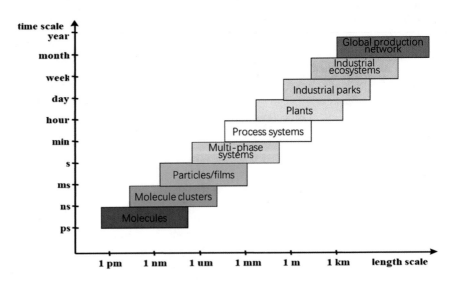

Fig. 1 Industry is a multi-scale complex system

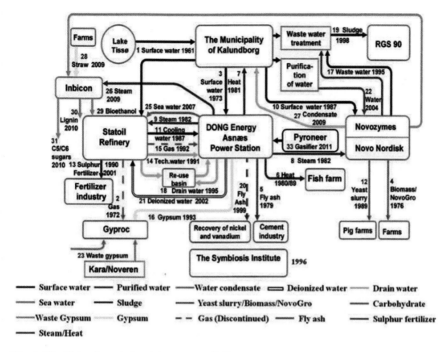

Fig. 2 The Kalundborg industrial ecosystem [10]

for environment and economy on the scale of industrial park. We call this type of industrial park "eco-industrial park."

"Example Ends"

2.2   The Circles of Industry

At every scale of the industrial system, there exist many opportunities to realize the reuse or recycle of the whole machinery, parts, materials, and chemical elements. Thus, we can thereof distinguish specific recycle and reuse circles including product circle, component circle, material circle, and chemical circle.

2.3   Product Circle

Being the carrier of service function, products can be reused or recycled at a whole level with the aid of some technical recovery measures when they enter the end of their lives. Generally speaking, the circulation at the product level has the following three forms: (1) on-site reuse, reusing in the original site after some repair or remanufacturing. For example, some factories usually reuse their machinery equipment or infrastructure products; (2) offsite reuse, reusing in different places. For example, some beverage bottles can be reused after simple cleaning; (3) cascading usage, refers to the usage to other degraded market users after restoration or without restoration. For example, clothing products can be made use of by different people through donation or other channels.

2.4 Component Circle

Most products in the assembly industrial sectors consist of a large number of components. Through replacement or repair of components or parts, the life of products can be extended, which generates the component cycle. Similar to the whole product cycle, there are also three types of component cycle, namely onsite reuse, offsite reuse and cascading usages. By innovating component recycling, new business models can be created. One of the famous component recycling examples is the reuse of printer cartridges by Xerox Company [6]. Another case is the reuse of carpet modules by Interface Company. In fact, it is the component reuse or recycling that supports the transition from product economy to service economy.

2.5 Material Circle

Physical products are composed of a variety of materials, such as biological materials, metallic materials and non-metallic materials. At the end of the product life cycle, they can be recycled at the material scale. Material recycling can be observed in all industrial sectors and in very diverse forms. For example, fibrous materials in clothing wastes can be recycled into carpets fillings; Aluminum materials in beverage cans can be used in aluminum casting production; Recycled aggregate materials can be used in new building structures. Material circle hosts not only degraded recycling but also equivalent or even advanced recycling. For example, waste PET bottles can be recycled for clothing production.

2.6 Chemical Circle

At material circles, there is a special kind of circulation that requires chemical reactions, which we call it chemical circulation or chemical recycling. Take plastic wastes as example, chemical recycling offers an innovative way to reuse mixed or contaminated plastic waste that can't be recycled at the moment: through thermochemical processes, the plastic can be used to produce syngas or oil, and the resulting recycled material can replace some fossil resources for chemical products. In fact, some chemical companies, such as BASF, have broken new ground in recycling plastic waste through its ChemCycling program. So far, BASF has worked with more than a dozen customers from different industries to develop pilot products for cheese packaging, refrigerator parts, and heat insulation panels.

Case Study The Ricoh's Comet Circle [11]

In 1994, Ricoh established the Comet Circle™ to express a comprehensive picture of how it can reduce its environmental impact, not only in its activities as a manufacturer and sales company but also from upstream to downstream—along the entire lifecycle of its products (Fig. 3).

The Comet Circle™ centers are based on the belief that all product parts should be designed and manufactured in a way that they can be recycled or reused. Ricoh management uses the Comet Circle™ as a real tool to plan its portfolio of products and activities. It is on this basis that Ricoh established the GreenLine label as a concrete expression of its resource recirculation business, with its priority on inner-loop recycling.

Its evolved relationship with its products in use is producing further results: optimizing the years that machines stay in operation at customer sites and generating annuity; generating additional revenue and margin by selling equipment more than once; and of course making a considerable contribution to resource conservation. Ricoh's objectives are to reduce the input of new resources by 25% by 2020 and by 87.5% by 2050 from the level of 2007; and to reduce the use of—or prepare alternative materials for—the major materials of products that are at high risk of depletion (e.g., crude oil, copper, and chromium) by 2050.

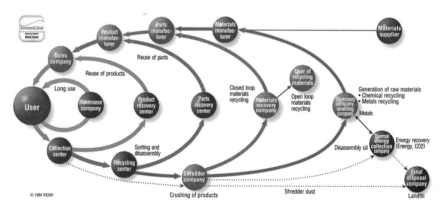

Fig. 3 The Ricoh's comet circle

3 Industrial Circular Transformation

3.1 Eco-Design at Product Scale

Products are the interface between production and consumption. Industry provides services and creates value through the production of products. Meanwhile, by adopting diversified measures to improve the circulation at the product level, industry can also achieve the purpose of circular economy. For example, industry can design and manufacture products that can be recycled more than once, products with longer life, products easier for recycling, or biodegradable, etc. In fact, both industry and academy have coined a term eco-design or design for environment to describe such an approach to designing products with special consideration for the environmental impacts of the product during its whole life cycle.

From a circular economy perspective, eco-design can save and recycle at maximum natural resources in a product's entire life cycle by considering multiple criteria in successive stages: raw material extraction, manufacturing, product use, and end-of-life management (recovery and recycling). Main criteria considered include raw materials consumption, energy consumption, water consumption, pollutant discharge, CO_2 emissions, biodiversity loss, etc.

The principles of eco-design were formally published in 2002 and they can be found in ISO/TR14062. Some goals and principles are specifically about

Using fewer materials, energy, and water for manufacturing products
Using materials and resources with a minimum environmental impact from a life cycle perspective
Producing waste and pollution as less as possible
Reducing the ecological impacts of distribution and packaging
Making remanufacturing, reusing, and recycling easier by intelligent design.

F-BOX of the SF Group [12]
Founded in 1993, SF Group currently is one of the leading integrated express logistics service providers in China, with annual operating income reaching 90.9 billion yuan (1.28 billion USD) and a delivery volume of 3.869 billion yuan (0.55 billion USD) in 2018. In terms of whole packing's waste reduction, SF's Sustainable Packaging Solutions Center designed F-BOX, a reusable packing box. F-BOX replaces the use of plastic tape with anti-theft zipper; the design of inner hooks and loop fasteners helps the fixation of express goods and replaces the use of plastic fillers. According to SF, each F-BOX can be recycled for 50 times. Each use of 100 F-BOX can reduce the use of 5,000 corrugated boxes, 154 rolls of plastic tapes, and 300 rolls of plastic inflated fillers. By October 2019, SF had invested 600,000 F-BOX, the total number of uses raised up to millions, with an average of 12 cycles each, while the estimated maximum number of recycles reaching 42. The F-BOX has covered all first-tier cities in China and some second-tier cities, reducing carbon emissions by about 1,600 tons.

3.2 Cleaner Production at Enterprise Scale

Enterprises are the main body of production. Thus, there exist a lot of opportunities for circular economy at the enterprise level. According to UNEP, cleaner production refers to continuous application of an integrated environmental strategy to processes, products, and services to increase efficiency and reduce risks to humans and the environment [13]. Since the implementation of 3P program (pollution prevention pays) by 3 M in 1974, cleaner production has been highly valued and systematically promoted at the global level. Cleaner production measures include

Waste-free or low waste processes and technologies
Substitution of raw materials and auxiliary materials (especially renewable materials and energy)
Reuse or recycle of waste (internal or external)
Improved control and automatization
Inventory and scheduling optimization.

Case Study 3 M's 3P program (Pollution Prevention Pay) timeline [14].
The 3P program (Pollution Prevention Pay) introduced by US Company 3 M is regarded as the first landmark of cleaner production. The 3P program introduced a new approach to pollution prevention by focusing on preventing pollution at the source—in products and manufacturing processes—rather than removing it after it has been created. Since the program's inception in 1975, 3 M employees worldwide have completed more than 8,000 3P projects in the following 35 years, which prevented more than 3 billion pounds of pollution and translated into nearly $1.4 billion in savings to 3 M.

1975: 3 M Pollution Prevention Pays (3P) Program is formed. First-year results: 19 projects prevented 73,000 tons of air emissions, 2,800 tons of sludge.

1977: 3 M and U.S. Environmental Protection Agency sponsor four conferences on pollution prevention.

1981: Environmental Audit Program begins.

1987: 3 M Corporate Safety, Health and Environmental Committee is formed. 3 M Air Emission Reduction Program is launched.

1988: 3 M Ozone-Depleting Chemical phase-out policy is adopted.

1990: Recycled paper Post-it® Notes are introduced. 3 M 16-point Safety Program is approved.

1993: 3 M goal of 70% reduction in worldwide air emissions is achieved.

1994: U.S. Environmental Protection Agency presents Achievement Award for 3 M's successful participation in the 33/50 Voluntary Air Emission Reduction Program.

1996: President Clinton's Council on Sustainable Development selects 3 M's 3P Program for the President's Sustainable Development Award.

1997: 3 M formalizes Life Cycle Management System to ensure the consideration of environmental, health, and safety issues in new product development.

2005: 3 M is awarded the 2005 Most Valuable Pollution Prevention (MVP2) Award from the National Pollution Prevention Roundtable.

2018: Listed on the Dow Jones Sustainability Index, recognizing Sustainability leadership, for the 19th consecutive year.

2019: 3 M embeds Sustainability Value Commitment in the new product launch process.

3.3 Circular Transformation at Industrial Parks' Scale

Due to the existing scale effect and scope effect, industries tend to develop in clusters, which form industrial parks in some specific areas. Meanwhile, in order to catch up with the development of industry, some countries, especially developing countries often introduce supporting policies on enterprises recruiting and infrastructure construction, which also stimulates the formation of industrial parks. From the perspective of industrial ecology, industrial parks create opportunities of formation of inter-firm industrial symbiosis which is described as engaging traditionally separate industries in a collective approach to competitive advantage involving physical exchange of materials, energy, water, and by-products. If a park realizes enough waste exchange and infrastructure sharing and obtain additional environmental and economic win–win effects, we can call this park an eco-industrial park. The worldwide classic example is the Kalundborg industrial system in Denmark. After the reporting of this case in 1989, some countries including the United States, the Netherlands, the United Kingdom, Japan, South Korea, and China began to plan or renovate industrial parks following the Kalundborg model. Up to now, there are more than 300 Eco-industry Park (EIP) cases reported.

Case Study Tianjin Economic and Technological Development Area (TEDA) EIP case.

TEDA was one of the first batches of industrial parks established in 1984 in China. In 2004, TEDA began to formulate and implement the EIP plan and then got approved the national pilot in 2008. Shi et al. had identified 81 inter-firm symbiotic relationships formed in TEDA involving utility, automobile, electronics, biotechnology, food and beverage, and resource recovery clusters [15]. Since 2010, TEDA implemented one EU Switch-Asia Project: Implementing Industrial Symbiosis and Environmental Management System in the Tianjin Binhai New Area. The objective of the project is to establish an industrial symbiosis network with 800-member SMEs. During a 4-year period (2010–2013), TEDA has organized 464 onsite visits, 22 quick-win workshops, and 14 sectoral seminars. Facilitated by the project, more synergies have been uncovered and realized (Table 1).

Table 1 Industrial synergies in TEDA (2010–2013) [16]

Sector	2010	2011	2012	2013
No. of membership	174	536	635	931
No. of synergies	10	27	43	87
CO_2 abatement (tons)	205	11,000	42,000	89,355
Landfill diversion (tons)	50	3000	257,000	321,076
Raw materials reduction (tons)	50	3000	872,000	936,388
Revenue increase (10,000 RMB)	7.23	552	8963	11,040

3.4 Sustainable Industrial Transition at Regional Scale

Today's industrial production has entered a new era of globalization. Products, together with their production factors including raw materials, labor, and capital, have been globalized and connected by trade networks in every corner of the world, forming the so-called global production network. This provides many opportunities for circular transformation at regional scale embodied in a global production network, such as

The change of business models. By changing product economy into services economy or function economy, the recycling rate of parts can be improved which means the decrease of wasted whole machines;

The emerge of sharing economy. With the help of the Internet and big data technology, more and more sharing economy opportunities have emerged in such industries as residence, travel, and tourism.

Green supply chain. Green supply chain management has been widely implemented in the automotive, electronics, and equipment industries. Within their supply chain system, the leading enterprises can exert environmental governance influence on their suppliers and customers.

Green trading. Green trade can prevent and stop the damage to ecological environment and human health caused by trade activities, so as to achieve sustainable development.

Green logistics. It can restrain the harm caused by logistics to the environment and human health by taking some measures, including green transport, green packaging, green circulation processing, etc.

Case Study: A new textiles economy—redesigning fashion's future [11] sourced from Ellen MacArthur Foundation.

As one of the typical representatives of the global production network, the current clothing system is extremely wasteful and polluting due to its nearly linear way. Based on circular economy principles, the clothing system should be leading a better economic model in which clothes, fabric, and fibers are kept at their highest value during use, and re-enter the economy after use, never-ending up as waste. The new textiles economy relies on four ambitions (see Fig. 4).

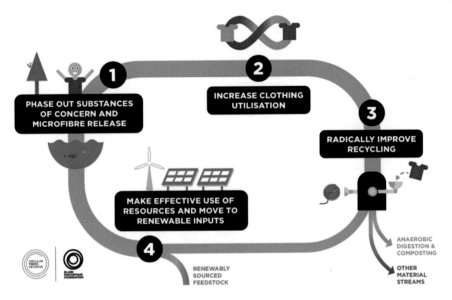

Fig. 4 Ambitions for a new textiles economy ©EMF

Issue of concern	Specific concerns related to environmental impacts, supply security, technology development			General environmental and economic concerns related to the throughput		
	within certain businesses, economic activities, countries, regions			of substances, materials, manufactured goods		
	associated with			at the level of		
Object of interest	Substances	Materials	Products (manufactured goods)	Businesses	Economic activities	Countries, regions
	chemical elements or compounds	raw materials, semi-finished goods	batteries, cars, computers, textiles	establishments, enterprises	mining, construction, chemical industry, iron & steel industry	total materials groups of materials, particular materials
Type of analysis	Substance Flow Analysis	Material System Analysis	Life Cycle Assessment	Business level MF Analysis	Input-Output Analysis	Economy-wide MF Analysis
	⇕	⇕	⇕	⇕	⇕	⇕
Type of measurement tool	Substance Flow Accounts	Individual Material Flow Accounts	Life Cycle Inventories	Business Material flow accounts	Physical Input-Output Tables, NAMEA-type approaches	Economy-wide Material Flow Accounts

Fig. 5 The spectrum of industrial metabolism and related methods

Phase-out substances of concern and microfiber release;

Transform the way clothes are designed, sold, and used to break free from their increasingly disposable nature;

Radically improve recycling by transforming clothing design, collection, and reprocessing;

Make effective use of resources and move to renewable inputs.

4 Industrial Metabolism Methods for Circular Economy

To support the circular transformation of industrial systems, we need not only to analyze the interaction between industrial systems and other systems but also to be able to construct eco-industrial systems which never exist before. Both of them need the support of systems methods and tools. Fortunately, the discipline of industrial ecology has developed a number of methodological tools, particularly on industrial metabolism. Along with methods and tools provided by other disciplines, the toolkits shown in Table 2 can help us to make industrial circular transformation happen.

Here, we focus on industrial metabolism which was proposed by Robert Ayres in analogy to the biological metabolism. Industrial metabolism, in the ontological sense, refers to "the whole integrated collection of physical processes that convert raw materials and energy, plus labor, into finished products and wastes…" [17]. From methodological perspective, industrial metabolism is a systems approach spectrum to quantitatively evaluate the stock and flow of substances in a socio-economic system with spatiotemporal boundaries according to the law of conservation of substances, so as to track the sources, paths, and sinks of substances flowing in the system [18].

The spectrum of industrial metabolism comprises substance flows analysis (SFA), material flows analysis (MFA), life cycle assessment (LCA), environmental input–output analysis (EIOA), etc., with the relationship shown in Fig. 4 [19]. SFA monitors flow of specific substances (e.g., Cd, Pb, Zn, Hg, N, P, CO_2, CFC) that are known for raising particular concerns as regards the environmental and health risks associated with their production and consumption. MFA is based on economy-wide material

Table 2 Methods and tools for industrial circular transformation

Activity	Targets group	Methods and tools
Systems analysis	Systematists, product analysts, environmental consultants, third-party assessors	Life cycle assessment, material flows analysis, substance flows analysis, environmental input–output analysis, scenario analysis, etc.
Design and plan	Systematists, Industrial designers, product designers, park planners, urban planners	Eco-design, design for X (environment, recycling, remanufacturing, disassembly, etc.), eco-industrial park plan, circular economy plan, etc.
Operation and management	Operators, managers, supply chain managers, third-party management	Cleaner production audit, green supply chain, responsible care, material stewardship, inventory management, etc.
Decision making	Government officials, financial investors, businesses, other policymakers	Agent-based modeling, scenario planning, decision-making supporting system, artificial intelligence, etc.

flow accounts that record all materials entering or leaving the boundary of the socioe-conomic system. LCA focuses on materials connected to the production and use of specific products (e.g., batteries, cars, computers, textiles), and analyse the material requirements and potential environmental pressures along the full life cycle of the products. EIOA is based on physical input–output tables that record material flows at various levels of detail to, from, and through the economy, and by economic activity and final demand category.

5 Extended producer's Responsibility

Industrial circular transformation is a typical complex system engineering task. Thus, it needs systems innovation to promote and implement industrial circular economy by integrating and coordinating technology, market, and policy. Characterizing industrial ecology as the science of sustainability, Allenby provided a policy framework to the transition in policy, regulation, and management that would be required to replace a brushfires approach to environmental protection with systematic management of the interaction between natural and industrial systems [20].

One important innovation is extended producer's responsibility (EPR) which refers to a policy approach under which producers are given a significant responsibility—financial and/or physical—for the entire life cycle of the product and especially for the take-back, recycling, and final disposal phases [9].

The concept was first formally introduced in Sweden by Thomas Lindhqvist in a 1990 report to the Swedish Ministry of the Environment. After several rounds of academic discussion and practical experience summary, it is recognized that EPR policy has the following features:

It is a product policy, not a waste policy.
Its priority is pollution prevention at the source, not end-of-life pollution treatment and control.
Its goal is to reduce the whole impact of the product itself and the corresponding production system during the life cycle of the product, rather than focusing on some specific pollution sources.
It promotes the polluter pays principle and seeks to endogenize the waste management costs into products.

A good example of EPR policy is the European packaging waste recovery and recycling schemes. EPR for packaging has delivered new innovations in packaging waste management and packaging design that have reduced the environmental impact of packaging and packaged goods. A EUROPEN analysis of EU data covering 1998–2011 shows that packaging production and packaging waste disposal have clearly been decoupled from economic growth in the EU-15.

In practice, EPR policy can be implemented by designing specific systems including deposit-refund system (DRS), green dot system, and so on. Taking the disposable packaging recycling as an example, the deposit system has been adopted extensively. Around the globe, more than 40 countries and regions have implemented the DRS for the disposable packages, showing the average recovery rate reached 85%.

6 Conclusion

Industry is a complex material and energy transformation system with multiple levels, which means that industrial circular transformation is also a complex system engineering involving multiple dimensions and multiple stakeholders. According to the hierarchical structure and circles of industry, we can summarize industrial circular transformation strategies.

At the product level, we can seek for opportunities and measures of recycling from the whole life cycle, and then take the LCA method to support eco-design and redesign of product–service systems. At the enterprises level, we can explore opportunities for recycling through cleaner production audit and other tools, to seek a win–win solutions through technological route change, raw material substitution, and process optimization. At the industrial park level, we promote circular transformation through waste exchange, infrastructure sharing, and material cascading utilization. At a larger regional scale, we can take measures like industrial restructuring, optimizing resource allocation, green supply chain management, and business model innovation.

In order to promote circular transformation at multiple levels, it is necessary to adopt, develop, and innovate policies or mechanisms to integrate technology, market, and policy measures such as the extended producer's responsibility (EPR).

Questions

1. What are the implementation paths of industrial circular transformation?
2. What are the differences and connections between life cycle assessment and substance flow analysis?
3. How do we design an eco-industrial park?
4. To eliminate plastics wastes, how should we design and implement the extended producer responsibility system?

Answers

1. Circular economy implementation paths can be observed at every scale of industrial systems, such as eco-design at product scale, cleaner production at firm scale, recycling transformation at industrial parks scale, urban mining at city scale, and green trade at regional and global scales.
2. The similarity between LCA and SFA is that they both take the life cycle perspective and include material preparation, production, consumption, and waste management four stages. The main difference is that SFA focuses on specific elements or substances, whereas LCA covers all material/energy inputs and outputs.

3. The design of eco-industrial parks is one of complex systems engineering projects. It needs to take into account the needs of enterprises, governments, and other stakeholders, and optimize and improve industrial development, infrastructure construction, and service operation from the perspective of circular economy.
4. The EPR system design can be considered from the whole life cycle of plastic products, such as the restrictive production in the production stage, the deposit system in the consumption stage, and the tax policy in the waste management stage.

Suggested Reading

Graedel, T. E., Allenby, B. R. (2003). *Industrial ecology*. Prentice Hall.
Brunner, P. H., & Rechberger, H. (2016). *Handbook of material flow analysis: For environmental, resource and waste engineers* (2nd ed.). Boca Raton, Florida: CRC Press—Taylor & Francis Group.
STAN—software for material flow analysis. https://www.stan2web.net/.
The open source life cycle and sustainability assessment software. https://www.opennlca.org/.
Ellen MacArthur Foundation. (2017). A new textiles economy: Redesigning fashion's future. https://www.ellenmacarthurfoundation.org/publications.
https://www.interface.com/US/en-US/sustainability/our-journey-en_US.
https://www.basf.com/global/en/who-we-are/sustainability/we-drive-sustainable-solutions/circular-economy/mass-balance-approach/chemcycling.html.

References

1. Grübler, A. (2003). Technology and global change. Cambridge University Press.
2. Gallopoulos, N. D. (1989). Strategies for manufacturing. *Scientific American, 1989*(261), 94–102.
3. United Nations Environment Programme. (2019). Global Environment Outlook: GEO 6, United Nations Environment Program.
4. Robert, U. A., & Leslie, W. A. (2002). A handbook of industrial ecology. *An-nals of Physics*. UK: Edward Elgar Publishing Limited.
5. Graedel, T. E. (2003). *Braden R*. Allenby. Industrial Ecology: Prentice Hall.
6. Charter, M. (2019). Designing for the Circular Economy, Routledge.
7. https://www.unep.fr/scp/cp/.
8. Lowe, E. A. (2001). *Eco-industrial park handbook for asian developing countries*. Environment Department, Indigo Development, Oakland, CA: A Report to Asian Development Bank.
9. OECD. (2016). Extended producer responsibility updated guidance for efficient waste management: Updated guidance for efficient waste management. OECD Publishing.
10. Chertow, M., & Ehrenfeld, J. (2012). Organizing self-organizing systems. *Journal of Industrial Ecology, 16*(1), 13–27.
11. Ellen MacArthur Foundation. (2017). A new textiles economy: Redesigning fashion's future. https://www.ellenmacarthurfoundation.org/publications.
12. SF Group, S.F. (2018). Corporate social responsibility report 2018. Retrieved from https://www.sf-express.com/cn/sc/download/20190316-2018-SF-CSR-Report.PDF.

13. Berkel, R. V. (2000). *Cleaner production perspectives for the next decade (II), UNEP's 6th international high-level seminar on cleaner production.* Canada: Montreal.
14. https://www.3m.com/3M/en_US/sustainability-us/goals-progress/#content2.
15. Shi, H., Chertow, M., & Song, Y. Y. (2010). Developing country experience with eco-industrial parks: A case study of the Tianjin economic-technological development area in China. *Journal of Cleaner. Production, 18,* 191–199.
16. Shi, L., & Yu, B. (2014). Eco-industrial parks from strategic niches to development mainstream: The cases of China. *Sustainability, 6*(9), 6325–6331.
17. Ayres, R. U. (1994). Industrial metabolism: Theory and policy. In R. U. Ayres & U. K. Si-monis (Eds.), *Industrial metabolism: Restructuring for sustainable devel-opment* (pp. 3–20). Tokyo: United Nations University Press.
18. Brunner, P. H., & Rechberger, H. (2003). *Practical handbook of material flow analysis.* New York: Lewis publishers.
19. OECD. (2008). Measuring material flows and resource productivity—synthesis report.
20. Allenby, B. (1998). Industrial ecology: Policy framework and implementation. Prentice Hall.

Dr. SHI Lei is associate professor of the School of Environment in Tsinghua University. His research interest lies in industrial ecology, focusing on industrial metabolism, industrial ecosystem complexity and eco-innovation. He contributes to the uncovering of structural and functional complexity of industrial ecosystems from complex network perspective. Dr. Shi also contributes to empirical studies on eco-industrial parks and circular economy models in China. He has so far led more than 30 projects and published more than 50 articles in Journal of Industrial Ecology, Journal of Cleaner Production, PNAS and Natural Climate Change. He was awarded the Research Frontier Award of the Chinese Society for Industrial Ecology (CSIE) in 2018. He served as the director of Asia Office of Journal of Industrial Ecology, associate editor of Journal of Cleaner Production and the chair of the 10th International Conference of Industrial Ecology (ISIE2019).

Industrial Symbiosis for Circular Economy: A Possible Scenario in Norway

Angela Daniela La Rosa and Seeram Ramakrishna

Abstract Interaction between industry and environment is crucial for industrial business performance as environmental impacts are constantly increasing pressure on industrial businesses. The creation of eco-industrial parks aims at transforming industrial systems into industrial ecosystems by including some measures like infrastructure and material/energy flows sharing. The introduction of industrial symbiosis scenario in which one firm's waste becomes another firm's feedstock represents a further development of eco-industrial parks design. This principle may be extended to cities and, doing so, an integration of socio-economic and ecological systems will be promoted. At the industrial park level, the practice of the circular transformation through waste exchange enhances circular economy. The application of the same principles to cities promotes the circular urban metabolism, where the conversion of natural resources into society occurs with zero-waste production.

Keywords Industrial symbiosis · Industrial ecology · Cleaner production · Eco-design · Eco-industrial parks · Urban metabolism

Learning Objectives

- Business, as usual, is not an option as we are facing an era of environmental and social changes. New economic and industrial scenarios are required dealing with the complexity of socio-ecological systems.
- Is circular economy possible? The creation of industrial symbiosis in industrial parks, in which the by-products produced by the park are consumed by companies in the park, maybe a way to follow.
- The principles of circular economy should be applied to cities, promoting a shift from linear to circular urban metabolism.

A. D. La Rosa (✉)
Department of Manufacturing and Civil Engineering, Norwegian University of Science and Technology, Gjøvik, Norway
e-mail: angela.d.l.rosa@ntnu.no

S. Ramakrishna
Circular Economy Taskforce, National University of Singapore, Singapore, Singapore
e-mail: seeram@nus.edu.sg

© Springer Nature Singapore Pte Ltd. 2021
L. Liu and S. Ramakrishna (eds.), *An Introduction to Circular Economy*,
https://doi.org/10.1007/978-981-15-8510-4_6

1 Introduction

Industrial symbiosis (IS) is an innovative and unique way of creating networks based on the ability of the partners of working together and exchanging materials, water, and energy streams with the purpose of increasing the resilience and the economic activities while reducing the environmental impact and production costs. The IS concept is in line with the recent Circular Economy (CE) principles referring to the challenges of resource scarcity, negative environmental impact, and economic development to promote a transition from a "Cradle to Grave" approach which means from materials extraction, manufacturing, use, and waste production to a "Cradle to Cradle" approach in a "closed loop," where the waste produced becomes itself nutrient for the next cycle. The CE concept is of great interest because it is a way for businesses to implement the much-discussed concept of sustainable development [13, 24, 25].

Since the beginning of the industrial revolution, mass production of goods was enabled by new manufacturing methods resulting in products with high availability and low costs. As a consequence, new consumer societies have risen, with increasing emission of pollutants to the environment, solid waste generation, and landfilling.

In addition, the growing world population demands a rising consumption of natural resources. Since planet earth's resources are limited, the requirements of exponential economic and population growth cannot be met. In this scenario, it is not only the challenge of environmental pollution that is becoming acute but the challenge of global resource scarcity as well.

In line with eco-industrial development, CE is understood as "realization of closed loop material flow in the whole economic system". In association with the so-called 3R principles (reduction, reuse, and recycling) the core of CE is the circular (closed) flow of materials and the use of raw materials and energy through multiple phases.

Definition

Industrial Symbiosis is an extension of the concept of eco-industrial park in which businesses and infrastructures cooperate with each other through exchanging wastes and sharing physical or non-physical resources, such as materials, water, energy, and information. The mechanism of industrial symbiosis (IS) is that one firm's waste becomes another firm's feedstock.

Circular economy is an economic system aimed at eliminating waste and the continual use of resources. Circular systems employ reuse, repair, refurbishment, remanufacturing, and recycling to create a closed-loop system, minimizing the use of resource inputs and the creation of waste. ISO/TC 323 is a new ISO technical committee that intends to develop requirements, frameworks, guidance, and supporting tools related to the implementation of circular economy projects.

Cradle to cradle. The term is a play on the popular corporate phrase "cradle to grave" (from birth to death, or "grave") implying that after products have reached the end of their useful life, they become either "biological nutrients" or "technical nutrients." Biological nutrients are materials that can re-enter the environment. Technical nutrients are materials that remain within closed-loop industrial cycles.

Urban metabolism. Urban metabolism is the study of material and energy flows arising from urban socio-economic activities and regional and global biogeochemical processes. The characterization of these flows and the relationships between anthropogenic urban activities and natural processes and cycles defines the behavior of urban production and consumption.

2 Some Examples of IS

The earliest example of industrial ecology was the symbiosis of industries at Kalundborg, Denmark [8]. Since then, the industrial symbiosis had successfully transformed existing industrial parks [29, 30] such as the National Industrial Symbiosis Program (NISP) in the UK, the regional synergies in the Australian mineral industries, and the Circular Economy program in Chinese industrial parks [17, 4, 11]. Other notable examples included the symbiotic alliance of Kymi pulp and paper mill and its allied industries in Kouvola, Finland [20], the waste management companies of Chamusca, Portugal [9], and the Tianjin Economic-Technological Development Area (TEDA) in China [26]. Boons et al. [5] formulated a conceptual framework demonstrating the dynamics of industrial symbiosis. The assessments mostly were qualitative and descriptive in nature. How did one quantify the benefits and impacts of industrial symbiosis and compare pros and cons across the industrial parks that took different forms and shapes? Several approaches had been employed to quantify the advantages of industrial symbiosis [31], [21], [10, 28]. Mattila et al. [23] compared process, hybrid, and input-output life cycle assessment (LCA) approaches in quantifying the environmental impacts of a forest industrial symbiosis in Kymenlaakso, Finland. The methods, however, did not cover entirely the impacts of industrial symbiosis in industrial parks as resource inputs from both natural and anthropogenic activities needed to be included.

Example

An early case study was a Swiss regional IS assessment (water, energy, metals, e.g., iron, copper, aluminum, wood, plastics, food, and building material are accounted) described by Massard and Erkman, in 2007. By-products exchanges create the IS. Together with utility sharing opportunities for supply and treatment, these elements grouped together to create resource synergies. By-product exchanges may be prohibited by national or local legislation intended to protect our environment. The legislation revealed that the Swiss laws on waste do not hinder by-product exchanges, in contrast with the EU policy.

By setting up very restrictive environmental laws, Swiss policy tends to charge the companies for the real costs of supply, effluent, and waste treatment [22]. Therefore, the companies themselves are often searching for alternative outlets for their by-products in order to reduce their costs. Implementing a new industrial park will change the local by-product exchange pattern and result in structural change, therefore, a macro-level decision support approach is necessary. Figure 1 highlights some key indicative points.

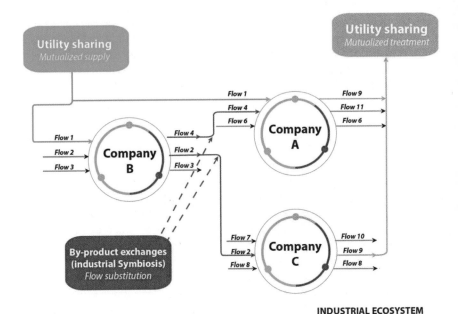

Fig. 1 An example of IS. The resources synergies include by-product exchanges and utility sharing

2.1 Weak Points of IS

The symbiotic relationship is a type of resource interdependence, which mainly involves the physical exchange of materials, energy, water, and by-products [7]. The mechanism of IS is that one firm's waste becomes another firm's feedstock therefore firms involved in IS relationships become resource interdependent. This type of interdependence may also extend from the operational level to the strategic one when a company uses wastes from the other company to generate new products for the market [1].

As firms become more and more embedded in the network of IS relationships, the degree of interdependence also rises and the need for coordination becomes high. In this regard, companies face interorganizational challenges and several inter-firm activities need to be planned to carry out the IS relationship [3, 15]. First, companies should agree on the quantity of waste that will be exchanged and the delivery time. Planning the right amount of waste that will be delivered to the right customer at the right time can be harder compared to similar activities in traditional businesses since waste is not produced upon demand but emerges as a secondary output of main production activities.

Furthermore, some wastes might require a treatment process before being used as inputs, removing impurities or contaminants from the waste, which can be operated by a third firm.

Such a practice increases the complexity of the IS relationship, which needs additional coordination. Accordingly, when inter-firm relationships require greater coordination, transaction costs increase [12]. Rather than to exchange waste with another company, waste producers might use wastes within their boundaries (e.g., by using a waste produced by a given production process as input for other production processes or simply selling wastes on the market, when a waste market exists) [27]. In such a case, there is no interdependence between firms within the system and the need for coordination is thus low, thus resulting in low transaction costs for the company. However, in order to use a given waste internally, companies need to operate production processes able to receive that waste as input.

The governance of the IS system is also characterized by centralization of control, i.e., the extent to which a central actor manages the entire system of relationships. A high centralization of control regards IS systems managed by a central actor who has disproportionate authority over which companies become part of the system, where and how symbiotic interactions take place.

3 A Possible Scenario in Norway

Scenario

The forest industry is very important in many areas of the country and profitable forestry industry is of great importance for settlement, employment, and sustainable business development within a specific region. The main tree species by volume and economic importance are spruce, pine, and birch, and analyses show that a significant amount of forest resources can be exploited in a sustainable and climate-friendly manner. The best practice would be to establish an industrial area in a strategic geographical location, covering a short distance between forests (the ecosystem) a city (the people), and the industries (the economy).

The following step would be to implement symbiosis and reduce the energy flows, by promoting waste/by-product exchanges between all the actors. Some possible ways are

1. Adopting and improving technologies that save resources and enhance waste reuse and recycling. Selecting materials that have lower embodied energy and reducing adverse environmental impacts material consumptions and productions. Technology innovations would be imperative in achieving goals.
2. Optimizing energy structure and promoting cascade utilization of energy would further improve energy efficiency and reduce emissions.
3. Restoring local ecosystem for sustainable developments of the industrial park. The ecosystem in which the industrial park was located being a fundamental life-support system not only provided essential services and resources but also was a resource base and a carrier for economic activities.

A schematic and simplified concept of symbiosis for the suggested scenario is drawn in Fig. 2.

The industrial symbiosis should consist primarily of infrastructure sharing and waste/by-products exchanges that lead to resource conservations. Industrial symbiosis reduces waste transportation and disposal.

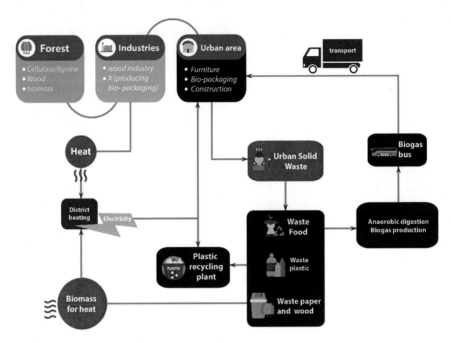

Fig. 2 An example of symbiotic network between forest, city, and industry (credit-illustration by Sawaros Thongkaew of STEAM Platform)

Three possible scenarios are considered, and the effects are summarized in Table 1:

– Scenario 1: the absence of industrial symbiosis.

The scenario assumes a park with more than two companies that manufacture products generating, e.g. heat, paper/cardboard, plastics, biomass waste. It is assumed that all these by-products are discarded outside the park.

– Scenario 2: the presence of partial industrial symbiosis.

The scenario assumes an amount of inbound by-product in the park, and an amount of outbound by-product discarded outside the park.

– Scenario 3: perfect symbiosis.

The scenario assumes a situation in which all the by-products produced by the park are consumed by companies in the park (a case of perfect symbiosis).

4 Contribution of the IS in the Sustainability of the Waste Management

The industrial symbiosis scenario includes the establishment of an anaerobic digester that will be able to treat the organic waste derived from crop cultivation and other

Table 1 Possible scenarios for eco-industrial parks. Benefits of IS compared to no IS

By-products/Wastes	Without IS		With IS	
	Action	*Effects*	*Action*	*Effects*
Heat	Heat released in the ecosystems. Wasted by-product	Pollution. GWP Waste of resources	Heat collected and provided to the Skjerven area	Benefits – Resources recovery – Avoided pollution
Biomass	Food waste and biomass waste collected from the municipality, the forest and the agro-industrial area	Economic costs Money required for treatment in specific service	Biomass collected and treated in the local anaerobic digestor Production of biogas.	Benefits – Public transport. Bus using methane from the municipal biogas plant – Biomass recovery. Wastes transformed into methane for public bus
Plastics	Waste collected from municipal wastes and from the industrial area. Packaging wasted	Economic costs Money required for treatment in specific service	Plastic wastes treated in the recycling plant located in the Skjerven industrial park	Benefits: – Money saving – New packaging produced by recycled plastic packaging
Infrastructure sharing	Use of private services and facilities	Economic costs Environmental burden.	Use of shared facilities and services	Economic saving from – Shared solid and liquid waste management – Shared training in new regulations and technologies – Shared emergency management services – Transportation services – Others

biomass from animal breeding as well as from food waste collection from the nearby municipality (and other possible neighbour municipalities) and from the industrial park area. Figure 3 reports a schematic diagram of a possible biogas plant for the selected area.

This process will contribute to bridge the gap between cities and industries by making a significant contribution to sustainable cities.

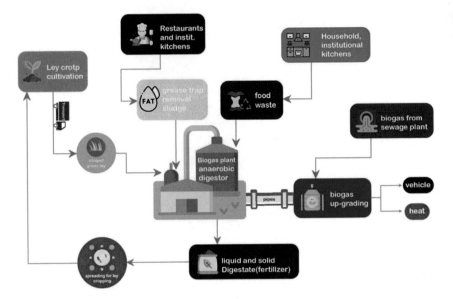

Fig. 3 Diagram of the biogas plant to establish in Industrial Park in symbiosis with the municipality (credit-illustration by Sawaros Thongkaew of STEAM Platform)

5 Future Trends

The study of the resource consumption and environmental pressure of urban areas in a systemic way falls within the scope of Urban Metabolism (UM). This field of research has become increasingly important in the last two decades, developing applications for Sustainability Indicators, Greenhouse Gas Accounting, Policy Analysis, Design and Material Flow Accounting. Kennedy et al. [18] and Zhang [32] provide reviews of the history, methodology, and applications of Urban Metabolism.

Urban metabolism is a multidisciplinary research domain focused on providing important insights into the behavior of cities for the purpose of advancing effective proposals for a more humane and ecologically responsible future. Material recycling should be a focus of urban policymaking and development due to its high potential to significantly reduce cities dependence on external and non-renewable resources. The principles of the industrial ecology approach are to be applied to the urban metabolism concept, in which it is seen as the conversion of nature into society. Girardet [14] coined and drew the difference between a 'circular' and 'linear' metabolism. In a circular cycle, there is nearly no waste and almost everything is reused.

Cities have a key role in the battle for sustainability. Cities are places of political contention [2, 16]. Cities represent the possibility to develop new regulatory structures and spaces of governance [6]. Furthermore, the social economy within each locality creates a dense fabric of relationships that allow local citizens to work together in identifying and acting on local problems or in taking local initiatives [19].

6 Conclusions

The approach of creating industrial symbiosis in industrial parks encounters the vision of an economy in loops (or circular economy) with a focus on job creation, economic competitiveness, resource savings, and waste prevention. This intends to promote a transition from a "Cradle to Grave" approach which means from materials extraction, manufacture use and waste production to a "Cradle to Cradle" approach in a "closed loop," where the wastes produced become itself nutrient for the next cycle. The closed-loop model can be extended to cities and to the urban metabolism with the intention of reaching the sustainable development goals (SDGs 2030) by interconnecting social-ecological and economic complex systems.

Questions

1. How the creation of industrial symbiosis brings benefits to industrial parks?
2. Is industrial symbiosis easy to plan and carry out?

Answers

1. It depends on the specific cases, but in general industrial symbiosis allows resources conservation due to waste recycling and reuse, avoided pollution and economic savings from.
2. Industrial symbiosis is a relationship that requires a certain degree of interdependence among the companies. In this regard, companies face interorganizational challenges and several inter-firm activities need to be well planned to carry out the IS relationship.

Suggested Reading

1. McDonough, W., & Braungart, M. (2002). *Cradle to cradle: Remaking the way we make things.* North Point Press.

References

1. Albino V., & Fraccascia, L. (2015). The industrial symbiosis approach: A classification of business models. *Procedia Environmental Science, Engineering and Management, 2,* 217–223.
2. Bakker, K. (2003). A political ecology of water privatization. *Studies in Political Economy, 70,* 35–58.
3. Bansal, P., & McNight, B. (2009). Looking forward, pushing back and peering sideways: Analyzing the sustainability of industrial symbiosis. *Journal of Supply Chain Management, 45,* 26–37.
4. Van Beers, D., & Biswas, W. K. (2008). A regional synergy approach to energy recovery: The case of the Kwinana Industrial Area, Western Australia. *Energy Conversion and Management, 49*(11), 3051–3062.

5. Boons, F., Spekkink, W., Mouzakitis ,Y. (2011). The dynamics of industrial symbiosis: A proposal for a conceptual framework. *Journal of Cleaner Production, 19*(9), 905–911.
6. Brenner, N. (2002). Decoding the newest 'metropolitan regionalism' in the USA: A critical overview. *Cities, 19*, 3–21.
7. Chertow, M. R. (2000). Industrial symbiosis: Literature and taxonomy. *Annual Review of Energy and the Environment, 25*(1), 313–337.
8. Chertow, M. R. (2007). "Uncovering" industrial symbiosis. *Journal of Industrial Ecology, 11*, 11–30.
9. Costa, I., & Ferrao, P. (2010). A case study of industrial symbiosis development using a middle-out approach. *Journal of Cleaner Production, 18*(10), 984–992.
10. Dai, T. J. (2010). Two quantitative indices for the planning and evaluation of eco-industrial parks. *Resources, Conservation and Recycling, 54*(7), 442–448.
11. Fang, Y. P., Côté, R. P., & Qin, R. (2007). Industrial sustainability in China: practice and prospects for eco-industrial development. *Journal of Environmental Management, 83*(3), 315–328.
12. Gereffi, G., Humphrey, J., & Sturgeon, T. (2005). The governance of global value chains. *Review of International Political Economy, 12*, 78–104.
13. Ghisellini, P., Cialani, C., & Ulgiati, S. (2016). A review on circular economy: The expected transition to a balanced interplay of environmental and economic systems. *Journal of Cleaner Production, 114*, 11–32.
14. Girardet, H. (2008). *Cities people planet: Urban development and climate change* (2nd ed.). England: Wiley.
15. Herczeg, G., Akkerman, R., & Hauschild, M. Z. (2018). Supply chain collaboration in industrial symbiosis networks. *Journal of Cleaner Production, 171*, 1058–1067.
16. Heynen, N., Kaika, M., & Swyngedouw, E. (Eds.). (2005). In *The Nature of Cities: Urban Political Ecology and the Politics of Urban Metabolism*. Routledge, Abingdon.
17. Jensen, P. D., Basson, L., Hellawell, E. E., Bailey, M. R., & Leach, M. (2011). Quantifying 'geographic proximity': Experiences from the United Kingdom's national industrial symbiosis programme. *Resources, Conservation and Recycling, 55*(7), 703–712.
18. Kennedy, C., Pincetl, P., & Bunje, P. (2011). The study of urban metabolism and its applications to urban planning and design. *Environmental pollution, 159*, 1965–1973.
19. Korten, D. (1995). Civic engagement to create just and sustainable societies for the 21st century. In M. Voula (Ed.), sustainable cities for the third millennium: The Odyssey of Urban Excellence, Springer Edt.
20. Lehtoranta, S., Nissinen, A., Mattila, T., & Melanen, M. (2011). Industrial symbiosis and the policy instruments of sustainable consumption and production. *Journal of Cleaner Production, 19*(16), 1865–1875.
21. Liu, Q., Jiang, P.P., Zhao, J., Zhang, B., Bian, H., & Qian, G. (2011). Life cycle assessment of an industrial symbiosis based on energy recovery from dried sludge and used oil. *Journal of Cleaner Production, 19*(15), 1700–1708.
22. Massard, G., & Erkman, S. (2007). A regional industrial symbiosis methodology and its implementation in Geneva, Switzerland. In *3rd International Conference on Life Cycle Management*. http://www.lcm2007.ethz.ch/paper/51_2.pdf.
23. Mattila, T. J., Pakarinen, S., & Sokka, L. (2010). Quantifying the total environmental impacts of an industrial symbiosis—A comparison of process-, hybrid and input output life cycle assessment. *Environmental Science and Technology, 44*(11), 4309–4314.
24. Murray, A., Skene, K., & Haynes, K. (2017). The circular economy: An interdisciplinary exploration of the concept and application in a global context. *Journal of Business Ethics, 140*(3), 369–380.
25. Patil, R. A., Seal, S., & Ramakrishna, S. (Jan, 2020). Circular economy, sustainability and business opportunities. *European Business Review*, 82–91. https://www.europeanbusinessreview.com/circular-economy-sustainability-and-business-opportunities/.
26. Shi, H., Chertow, M., & Song, Y. (2010). Developing country experience with ecoindustrial parks: A case study of the Tianjin economic-technological development area in China. *Journal of Cleaner Production, 18*(3), 191–199.

27. Shi, L., & Chertow, M. (2017). Organizational boundary change in industrial symbiosis: revisiting the Guitang Group in China. *Sustainability, 9*, 1–19.
28. Soratana, K., & Landis, A. E. (2011). Evaluating industrial symbiosis and algae cultivation from a life cycle perspective. *Bioresource Technology, 102*(13), 6892–6901.
29. Tudor, T., Adam, E., & Bates, M. (2007). Drivers and limitations for the successful development and functioning of EIPs (eco-industrial parks): A literature review. *Ecological Economics, 61*(2–3), 199–207.
30. Veiga, L. B. E., & Magrini, A. (2009). Eco-industrial park development in Rio de Janeiro, Brazil: A tool for sustainable development. *Journal of cleaner production, 17*(7), 653–661.
31. Yang, S. L., & Feng, N. P. (2008). A case study of industrial symbiosis: Nanning Sugar Co., Ltd. China. *Resources, Conservation and Recycling, 52*(5), 813–820.
32. Zhang, Y. (2013). Urban metabolism: A review of research methodologies. *Environmental pollution, 178*, 463–473.

Angela Daniela La Rosa is associate professor at the Norwegian University of Science and Technology (NTNU), in the Group of Sustainable Composites of the Department of Manufacturing and Civil Engineering, Gjøvik site. She obtained a Laurea degree in applied chemistry in 1994 from the University of Catania (Italy). Since 2001 she has a PhD in Polymer Science, with most of the work done in the Strategic Technology Group of ICI (Imperial Chemical Industries) UK. She is also currently manager R&D in remote of the company All Green srl (Romania). Her main research activity is focused on the evaluation of life cycle assessment (LCA) of new formulations of polymer composites and bio-composites. Her research and teaching interests include LCA, sustainable development and circular economy as well as the production and end-of-life processes of polymers and polymer composites.

Professor Seeram Ramakrishna *FREng, Everest Chair* (https://www.eng.nus.edu.sg/me/staff/ramakrishna-seeram/), is among the top three impactful authors at the National University of Singapore, NUS (https://academic.microsoft.com/institution/165932596). NUS is ranked among the top five best global universities for engineering in the world (https://www.usnews.com/education/best-global-universities/engineering). He is the Chair of Circular Economy Taskforce. He is a member of Enterprise Singapore's and ISO's Committees on ISO/TC323 Circular Economy and WG3 on Circularity. He also the Chair of Sustainable Manufacturing TC at the Institution of Engineers Singapore and a member of standards committee of Singapore Manufacturing Federation (http://www.smfederation.org.sg). He is an advisor to the Ministry of Sustainability & Environment—National Environmental Agency's CESS events, (https://www.cleanenvirosummit.sg/programme/speakers/professor-seeram-ramakrishna; https://bit.ly/catalyst2019video; https://youtube.com/watch?v=ptSh_1Bgl1g). European Commission Director-General for Environment, Excellency *Daniel Calleja Crespo, said, "Professor Seeram Ramakrishna should be praised for his personal engagement leading the reflections*

on how to develop a more sustainable future for all", in his foreword for the Springer Nature book on Circular Economy (ISBN: 978-981-15-8509-8). He is a member of UNESCO's Global Independent Expert Group on Universities and the 2030 Agenda (EGU2030). He is the Editor-in-Chief of the Springer NATURE Journal Materials Circular Economy—Sustainability (https://www.springer.com/journal/42824). He is an Associate Editor of eScience journal (http://www.keaipublishing.com/en/journals/escience/editorial-board/). He is an opinion contributor to the Springer Nature Sustainability Community (https://sustainabilitycommunity.springernature.com/users/98825-seeram-ramakrishna/posts/looking-through-covid-19-lens-for-a-sustainable-new-modern-society). He teaches ME6501 Materials and Sustainability course (https://www.europeanbusinessreview.com/circular-economy-sustainability-and-business-opportunities/). He also mentors Integrated Sustainable Design ISD5102 project students. Microsoft Academic ranked him among the top 25 authors out of three million materials researchers worldwide based on H-index (https://academic.microsoft.com/authors/192562407). He is named among the World's Most Influential Minds (Thomson Reuters) and World's Highly Cited Researchers (Clarivate Analytics). Listed among the top three scientists of the world as per the Stanford University researcher study on career-long impact of researchers or c-score (https://drive.google.com/file/d/1bUJrvurVVBbxSl9eFZRSHFif7tt30-5U/view). He is an Impact Speaker at the University of Toronto, Canada Low Carbon Renewable Materials Center (https://www.lcrmc.com/). He is a judge for the Mohammed Bin Rashid Initiative for the Global Prosperity (https://www.facebook.com/Make4Prosperity/videos/innovation-inclusive-trade/479503539339143/). He advises technology companies with sustainability vision such as TRIA (www.triabio24.com), CeEntek (https://ceentek.com/), Green Li-Ion (www.Greenli-ion.com) and InfraPrime (https://www.infra-prime.com/vision-leadership). He is a Vice-President of Asian Polymer Association (https://www.asianpolymer.org/committee.html). He is a Founding Member of Plastics Recycling Association of Singapore (PRAS). His senior academic leadership roles include University Vice-President (Research Strategy), Dean of Faculty of Engineering; Director of NUS Enterprise and Founding Chairman of Solar Energy Institute of Singapore (http://www.seris.nus.edu.sg/). He is an elected Fellow of UK Royal Academy of Engineering (FREng), Singapore Academy of Engineering and Indian National Academy of Engineering. He received PhD from the University of Cambridge, UK, and The TGMP from the Harvard University, USA.

Agriculture and Food Circularity in Malaysia

Hung Teik Khor and Gary Kiang Hong Teoh

Abstract The area of circular economy in terms of organic waste in agriculture and food sector has always been challenging without an effective and efficient mechanism of collection and processing into value-added products. There is no circularity in food and agriculture unless the waste management process (for both production and consumption) is able to produce value-added and financially viable products, services and cash. Natural resources are used to grow agriculture products which are processed into food, bio-based materials and energy for consumption. In this process of production and consumption, waste is generated from post-harvesting, post-processing, pre-consumption and post-consumption. In the context of food, there is pre- and post-consumption food waste. Pre-consumption food waste can be edible food which can be recollected and resold or distributed to the needed consumer. The post-consumption food, the leftovers, can be processed together with the green waste into compost and soil enhancer which is then used by the agriculture sector to produce more food, and in so doing, the economic cycle is complete from food back to food.

Keywords Agriculture and food circular economy · Post-harvesting · Pre-consumption · Post-consumption · Post-consumer discards · Food to Food Programme · Waste food slurry · XLR8® · Microbial photosynthesis · Groundswell® Continuous Fermentation Process · Hydrosynthesis · Humisoil® · Soil enhancers · Edible food · Green waste · In-vessel composting · Hydrosoil · Farmers' market · Collection and discount selling · Food waste diversion

H. T. Khor (✉)
International Sustainable Environment Networking, Office of the Penang State Minister for Welfare, Caring Society and Environment, Penang State Government Malaysia, George Town, Malaysia
e-mail: htkhor@gmail.com

G. K. H. Teoh (✉)
Golden Highway Auto-City Sdn Bhd, Perai, Penang, Malaysia
e-mail: garyteoh1225@gmail.com

© Springer Nature Singapore Pte Ltd. 2021
L. Liu and S. Ramakrishna (eds.), *An Introduction to Circular Economy*,
https://doi.org/10.1007/978-981-15-8510-4_7

Learning Objectives
Understand the agriculture and food circular economy
Learn about the key concepts of agriculture circularity and edible waste food circularity

- Circular economy is a transformation of forms from one to another which creates value without wastes.
- Food waste can be transformed into nutrient-rich compost for organic farming.
- Social business models on youth social enterprises and government food bank.

1 Introduction

In the area of circular economy for agriculture and food, organic waste has always been hard to manage without an effective and efficient mechanism for collection, processing into products with value and there is no circularity unless something comes back. Organic waste can be transformed into useful raw resource to grow a new product and thereafter transformed again the waste from the new product into raw resource for another new product.

There is no circular economy unless there is a value-add during the recycle process to make it financially viable to turn it into products, services and cash.

Natural resources (such as forests, land, water and sunlight) contribute to the growing of food through agricultural practices and from these food and other organic products (e.g. wood, paper) are manufactured for consumption or use. Manufactured products are then sold to consumers at the consumption site. The consumption of such products results in post-consumer discards such as food waste which includes kitchen food preparation and trimmings, processed or manufactured food stuff (e.g. canned food, cooked food, bakery products, etc.) or green waste (e.g. garden and yard trimmings, roadside trimmings). Most of these are sent to disposal sites such as landfills or incinerators or incinerated.

Figure 1 shows the circular economy for agriculture and food. This chapter focuses on the post-harvest activities mainly at the consumption site (pre-consumption and post-consumption) and part of it will cover green waste which is also a contributor of organic waste in all municipalities in Malaysia.

However, some of these post-harvest, pre- and post-consumption food discards are edible or reusable and thus possess an added value during the recycling process whereby they are no longer considered waste.

Edible food (fresh produce or cooked food) can be recollected and resold or distributed to the needy or consumers. Even the leftover food waste can be processed with green waste to be transformed into compost and soil enhancers for the agriculture sector to produce more food and in so doing complete the circular economy from 'food back to food'. The recycling of these post-consumption organics will be diverted from landfills and further mitigate greenhouse gases such as methane and carbon dioxide which are generated in the process of landfilling. Eventually, the life

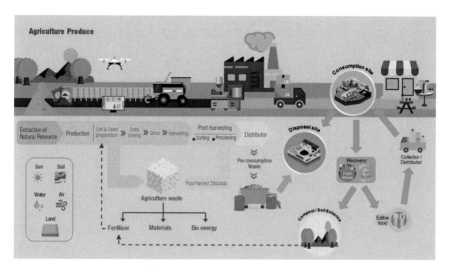

Fig. 1 Circular economy for food and agriculture—Graphics by Ms. Sawaras Thongkaew (STEAM Platform)

span of landfills will be prolonged saving much needed space for other forms of land use.

Malaysians generated an average of 1.17 kg each day per person in 2018. Food waste constituted 44.5% of the waste collected according to the Solid Waste Management and Public Cleansing Corporation (SWCorp) and this can be separated and composted and these reuse, recycling and composting practices could reduce waste generation by 60–70% if there was mass adoption [4].

Waste generated in Penang state-wide is estimated to be about 1,800 metric tonnes/day. From this, about 600–800 tonnes/day are generated from Penang Island and about 400–600 metric tonnes/day from Seberang Perai. The per capita generation is about 1.1 kg. From data available for Penang, organic waste constitutes about 40–60% of total waste and a large proportion of this is from food and garden waste [9].

The Penang State Government has been carrying out several organic waste minimisation and diversion from the landfill since 2011. These fall into three main categories, i.e. food and kitchen waste, green waste (garden and yard trimmings) and animal waste (manure). Some of the approaches for the different stakeholders in Penang are shown in Table 1 [9].

The Penang Organic Waste Policy (2016) seeks to encourage the separation and treatment of organic waste at source in order to divert the waste away from the landfill so as to prolong its lifespan and reduce municipal costs. This is part of Penang's local action towards a global commitment to mitigate the effects of global warming and climate change [9].

The Penang State Government has formulated several long-term objectives and strategies as shown below.

Table 1 Strategies for reducing organic waste

No	Stakeholders	Approach	Remarks
1	Hotels, hospitals, restaurants and coffee shops, food courts and hawker centres	• Waste separation at source • Onsite composting • Collect food waste and used cooking oil	Can use composting machines
2	Community/residential area committees/village committees/highrise dwellings/schools	• Waste separation at source • Two-stream system • Community composting • Individual household composting	Collection on alternate days
3	Households (individual landed properties)	• Individual household composting	
4	Municipal level	Large-scale composting	Have smaller district-level composting plants if possible

Source Integrated Solid Waste Management (ISWM) for Penang—implementation strategies [8]

Box 1: Penang Organic Waste Reduction Objectives and Strategies

OBJECTIVE 1:

To divert the amount of organic waste (putrescibles) from the Pulau Burung Sanitary Landfill and moving towards a total ban in the long term.

Strategy 1.1: Impose separation of organic waste at source.

Strategy 1.2: Develop relevant policies for different waste generators.

Strategy 1.3: Increase community awareness and understanding of separating organic waste at source.

OBJECTIVE 2:

To reduce the costs of collection, transfer and treatment of organic waste for the local authorities by treating organic waste at source.

Strategy 2.1: Encourage treatment of organic waste into useful by-products at source where possible.

OBJECTIVE 3:

To incentivise organic waste treatment by private and community efforts through cost savings by local authorities.

Strategy 3.1: Develop incentive systems to reward efforts such as Neighbourhood Watch (Rukun Tetangga) that treat and process organic waste at source.

Strategy 3.2: Set the stage for future voluntary carbon offset schemes.

OBJECTIVE 4:

To emulate nature and return all organic outputs to food production, parks and gardens and energy production, thereby completing the nutrient cycle to ensure a sustainable food supply and security.

Strategy 4.1: Develop and promote new linkages to return organic outputs to the food production process in agricultural, horticultural and agro-forestry sectors.

Strategy 4.2: Encourage a food waste to food policy to maintain food security through urban and peri-urban agriculture.

The success of these objectives and policies needs the cooperation and involvement of the residential, commercial, industrial and institutional sectors in collaboration with the state government.

Four case studies that are worthy of mention are shown in the list below:

Case Study 1: Food to Food Programme by VRM Biologik® (Waste food slurry to soil enhancer and increased nutrient uptake for plants like probiotics) **and Groundswell® Continuous Fermentation Process** (Organic/green waste to via methane-free continuous static fermentation pile to high-content humic organic media for soil improvement);

Case Study 2: From Food Waste to Farm Produce: Circular Economy by Auto-City, Penang (Food waste from commercial F&B outlets is transformed using in-vessel composting into compost for use in farming of organic produce);

Case Study 3: Green Hero—Social Enterprise Combatting Food Wastage (Collection and discount selling of edible food from food outlets via apps by a Youth Enterprise);

Case Study 4: Mutiara Food Bank—Saving Food, Feeding Lives (Diverting edible food to the needy by the Penang State Government).

These four case studies will be individually elaborated in the sections below.

2 Case Studies

2.1 Case Study 1: Food to Food Programme and Groundswell® Continuous Fermentation Process by VRM Biologik® (vrm.com.au)

VRM Biologik® is an Australian company with a local Malaysian branch which collaborated with the Penang Island City Council in 2011 to divert green waste, food and market waste in Penang, Malaysia given the strong local interest organic waste from the only landfill in Penang to help extend its lifespan.

VRM has designed and implemented several programmes that provide a triple bottom line impact and in doing so add significant value to organic residues.

Described below are two examples of their processes which resulted in two products called XLR8® which is a soil enhancer using their Bio-Regen® conversion machine and the production of HumiSoil® using their Groundswell® process that has engendered a circular economy.

The 'Food to Food' Programme developed by VRM Biologik® immediately diverts food waste/organic waste away from the landfill at the source of generation. VRM Biologik® is a leading provider of high-quality formulations that catalyse natural biological reactions. These formulations are used in organic recycling, advanced soil and water remediation, agricultural support programmes and industrial and domestic cleaning.

VRM Biologik® has also developed a food waste processing machine which is used to collect organic/food waste from wet markets, food courts and canteens. These units are installed at the source of organic residues. Bio-Regen® conversion units are installed on site and supplied with catalyst additive which is added to a resultant slurry after grinding and mixing of the organic material. The slurry is fermented in a patented perpetual fermentation process and converted to a soil enhancer which carries significantly more nutrient and other values than were present in the organic residue itself. The material is no longer a waste product when it leaves the site. Ferment is collected by VRM Biologik® staff or licensees to be further processed and bottled as a marketable product. This resulting product is a soil conditioner that stimulates natural reactions which improve soil structure, unlock nutrients, manufacture water and mobilise phosphorus.

VRM Biologik® has pioneered the manufacture and deployment of formulations which stimulate bacterial photosynthesis as a way to energise depleted soils. An important consequence of bacterial photosynthesis is hydrosynthesis or the manufacture of water! This means that by-products of the conversion of organic residues are suitable for assisting growth in dry or arid soil conditions.

Use of these products helps build the basic biomass which, with good management, will help maintain water availability in soil. VRM Biologik®'s XLR8® products deliver a range of micro-nutrients and biological substrates which combine to trigger natural reactions in soil which help to feed plants. These reactions also help stimulate carbon sequestration and nitrogen fixation (Fig. 2).

VRM's XLR8 is used together with another product in another process pioneered by VRM Biologik® called the Groundswell® process and its resultant product HumiSoil®. Unlike composting, the Groundswell® process is a static continuous fermentation process that utilises *bacterial photosynthesis* to generate energy needed to convert organic material to humus and pre-humic substances rather than digestion to CO_2 and water vapour as happens in compost. The Groundswell® process is designed to retain and enhance all nutrient available in the organic material and as such does not have a net carbon emission during the process. The resulting Humisoil® has a nutrient and catalytic value which makes it far more valuable than either the original organic material or a comparative compost (Photo 1; Fig. 3).

Fig. 2 Bio-Regen
food-to-food
flowchart—Graphics
Courtesy of Ken Bellamy,
Founder of VRM

Photo 1 VRM XLR8 sold
in 1000 litre
tanks—Graphics Courtesy of
Ken Bellamy, Founder of
VRM

The process is completed over 26 weeks and completes long-term pasteurisation as well as conversion of organics without the requirement for mechanical turning or other energy supply. Incoming feedstock is usually shredded, mixed with catalysts supplied by VRM Biologik® and covered with large tarpaulins for fermentation to occur. At 6 weeks, the pile is opened and resprayed, repiled, recovered and left for another 20-week period. Incoming feedstock is not limited to green waste only but also covers other types of materials such as animal carcasses and dung.

HumiSoil® is an organically based product in which the biological reactions which result in the formation of humus have already been started. This pre-fostering of humus manufacture allows farms to speed up the natural process of humus formation in the soil. HumiSoil® is a fully matured topsoil enhancer made from totally organic inputs which contains high levels of humic materials together with a range of other

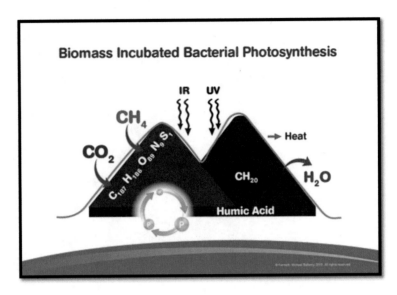

Fig. 3 Biomass-incubated bacterial photosynthesis process—Graphics Courtesy of Ken Bellamy, Founder of VRM

biological substrates. It has the characteristics of a humus-rich topsoil. It contains high levels of active humic materials, fosters carbon fixation and natural humus formation, supports nitrogen fixation from air and allows reduced reliance on nutrient application. The mobilisation of phosphorus and a higher sustained available phosphorus are key tangible values offered by the use of Humisoil® on farms. The promotion of beneficial fungi responsible for soil formation and nutrient transfer gives added value. Importantly, when used with its companion product XLR8 Bio®, hydrosynthesis is stimulated in soil. This manufacture of water via bacterial photosynthesis is a significant value-add over the properties of the original material.

Reactions responsible for the fixation and mobilisation of additional carbon, nitrogen, phosphorus and numerous trace elements from their environment through bacterial photosynthesis thrive in the Groundswell® Process (Fig. 4).

Studies in China have shown that natural reactions stimulated in soil following the regular application of HumiSoil® and its companion product XLR8 Bio® resulted in the

- Manufacture of water in the soil;
- Fixation of nitrogen from the air;
- Mobilisation of phosphorus reserves in soil;
- Capture of CO_2 from the air and use of it to make sugars and
- Manufacture of energy storage compounds like ATP.

All of these functions help build a healthy soil and can assist plant growth. Healthy soil reactions have been linked with reduced incidence of disease outbreak as well as with much more efficient nutrient transfer to plants (Fig. 5).

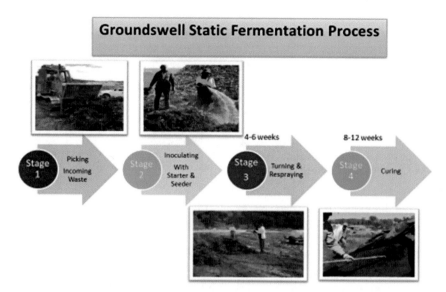

Fig. 4 Groundswell static fermentation process—Graphics courtesy of Ken Bellamy, Founder of VRM

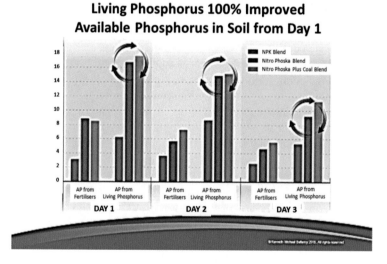

Fig. 5 Improved soil phosphorus availability—Graphics courtesy of Ken Bellamy, Founder of VRM

Photo 2 Humisoil—Courtesy of Ken Bellamy, Founder of VRM

HumiSoil® also makes phosphorus up to 100% more available to plants and in doing so reduces the need of chemical fertilisers which are in the long run detrimental to soil health (Photo 2).

The use of VRM Biologik® catalysts made in place with local organic residues can enhance the impact of chemical fertilisers and when used alone can provide at least equivalent impact on phosphate availability as through the use of chemical fertilisers. This reduces the amount and costs of chemical fertilisers required.

The two (2) VRM processes and the resultant products (XLR8® and Humisoil®) close the circularity loop. This is a type of circular economy that utilises food waste and then recirculates it back into the soil with a much greater tangible value to further produce food. The Groundswell® process not only enriches soil but also increases the soil phosphate content in turn reducing the need to apply chemical fertilisers.

The VRM Biologik® processes have been applied to many crop types and planting circumstances including fruit, vegetables, root crops, rice and other grains, pasture enrichment and plantation crops such as banana, successfully.

From its inception in August 2011 in the State of Penang, Malaysia to the end of 2016, a total amount of 398,797 metric tonnes of food and organic waste have been diverted from the landfill using this process. This translated to RM 93,222 (approximately USD23,305) savings in tipping fees alone and an estimated 358,908 tonnes of CO_2.eq (conversion factor = 0.9 per tonne of food waste) prevented from being released into the atmosphere from the landfill. Bio-Regen® units are also used in Australia, China, USA, Great Britain, Papua New Guinea and Singapore.

2.2 Case Study 2: From Food Waste to Farm Produce: Circular Economy by Auto-City, Penang (https://www.aut ocity.com.my/)

Since 2014 the Management of Auto-City (Management) embraces the circular economy whereby the Raw Food Waste (RFW) generated in Auto-City is no longer regarded as waste but as raw material of value that is transformed into Matured Food

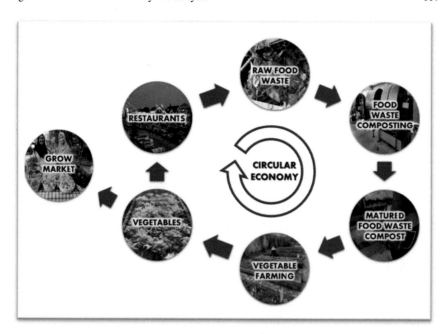

Fig. 6 Auto-City's circular economy

Waste (MFW) compost which is used in farming of organic vegetables for human consumption. This circular economy is shown in Fig. 6.

2.2.1 From Waste to Compost

The management uses a Natural Fermentation Method (NFM) of food waste composting and a semi-automated system to transform the RFW comprising of meat, bones, shells, skins, vegetables, fruit peels, oils, sauces and other organic matters into MFW compost within days. The process flow from RFW to MFW compost is shown in Fig. 7. The semi-automated system comprises several equipment: (a) a mixer to evenly mix the RFW with coco peat and in-house cultured 'Oommi' beneficial microbes, (b) a shredder to break down the RFW into smaller pieces so as to accelerate the composting process and (c) a rotary compost drum to continuously turn and evenly transform the shredded RFW into MFW compost. From 2015 to 2018, a total of about 105 tonnes of MFW compost is produced. The NFM is properly documented and published in a technical paper entitled 'Food Waste Composting: Natural Fermentation Method' in the International Journal of Recent Technology and Engineering [6].

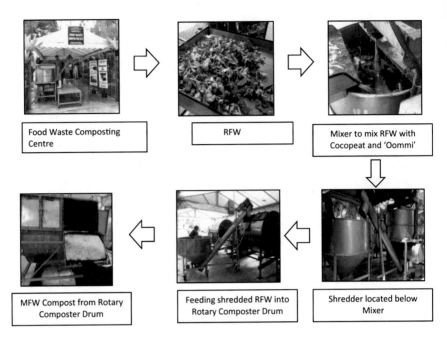

Food Waste Composting Centre | RFW | Mixer to mix RFW with Cocopeat and 'Oommi'

MFW Compost from Rotary Composter Drum | Feeding shredded RFW into Rotary Composter Drum | Shredder located below Mixer

Fig. 7 Process flow from RFW to MFW compost

2.2.2 Compost for Farming

The management uses the MFW compost for its organic soil-based farming to grow six types of lettuces. It develops its in-house Hydrosoil (HS) beds and Standard Operating Procedures (SOP) to grow vegetables: germinate seeds into seedlings, transplant the seedlings onto the HS beds, grow the seedlings into mature vegetables, harvest the vegetables, replenish the soil with MFW compost and prepare the HS beds for replanting. The SOP to grow vegetables using HS beds is shown in Fig. 8.

2.2.3 From Farm to Consumers

The freshly harvested vegetables are sold to selected restaurants and at a Grow Market in Auto-City. Photo 3 shows the Grow Market which serves as a platform for local farmers to sell their fresh organic produce directly to health-conscious consumers. The local farmers use the organic soil-based farming method and the challenge they face is that once they grow and harvest their vegetables and fruits, they need to quickly find consumers to buy their fresh produce. Whereas the challenge faced by the health-conscious consumers is to ensure the vegetables and fruits they buy are farmed without chemical fertiliser and pesticide. Thus, the Grow Market is an opportunity for the consumers to get to know and understand the farmers and this helps to build good relationship and trust.

HS bed ready
for seedlings

3 days old seedlings ready
for transplant

Transplant seedlings

30 days old vegetables and
ready for harvest

15 days old vegetables

Seedlings fully
transplanted

Fig. 8 Growing vegetables in hydrosoil beds

2.3 Case Study 3: Green Hero—Social Enterprise Combatting Food Wastage (www.greenhero.net)

This social enterprise was started by a group of young people who saw the need to combat food wastage in the City of George Town, Penang Malaysia and has expanded to Kuala Lumpur and soon to be in Johor. Green Hero redistributes food by convincing merchants to go Happy Hour online to students who are on a tight budget, the lower income group, orphanages, old folks' homes and even charitable organisations. Still edible food such as sushi, bread, cakes, pastries, bento boxes and even groceries (Fig. 9).

It was started by Penang-born Calvin Chan who studied International Business Management started off selling surplus food at F&B outlets in his college's cafeteria at discounted prices to students during breakfast and tea time. The online platform had humble beginnings as a WhatsApp group of a bunch of students in DISTED College and later spread to other colleges in Penang.

Chan was inspired by his research and found that France has legislation dealing with this issue while United Kingdom has a food surplus app, and Hong Kong, a group of food rescuers who use blaster freezer to freeze the food they save and bring it back to their central kitchen to redistribute over the days.

Photo 3 AutoCity's grow market

Together with five other friends, Chan launched social enterprise Food Plus Life (recently rebranded as Green Hero) in mid-2017 as an online platform where F&B businesses can channel edible food that remains unsold after operating hours and which would otherwise go to waste. The young, enterprising team of college-goers began their business with RM200 capital (Photo 4).

They insert the information about their food items onto the platform and a notification will be sent to the youth's customer base. From this platform, their customers, in turn, can pick and choose what they want to buy and the selected items will either be delivered or picked up at an agreed place. The online platform is only open to customers at 8 pm daily back in 2017 and today due to more merchants they go as early as 7 pm when most of F&B outlets are about to close in an hour time by going Happy Hour online on a daily basis.

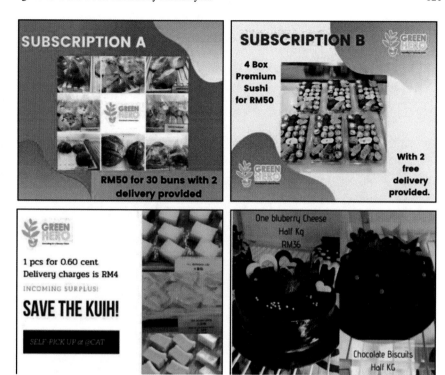

Fig. 9 Examples of edible surplus food offered by Green Hero—Courtesy of Green Hero

Their customer base has currently expanded at least six WhatsApp groups in Penang. Fast forward 2 years, and the social enterprise spread its wings to Kuala Lumpur in April. With the number of merchants on the platform totalling to over 100. Green Heroes has launched a mobile app for beta tester first. It will then be available for both Android and IOS by December 2019 (Fig. 10).

The process starts with participating merchants updating Green Hero in WhatsApp on the surplus they have or even in Google Sheets for the food they have. Once Green Hero has been updated with the food surplus, Calvin and his admin team will then start to post it out in the WhatsApp groups and Facebook page to collect order. After collecting order and confirmation, the order details are then be given to a driver for pick up. After picking up the food, every driver who start the trip with works with Green Hero will have to share their live location with Green Hero so that they can keep track on where the drivers are and keep customer updated on the time and delivery status (Fig. 11).

The operational flowchart for Green Hero is as follows as shown in Fig. 12. It involved the F&B outlets as the supply or food source at the consumption site which is then resold at discounted prices or given free via the use of Internet technology.

Figure 13 shows that since April 2019, Green Hero has diverted around 600 meals from the landfill.

Photo 4 Photograph of Calvin Chan (seated far right blue jumper) with his team—Courtesy of Green Hero

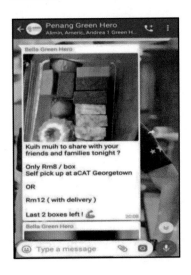 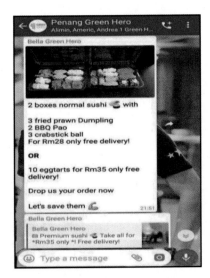

Fig. 10 Examples of participating merchants updating their food surplus through WhatsApp—Courtesy of Green Hero

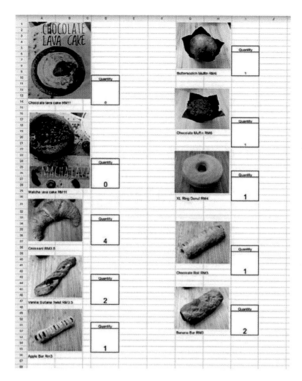

Fig. 11 Examples of participating merchants updating their food surplus through Google Sheets—Courtesy of Green Hero

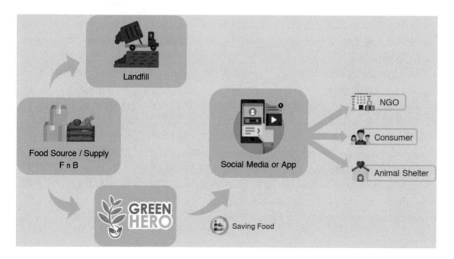

Fig. 12 Green Hero operation flowchart—Graphics by Ms. Sawaras Thongkaew, STEAM Platform

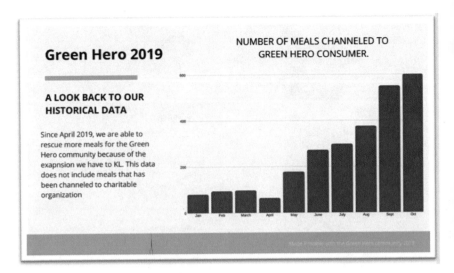

Fig. 13 Number of meals diverted from landfill since April 2019—Courtesy of Green Hero

Most of the food items sold on this platform are without branding and even the pick-up points are not at the outlets themselves. This is because most merchants would prefer not to reveal the brand. They are worried that the food quality and taste may deteriorate over time if the food is consumed on or too near to the date of expiry may drive away customers.

Food outlets are also worried that less customers may want to buy for the full price during other times.

Green Heroes also allows users to sponsor surplus food for children living in shelter homes, in addition to arranging surplus food from events to be distributed to those in need.

The last resort for any unsold food is to send the food to animal shelters that are willing to accept them.

The innovative social enterprise also benefitted from Hong Leong Bank Jumpstart, the CSR platform of Hong Leong Bank, which played a crucial role in its rebranding, media and public relation, as well as in developing the iOS version of the app. Their plan is to achieve 30% of preventing good food from going to the landfill but feed the tummy instead in Malaysia within the next 5 years with a sustainable model and eventually go international. Green Hero also aims to work with other relevant parties who shares the same vision so that the number can be achieved in 5 years' time because teamwork makes the dream work [3].

2.4 Case Study 4: Mutiara Food Bank—Saving Food, Feeding Lives (https://www.facebook.com/MutiarafoodB ank/)

Malaysians are throwing away about 15,000 tonnes of excess food every day and all this goes to the landfill [10]. In order to address this wasteful practice and avoid such perfectly edible food from going to the landfills, the Mutiara Food Bank programme in Penang was launched in May 2018. This excess food will provide sustenance for the underprivileged, arrest methane emissions and slash carbon footprint.

Food collected from contributors would be transported to the central kitchen for it to be reheated and packed before being distributed to the needy recipients on the same day.

The Mutiara Food Bank programme has also grown through collaboration with strategic partners such as TESCO Hypermarket. The excess food such as bread, fruits and vegetables from hypermarkets every day.

The Mutiara Food Bank's central kitchen is based at Penang's Caring Society Complex and volunteer chefs from Penang Chefs' Association who will inspect cooked excess food delivered from resorts and hotels. The chefs will reheat or blend the various deliveries into meals, which are delivered every day to the poor throughout the state.

Data from the Mutiara Food Bank shows that Bakery products such as doughnuts, cakes, pizzas and other pastries amounted to 154 tonnes since the programme was launched in May 2017 up to October 2019. Fresh produce such as vegetables and fruits totalled up to 259 tonnes. The total number of recipients amounted to (May 2017 to October 2019 since) is 35,276 needy people [8]. If each kilo is valued at USD1.20 (RM5), it would mean that about USD500, 727 (RM1.06 million) worth of food has been distributed [10] (Photo 5).

The Domestic Trade and Consumer Affairs Ministry impressed with the Mutiara Food Bank project has also duplicated Penang's good example and started the Malaysia Food Bank programme nationwide starting from January 2019.

Photo 5 Mutiara Food Bank delivery van

Photo 6   Vegetables for distribution to the needy—Co-Courtesy of Mutiara Food Bank

The Malaysia Food Bank Foundation is an extensive network of hypermarkets, hotels, food producers and volunteers to make sure that excess food—from fruits and vegetables to bread and even a five-star resort's mutton curry—will go to the poor (Photo 6).

The Malaysia Food Bank programme, under the purview of the Domestic Trade and Consumer Affairs Ministry, coincides with the Government's intention to reduce financial burden of the people especially those in the bottom 40% of the low-income group (B40) category. Food takes up about 30% of our living expenses so for poor families and the recipients would be households.

The Food Bank Malaysia programme has benefited 48,850 households nationwide since its launch in January 2019 and during that period, 1,055 metric tonnes of surplus food were saved and channelled to deserving households including students at public universities [13].

3   Summary

This chapter introduces the basics of what can be practised on the ground for both agriculture and food economy circularity. It shows practical case studies of agriculture circularity from the Food to Food Programme by VRM Biologik® (waste food slurry to soil enhancer and increased nutrient uptake for plants like probiotics) and its complementing Groundswell® Continuous Fermentation Process (organic/green waste to via methane-free continuous static fermentation pile to high-content humic organic media for soil improvement) as an example of both green waste and post-consumption food waste technologies.

The Auto-City' Commercial F&B outlets recycle their food waste for in-vessel composting to grow organic vegetables and sells them at the Grow Market which also provides a marketing venue for other farmers in the area.

The Green Hero which is a Youth Social Enterprise Combatting Food Wastage by collecting and discount selling of edible food from food outlets via apps is an example of young entrepreneurs using social media and to ensure circularity of edible food which would otherwise go to waste.

The Mutiara Food Bank Diverting Edible Food to the Needy programme is another good example on how government collaborates with the private sector to ensure that food is not wasted.

However, efforts by the non-government sectors are not sufficient without the important role of government as the policymaker and facilitator in the whole circularity system. This is to ensure correct policies and regulations are formulated to ensure a sustainable environment. Ultimately, the policy should be able to bring about a change in attitude of the public in attitude and behaviour on solid waste management.

Good examples of government interventions are food banks and 'Empty Plate—Love food don't waste it' policies to curb over consumption and wastage especially hotels, restaurants and other eating outlets as well as collection of waste food. A good tie up would be the **collection of fresh food or edible food** from supermarkets, fresh markets and food manufacturers **for redistribution** to the needy. The government should allocate funds on an annual basis for public education and awareness programmes for sustainability.

The management of Municipal Solid Wastes (MSW) still faces many challenges as the population grows. The current regulation system is not perfect and the existing management system and the collection facilities need to be fine-tuned towards **mandatory waste separation at source**. Most municipal solid wastes are still collected without proper separation at the source, and treatment facilities are limited and much of the collected wastes are still unrecovered. Government, NGOs, Community-Based Organisations (CBOs) and private sectors must continue to collaborate as there is still much needs to be done. The main management strategies to remedy this should include amendment of current laws and regulations, improve current management systems and introduce classified collections. The effective implementation of these strategies will help to reduce the problem of waste generation and divert waste from the landfill and through the effective and widespread practice of the 3Rs.

There must be more aggressive resource recovery and diversion of putrescible materials from the landfill. The rationale behind the objective is that removing putrescibles from the waste stream will enhance recycling because it reduces the co-mingled waste portion to the landfill.

There is a need to promote public education and also to raise awareness on SWM issues on a sustained basis.

4 Exercises

Questions
1. Can post-harvest agriculture waste or waste food be recycled instead of disposing them?
2. What common technologies are available to process agriculture waste products or green waste (roadside and garden trimmings)?
3. What can the commercial sector do on the agriculture and food circularity?
4. What can the government do to reduce food waste?
5. What circular economy is being practiced by the Management of Auto-City?
6. How does the Management of Auto-City create value on the matured food waste compost?

Answers
Answer 1: Post-harvest agriculture waste and waste food can be recycled or processed into useful agriculture inputs such as compost, soil enhancers thus completing the circularity.

Answer 2: Commonly used technologies include in-vessel processing or open composting technologies. There are a variety technologies under each of these technologies which are developed for appropriateness and adoptability to the situation at hand.

Answer 3: The commercial sector can be encouraged to collect waste food at source to produce useful products so that these waste can immediately be diverted away from disposal sites.

Answer 4: Governments can come up with policies like starting food banks to collect edible food for redistribution, have 'Empty plate' policies to curtail excessive or over consumption especially in the food and beverage industry such as eating outlets, hotels and restaurants.

Answer 5: The circular economy is 'From Food Waste to Farm Produce'.

Answer 6: The Management of Auto-City develops its in-house hydrosoil farming system which uses the matured food waste compost for organic farming to grow good quality and yield of vegetables for human consumption.

Acknowledgements The author would like to thank the following people for their inputs of data, photos and graphics in the various sections:

Case Study 1: Food to Food Programme by VRM Biologik® by Mr. Ken Bellamy, Founder of VRM Biologik®.

Case Study 2: From Food Waste to Farm Produce by Auto-City, Penang, Malaysia, authored by Mr. Gary Teoh King Hong, Managing Director, Golden Highway Auto-City Sdn Bhd, Penang, Malaysia.

Case Study 3: Green Hero—Youth Social Enterprise Combatting Food Wastage by Calvin Chan, Chief Environmental Officer (CEO) and Founder of Green Hero.

Case Study 4: Mutiara Food Bank Diverting Edible Food to the Needy (Government) by Mr. Muhammad Zakwan Mustafa Kamal. Chief Executive Officer, Mutiara Food Bank.

Special thanks to Ms. Sawaras Thongkaew, STEAM Platform for her hard work in providing the illustrations for Fig. 1: Circular economy for food and agriculture and Fig. 9: Green Hero operation flowchart.

References

1. Bellamy, K. (2019, December). *Food to food program by VRM Biologik®* (H. T. Khor, Interviewer).
2. Bellamy, K. (2020, December). *Groundswell® continuous fermentation process* (H. T. Khor, Interviewer).
3. Chan, C. (2019, December). *Green Hero—Social enterprise combatting food wastage* (H. T. Khor, Interviewer). George Town, Penang, Malaysia.
4. Chu, M. (2019, July 30). *The star nation.* Retrieved December 18, 2019, from Generation more waste than ever: https://www.thestar.com.my/news/nation/2019/07/30/generating-more-waste-than-ever.
5. Green Hero Team. (2019, December). *Green Hero—Innovating for a greener future.* Retrieved from Green Hero: www.greenhero.net.
6. Hong, G. T., Samah, M. A., Nowroji, K., & Chet, S. S. (2019). Food waste composting: Natural fermentation method. *International Journal of Recent Technology and Engineering (IJRTE), 8*(1C2).
7. Khor, H. T. (2012). *Integrated Solid Waste Management (ISWM) for Penang—Implementation strategies.* Submitted Penang State Government. George Town, Penang, Malaysia: Unpublished.
8. Kamal, M. Z. (2020, March 29). *Chief Executive Officer, Mutiara Food Bank. Mutiara* (H. T. Khor, Interviewer).
9. Khor, H. T. (2016). *Penang organic waste management plan and policy: Climate & clean air coalition (CCAC).* George Town, Penang. Retrieved from https://www.waste.ccacoalition.org/sites/default/files/files/1_report_on_penang_organic_waste_management_plan_plan_and_policy_0.pdf.
10. Low, A. (2018, December 23). *Food bank goes nationwide.* Retrieved from The Star: https://www.thestar.com.my/news/nation/2018/12/23/food-bank-goes-nationwide/.
11. Sekaran, R. (2017, November 20). *Penang sets up food bank for the poor.* Retrieved from The Star: https://www.thestar.com.my/news/nation/2017/11/20/penang-sets-up-food-bank-for-the-poor/.
12. Tan, C. (2018, December 22). *Government launches food bank to aid the needy.* Retrieved from Buletin Mutiara: https://www.buletinmutiara.com/govt-launches-food-bank-to-aid-the-needy/.
13. The Sun Daily. (2019, March 23). *48,850 households benefit from national Food Bank.* Retrieved from https://www.thesundaily.my/local/48-850-households-benefit-from-national-food-bank-KY717059.
14. VRM international Pty. (2019, December). *VRM Biologik.* Retrieved from VRM Biologik: www.vrm.science.

Hung Teik Khor is a Senior Advisor for International Sustainable Environment Networking to the Penang State Minister for Welfare, Caring Society and Environment, Malaysia and freelance consultant. Prior to his retirement in 2012, he has worked under the Malaysian Agriculture of Ministry for 18 years before working a Senior Research Analyst at the Penang Institute (formerly known as the Socio-economic and Environmental Research Institute—SERI) which is a State Government Think Tank for another 14 years. He is trained in Agriculture Science and Business Administration and is also a qualified TESOL teacher. He is involved as in the fields of environment, food and agriculture, solid waste management and disability issues. He was also involved in the formulation of the environment and waste management policies for Penang State. He was also external consultant to various international agencies such as UNEP, UNDP, UN ESCAP, DANIDA, CIDA and other private sector companies. He also advises the Penang Natural Green Organisation (PNGO) on environmental and waste management issues and also acts as Advisor and facilitator on LA21 (Local Action 21) community engagement to Penang Island City Council.

Gary Kiang Hong Teoh is the Founder and Managing Director of the AUTO-CITY Group whereby he and his wife Ong Bee Lee, are responsible for the conceptualization, marketing, development and management of AUTO-CITY in Penang, Malaysia. Known for being the first of its kind development in Malaysia, AUTO-CITY is being developed as the 1-Stop for Automobile, Food, Entertainment, Banking, Shopping and Eco-tourism and is endorsed as a tourism destination by the Penang State Government. Gary Teoh has over 30 years of working experience in various businesses, including research and development in green practices and farming, agriculture and plantation, real estates, property development and management, project management, construction, event management, manufacturing, wholesaling, retailing, food and beverages, entertainment, education, training, consultancy and tourism. In his academic achievements, Gary Teoh was educated as a Civil Engineer in Canada with a Bachelor's Degree in Civil Engineering and a Master's Degree in Building Engineering in the early 1980s. In the late 1980s, he was also trained as an Accredited Gemologist in Thailand and he was awarded a Diploma in Gemological Science. In the mid-90s, Gary graduated with a Master's Degree in Business Administration from University of Portsmouth, UK and a postgraduate Diploma from the Chartered Institute of Marketing, UK.

Material Passports and Circular Economy

Mohamed Sameer Hoosain, Babu Sena Paul, Syed Mehdi Raza, and Seeram Ramakrishna

Abstract A set of trusted information in the form of material passports is necessary in order to understand the circular value of systems and materials in the built environment. Innovations such as digital technologies and material passports are useful circular economy transitional tools in managing the materials flows and decarbonizing the built environment. Material ppassports are datasets and reliable information consisting of the entire value chain—specifications of the materials used, and specific supply chains involved from the sources to producers, distributors and consumers or users; and technical facts to improve the re-use or recycling of materials and to enhance their residual value. This paper presents the material passports in the built environment for the transition towards a circular economy as well as the latest trends in research and practice.

Keywords Material passports · Circular economy · Digital marking · Fourth industrial revolution (4IR) · Digital technologies

Learning Objectives

- Understand the principles of material passports and circular economy.
- Understand material passports methodology.
- Identify and utilize software tools and digital technologies that are available to create a material passport.

M. S. Hoosain (✉) · B. S. Paul
University of Johannesburg, Johannesburg, South Africa
e-mail: sameer.hoosain@gmail.com

B. S. Paul
e-mail: bspaul@uj.ac.za

S. M. Raza · S. Ramakrishna
National University of Singapore, Singapore, Singapore
e-mail: syedmehdiraza@u.nus.edu

S. Ramakrishna
e-mail: seeram@nus.edu.sg

© Springer Nature Singapore Pte Ltd. 2021
L. Liu and S. Ramakrishna (eds.), *An Introduction to Circular Economy*,
https://doi.org/10.1007/978-981-15-8510-4_8

- Identify and utilize current global applications and material passports databases.
- Understand and implement material circularity indicators, within material passports.
- Be able to create a material passport of your own in the built environment as a tool for transition towards a circular economy.

1 Introduction

Thomas Rau, a Dutch architect and pioneer had quoted: "Every building is a material depot." Keeping this in mind, if material passports are utilized, therefore, a building about to be destroyed becomes a storage warehouse for useful materials or buildings as material banks (BAMB) [15].

Material passports can serve as a tool to bring residual value back to the market, using its value tracking capabilities. Imagine knowing the exact worth of every material used in a building and importantly what can be recycled or re-used. Parts of Europe, some of which are mentioned in Table 1 [17], have piloted the concept, which will have a global effect in the years to come. This follows upon the European Union (EU) commission unveiled vision in 2015 to make the economy more sustainable [23]. The full list of circular economy policies in EU countries can be found in Ref. [7].

There are other initiatives in place in other sectors:

The food processing or manufacturing industries—Hazard analysis and critical control points (HACCAP) is a preventative food safety approach to reduce risks and hazards. More recently, Blockchain and Artificial Intelligence (AI) is being utilized as a technique in the fourth industrial revolution (4IR), to track and solve issues such as food fraud, safety recalls, supply chain inefficiency and food traceability. Therefore, the record of a manufactured item's journey, from the source to the consumer, is able to be traced in real-time [29].

The health and environmental sectors—Life cycle analysis (LCA) or Life Cycle Impact Assessment (LCIA) over cycles of cradle-to-grave or cradle-to-cradle is a system that tracks processes and transport routes used for raw materials extraction, production and waste management [27]. They also leverage other already available instruments such as technical data sheets (TDS) and material safety data sheets (MSDS), these contain information about product composition, applications, the potential hazards of these products and how to work safely with them.

Passports have the potential to incorporate the above standards and more, but in this case material passports would serve the built environment.

Table 1 Circular economy and material passport examples by some EU countries

Countries	Material passport implementations	Circular economy policies	Government websites
UK	<u>Queen Elizabeth Olympic Park</u> An Asset Disposal scheme was created to assist contractors to re-use materials after the games by selling them or gifting them to charities. Sustainability was a standout feature as the surrounding community have benefited from the complex	London's circular economy route map	https://www.lwarb.gov.uk/what-we-do/circular-lon don/circular-economy-route-map/
Germany	<u>New office building in Essen</u> Was previously a coal mine industrial complex. The project focused on sustainable design, especially on cradle-to-cradle design. Importantly this leads to a material passport {in this case the passport focuses on the entire building "Buildings' Material passport" (BMP)}, healthy materials and recyclability [4]	Resource efficiency	https://www.bmu.de/en/top ics/economy-products-res ources-tourism/resource-eff iciency/overview-of-ger man-resource-efficiency-programme-progress/
Netherlands	<u>1. Brummen Townhall</u> Was the first building in the world equipped with a material passport. It served as a raw materials depot, the material passport made it possible to re-use the material after the building is dismantled <u>2. Liander head office</u> It is the first building in the Netherlands that is sustainable and energy positive. A material passport was developed. Less materials were used, re-use of materials, therefore, materials can continue their full life cycle and a reduction in carbon emissions	The government-wide circular economy program. No more waste as resources will be re-used across a number of sectors. They plan to achieve this by the year 2050	https://www.government.nl/topics/circular-economy

2 Background

In 2017, the world resource use was approximately 20 tons/person/year. The building sector produces the largest amount of waste and consumes the most resources. It is estimated that these amounts will double by the year 2050. Not forgetting the environmental impact such as CO_2 emissions. For this, the European Union (EU)'s commission in 2015, planned to make their economy more sustainable by implementing a circular economy approach.

What is circular economy? A circular economy basically entails materials being kept in use for longer periods of time.

Ken Webster who is the head of innovation at the Ellen MacArthur Foundation, had argued in the book publication, "The circular economy: A wealth of flows," that our linear economy (take–make and dispose) is a nineteenth-century heritage unmoored in the twenty-first-century reality [33]. In previous years, building materials were re-used when constructing new buildings, but in the last 70 years this has decreased. This is due to the fact that newer material compositions make recycling or re-use complex [18]. For this reason, material ppassport is a tool that can be used for a transition from a linear model to a circular economy, both these models are depicted in Fig. 1 a and b [26].

3 Material passports

Material passports? Material passports can be defined as a value tracking tool and brings back residual value to the market. Material ppassports make information available and relevant from production to purchase to use to maintenance. A single material passport contains a set of information about a particular material, product or system.

The process of generating a material passport may involve a number of persons or companies. Therefore, they are able to contain information from a number of sources and provide a number of stakeholders with this information, this structure of material passports can be seen in Fig. 2, and for easy comprehension, we break down as each section below, i.e. Materials, Information and Stakeholders [24].

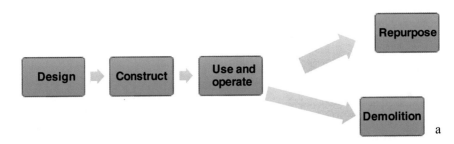

Fig. 1 **a** Linear model. **b** Circular model

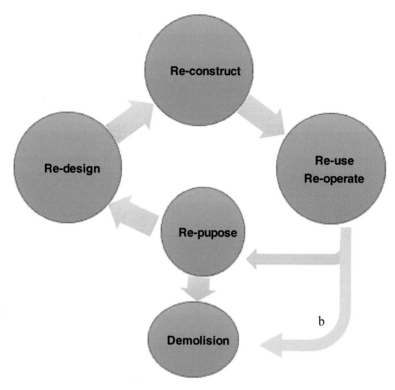

Fig. 1 (continued)

Fig. 2 The structure of material passports

3.1 Materials

The materials spectrum is based on different levels of hierarchy, Fig. 3. This is required in order to define the characteristics of a particular product/material and its level of recovery.

One example using the above approach would be the lighting in a building [11]:

- The finished lighting assembly in a "Building" would fall under "Product" and "System."
- The parts of the lighting assembly would fall under "Component."
- The chemicals needed to manufacture the components of the lighting assembly would fall under "Material."

Material passports in the built environment require large amounts of data, some of which are already available in other sectors, a few examples were listed in the previous sections. The idea is to centralize all this data. Some of the most important product/material characteristics required for a material passport in the built environment are [18]:

- *Physical properties*—Dimensions and weight, density, energy and strength.
- *Chemical properties*—Chemical composition, health and safety, is it recyclable or not and lifespan.
- *Biological properties*—Renewable or not and is it recyclable or not.
- *Health of materials*—Emissions, what does the law say, certifications and MSDS.
- *Design and production*—MSDS, TDS, certifications, Bill of materials (BOM) and logistics.

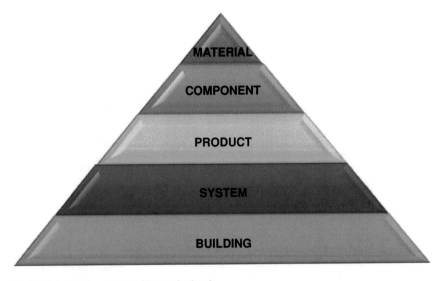

Fig. 3 Materials spectrum and hierarchy levels

- *Logistics*—Weight, dimensions, traceability and correct documentation.
- *Disassembly*
- *Recyclable.*

The choice of materials plays an important role in the design of a material passport for a circular economy, and to ensure that they can be re-used and there's safety for humans and the environment. Let us look at the steps to follow in order to generate one [12]:

1. Create a Bill of materials (BOM) or a full product/material list for the particular project. Explore the properties of each product/material.
2. Assign a product/material type for each homogeneous product/material, e.g. metal, plastic, textile, glass, etc.
3. Classify each product/material with a group or cycle. This can be done by designating each product/material either as biological or technical. Biological means that the product can be returned to the environment whilst in use or after (wood, cotton, paper, etc.), whereas technical implies that products/materials that cannot be returned to the environment (metals and plastic, etc.).
4. Receiving information from suppliers about the chemical properties of the product/material. Request for MSDS and TDS and other certifications about the product/material as well as the supplier.
5. Understanding the hazards or health and safety of the product/material, as well as finding a suitable replacement if any.
6. Check if the product/material was part of a previous circular process.
7. Check the lifespan of the product/material, can we recover value in the future.
8. Finally, is our product/material recyclable?

3.2 Information

The construction industry is one of the largest sectors in the world's economy. It employs approximately 7% of the world's working population and spends around $10 trillion on construction goods and services per year. One of these services and goods are software-based design items. Computational thinking such as Building Information Modeling (BIM) software and material databases allow engineers and designers to create three-dimensional simulations and the ability to view the reality of the structure. Unlike the traditional computer-aided design (CAD), which represent flat shapes and two-dimensional representations.

Software models and databases allow us to create digital representations of real-life structures. From the geometry of the building, spatial and geographic information, and importantly, the exact quantities and the properties of the materials are required [13]. In order to allow high recycling rates and low environmental impacts of buildings, detailed information databases of the construction materials, as well as their building stock characteristics, are essential [20].

We are living in an era of digitization which is ideal for analytical purposes. Big data analytics enabled by machine learning using classification algorithms for anomaly detection and time series forecasting, among others, would allow maintenance predictions, how well the property may sell, modeling errors, recyclability, and, importantly, life cycle analysis of materials [6].

There have already been software and database developments and prototype models within the industry for the industry, some of which are mentioned in Table 2 [36].

3.3 Stakeholders

In order to complete a material passport and a transition toward a circular economy, all relevant stakeholders in Fig. 4, need to play their parts. Data needs to be shared via a material passport database, which allows for regular updates within a centralized location which is accessible to all stakeholders. Some of these databases already exist, namely BAMB and Madaster.

4 Goals and Benefits of Material passports

Below goals and benefits of material passports are summarized [18]:

- Resource for switching from a linear system to a circular economy
- Improving the importance, quality, and safety of material supplies
- Waste reduction
- Eco-footprint reduction
- Supply and demand management
- Managing resources rather than waste will lead to significant cost reduction
- An increase in residual value
- Sufficient data for all stakeholders.

5 Current Applications and Projects

In September 2015, 15 partners from seven European countries worked on a project called Buildings as Material Banks (BAMB), under the Horizon 2020 research programme, WASTE-1–2014 Moving toward a circular economy through industrial symbiosis. It is led by Brussels Environment, with an approximate budget of 10 million Euros.

BAMB's aim is to develop a circular model to the use of building, designs, and materials, thereby increasing the value of materials. Material passports and reversible building designs are tools that are used in this shift to a circular economy [3].

Table 2 Various databases and software, implementations and prototypes

Company and software	Description	Website
Substance flow analysis (STAN)	Software which is available online, analyses material flow by importing material data into the software, the software follows the Austrian standard	https://www.stan2web.net/
OpenLCA	Another freeware available and still in development is openLCA which can be used for Sustainability and Life Cycle Assessment of materials data	https://www.openlca.org/ https://nexus.openlca.org/databases
TOMRA expert line	TOMRA Expert Line in Canada, have developed algorithms for material recycling and waste solutions	https://www.tomra.com/en
BigML	Apply Machine Learning to BIM	https://bigml.com/
OPTORO	Have also developed algorithms for material recycling and waste solutions	https://www.optoro.com/
CEP-AMERICAS	This Circular Economy Platform (CEP) of the Americas is an initiative powered by the Americas Sustainable Development Foundation (ASDF). It fills the vacuum for an easy-access one-stop-shop portal where information about Circular Economy from and for the Americas is made available	https://www.cep-americas.com/
Eco invent	Database for the Life Cycle Assessment on energy supply, resource extraction, material supply, chemicals, metals, agriculture, waste management services, and transport services	https://www.ecoinvent.org/home.html
Eco2soft	Eco and recycling data	https://www.baubook.info/eco2soft/?lng=2
SimaPro	Database for Life Cycle Assessment	https://simapro.com/
USEtox	Characterizing human and Ecotoxicological impacts of metals, calculation of emission fractions, and characterization factors	https://usetox.org/

(continued)

Table 2 (continued)

Company and software	Description	Website
Thinkstep GaBi	Life cycle assessments, product and organizational carbon footprints. This database spans over multiple sectors	https://www.gabi-software.com/international/databases/gabi-dat abases/
Circularise	An open, distributed and secure communications protocol for the circular economy. The platform allows information exchange between stakeholders throughout the value chain, creating transparency around product histories and material destinations	https://www.circularise.com/
Buildingone	Efficient planning, management and analysis for building life cycle	https://www.onetools.de/en/
Buildings as material banks (BAMB)	Material passport platform basically is digitally marking materials. It allows users to monitor the building cycle, from planning to construction, occupancy, repairs, renovations, repurposing and decommissioning, and the capacity to track materials quality and changes and track materials health	https://www.bamb2020.eu/
MADASTER	Material passport platform for the public, which acts as an online library of materials in the built environment. Madaster utilizes 3D scans and building information modeling (BIM) to register or digitally mark the parts of buildings, all this information is put into a passport	https://www.madaster.com/en/our-offer/Madaster-Platform
Circular economy toolkit (CET)	An assessment tool, which identifies improvements in products circularity	https://circulareconomytoolkit.org/
Material circularity indicator (MCI)	Described by the Ellen MacArthur Foundation as a tool used to assess European products in regards to a circular economy	https://www.ellenmacarthurfoun dation.org/our-work/activities/ce100/co-projects/material-circul arity-indicator

(continued)

Table 2 (continued)

Company and software	Description	Website
Circularity calculator	Supports manufacturers in product designs for a circular economy	https://www.circularitycalculator.com/

Fig. 4 Diverse stakeholders who would contribute to a material passport

BAMB's Material passport platform basically is digitally marking materials. It allows users to monitor the building cycle, from planning to construction, occupancy, repairs, renovations, repurposing and decommissioning, and the capacity to track materials quality and changes and track materials health. BAMB has succeeded in the goal of generating 300 material passports, materials are developed together with a software solution, information is easily accessible by all stakeholders during each process [8].

The approach was applied to the wood frame system in Brazil, in order to test the application and feasibility of the system. The project had proved that stimulates circular thinking [25].

A similar project was investigated in the United Kingdom in the steel industry. Steel that is re-used is approximately 8–10% more expensive as compared to new steel. This is due to the reconditioning process that is required. The project had shown that, by utilizing material passports and the BAMB process, there was a reduction in financial barriers ranging from 150 to 1000 £/t [31]. The re-use of metal in ships have also come under the spotlight, with that in mind, Maersk Shipping Line had implemented its own material passport called 'cradle-to-cradle passport.' Every nut and bolt on a 60,000-ton ship can be identified using this passport, materials are numbered, thereby separating the quality of metals. This creates better recycling possibilities and safety [9].

The Ellen MacArthur Foundation (EMF) collaborates with global partners such as Cisco, Google, H&M to name a few and The Circular Economy 100 (CE100) Network. Its ultimate goal is to accelerate the transition to a circular economy. The Built Environment Case Studies Co. Project is a collaboration between BAM, BRE, cd2e, London Waste & Recycling Board, Ouroboros, Tarkett, and Turntoo to provide the CE100 Network with relevant circular information with regards to the built environment. The projects that were carried out are listed below [10]:

- Rehafutur Engineer's House
- Olympic Park
- Resource Efficient House
- Brummen Town Hall
- Renovation offices, workshops and storage: Liander
- Bus Boarder
- Pôle de Police Judiciaire/Judicial Police Compound
- BioBuild
- Buildings as Material Banks (BAMB)
- Construction Re-use Platform: Bexleyheath
- ROC A12 School: Carpet Lease
- Reviva shelving.

Taiwan has increased its efforts toward a circular economy. In 2016, the government initiated several measures to implement a circular economy. In a recent publication, there were a total of 66 circular implementations in Taiwan, where over 360 partners are involved. Taiwan being the leading supplier and manufacturer of electronic equipment, the stand out initiative was they developed a circular economy for economic development with environmental protection [21, 32].

In 2017, at the real estate expo in Amsterdam, Madaster launched a material passport platform for the public, which acts as an online library of materials in the built environment. Thomas Rau described Madaster as being a 'land registry for materials.' This project inspired others in the industry to follow suit, ABN AMRO partnered with CAD & Company, Rendemint and the architects' (specialists in circular economy) firm Architekten Cie to develop its own material passport [5]. Madaster utilizes 3D scans and BIM to register or digitally mark the parts of buildings, all this information is put into a passport.

There are other companies and organizations that have implemented or in the process of implementing the circular economy, namely:

- Timberland—From tires to shoes.
- Johnson controls—Recycled batteries.
- Aquazone—Water waste into fertilizer.
- Schneider Electric—Increase product lifespan through leasing and pay per use.
- AB Inbev—Returnable glass bottles.

5.1 A Material passport Case Study

We use a case study published in 2019, by three researchers from TU Wien, which is one of the major universities in Vienna, Austria. They had demonstrated how to generate a material passport, as well as assess the recycling potential and environmental impact of materials and an entire building [19].

The material passport concept was tested using a concrete office building. The building consisted of three stories. The following building components were considered for the assessment:

- Exterior and interior walls
- Roof
- Flooring
- Ceilings
- Basement
- Glass facade
- Concrete columns.

The first step was to model the building using BIM software. The above-listed components had to be inserted into the BIM software. Once the building has been modeled, the second step is to utilize some of the databases or software mentioned in Sect. 3.2. In this example, the modeled components were linked to BuildingOne for data management and assessment, and Eco2soft was used to determine the eco and recycling data of each component. Once all the suitable information had been acquired regarding each component, a material passport is developed. A graphical representation is depicted below in Fig. 5.

The case study had shown that Material passports can work together with current software and prototypes, the building was recyclable to around 50%, and the main environmental impact was caused by the concrete. This had proved that by utilizing material passports in the early stages of project developments, the correct choices of materials with regards to recyclability and eco-friendliness can be made. Which is an important step toward a circular economy.

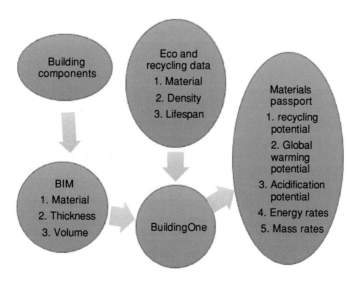

Fig. 5 A graphical representation of the development of a material passport, based on a case study

6 Barriers and Solutions of a Circular Economy and Material passports

In summary, most of our global economy is modeled around linearity rather than circularity. Humanity and their habit of wasting and polluting have brought the practice of a circular economy into the spotlight, in all sectors. As much as we develop tools such as material passports or digital marking of materials as well as many other so-called 'ingredients lists,' there are still some hurdles. In order to get over these hurdles, certain rules need to be put in place [35]:

- Stricter government policies and laws.
- Government support in form of incentives and funding to companies and public and private partnerships.
- Production of products or materials in a more circular fashion with innovation.
- Development of better recycling technologies and waste management with the introduction of 4IR.
- Finally, changing the mindset of humanity and consumption behavior.

7 Advances to Be Made in Terms of Digital Technologies and Software to Fully Realize 'Material passport' in a Circular Economy

Digitization seems to be lagging within the built environment, particularly in the construction sector. Implementation of this important aspect has been a challenge in the past years. As we stand at the cusp of a new industrial revolution (4IR), technologies such as Artificial intelligence (AI) and Cloud IoT platform. The tools for implementing material passports and a circular economy are now readily available, the idea of digitally marking materials is now a reality. This would also allow for an increase in profit for supply chains.

We are now able to integrate material passports within BIM, Geo-information (GIS), and Unified building modeling (UBM) systems. This will allow for capturing and accessibility of material data and their properties within a circular design.

With the use of AI, material properties can be viewed using automated methods. Machine learning algorithms can analyze the above material data patterns. Or imagine material passports communicating with each other via wireless sensors, this can be achieved using IoT platform. This would allow for smoother and faster data exchange and input changes across different networks and stakeholders, lifespan monitoring of materials, and early maintenance warnings.

Digitization, marking, capturing, storing, and analyzing materials data for material passports allow for the introduction and implementation of a number of technology applications and technological advancements in this research area within a circular economy. We list some of the technologies below [18]:

- Blockchain
- AI (Advanced Robotics and Machine Learning)
- Virtual Reality
- Augmented Reality
- IoT
- Big Data
- Radio-frequency Identification (RFID)
- 3D Printing and Scanning.

These technological advancements catalyze a circular economy implementation. The construction industry will change drastically in the coming years as a number of technologies will be adopted from the autonomous operation of construction equipment to digital marking of products. Digital marking is a link between the physical product and the performance of a product or material {Declaration of Performance (DoP)}. It gives access to product or material information in a digital format in a harmonized way. This info can be remotely accessed via cloud-based platforms or applications. This contributes in reducing administrative burdens, traceability of products or materials, and serves as a link between manufacturers and consumers use of the product.

To date, there have already been applications and innovations from global leaders in other sectors. Apple's iPhone disassembly robot is capable of dismantling an iPhone in 11s and sorts its components into re-usable materials. By doing this, Apple has captured materials that are re-usable for future products to the value of approximately $40 million [34].

8 Conclusion

Material passport will increase the value of materials and incentivize suppliers and manufacturers to produce circular materials. It enables the easy acquisition of sustainable materials and reversed logistics. Given the number of ongoing initiatives globally, it is reasonable to say that there is room for possibilities in this research area. The concept of circular economy in the built environment, as well as material passports and digital marking tools, provide opportunities for innovation and value creation.

9 ANNEX: A Case Study

In this second case study, we used a student-oriented construction project report in order to come up with a material passport, and demonstrate the quantitative benefits before and after material passport analysis.

Step 1

Identify a project: in this case, the project chosen was the Delft University of Technology, Faculty of Architecture, Urbanism and Building sciences, Track Building Technology, where a student had worked on a project located in Rotterdam, Netherlands. A typical old office building is refurbished into livable housing [1].

Step 2

Make up a list of the materials used in the construction process or create a BOM, their quantities and specific use in the construction process. In this case, we had selected some of the most important components.

- Concrete
- Aluminium
- Glass
- Wood.

Step 3

Once all the components and their respective quantities are determined, develop a BIM model using design software or acquire the building design. The site plans and graphical BIM models of this project can be viewed in Fig. 6a before design and Fig. 6b after design [1].

Step 4

Acquire each materials eco and recyclable properties, as well as the material-specific fixed data such as MSDS, TDS, LCAs, etc., or whether it has been used in previous cycles, all these data can be acquired from one of the databases in Table 2; in this case, we had used Thinkstep Gabi Database.

Step 5

We create our own material passport in Table 3; this can be done by utilizing the information in the previous steps. For the purpose of this paper, we had created a tabular design. Since the material data is of large amounts, links have been provided in the table. Newer trends are based on online material passport databases where passport information can be accessed and adjusted accordingly by the building's stakeholders.

In order to see the benefits of this material passport, we will look at the circularity index of the above-listed materials. Circular indicators are decision-making tools for developers, it assesses how well a company or product performs in the transition from a linear to a circular economy. Material circularity indicator (MCI) of a product gives a value between 0 and 1 (or 0%–100% recirculated parts), a value that is higher than 1 means a higher circularity. In order to calculate the indicator value, complex mathematical calculations and input values are required, these involve Mass, Recycled feedstock, and Recycling efficiency.

$$Product\,Level\;Circularity\;=\;\frac{Economic\;Value\;of\;Recirculated\;Parts}{Economic\;Value\;of\;All\;Parts}$$

MCI data requirements are as follows:

- Source of the material (virgin, recycled, re-used)
- Manufacturing process losses
- How the manufacturing losses treated?
- How is the waste of the product or end of life treated?
- Recycling process efficiency
- The mass of the product
- The lifetime of the product

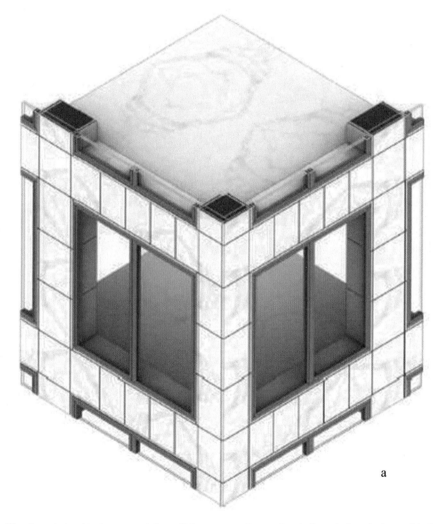

a

Fig. 6 **a** A graphical representation of the construction project before design. **b** A graphical representation of the construction project after design

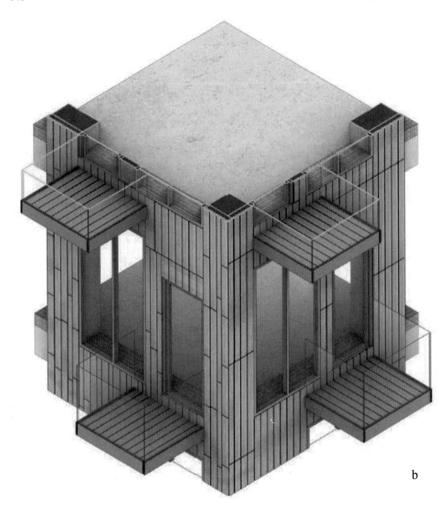

b

Fig. 6 (continued)

- Intensity of use of the product
- The average lifetime of the average product
- The average use intensity of the average product.

The same datasets are required for life cycle analysis (LCA), except for the last two points, i.e., the average lifetime and average use intensity of the average product.

For the purpose of MCI calculations, online tools are available such as Circular economy toolkit (CET), Circularity calculator, Ellen MacArthur's Material circular indicator (MCI) tool, OpenLCA, and GabI software, the links to these platforms and many others are available in Table 2. In this case study, we used Ellen MacArthur's Material circular indicator (MCI) tool. Tables 4 and 5 shows the material circularity

Table 3 Material passports

Materials	Mass (kg)	Life expectancy (years)	CO_2 per year	Fixed material data, manufacturers, LCA	Eco and recycling properties
Concrete	33,600	100	2657	https://gabi-doc umentation-2020. gabi-software.com/ xml-data/proces ses/606a5ad9-3317-4311-91b7-6c2147955729.xml	The concrete structure and brick work is considered to have a lifespan of 100 years and is re-usable at end of use phase of the building. 1ton concrete = 1ton of CO_2
Aluminium	810	50	145	https://gabi-doc umentation-2020. gabi-software.com/ xml-data/proces ses/695f3519-ea45-414b-9f18-a1dd92 00b765.xml https://gabi-doc umentation-2020. gabi-software.com/ xml-data/proces ses/76a28c07-f7d7-40f9-8fcd-1f2323 7daf2e.xml	Aluminium and steel are high embodied, due to the energy required to produce, mining, heating, manufacturing, and transportation. The more they are recycled their embodiment reduces making it more recyclable and eco-friendly. Long-lasting, durable, and able to withstand ambient conditions
Wood for balconies	89	100	70	https://gabi-doc umentation-2020. gabi-software.com/ xml-data/proces ses/3e4bd27e-caa6-42b4-8e4e-95da9a 96eed3.xml	Wood, just like steel and aluminium, is a good recyclable material but with less strength. Perfect for structural framing, flooring, furniture. Cork and bamboo are good alternatives

(continued)

Table 3 (continued)

Materials	Mass (kg)	Life expectancy (years)	CO_2 per year	Fixed material data, manufacturers, LCA	Eco and recycling properties
Glass for balconies	211	50	210	https://gabi-doc umentation-2020. gabi-software.com/ xml-data/proces ses/6d58cca8-213a-447e-942c-89701a609005.xml	Recycling glass reduces air and water pollution. Environmentally and economically friendly

Table 4 Material circularity indicator (MCI) before the implementation of the Material passports

Materials	Virgin %	Re-used %	Recycled %	Re-usable %	Recyclable %	Waste %
Concrete						
Aluminium	0	8	0	3	0	0
Glass	7	0	35	0	28	28
Wood	56	41	0	60	13	9
MCI value of building before material passport is 69%						

Table 5 Material circularity indicator (MCI) after the implementation of the Material passports

Materials	Virgin %	Re-used %	Recycled %	Re-usable %	Recyclable %	Waste %
Concrete	0	69	0	95	0	0
Aluminium	0	0	0	0	0	0
Glass	7	1	35	0	6	7
Wood	56	1	0	3	1	1
MCI value of building before material passport is 73%						

indicator of the building in this case study, before and after developing the material passport [1].

For further reading on materials circularity indicators, measuring and mapping circularity, as well as the International Organization for Standardization (ISO) TC 323 on Circular Economy, refer to these references [2, 16, 22, 30].

The original glass can be broken down for recycling, the aluminium can be re-used in the building renovation process or recycled, and glass and stone are recycled.

Finally, the concrete which is 70% of the structure is expected to last for 100 years, therefore re-usable. Based on this information, we were able to select the correct materials and we had developed our material passport in Table 3 above. The new MCI value is shown in Table 5.

We find that there is an increase in the MCI value from 69 to 73% based on the selection of materials in this case study. This change in MCI is due to the fact that we had kept the specific materials in a circular flow (re-used or recycled or recyclable) for as long as possible, with a reduction in waste.

Material selection plays a huge role in circular design processes. In this case, standardization of components, cleaner material flows, material cost factors, and simpler disassembly or modular designs were taken into account when selecting materials. By comparing the material indicators based on the selected components for the renovation of this building, we can conclude that Material passports generation from the beginning of the building's lifespan proves to be an important feature in the construction process, and would assist in creating a perfect balance in selecting the correct materials and their quantities in order to create a more circular economy.

10   Questions and Answers

1. **Define a Material passport**
   Material passports can be defined as a value tracking tool and brings back residual value to the market. Material passports make information available and relevant from production to purchase to use to maintenance. A single material passport contains a set of information about a particular material, product, or system.
2. **Define a Circular economy**
   Materials being kept in use for longer periods of time.
3. **What is the difference between a linear economy and a circular economy?**
   In a linear economy, raw materials are transformed into a required product and, at the end of its life cycle, it is thrown to waste.
   Whereas in a circular economy, the product is not thrown to waste, rather it is recycled or re-used.
4. **What is the basic structure of material passport?**
   Input material data.
   A material database.
   Output material data accessible by stakeholders.
5. **What are the properties or characteristics one should look out for when selecting materials?**
   Physical properties.
   Chemical properties.
   Biological properties.
   Health of materials.
   Production and design.
   Logistics.

Disassembly.

Recyclable.

6. **List and explain the types of data we can get from the above Material properties and characteristics**

   Life cycle analysis (LCA) or Life Cycle Impact Assessment (LCIA) is a system that tracks processes and transport routes used for raw materials extraction, production, and waste management.

   Technical data sheets (TDS) and material safety data sheets (MSDS), these contain information about product composition, applications, the potential hazards of these products and how to work safely with them.

   Bill of materials (BOM), list all the required materials, their quantities, and their use.

7. **List the steps to follow to generate a Material passport.**

   1. Create a Bill of materials (BOM) or a full product/material list for the particular project. Explore the properties of each product/material.
   2. Assign a product/material type for each homogeneous product/material, e.g., metal, plastic, textile, glass, etc.
   3. Classify each product/material with a group or cycle. This can be done by designating each product/material either as biological or technical. Biological means that the product can be returned to the environment while in use or after (wood, cotton, paper, etc.), whereas technical implies that products/materials that cannot be returned to the environment (metals and plastic, etc.).
   4. Receiving information from suppliers about the chemical properties of the product/material. Request for MSDS and TDS and other certifications about the product/material as well as the supplier.
   5. Understanding the hazards or health and safety of the product/material, as well as finding a suitable replacement if any.
   6. Check if the product/material was part of a previous circular process.
   7. Check the lifespan of the product/material, can we recover value in the future.
   8. Finally, is our product/material recyclable?

8. **Name two current Material passport databases**

   BAMB and MADASTER.

9. **Who are the typical stakeholders involved in the development of Material passport? (Name five or more)**

   Engineers, architects, owners, builders, material suppliers.

10. **Name three benefits of Material passports**

    Eco-footprint reduction.

    An increase in residual value.

    Sufficient data for all stakeholders.

11. **What is digital marking of materials?**

    Digital marking is a link between the physical product and the performance of a product or material {Declaration of Performance (DoP)}. It gives access to product or material information in a digital format in a harmonized way. This info can be remotely accessed via cloud-based platforms or applications.

12. **Which of the approach creates the highest value—Recycling, Re-use, or remanufacturing?**
Re-use.

13. **What are Material circular indicators?**
Circular indicators are decision-making tools for developers, it assesses how well a company or product performs in the transition from a linear to a circular economy. Material circularity indicator (MCI) of a product gives a value between 0 and 1 (or 0%–100% recirculated parts), a value that is higher than 1 means a higher circularity.

14. **True or false—Design for endless RECYCLING is a key principle in a circular design?**
False.

15. **Develop a Material passport of your own, based on a construction project you are currently busy with.**
(Example: ANNEX: A Case Study)

16. **Come up with a circular designed product of your own, not necessarily in the built environment.**

References

1. Ankur, G. (2019). Accelerating circularity in built-environment through "Active-Procurement": An aggregated assessment framework to make sustainable choices while using secondary material at early design phase. TU Delft Architecture and the Built Environment.
2. Assets.website-files.com. (2020). Retrieved March 19, 2020, from https://assets.website-files.com/5e185aa4d27bcf348400ed82/5e318d51e9eac45e7a658aac_Measuring%20Circularity%20-%20technical%20methodology%20document.docx.pdf.
3. BAMB. (2020). About bamb—BAMB. Retrieved February 1, 2020, from https://www.bamb2020.eu/about-bamb/.
4. BAMB. (2020). New office building—BAMB. Retrieved March 16, 2020, from https://www.bamb2020.eu/topics/pilot-cases-in-bamb/new-office-building/.
5. Bikker, A. (2020). MADASTER-materials passport opens the way for waste-free building culture. Abnamro.com. Retrieved February 5, 2020, from https://www.abnamro.com/en/newsroom/press-releases/2017/madaster-materials-passport-opens-the-way-for-waste-free-building-culture.html.
6. The Official Blog of BigML.com. (2020). Building information modeling (BIM): Machine learning for the construction industry. Retrieved March 12, 2020, from https://blog.bigml.com/2018/08/07/building-information-modeling-bim-machine-learning-for-the-construction-industry/.
7. Böhme, K., Holstein, F., & Salvatori, G. (2019). Circular economy strategies and roadmaps in Europe. Visits and Publications.
8. C2c-centre.com. (2020). Materials passports platform prototype for materials banking now live I C2C-centre. Retrieved February 2, 2020, from https://www.c2c-centre.com/news/materials-passports-platform-prototype-materials-banking-now-live.
9. C2c-centre.com. (2020). Maersk cradle to cradle® passport I C2C-centre. Retrieved February 4, 2020, https://www.c2c-centre.com/library-item/maersk-cradle-cradle%C2%AE-passport.

10. CE100. (2016). Circularity in the built environment: Case studies. Compilation of case studies from the CE100. Ellen Macarthur Foundation.
11. Charter, M. (2019). *Designing for the circular economy*. London: Routledge.
12. Circulardesignguide.com. (n.d.). Material selection. Retrieved February 18, 2020, from https://www.circulardesignguide.com/post/material-selection.
13. Community, B. (2020). Could machine learning be the new scope for BIM? | Bimcommunity. BIM Community. Retrieved March 12, 2020, from https://www.bimcommunity.com/news/load/948/could-machine-learning-be-the-new-scope-for-bim.
14. Construction-products.eu. (2020). Smart CE marking: Construction products Europe AISBL. Retrieved March 19, 2020, from https://www.construction-products.eu/services-jobs/smart-ce-marking.
15. Material District. (2020). The material passport as next step in circular economy—Material District. Retrieved January 23, 2020, from https://materialdistrict.com/article/material-passport-next-step-circular-economy/.
16. Ellenmacarthurfoundation.org. (2020). Retrieved March 18, 2020, from https://www.ellenmacarthurfoundation.org/assets/downloads/insight/Circularity-Indicators_Project-Overview_May2015.pdf.
17. Finamore, M. (n.d.). Circular Economy in the Built Environment. Oneplanetnetwork.org. Retrieved February 11, 2020, from https://www.oneplanetnetwork.org/sites/default/files/circular-economy-versio-7.11.2017-pdf.pdf.
18. Heinrich, M., & Lang, W. (2019). *Materials passports—Best practice*. München: Technische Universität München.
19. Honic, M., Kovacic, I., & Rechberger, H. (2019). Assessment of the recycling potential and environmental impact of building materials using material passports - a case study. In *Energy efficient building design and legislation*. Croatia: SMSS2019 (pp. 172–179). Retrieved February 29, 2020, from https://www.researchgate.net/publication/334611379.
20. Honic, M., Kovacic, I., & Rechberger, H. (2019). Concept for a BIM-based material passport for buildings. *IOP Conference Series: Earth and Environmental Science., 225,* 012073. https://doi.org/10.1088/1755-1315/225/1/012073
21. Ibitz, A. (2020). Assessing Taiwan's endeavors towards a circular economy: The electronics sector. *Asia Europe Journal*. https://doi.org/10.1007/s10308-019-00568-w
22. International Organization for Standardization. (2019). ISO/TC 323—Ad-Hoc Group 3 'Measuring Circularity' Explanatory note as part of the New Work item proposal (NWIP). Retrieved March 19, 2020, from https://www.iso.org/committee/7203984.html.
23. Architects Journal. (2020). Material passports: Finding value in rubble. Retrieved January 26, 2020, from https://www.architectsjournal.co.uk/news/material-passports-finding-value-in-rubble/10043989.article.
24. Luscuere, L. (2017). Materials passports: Optimising value recovery from materials. *Proceedings of the Institution of Civil Engineers—Waste and Resource Management, 170*(1), 25–28.
25. Munaro, M., Fischer, A., Azevedo, N., & Tavares, S. (2019). Proposal of a building material passport and its application feasibility to the wood frame constructive system in Brazil. *IOP Conference Series: Earth and Environmental Science, 225,* 012018.
26. Peters, M., Oseyran, J., & Ribeiro, A. (2016). BAMB value network by phase. BAMB—Buildings as material banks consortium.
27. DEAT. (2004) Life cycle assessment, integrated environmental management, information series 9, department of environmental affairs and tourism (DEAT), Pretoria.
28. Pyzyk, K. (2020). 5 of the world's most eco-friendly building materials. Smart Cities Dive. Retrieved March 14, 2020, from https://www.smartcitiesdive.com/news/most-eco-friendly-building-materials-world-bamboo-cork-sheep-wool-reclaimed-metal-wood/526982/.
29. Qiu, E. (n.d.). How blockchain will transform the food supply chain. Retrieved February 11, 2020, from https://www.plugandplaytechcenter.com/resources/how-blockchain-will-transform-food-supply-chain/.
30. Saidani, M., Yannou, B., Leroy, Y., & Cluzel, F. (2017). How to assess product performance in the circular economy? Proposed requirements for the design of a circularity measurement framework. *Recycling, 2*(1), 6.

31. Smeets, A., Wang, K., & Drewniok, M. (2019). Can material passports lower financial barriers for structural steel re-use? *IOP Conference Series: Earth and Environmental Science, 225,* 012006.
32. Towards a circular Taiwan. (2019). 1st ed. Taiwan: Taiwan circular economy network.
33. Webster, K. (2017). *The circular economy.* Isle of Wight: Ellen MacArthur Foundation Publishing.
34. World Economic Forum. (2020). These 5 disruptive technologies are driving the circular economy. Retrieved March 6, 2020, from https://www.weforum.org/agenda/2017/09/new-tech-sustainable-circular-economy/.
35. World Resources Institute. (2020). Barriers to a circular economy: 5 reasons the world wastes so much stuff (and why it's not just the consumer's fault). Retrieved February 5, 2020, from https://www.wri.org/blog/2018/05/barriers-circular-economy-5-reasons-world-wastes-so-much-stuff-and-why-its-not-just.
36. Wright, S. (2020). AI and a circular economy | Stephen J. Wright. Retrieved February 23, 2020, from https://www.stephenjwright.com/ai-and-a-circular-economy.

Mohamed Sameer Hoosain is a post-graduate member of the University of Johannesburg (UJ) in South Africa. He has qualified with M-tech and B-tech in electrical and electronic engineer. His master's thesis in electrical and electronic engi-neering dissertation—is in the area of "energy efficiency in smart homes based on demand side management, with further research based on Engineering education and Industry 4.0". His manuscripts have been presented and published at local and international conferences and journals in his research field. He is currently a PhD candidate working with; the National University of Singapore (NUS) and UJ, his interest lies in industry 4.0 topics and Circular economy. Mohamed Sameer has gained experience working in the academic sector and industrial sector. He also does work related his field in Engineering, Production management, Internal auditing ISO 22000 FSSC 22000 including HACCAP at an FMCG company. Among oth-ers, he has successfully completed EPICS-in-IEEE projects, where he had developed ways to provide energy for rural communities as well as in-novative farming techniques at UJ. He has fostered collaborations with external parties such as Engineers without borders, Growing up Africa and other private, public and non-profit organizations. Other areas in-clude research and development at the Institute for Intelligent systems (IIS) at UJ, where we had found solutions for the university and external companies such as shortest path algorithms for underground cars in the mining sector for Komatsu, others include; IoT based management sys-tems, to name a few. Also the development of a Makerspace where 3D printing technologies and robotics are now available at the university. He is a member of ECSA as a candidate engineering technologist en-route professional regis-tration and a graduate student member with the IEEE. (https://www.linkedin.com/in/mohamed-sameer-hoosain-16a70a64).

Babu Sena Paul received his B. Tech and M. Tech degree in Radio physics and Electronics from the University of Calcutta, West Bengal, India, in 1999 and 2003 re-spectively. He was with Philips India Ltd from 1999–2000 as a supporting engineer. From 2000–2002 he served as a lecturer at Electronics and Communication Engineering Dept., Sikkim Manipal Institute of Technol-ogy, Sikkim, India. He completed his Ph.D. in the area of wireless com-munication, with thesis titled "Modelling of Multi-Antenna Wireless Channels & Relay Based Communication Systems", from Indian Institute of Technology (IIT) Guwahati, Assam, India in 2009. He joined the De-partment of Electrical and Electronic Engineering Technology, University of Johannesburg in 2010. He served as the Head of the Department from April 2015 until March 2018. He is currently serving as an Associate Professor and Director of the Institute for Intelligent Systems (IIS) at the University of Johannesburg. He has successfully supervised/co-supervised and is currently supervising/co-supervising post-graduate, doctoral and post-doctoral research fellows. His research interests are in the area of wireless communication, channel modelling, MIMO systems, relay based communication, mobile-to-mobile communication, wireless sensor networks, Data Science, Machine Learning etc. He has attended and published several papers in ISI indexed journals, international and national conferences and sympo-siums. He was awarded the IETE Re-search Fellowship during his Ph.D. studies. He is a member of IEEE and a life member of IETE.

Syed Mehdi Raza is a Professional Engineer (PEC No. Civil/33902) registered from Pakistan Engineering Council with over 7 years Civil Engineering experience. He completed his Bachelors in Civil Engineering from NED University of Engg & Tech. He is currently pursuing his Msc in Integrated Sustainable Design at the School of Design & Built Environment, National University of Sin-gapore. His studies are based on BIM, GIS and Circular economy in De-sign-Construction sector of Singapore.

Professor Seeram Ramakrishna *FREng, Everest Chair* (https://www.eng.nus.edu.sg/me/staff/ramakrishna-seeram/), is among the top three impactful authors at the National University of Singapore, NUS (https://academic.microsoft.com/ins titution/165932596). NUS is ranked among the top five best global universities for engineering in the world (https://www. usnews.com/education/best-global-universities/engineering). He is the Chair of Circular Economy Taskforce. He is a member of Enterprise Singapore's and ISO's Committees on ISO/TC323 Circular Economy and WG3 on Circularity. He also the Chair of Sustainable Manufacturing TC at the Institution of Engineers Singapore and a member of standards committee of Singapore Manufacturing Federation (http://www.smfederation.org.sg). He is an advisor to the Ministry of Sustainability & Environment—National Environmental Agency's CESS events, (https:// www.cleanenvirosummit.sg/programme/speakers/professor- seeram-ramakrishna; https://bit.ly/catalyst2019video; https:// youtube.com/watch?v=ptSh_1Bgl1g). European Commission Director-General for Environment, Excellency *Daniel Calleja Crespo,* said, *"Professor Seeram Ramakrishna should be praised for his personal engagement leading the reflections on how to develop a more sustainable future for all"*, in his foreword for the Springer Nature book on Circular Economy (ISBN: 978-981-15-8509-8). He is a member of UNESCO's Global Independent Expert Group on Universities and the 2030 Agenda (EGU2030). He is the Editor-in-Chief of the Springer NATURE Journal Materials Circular Economy—Sustainability (https://www.springer.com/journal/42824). He is an Associate Editor of eScience journal (http://www.keaipublishing.com/en/ journals/escience/editorial-board/). He is an opinion contributor to the Springer Nature Sustainability Community (https://sus tainabilitycommunity.springernature.com/users/98825-seeram- ramakrishna/posts/looking-through-covid-19-lens-for-a-sustai nable-new-modern-society). He teaches ME6501 Materials and Sustainability course (https://www.europeanbusinessreview. com/circular-economy-sustainability-and-business-opportuni ties/). He also mentors Integrated Sustainable Design ISD5102 project students. Microsoft Academic ranked him among the top 25 authors out of three million materials researchers worldwide based on H-index (https://academic.microsoft.com/ authors/192562407). He is named among the World's Most Influential Minds (Thomson Reuters) and World's Highly Cited Researchers (Clarivate Analytics). Listed among the top three scientists of the world as per the Stanford University researcher study on career-long impact of researchers or c-score (https://drive.google.com/file/d/1bUJrvurVVBbxSl9eFZRSHFi f7tt30-5U/view). He is an Impact Speaker at the University of Toronto, Canada Low Carbon Renewable Materials Center (https://www.lcrmc.com/). He is a judge for the Mohammed Bin Rashid Initiative for the Global Prosperity (https://www. facebook.com/Make4Prosperity/videos/innovation-inclusive- trade/479503539339143/). He advises technology companies with sustainability vision such as TRIA (www.triabio24.com),

CeEntek (https://ceentek.com/), Green Li-Ion (www.Greenli-ion.com) and InfraPrime (https://www.infra-prime.com/vis ion-leadership). He is a Vice-President of Asian Polymer Association (https://www.asianpolymer.org/committee.html). He is a Founding Member of Plastics Recycling Association of Singapore (PRAS). His senior academic leadership roles include University Vice-President (Research Strategy), Dean of Faculty of Engineering; Director of NUS Enterprise and Founding Chairman of Solar Energy Institute of Singapore (http://www.seris.nus.edu.sg/). He is an elected Fellow of UK Royal Academy of Engineering (FREng), Singapore Academy of Engineering and Indian National Academy of Engineering. He received PhD from the University of Cambridge, UK, and The TGMP from the Harvard University, USA.

Plastics in Circular Economy: A Sustainable Progression

Anand Bellam Balaji and Xiaoling Liu

Abstract Extensive usage of plastics causes environmental deterioration, global warming and health imperilments, costing the economy around $139 billion annually. Conversely, from childcare products to coffins manufactured with plastics, they have become an integral part of our life. At this situation, banning the use of plastics is not sustainable. Therefore, it is necessary to espouse circular economy (CE) in plastic sector. Meaning, preventing waste by manufacturing products that are efficiently reusable, recyclable or recoverable and gradually replacing non-degradable with degradable plastics. This chapter focuses on the concept of CE in plastic industries, recycling and recovering methods of plastics, government frameworks and challenges faced for implementation of circularity. From the case studies reported in this work, though there are successful execution of circularity in few scenarios, it can be noted that the implementation of CE is still at infant stages, as large proportion of companies have yet not committed 100% circularity until 2025. This work also identifies that more advancements in research and technologies, more tax benefits and funding allocation, need for collaborative business models, boosting and advertising the demand for recovered products and increased awareness on social responsibility of consumers and manufacturers are still necessary for achieving efficient circularity of plastics.

Keywords Plastics circularity · Key indicators for CE · Plastics recycling techniques · Challenges

Learning Objectives

- Understanding the concept of plastic circularity.
- Necessity to adopt plastic CE.
- Understanding the key challenges and indicators involved in CE implementation.
- Case studies, where plastic circularity is implemented.
- Knowing the status of global initiatives and legal frameworks.

A. B. Balaji · X. Liu (✉)
University of Nottingham, Ningbo campus, Ningbo, China
e-mail: Xiaoling.liu@nottingham.edu.cn

© Springer Nature Singapore Pte Ltd. 2021
L. Liu and S. Ramakrishna (eds.), *An Introduction to Circular Economy*,
https://doi.org/10.1007/978-981-15-8510-4_9

1 Introduction

In recent years, a number of products are being manufactured using plastics, mainly due to it's low cost, less weightness and ease of processing. It is estimated that 359 million metric tonnes of plastics were produced in 2018 alone to meet the global demands [1]. In 2019, the value of global market for plastics is approximately $568.7 billion with an average increasing rate of 3.5% until 2027 [2]. The consumption of plastics per person per year ranges from 9 to 108 kg. Figure 1 shows the per capita consumption of plastics across various countries in 2015. Polyolefin (PO) makes up the biggest share thereof, with low-density polyethylene (LDPE) being the most prominent, followed by polypropylene (PP), high-density polyethylene (HDPE) and polyethylene terephthalate (PET). Smaller shares are contributed by polyvinyl chloride (PVC), polystyrene (PS), polyamide (PA) and other plastics. Figures 2 and 3 show the consumption of general plastics and usage of these common plastics in different sectors. Most of the common plastics used in the industry are non-biodegradable. With its extreme durability, the life span of plastics and polymers

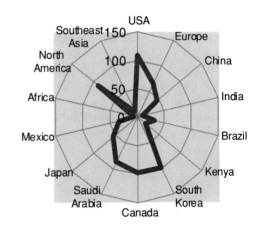

Fig. 1 Per capita consumption of plastics (kg/person) [4–7]

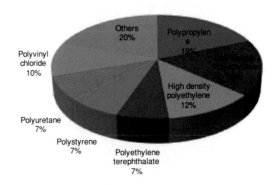

Fig. 2 Global consumption of common plastics [8]

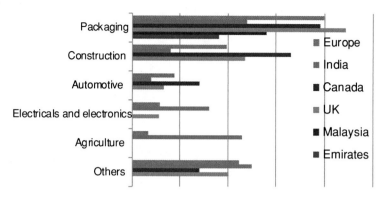

Fig. 3 Consumption of plastics by sector (%) [9–13]

is around some centuries. With increase in global usage of plastics, they often end up in landfills after their consumption. The usage of plastics has become one of the most pressing issues for the globe, as they are the leading source of global warming, reduction in groundwater due to clogging of river/canal streams and health hazards. Management of plastic waste (as they reach their end of life) has turned out to be so omnipresent that it has triggered the efforts to frame a worldwide treaty under the witness of United Nations. To tackle the issues, the stakeholders are also marching towards the adoption of circularity of the plastics. Circularity is adopted through developing and pursuing a concept of 'circular economy'. Circular economy is briefly defined by European parliament as 'a model based inter alia on sharing, leasing, reusing, repairing, refurbishing and recycling, where products and the materials they contain are highly valued, and where waste is reduced to a minimum' [3].

2 Circular Economy of Plastics

In recent trends, all the stake holders are moving towards more sustainable approach of circular economy from linear economy, where the products are used and disposed. 'Circular economy aims to keep resources in use for as long as possible, to extract the maximum value from them while in use, and to recover and regenerate products and materials at the end of their service life. It offers an opportunity to minimize the negative impacts of plastics while maximizing the benefits from plastics and their products, and providing environmental, economic and societal benefits'. The concept of progression from linear economy to circular economy is represented in Fig. 4.

In short, three important principles to manage the circularity of plastics or circular economy, in general, are as follows:

- Preserve and enhance the natural capacity by controlling finite stocks and balancing the flow of renewable resources.

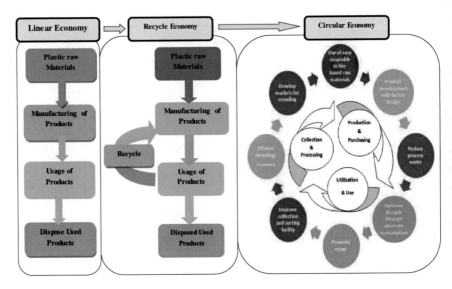

Fig. 4 Progression of polymer usage from linear model to circular model

- Optimizing the raw materials and yield through product circularity and using the materials with high value at all times.
- Make the system more effective by removing the negative externalities.

3 Recycling Rates, Policies and Initiatives for Circularity of Plastics Undertaken in Various Countries

Recyclability and reusing are parts of initiatives taken for promotion of circularity of plastics. In early days, many countries were exporting the plastic wastes to other countries, which were equipped with state-of-the-art recycling technologies. Until recently, China was the leading country to import the plastic wastes from various countries. For instance, Australia alone exported around 1.25 million tonnes of plastic wastes during a period of 2016–2017.

In early 2018, China began enforcing its National Sword policy, where stringent rules and regulations were imposed on the quality of plastic wastes to be imported. Further, in recent years, China minimized the import plastic wastes from major countries. Since the ban imposed by China, many countries started recycling of plastics within its territory. However, the recycling rates are still quite lower. Figure 5 shows the recycling rates of plastic wastes across various countries in 2018. India, South Korea, Japan and China are the leading countries in recycling of plastic waste.

From Fig. 5, it can be noted that the recycling rates of countries such as Singapore, USA, Brazil, etc. are still only at very low rate of 4–20%. To fight against the waste management, several governments and organizations are taking leaping steps towards

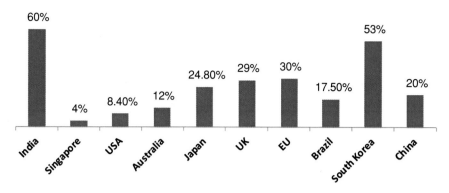

Fig. 5 Recycling rates of plastic waste of different countries [12–19]

the adoption of circularity of materials. Taking the lead, China adopted the Circular Economy Promotion law in 2009. In the following years, several detailed action plans (such as State Council 2013) have briefed the clear strategies and legal frameworks for the successful execution of CE in China [20].

Down the timeline, in 2012, Germany enforced Circular Economy Act also called as 'kreislaufwirtschaftgesetzKrWG' or German Circular Economy Law. Further, in December 2015, European Union came forward in adoption of CE. The 'Circular Economy Package' of European Union emphasizes on the new path of closed-loop cycle for production, usage and recycling of plastics. The framework also serves as a reference point for drafting of circularity principles for the countries, which focused on CE in later years. With respect to strategies for plastics in focus, the framework came into force from 2018. Later, Netherlands, in September 2016, passed on new legislation called 'A Circular Economy in the Netherlands by 2050'. Apart from this, several other programmes and initiatives such as Green Growth, From Waste to Resource (VANG) and the Bio-based Economy are also encouraged to embrace circularity of plastics. Later, the government of Spain elaborated a new initiative in 2017 for reduction of landfills and to promote CE. It is stated that around 300 signatories and stakeholders have indulged in the new pact of Government of Spain. The Government of India has recently drafted a National Resource Efficiency Policy in 2019, aiding the country to move towards the circular economy and approach zero landfill strategy.

Apart from government institutes, several non-governmental entities have come forward with a mission to progress towards circularity of plastics. For instance, American Chemistry Council's (ACC) Plastics Division has released a new obligation for its members. The new proposal has set a target to recycle and recover the plastics used in packaging sector, completely by 2040. In addition, the plastic pellet stewardship is to be encouraged by 2022. Further, in 2018, around 200 leading companies and business organizations, 16 governments of various countries, 26 financial institutes and 50 academic and research institutes have collaborated in a program called 'New

Plastics Economy Global Commitment' to make the virtual vision of CE into reality. The Ellen MacArthur Foundation heads this new initiative in collaboration with UN Environment [21].

4 Challenges Faced in Embracing Circularity of Plastics

In the holistic progression from open-loop/linear usage of plastics to implementation of close-loop/circularity of plastics, all the participants from industrialists to consumers need to face a significant challenge. Several strategies are reviewed and available for implementation of circular economy, especially in packaging industries. Researchers and analysts have constructed a database of the key factors and challenges that are to be considered in successful operation for adoption of CE [22], which is briefly summarized in Fig. 6. A number of important challenges are discussed as follows.

4.1 Finding Value

The recycled plastics are sometimes expensive than the virgin plastic raw materials. Though recycled plastics are up expensive, the quality is often questionable. Therefore, there is a need to find the addition of value to the recycled plastic materials.

4.2 Redesigning

Many goods are manufactured such that plastic content is made complicated to recycle. Many products such as bottles, packaging covers etc., in order to meet the market needs, are often produced with other additives and materials such as glues and bonds. However, these lead to difficulty in dissembling the products into waste separates. Therefore, a sustainable approach must be adopted to design the products with focus of easy recycling. In addition, new eco-based materials should be designed which not only offers recyclability but also are eco-friendly. For instance, polyethylene furanoate (PEF), a new bio-based plastic, is being developed, which serves as an alternate to existing PET bottles. In 2016, AVANTIUM, BASF with partnership of other leading companies such as Coca-Cola Company, Danone and ALPLA came forward to start a joint venture called 'Synvia'. They established a manufacturing unit in Belgium for the production of these bio-based materials which also offer opportunities to be recycled [24].

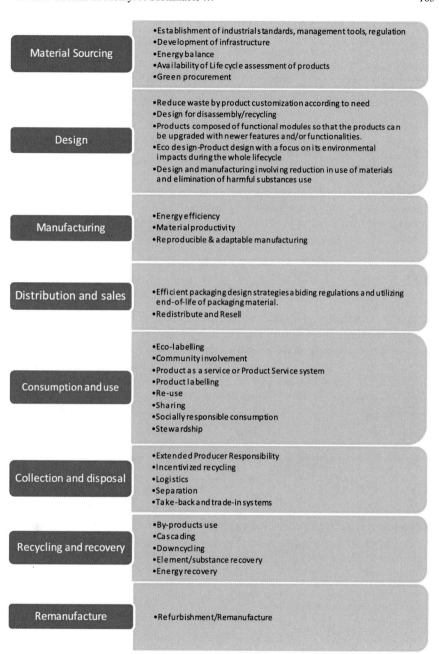

Fig. 6 Challenges and key indicators in implementation of circular economy [23]

4.3   Increasing Availability

Finding an alternate sustainable material is a key aspect in implementation of CE. Further, it is also equally important to make sure that new raw materials need to possess a stable supply, for uninterrupted manufacturing process.

4.4   Respecting Manufacturing Processes

Redesigning and usage of new alternate materials may affect the existing manufacturing process. Therefore, it is also necessary to find efficient approaches that do not interfere with the existing production processes and cost.

4.5   Sorting

Since different plastic goods have variable attributes such as shape, design, structure, different chemical properties and colour, it becomes very difficult for the recycling units to sort and categorize them to retrieve a good yield of recycled plastics. Manual sorting, induction sorting, eddy current sorting, sink float separation, tribo-electric separation and speed accelerators are some of the state-of-the-art sorting technologies used in current recycling facilities. X-ray, infrared and laser-induced breakdown spectroscopy are some of the technologies used for quality control of the sorting process.

4.6   Recycling

Recycling is one of the important aspects in CE to curb the leakage of plastics from the loop. The plastics are generally recycled either by mechanical or chemical recycling. A brief note on the types of plastic recycling is explained in the following sections.

4.6.1   Mechanical Recycling

Using mechanical recycling, the plastic wastes are processed into secondary/recycled raw materials without altering the chemical structures of the polymer. Mechanical recycling is more suitable and limited to thermoplastic polymers. It is often not suitable for thermoset polymers mainly due their cross-linked structures. The steps involved in mechanical recycling include collection, sorting, chipping, washing and pelletizating. The main challenge faced in mechanical recycling is that the plastic wastes need to be sorted, as they are susceptible to contamination. Another drawback is that the heat occurred during the process tends to soften the polymer or even cause degradation resulting in poor properties of the recycled yield.

4.6.2 Chemical Recycling

Chemical recycling (also called as feedstock recycling) is used for converting the larger polymeric chains into smaller units that can be further used into a variety of valuable resources. For instance, the plastic wastes are converted into gases, fuels and other chemicals through a number of processes such as pyrolysis, gasification and hydrogenation. On other hand, chemical depolymerization or chemolysis technology is used to convert the plastic wastes into monomers, oligomers and higher hydrocarbons that can be used to reproduce plastics that has been recycled or can be used for producing new plastic materials. The processing conditions, advantages and challenges of the process and examples of plastic treated and their yield are summarized in Table 1.

4.7 Collaboration

The adoption of circularity of plastics is a tedious progression, which is often not possible by a single stakeholder. It is necessary that all the stakeholders involved in the product and supply chain must corporate and collaborate for a smooth progression from linear to circular economy.

5 Case Studies of Circularity of Plastics

With the above-mentioned challenges, there are many designers, economists, activists, industrialists and manufacturers coming up with novel solutions to design the circularity of consumer used plastic products. Many smarter solutions for recycling, redesigning and minimal disposal of plastic wastes are being implemented in the day-to-day life. Figure 7 summarizes the consumption of some of established consumer brands, packaging companies, retail and hospitality companies. In addition, the organization's commitment to recycle and use the post-consumer recycled plastics by 2025 are listed in Fig. 7. A few examples of plastic waste circularity with respect to both upcycling and downcycling are briefed.

Table 1 Summary of chemical recycling methods [25–28]

Recycling method	Process Description	Processing conditions	Advantages	Challenges and drawbacks	Polymer Type	Products
Chemolysis/Chemical depolymerization	Depolymerization of plastic wastes with various chemical reagents to produce monomers	Temperature: 25–280 °C Pressure: 1–40 atm	Generates recycled monomers of plastics with no significant compromise on the quality	• Limited to recycling of condensation polymers and not suitable for additional polymers such as PP, PE and so on • Needs larger volume to be cost efficient	PET	• Terepthalic acid (TPA), bis(hydroxyethyl) terephthalate (BHET), Dimethyl terephthalate (DMT) (monomers for PET) • Polyols (monomers for PE, PU and epoxy)
					PA, PC	Caprolactam, hexamethylene diamine and adipic acid (monomers for PA and PC)
					PU	• Polyols and isocyanate (monomers for PU) • Urea

(continued)

Table 1 (continued)

Recycling method	Process Description	Processing conditions	Advantages	Challenges and drawbacks	Polymer Type	Products
Gasification	Partial oxidation process (using air, steam orsub-stoichiometric oxygen) for conversion of plastic wastes into gases.	Temperature: 1200–1500 °C; Pressure: 10–50 atm	• Polymer separation into different categories is not necessary • Can also be gasified with non-plastic contaminants in the waste	• Release of toxic gases such as NOx, H2S and Carbonyl sulfide • Still lacks technological advancements and research for gasification of mixed plastic waste	All plastics	• Syngas (A flammable gas consisting mainly of CO, H_2, with smaller amounts of CH_4, CO_2, H_2O and some inert gases) can be used as fuel or for production of methanol and paraffinic hydrocarbons
Pyrolysis	Degradation of polymeric chains in presence of inert atmosphere	Temperature: 500–800 °C	• Simple technology • Relatively no significant release of toxic gases • Can be used to produce electricity and heat	• Complex chemistry involved depending upon nature of plastics • Requires high volumes to be cost effective • Less easiness for PVC processing	PE, PP, PS	• Gases such as butene and propylene • Hydrocarbon liquids ranging from C_5–C_{16}
					HDPE, LDPE	Gases such as isobutane, isobutene, propylene, propane and isopentane
					PET	Benzoic acid, benzene and terephthalic acid
					PVC	HCl, benzene, toluene, and naphthalene
Hydrogenation	Plastics degraded using hydrogen in presence of heat	Temperature: 100–300 °C; Pressure: 50–100 atm	Fuels obtained can be used directly without further vigorous treatments	• High cost, due to usage of hydrogen • Operated under high pressure	All plastics	Hydrochloric acid, halogenated solid residue and gas

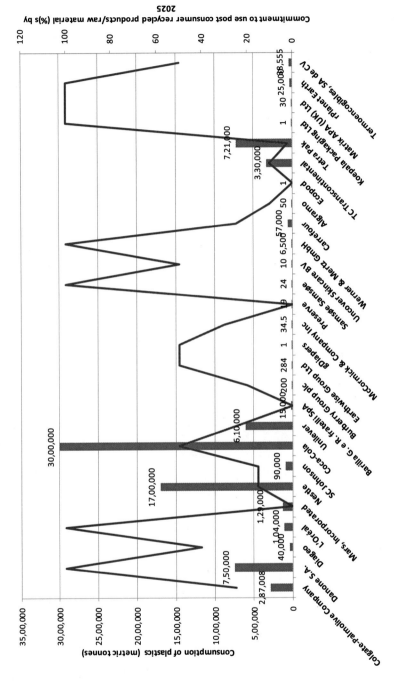

Fig. 7 Consumption of plastics by companies and their commitment to circularity of plastics (*Data Source* New Plastics Economy Global Commitment Report—2019)

5.1 Plastic Wastes in Construction and Maintenance of Roads

As an initiative to adopt CE, a new trend towards using the recycled plastics in laying roads and pavements has been tested and implemented across various countries. In general, the collected plastic wastes are first segregated into thermoplastic and thermosetting plastics. Mostly, thermoplastics are preferred because for this concept, as the softening point of thermoplastics is in acceptance level with the processing temperatures of bitumen and tar (155–165 °C). Further, plastic waste is melted and mixed with bitumen in a prescribed ratio (from 5 to 20%) based on the quality of tar or asphalt. Normally, blending takes place when temperature reaches 45.5 °C [29]. The mixture is then heated to higher temperatures of 155 °C, for enhancing better bonding between the bitumen and plastics. These heated mixtures are used for laying roads.

Other benefits of plastic roads include [30–32] the following:

- Approximately, cost of laying up roads reduces up to Rs. 45,000 per kilometre.
- The strength of the roads is almost doubled with respect to conventional roads.
- Since seepage during rainwater stagnation is reduced, the problems of frequent potholes occurrence are most likely reduced.
- The problem of bleeding that occurs in hot weather is also reduced.
- The use of plastic roads does not need any additional or specialized machineries and technologies. Hence, no significant change in the operation cost.
- Reduces the usage of initial raw materials such as bituminous, tar or asphalt leading to reduction of procurement cost.

India is one of the pioneer countries to have laid roads made up of plastic waste. It all started in 2015, when the government of India made it compulsory for all the road builders to implement the plastic wastes in roads. So far, the country has around 1 lakh kilometres of road laid using plastic across 11 states [33]. Further, the roads are being laid in UK, Canada, Australia, New Zealand and gulf nations using the patented product MR6, which consists of pellets manufactured with the blends of plastics and 20% bitumen. These patented plastic waste/bitumen blends are supplied by a road product specialists, Macrebura. These specialized products were also used as pot hole fillers during the repairing and maintenance of roads [34].

Apart from construction of roads and repairing potholes, the plastic wastes are also being used in manufacturing tools for reducing the cracks and buckling of the roads. For instance, in Texas (USA), the frequency of occurrence of cracks and buckling is higher due the nature of the soil, which often contacts or expands at different climatic conditions. To combat the issue, Dr. Sahadat Hossain, University of Texas at Arlington has projected a new technology of using 'pins' manufactured using plastic wastes, drilled into the roads. These pins keep the materials in contact and reduce the occurrence of cracks and the buckling. Approximately, 500 used plastic bottles are used in producing one such pin [35].

Key points: (1) Plastic roads not only offer an alternate for reducing the plastic wastes to landfill, but also the value of the plastic waste is upcycled as plastics in form of roads stay longer in the circular loop. (2) Use of plastic waste reduces the cost of roads thus adding potential contribution to the circular economy. (3) Plastic in roads increases the lifespan of the roads.

5.2 Circularity of Plastic Bottles and Liners

Procter & Gamble (P&G) collaborates with Purecycle™ to reuse the PET bottles. A recycling plant has been set up in Ohio, USA to have a commercial capacity to recycle 105 million pounds of PET and PP bottles with operation resuming in 2020. They use a patented depolymerization technology, which is less energy consuming to make the recycled materials economically viable [36]. P&G has also collaborated with other recycling specialists such as TerraCycle and SUEZ to make use of ocean/beach recycled plastics in manufacturing of shampoo bottles. The companies have committed to use up to 25% of beach plastics in manufacturing of shampoo bottles [37].

Coca-Cola European Partners (CCEP) along with Avery Dennison, Viridor and PET UK have collaborated with a mission to recycle the PET bottle liners rather than incinerating. In 2016, they upcycled 70 tonnes of PET bottle liners. The liners were shredded mechanically and reused into PET staple fibres and strappings. Also, thermo-foam sheets are produced from these plastic wastes which will be used in the production of trays [38].

Key points: (1) Collaboration between the manufacturers and plastic recycling experts is more efficient for adopting circularity. (2) New product design and materials must be implemented in order to avoid contamination caused by labelling, glues and so for easy processing during recycling.

5.3 Circularity of Plastics Used in Absorbent Hygiene Products

Single use disposable absorbent hygiene products such as nappies, diapers and sanitary napkins are often manufactured using absorbent pads covered with plastics such as polypropylene and polyethylene. Polyester fleeces are also used in manufacturing of the cloth diapers. It is estimated that around 300,000 tonnes of used diapers end up in Netherlands and 900,000 tonnes in Italy. Around 5% of landfills in U.S.A consists of absorbent hygiene products (AHP) [39]. Most commonly, these wastes are sterilized and incinerated for energy recovery. However, recently many novel ideas have been developed to recycle and reuse the plastics recovered from such wastes. P&G in Russia has started an initiative to use the recycled plastics from such product in cement applications. Further, the company has also collaborated with Angelini Group and Fater Spa to develop more circularity options. The plastics recovered from these wastes is proposed to be recycled into various applications such as bedding pads for pets, automotive, bottle caps and plastic park benches [40].

Key points: (1) Separate collection of these AHP wastes, separate sterilization process and quality control are serious challenges involved in developing circularity ideas for such wastes. (2) New disposal system and specialized collection must be implemented.

6 Conclusion

With respect to the increasing pressure of plastic waste management, there is an urgent need for the implementation of plastics circular economy. To help all the stakeholders move towards execution of CE, several government and non-governmental agencies have put forward new legislations and framework. Nevertheless, it is to be noted that, not all the leading industrialists, companies and even some smaller countries have yet committed for endorsement of complete circularity of plastics. These could be mainly because of the challenges such as

(a) lack of more evidence on the merits and demerits of recycled plastics,
(b) lack of more specialized waste collection and processing infrastructure, low progress in setting eco-design requirements,
(c) lack of advanced technologies, high cost of recycled plastics in comparison to virgin plastics and uncertainty in addition of value to recycled plastics.

In spite of these challenges, a few case studies have shown that circularity of plastics is possible. However, the implementation has not reached its maturity. From the market study, it can be noticed that all the entities are focusing on dealing with the challenges mentioned. However, the effect of absolute transformation is time-consuming and needs to be assessed regularly down the time. Focusing more and addressing the following priorities can make a significant progression from linear economy to circular economy of plastics easier:

(a) accessing ways to provide better transportation/collection/sorting of plastics,
(b) expanding the eco-design rules,
(c) facilitating and boosting the demand for circular products,
(d) providing funds for research advancements and
(e) use of new materials and providing more tax benefits.

From the case studies, it can also be seen that encouraging and developing more effective business models for collaboration between the companies, experts and speciality commodity manufacturers will also help the progression to be better and faster. Moreover, the concept of plastic circularity should not be burdened on the shoulders of established companies and government alone, but should be cultivated as a social responsibility from the grass root levels of every entity.

Questions

(Q1) What is the feasibility of chemical recycling methods for recycling plastic waste?
(Q2) Is plastic waste recycling energy efficient when compared to incineration?
(Q3) Is chemical recycling better than mechanical recycling?
(Q4) Can 100% circularity of plastics be achieved by enforcing government legislations?

Answers

(A1) With chemical recycling methods, the long polymer chains of the plastics into small units when subjected to high temperatures and with presence of solvents or catalysts. Chemical recycling in most cases is used for converting plastic wastes into fuels and energy. Except for chemolysis technique of PS and PET where the raw materials can be produced through repolymerization of monomers produced through chemical recycling. In general, the operational costs are higher. However, successful development of these technologies can help the plastic slowly decouple from petroleum dependence and self-potentially close the circular loop.

(A2) Plastic waste is commonly incinerated to tackle the problem of plastic waste management. However, new technologies are being developed to recover the raw materials through various recycling methods. On the other hand, it is speculated the recycling methodologies are very energy consuming making the process not sustainable. But in reality, for incineration the heating value for plastics is ~36,000 kJ kg^{-1}, whereas mechanical recycling conserves ~60,000–90,000 kJ kg^{-1}. Therefore, recycling plastic waste is more energy efficient than incineration.

(A3) Though both methods have its own pros and cons, the waste recycled using chemical methods can recover raw materials repeatedly for a number a times. In chemical recycling methods, the plastics are broken down into their molecular level and repolymerized from the monomers yielded from the process. The loss of integrity of the properties of plastics is lesser when compared to mechanical recycling. As in mechanical recycling, the heat and shredding of plastics causes an impact on the properties of the materials recovered.

(A4) Several governments have enforced the frameworks for implementation of circular economy of plastics. Nevertheless, these legislations more often provide a detailed framework and monitoring process for encouraging the plastic sector to progress towards CE. They have also provided tax benefits and funding opportunities for development of technologies and materials for improving plastic circularity. However, they do not impose fines or any other legal punishments for not adopting 100% circularity of plastics. From market study, it can be seen that not all the companies and brands have committed to 100% plastics circularity for a period until 2025. Hence, these rules and legislations cannot help achieve complete circularity of plastics but they encourage and push the economy to progress towards circularity.

References

1. Garside, M. (2019). Global plastic production statistics Published by M. Garside, Nov 8, 2019. This statistic displays the production volume of plastics worldwide (and in Europe) from 1950 to 2018. In 2018, world plastics production totaled around 359 million metric tons. Wor. *Statista* [Online]. Retrieved from https://www.statista.com/statistics/282732/global-production-of-plastics-since-1950/.
2. Grand View Market Research Report. (2020). Plastics market size, share & trends analysis report by product (PE, PP, PU, PVC, PET, Polystyrene, ABS, PBT, PPO, Epoxy Polymers, LCP, PC, Polyamide), By application (Packaging, Construction), By region, and segment forecasts, 2020–2027.
3. Breifing of European Parliament. (2017). *Plastics in a circular economy.*
4. IEA. (2020). *Per capita demand for major plastics in selected countries in 2015* [Online]. Retrieved from March 06, 2020, from https://www.iea.org/data-and-statistics/charts/per-capita-demand-for-major-plastics-in-selected-countries-in-2015.
5. FICCI. (2014). *Potential of plastics industry in northern India with special focus on plasticulture and food processing.*
6. Dubai Business Pages. *The plastics industry in Africa* [Online]. Retrieved from March 06, 2020, from https://dubai-business-pages.com/features/plastics.html.
7. Cleetus, C., Thomas, S., & Varghese, S. (2013). Synthesis of petroleum-based fuel from waste plastics and performance analysis in a CI engine. *Journal of Energy, 2013.*
8. Yeo, J. C. C., Muiruri, J. K., Thitsartarn, W., Li, Z., & He, C. (2018). Recent advances in the development of biodegradable PHB-based toughening materials: Approaches, advantages and applications. *Materials Science and Engineering C, 92,* 1092–1116.
9. British Plastic Federation. (2020). *About the British plastics industry* [Online]. Retrieved March 04, 2020, from https://www.bpf.co.uk/industry/default.aspx.
10. Government of Canada. (2017). *Industry profile for the Canadian plastic products industry* [Online]. Retrieved March 06, 2020, from https://www.ic.gc.ca/eic/site/plastics-plastiques.nsf/eng/pl01383.html.
11. NPCS. (2017). *Plastic products manufacturing: Profitable plastic industries.* Spectacle Frames, P.V.C. Rexine Cloth, Plastic Granules.
12. European Union. (2017). *Plastics in a circular economy.*
13. Johnson, J. US plastics recycling rate continues to fall. *Plastic News* [Online]. Retrieved from https://www.plasticsnews.com/news/us-plastics-recycling-rate-continues-fall.
14. The Energy and resources Institute. (2018). *Creating innovative solutions for a sustainable future* [Online]. Retrieved March 04, 2020, from https://www.teriin.org/sites/default/files/files/factsheet.pdf; https://www.unido.org/sites/default/files/files/2018–11/Plenary2-Plastics-Mohanty.pdf.
15. Heijmans, P. (2019). Singapore is only recycling a tiny fraction of its plastic waste. *Bloomberg* [Online]. Retrieved from https://www.bloomberg.com/news/articles/2019-10-11/singapore-is-only-recycling-a-tiny-fraction-of-its-plastic-waste.

16. The Canberra Times. (2019). *Only 12% of plastic waste is recycled* [Online]. Retrieved March 04, 2020, https://www.canberratimes.com.au/story/6320516/only-12-of-plastic-waste-is-recycled/?cs=14231.
17. Inoue, Y. (2018). *Japan's resource circulation policy for plastics.*
18. Koh, T. (2019). Plastic pollution: Greenest countries in Asia. *Asian Geographic Magazines* [Online]. Retrieved from https://www.asiangeo.com/environment/plastic-pollution-greenest-countries-in-asia/.
19. Morgan, D. (2019). How China is trying to stem its massive plastic pollution problem. *Huffpost* [Online]. Retrieved March 13, 2020, from https://www.huffingtonpost.in/entry/shanghai-trash-sorting-china-plastic_n_5d35fc12e4b020cd99478d8b?ri18n=true.
20. McDowall, W., et al. (2017). Circular economy policies in China and Europe. *Journal of Industrial Ecology, 21*(3), 651–661.
21. Ellen MacArthur Foundation. (2019). *New plastics economy global commitment.*
22. Kalmykova, Y., Sadagopan, M., & Rosado, L. (2018). Circular economy—From review of theories and practices to development of implementation tools. *Resources, Conservation and Recycling, 135,* 190–201.
23. Kalmykova, Y., Sadagopan, M., & Rosado, L. (2017). Circular economy—From review of theories and practices to development of implementation tools. *Resources, Conservation and Recycling,* 1–13. https://doi.org/10.1016/j.resconrec.2017.10.034.
24. Avantium. (2016). *Synvina: Joint venture of BASF and Avantium established* [Online]. Retrieved March 04, 2020, from from https://www.avantium.com/press-releases/synvina-joint-venture-basf-avantium-established/.
25. Al-Sabagh, A. M., Yehia, F. Z., Eshaq, G., Rabie, A. M., & ElMetwally, A. E. (2016). Greener routes for recycling of polyethylene terephthalate. *Egyptian Journal of Petroleum, 25*(1), 53–64.
26. Ragaert, K., Delva, L., & Van Geem, K. (2017). Mechanical and chemical recycling of solid plastic waste. *Waste Management, 69,* 24–58.
27. Aguado, J., & Serrano, D. P. (2007). *Feedstock recycling of plastic wastes.* Royal Society of Chemistry.
28. Beyene, H. D. (2014). Recycling of plastic waste into fuels, a review. *International Journal of Science, Technology and Society, 2*(6), 190–195.
29. Chavan, M. A. J. (2013). Use of plastic waste in flexible pavements. *International Journal of Application or Innovation in Engineering & Management, 2*(4), 540–552.
30. Pandi, G. P., Raghav, S., & Selvam, D. T. (2017). Utilization of plastic waste in construction of roads. *International Journal of Science and Research, 5804.*
31. Patel, V., Popli, S., & Bhatt, D. (2014). Utilization of plastic waste in construction of roads. *International Journal of Science and Research, 3*(4), 161–163.
32. Trimbakwala, A. (2017). Plastic roads: Use of waste plastic in road construction. *International Journal of Science and Research Publications, 7,* 137–139.
33. Karelia, G. One lakh kilometres of roads in India are being made from plastic waste, is this the Solution to end plastic crisis? *NDTV.*
34. McCarthy, J. (2018). Recycled plastic is being used to repave roads around the world. *Global Citizen Events.*
35. Dykes Paving. (2020). *Texas roads made from plastic* [Online]. Retrieved March 04, 2020, from https://www.dykespaving.com/blog/texas-roads-made-from-plastic/.
36. Clean Technica. (2019). *Procter & gamble and pure cycle collaborate on polypropylene recycling process* [Online]. Retrieved March 04, 2020, from https://cleantechnica.com/2019/09/27/proctor-gamble-and-purecycle-collaborate-on-polypropylene-recycling-process/.
37. P&G. (2017). *P&G's head & shoulders creates world's first recyclable shampoo bottle made with beach plastic* [Online]. Retrieved from https://news.pg.com/press-release/head-shoulders/pgs-head-shoulders-creates-worlds-first-recyclable-shampoo-bottle-made.
38. SB. (2017). Coke, Avery Dennison drive smartwater towards circularity with recycled PET waste [Online]. Retrieved March 04, 2020, from https://sustainablebrands.com/read/chemistry-materials-packaging/coke-avery-dennison-drive-smartwater-towards-circularity-with-recycled-pet-waste.

39. C. of A. M. Circularity. (2019). The Netherlands leads the way in sustainable diapers. *Circularity, Chemistry of Advanced Materials* [Online]. Retrieved fromhttps://hollandchemistry.nl/case/the-netherlands-leads-the-way-in-sustainable-diapers/.
40. McIntyre, K. (2017). *Giving new life to old diapers* [Online]. Retrieved from https://www.nonwovens-industry.com/issues/2017-01-01/view_features/giving-new-life-to-old-diapers/.

Dr. Anand Bellam Balaji received his Ph.D. degree in Chemical Engineering from University of Nottingham, Malaysia in 2019. Currently, he pursues as Postdoctoral research fellow in New materials Institute, University of Nottingham Ningbo campus (UNNC). He currently works on development of green, sustainable and flame retardant composites for aerospace, automobile and high-speed train applications. He works actively along with his team, whose current interests' focuses on developing and application of composite materials with sustainability and recyclability as one of the key aspect. His other area of research includes thermoplastic polymer applications for biomedical applications, radiation processing of polymers and nano-composites.

Dr. Xiaoling Liu graduated from University of Nottingham, UK and joined Ningbo campus (UNNC) in 2015. Within only a little over 2 years, Dr. Liu has built up a research platform from scratch to train undergraduate students, Ph.D.s and support industry with 2 key labs, collaborating a joint lab with AVIC Composite group, China. Her research focuses on developing polymer composites with interests towards biomedical, fire retardant, conductive, green and thermoplastic composites. Until today, she has been successful in securing various grants (apprx. RMB 25M) from multiple sources including international companies and prestigious state-level SoEs. It is noteworthy that in recent years, she has also extended her interests in research and development of sustainable materials, recyclability and adopting principles of circular economy to keep the carbon fibres and polymer composites at their highest utility and value at all times. Below are few notable undergoing projects to generate solutions, develop competence and technology through cross-disciplinary cooperation for value creation, sustainability, recyclability and resource management in the field of composites.

- ACC TECH and UNNC Joint Laboratory in "Sustainable Composite Materials", Industry project; 2017–2022
- "Composite recycling and reuse"—Collaboration project; 2017–2021
- "Eco-friendly, high performance and multi-functional green building composites and its application technology" — Commonwealth project; 2019–2022
- "Development of Property-Improvement of High Temperature Cure, Full Green composites"—Airbus (Beijing) Engineering technology centre co. ltd; 2019

- Zhejiang Innovation Team—"Multi-functional green composite for the next generation aerospace application"; 2018–2021

With her keen interest, proactive nature and hard work, she has played an important part in taking UNNC composites technology from start-up status to national recognition within China's fast-moving composites industry.

Circular Economy Enabled by Community Microgrids

Deva P. Seetharam, Harshad Khadilkar, and Tanuja Ganu

Abstract Electricity is at the heart of modern economies and it is providing a rising share of energy services. Global demand for electricity is set to increase further as a result of rising household incomes, with the electrification of transport and heat, and growing demand for digital connected devices and air conditioning. Rising electricity demand was one of the key reasons why global CO_2 emissions from the power sector reached a record high in 2018. In fact, as per the projections of International Energy Agency (IEA), electricity's share in total final energy consumption is expected to increase from 19% in 2018 to at least 24% in 2040. One of the potentially effective approaches for providing universal access to electricity is community microgrids built on distributed energy resources (DER) such as photovoltaic (PV) systems and batteries. As DER can generate and store electricity locally, they can power microgrids. A microgrid is a group of interconnected loads and distributed energy resources within clearly defined electrical boundaries that acts as a single controllable entity. A set of microgrids can be interconnected together to form community microgrids. In a community microgrid (CM), prosumers and consumers can cooperate to generate, share and consume electricity. Such CMs **minimise** the use of fuels required by conventional power plants, **reduce** creation of waste, pollution and carbon emissions. Additionally, DERs such as solar panels reduce the need for fuels and manufacturing supplies for long periods as their expected lifetime is about 20 years. Moreover, in a CM, the excess energy generated by a producer can be shared with the other members of that CM thereby **recycling** the excess energy. In summary, community microgrids incorporate the principles of circularity or circular economy to enable universal access to electricity while reducing air pollution and thereby addressing the climate change.

D. P. Seetharam (✉)
Independent Researcher, Bangalore, India
e-mail: deva@alum.mit.edu

H. Khadilkar
Tata Consultancy Services Ltd, Thane, India
e-mail: harshad.khadilkar@tcs.com

T. Ganu
Microsoft Research, Bangalore, India
e-mail: tanuja.ganu@microsoft.com

© Springer Nature Singapore Pte Ltd. 2021
L. Liu and S. Ramakrishna (eds.), *An Introduction to Circular Economy*,
https://doi.org/10.1007/978-981-15-8510-4_10

Keywords Distributed energy resources · Community microgrids · Circular economy · Solar PV · Batteries · Energy storage systems

1 Introduction

Electricity[1] is one of the most widely used forms of energy, and it has been adapted to power a huge and a growing number of applications. Electricity is a practical energy source for many residential, commercial and industrial use cases such as lighting, heating, cooling, telecommunications and transportation. As described by the International Energy Agency (IEA) report, electricity is at the heart of modern economies and it is providing a rising share of energy services. Demand for electricity is set to increase further as a result of rising household incomes, with the electrification of transport and heat, and growing demand for digital connected devices and air conditioning. Rising electricity demand was one of the key reasons why global CO_2 emissions from the power sector[2] reached a record high in 2018 [2].

The IEA analyses the future of electricity and electricity-related emissions under two scenarios:

1. **Stated policies scenario**—The aim of the stated policies scenario is to provide a detailed sense of the direction in which existing policy frameworks and today's policy ambitions would take the electricity sector out to 2040. In this scenario, global electricity demand grows at 2.1% per year to 2040, and this raises electricity's share in total final energy consumption from 19% in 2018 to 24% in 2040 [2].

2. **Sustainable development scenario**—The sustainable development scenario takes its inspiration from the UN sustainable development goals (SDGs) most closely related to energy: achieving universal energy access, reducing the impacts of air pollution and tackling climate change. In this scenario, electricity plays an even larger role, reaching 31% of final energy consumption. That is, the share of electricity in final consumption, less than half that of oil today, overtakes oil by 2040 [2].

One of the potentially effective approaches for achieving SDGs in power sector is by generating electricity using distributed energy resources (DER) (such as solar photovoltaic (PV) systems and batteries), and by distributing and sharing electricity locally using the principles of circular economy. As DER can generate and store electricity at the premises of consumers, they can power microgrids. A microgrid is a group of interconnected loads and distributed energy resources within clearly defined electrical boundaries that acts as a single controllable entity. A set of microgrids can be interconnected together to form a community microgrid. In a community microgrid (CM), prosumers and consumers can cooperate to generate, share and consume

---

[1] A prosumer is a person (or entity) who consumes and produces a product [1].

[2] The phrases 'electricity sector' and 'power sector' are used interchangeably in this chapter.

electricity. Such CMs **minimise** the use of fuels required by conventional power plants, **reduce** creation of waste, pollution and carbon emissions. Additionally, DERs such as solar panels reduce the need for fuels and manufacturing supplies for long periods as their expected lifetime is about 20 years. Moreover, in a CM, the excess energy generated by a producer can be shared with the other members of that CM thereby **recycling** the excess energy. Thus, community microgrids incorporate the principles of circularity or circular economy to enable universal access to electricity while reducing air pollution and thereby tackling climate change.

Given the importance of circular economy to electricity systems, there is considerable interest in applying circularity to energy systems [3]. However, the main focus has been on recycling material and reusing assets. To list a few such efforts:

1. Reusing facilities—for example, Enel, an Italian multinational energy company, has started Futur-e [4] to repurpose 23 thermoelectric power stations and a disused mining site. For instance, Porto Tolle, a decommissioned thermoelectric power plant, is being transformed into an open-air tourist hub; Carpi, a retired power station, is being converted into an innovative logistics hub.
2. Recycling material—for instance, TransAlta, Canada's largest publicly traded power generator and marketer of electricity and renewable energy, decommissioned Canada's oldest wind farm and recycled about 90% of the wind farm equipment by weight [5]. That is a total of approximately 1,250 tonnes of metal from gearboxes, generators, nacelles, towers, wiring and pad-mount transformers, plus 44,600 litres of oil.
3. Reusing assets—Circusol, an innovation action project from the European Union, has developed a novel business model known as product-service system (PSS) business model. In a PSS model, a supplier provides solar power generation and storage to a user as a service. The PV and batteries are installed at the user's site, but the supplier remains as the owner and is responsible for their optimal functioning. When the equipment reach their end of life at the sites, the supplier takes them back and decides whether they can be given a second life and reinstalled somewhere else, or must be sent for recycling. The importance of this business model can be understood given the fact 8 million tonnes of used PV units and 2 million old electric vehicle batteries are expected to be discarded by 2030.

Compared to the aforementioned initiatives that focus on reusing/recycling of physical assets, community microgrids primarily focus on operational efficiency— efficient generation, consumption and sharing of clean electricity. This chapter describes how principles of circular economy can be applied to community microgrids. The rest of the chapter is organised as follows. First, Sect. 2 presents a quick overview of conventional power grids. Next, Sect. 3 provides an overview of distributed energy resources (DER). It is followed by Sect. 4 that discusses microgrids built using DER, and the corresponding operational challenges. Section 5 describes how community microgrids work and address some of those challenges with the contributions from prosumers, producers and consumers. Section 6 discusses a few case studies of community microgrids. Section 7 summarises the key points covered in this chapter.

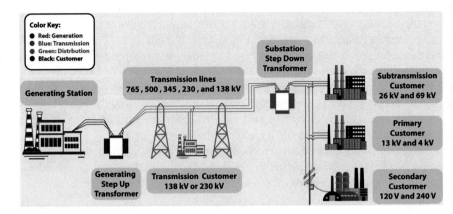

Fig. 1 Simplified diagram of the topology of electricity grids in North America. (*Source* https://www.ferc.gov/industries/electric/indus-act/reliability/blackout/ch1-3.pdf)

2 Power Grids

As shown in Fig. 1, traditional power grids are based on centralised architectures[3] in which large-scale power plants (nuclear, thermal, hydroelectric, etc) generate electricity. For example, in the USA, about 20% of electricity is generated by 61 large (capacities ranging from 582 MW to 3,937 MW [6]) nuclear power plants [7].

The electricity generated by such large plants gets transported to the loads (consumption points) by transmission and distribution networks. Former transports energy over long distances at hundreds of Kilo volts, and the latter, using distribution transformers, steps the voltage down (and correspondingly steps the current up) to deliver electricity to end consumers. Most of the residential and commercial consumers connected to distribution networks at about 120 V or 240 V.

Although such centralised AC power systems have been effective for more than a century, they suffer from certain distinct disadvantages as explained below.

- **Energy losses**—As electrical energy must be transmitted from power plants to the consumers via extensive networks, about 8–15% of energy is lost as heat in transformers and power lines due to Joule effect[4] [8]. Based on the quality of grid construction, ambient temperature, cable length and the effectiveness of maintenance procedures, these losses can be higher or lower. As per the World Bank's analysis [9], in 2014,[5] these losses were amounting to 5% in China, 19% in India, 9% in Indonesia and 6% in Thailand.

---

[3]As this AC power system architecture is proven to be robust and reliable, power grids across the world are built based on this exactly same or similar architecture.

[4]When electric current flows through a material with finite conductivity, electrical energy is converted to heat. Joule's first law states that the heat generated by such a material is proportional to the product of its ohmic resistance and the square of the current.

[5]Most recent data available is from the year 2014.

- **Peak capacity under utilisation**—Unless grid-scale storage is deployed, electricity must be generated and transported, in real time, as and when required by consumers. So if the magnitude of peak load is much higher than the magnitude of average load, then the peak capacity (for generation and transportation) will be be underutilised during the off-peak hours. For instance, during 2006, in the New England region of the USA, 15% of all capacity ran 0.90% of the time or less, 25% of all capacity ran 2.92% of the time or less. Furthermore, the peak-to-average ratio has been rapidly increasing over the last 4 decades [10].

Environmental impact of centralised generation—Based on the source of generation (fossil fuel or renewable resource), centralised power plants can contribute to large-scale regional environmental concerns as well as localised concerns that affect the area directly surrounding a power plant. They can affect the environment in the following ways [11]:

- **Air pollutant emissions**:

  • The amount and type of emissions will vary by fuel burned and other plant characteristics.
  • Air pollution from burning fuel often includes carbon dioxide, sulphur dioxide, nitrogen oxides, mercury and particulate matter.

- **Water use and discharge**:

  • Water used for steam production or cooling may be returned at warmer temperatures to water bodies and may contain contaminants.
  • Some water may also be lost to evaporation.

- **Waste generation**:

  • Burning certain fuels results in solid waste such as ash, which must be stored and eventually disposed of properly.
  • Some wastes contain hazardous substances. For example, nuclear power generation produces radioactive waste, while coal ash can contain heavy metals like mercury.

- **Land use**:

  • Large power plants require space for their operations.
  • Centralised generation requires transmission lines, which also use land.

In many countries, due to aforementioned disadvantages, conventional power grids are being rapidly replaced by decentralised energy systems built from distributed energy resources (DER) that can generate and store electricity near the points of consumption.

3 Distributed Energy Resources

Distributed energy resources (DER) are small, modular, energy generation and storage technologies that supply electricity where needed. Typically producing less than 10 megawatts (MW) of power, DER systems can usually be installed in the premises of consumers and can be sized to meet their particular needs [12].

3.1 Solar PV Systems

Although there are many different types of DERs such as back-up generators, wind turbines and combined heat and power (CHP) systems, we would mainly focus on photovoltaic (PV) generation systems—both roof-mounted and ground-mounted—installed in premises. As they are clean and do not generate any noise, they can be conveniently installed within the premises of consumers. They have the following attractive properties:

– **Scale-free**—As solar panels come in a wide range of capacities such as 10W, 20W, 100W, 250W, 300W, etc., generation units of arbitrary capacities can be created by assembling the required number of solar panels.
– **Falling prices**—As illustrated in Fig. 2, the prices of silicon PV cells and the prices of installed solar PV systems have been falling. Further, as per Swanson's law,[6]

Fig. 2 Plummeting prices of solar PV cells. (*Source* https://commons.wikimedia.org/wiki)

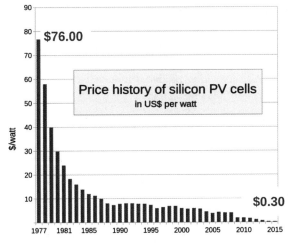

[6]Swanson's law is the observation that the price of solar photovoltaic modules tends to drop 20% for every doubling of cumulative shipped volume [13]. It is important to note that Swanson's law does not address the other costs such as labour costs and equipment costs.

Fig. 3 Falling prices of solar panels in India. (*Source* https://saysolar.in/home-solar-system/*)

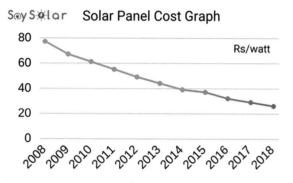

based on the current volume of panels shipped, costs are expected to decrease by 75% about every 10 years. For instance, as shown in Fig. 3, price per watt of solar panels in India has fallen by almost 600%.

- **Relatively long expected life**—Once installed, solar panels can operate and produce electricity for more than 20 years. In fact, some of the panel manufacturers assure that generation capacity will not reduce by more than 20% for the first 20 years of operation.
- **Relatively low maintenance**—As they do not include any moving parts, the in-premise installations do not require any regular maintenance other than periodic cleaning of solar panels and tuning of inverters.

As a result of such attractive properties, solar PV systems are becoming popular across the world. As shown in Fig. 4, the installed capacity of PV systems has increased by more than 500-fold since the early 1900s and is continuing at a robust pace. More importantly, the cumulative installed capacity of smaller PV systems that are installed on premises forms one-third of the total installed solar capacity.

3.2 Batteries

As solar is an intermittent (generated energy fluctuates due to cloud movements, weather changes, etc) energy source, solar PV systems, to smooth out the fluctuations, usually are either connected to power grids and/or augmented with batteries. Batteries are an integral component of off-grid systems. While selecting a battery, the following critical characteristics[7] must be considered.

- Capacity or nominal capacity—the total ampere hours available when the battery is discharged at a certain discharge current (specified as a C-rate) from 100% state-of-charge to the cut-off voltage (point at which a battery cannot supply any more

---

[7]The comprehensive details of these and other characteristics are presented in the guide from MIT [14].

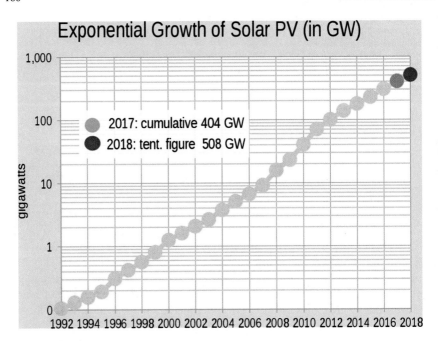

Fig. 4 Exponential growth curve on a semi-log scale of worldwide installed photovoltaics in gigawatts. (*Source* https://commons.wikimedia.org/wiki)

current). Capacity is calculated by multiplying the discharge current (in amperes) by the discharge duration (in hours). Capacity decreases with increasing C-rate.[8] C-rate is used to indicate the maximum current that a battery can safely deliver on a circuit.

- Depth of discharge (DoD)—the percentage of battery capacity that has been discharged expressed as a percentage of maximum capacity. A discharge to at least 80% DoD is referred to as a deep discharge. A deep-cycle battery is a battery designed to be regularly deeply discharged using most of its capacity.
- Cycle life—the number of discharge-charge cycles a battery can experience before it fails to meet specific performance criteria. Cycle life is estimated for specific charge and discharge conditions. The actual operating life of the battery is affected by the rate and depth of cycles and by other conditions such as temperature and humidity. The higher the DoD, the lower the cycle life.
- Cost—in addition to the initial investment, the operating cost—cost per kWh—as a function of the upfront investment and all the electricity stored in and discharged from that battery can be analysed using the following formulas:

---

[8]It is defined as the current through the battery divided by the theoretical current draw under which the battery would deliver its nominal rated capacity in 1 h. The capacity of a battery is commonly rated at 1C, meaning that a fully charged battery rated at 1Ah should provide 1 A for 1 hour. The same battery discharging at 0.5 C should provide 500 mA for 2 hours, and at 2C it delivers 2 A for 30 min.

$$total - kWh = (C * V * D * CL)/1000 \qquad (1)$$

$$cost/kWh = Battery - Price/total - kWh \qquad (2)$$

where
C—capacity of battery expressed in ampere hours,
V—operating voltage of the battery,
D—recommended depth of discharge
and CL—number of charge/discharge cycles supported by the battery.

The following two types of batteries are commonly used with solar installations [15]:

1. **Lead-acid batteries**—they have been used in energy storage applications for more than 150 years. Their main benefit is their lower cost compared to other two types of batteries. However, their cycle life is usually lower and can support only shallow (about 50%) DoD.
2. **Lithium-ion batteries**—these batteries have many advantages over lead-acid batteries. First, their energy density (kWh/Litre) is much higher, allowing lithium batteries to be smaller and lighter than lead-acid batteries with similar capacity. Secondly, and more importantly, they have long cycle lives (3,000–10,000 cycles) at a high DoD (80%). However, these batteries are routinely 500–700% costlier than similarly sized lead-acid batteries. Fortunately, as seen in Fig. 5, as

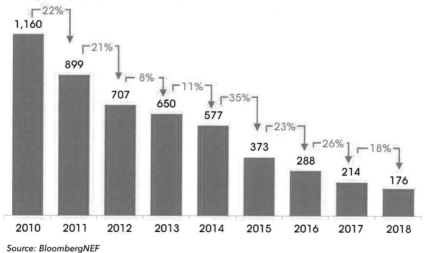

Fig. 5 Lithium-ion battery prices

per Bloomberg New Energy Finance [16], the prices of lithium-ion batteries have been rapidly falling.

The use of batteries must be carefully considered as both the aforementioned batteries (are based on electrochemical cells[9]) can severely impact the environment, if they are not carefully recycled and/or disposed [17].

4 Microgrids

A **Microgrid** is defined as a group of interconnected loads and distributed energy resources within clearly defined electrical boundaries that acts as a single controllable entity with respect to the grid. A microgrid can connect and disconnect from the grid to enable it to operate in both grid-connected or islanded modes [18].

Microgrids are broadly categorised as grid-connected (on-grid) and isolated microgrids (off-grid).

- **Grid-connected microgrids**—the entities which are equipped with one or more of DERs (such as solar, batteries or wind turbines) in their premises but are also connected to the central grid are examples of grid-connected microgrids. Based on the availability of locally generated energy, the loads can be either served by the electricity drawn from the grids or from the locally generated electricity. Moreover, if local DERs generate more electricity than required by the local loads, the excess energy can be exported to the grid.
- **Isolated microgrids**—the microgrids which are totally disconnected from central grid and work as independent and self-sufficient systems. They are often found in remote areas without access to reliable power grids.

Figure 6 shows the exemplary structure of grid-connected and isolated microgrids. In last few years, microgrids are becoming popular in southeast Asia [19, 20], India and Africa [21, 22]. These microgrids not only provide energy for a community that currently has no access to power grids, but they also can function as a reliable and economic source of electricity to urban and rural populations.

The microgrids, based on on-premise solar installations, avoid the disadvantages of conventional power grids since they supply clean energy without network losses, fuels and emissions. However, they have the following shortcomings:

- **Intermittent generation**:
  - Solar panels can generate electricity only during the daytime. Further, the generation follows a bell curve pattern that peaks when the sun is close to its local zenith. This pattern could lead to a timing imbalance between peak demand and solar production. For instance, in many energy markets, the peak demand occurs after sunset, when solar power is no longer available [23].

---

[9]An electrochemical cell is a device capable of either generating electrical energy from chemical reactions or using electrical energy to cause chemical reactions.

Microgrids

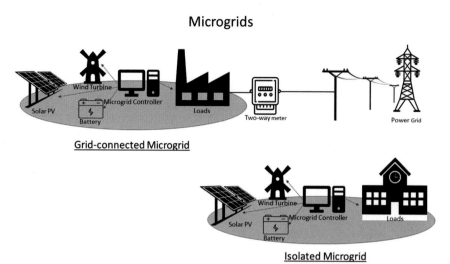

Fig. 6 Grid-connected and isolated microgrids

- Solar energy generation can also fluctuate on a minute-by-minute basis due to various environmental factors such as cloud movements, smog, fog, etc.

- **Low-generation power density**[10]: After the atmospheric absorption and the Earth's albedo,[11] the total solar radiation that reaches the Earth's surface can be prorated to about 170 W/m^2, and ranging from less than 100 W/m^2 in cloudy northern latitudes to more than 230 W/m^2 in sunny desert locations [25]. As a result, the specific yield (kWh/kWp) is different in different parts of the world. For instance, a one kW solar installation in Windhoek, Namibia, would yield about 5.4 kWh per day, and the same installation in Stockholm, Sweden, would yield only 2.8 kWh per day [26]. Due to low power density of PV systems, even the largest solar PV parks generate electricity with power densities that is at best 1/10 and at worst 1/100 of the power densities of coal-fired electricity generation [25].

From these shortcomings, it can be seen that standalone microgrids may not be adequate to satisfy the demands of electricity consumers. Moreover, some of the consumers may not be able to adopt solar due to unavailability of suitable (orientation, shadow-free, etc) space and/or initial capital funds. To support such resource-constrained consumers, it is important to implement 'Community Microgrids' in which producers, consumers and prosumers can cooperate to address the collective economic objectives and energy requirements of the community. Further, given the inherent intermittent nature of solar energy, CMs must be augmented with energy storage systems to smooth out the fluctuations. In the next section, we shall discuss community microgrids in detail.

---

[10]Power density is the amount of power (time rate of energy transfer) per unit volume [24].
[11]Fraction of radiation reflected to space by clouds and surfaces.

5 Community Microgrids

As discussed in earlier sections, **community microgrid** provides energy to a community or a group of customers.

As shown in Fig. 7, a community microgrid can be constructed by electrically interconnecting the participants. A participant in a CM will possess some or all of the following: (i) Solar PV (source), (ii) rechargeable batteries (storage) and (iii) local loads. The sources always provide energy, the loads always consume energy and the batteries, on demand, can either supply (discharge) or consume energy (recharge). So, a participant can be classified as a producer (owns only the source and optionally, batteries) or as a consumer (owns only the loads) or as a prosumer (has both source and loads, and, optionally, storage).

Based on the consumption and generation of these participants, the net current in a CM, at any instant, can be positive or negative or zero. However, for a CM to be self-sufficient and stable, the amount of electricity supplied and consumed in that microgrid must be continuously matched, as stated in Eq. 3.

$$\sum_{i=1}^{n}(E_{ig} + E_{id}) - \sum_{i=1}^{n}(E_{ic} + E_{ic}) \approx 0 \qquad (3)$$

$n-$ number of premises
$E_{ig}-$electricity generated by the ith premise,
$E_{id}-$electricity discharged by batteries at the ith premise,
$E_{il}-$electricity consumed by the loads at the ith premise and
$E_{ic}-$electricity used for charging batteries at the ith premise.

A CM controller can coordinate with microgrid controllers within each premise to increase/decrease generation, alter charging/discharging patterns and adjust consumption levels and schedules. The following subsections describe how different types of participants can contribute to continuous balancing of a CM.

5.1 Consumer

A consumer can use electricity for powering loads (lighting, cooling, heating, computing, etc.) and/or for charging energy storage systems (batteries, electric vehicles, etc.).

5.1.1 Powering Electrical Loads

Theoretically, CMs must be able to power all the loads concurrently operating in the network. However, the total generation capacity required for achieving that can be

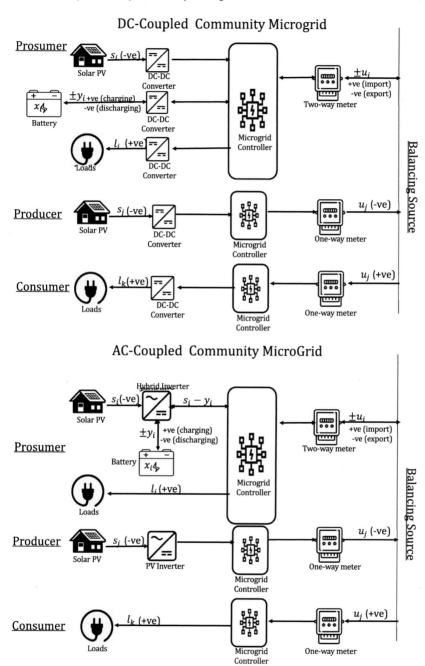

Fig. 7 Flow of energy with different types of participants

Table 1 Types of electrical loads

	Schedule	Level
Inflexible	Lighting, fans,	Incandescent bulbs, conventional
	Entertainment centres, etc.	Water heaters, etc.
Flexible	EV chargers, water heater, etc.	Air conditioners, EV chargers,
		Fans, LED lights, etc.

expensive. So, the total concurrent consumption can be reduced, if there is flexibility in schedules and/or operating levels of those loads. As illustrated in Table 1, while loads such as incandescent bulbs are completely inflexible, electric vehicle (EV) chargers can be operated at different charging rates and can be scheduled when the CM is not heavily loaded.

An entity that does not have the capacity to store or generate electricity can also contribute to CMs by participating in **demand response (DR)** programmes. That is, when the total generation of a CM is high, they can increase their consumption, and, conversely, when the generation is low, they can decrease.

As presented by Zhang et al. [27], several different types of DR programmes have been designed and deployed. In these programmes, DR messages are sent to consumers to decrease (or increase) their consumption in response to grid conditions. The authors classify DR programmes into two types based on the economic means employed by utilities to effect such a change:

1. **Price-based DR**—consumers are induced to minimise their consumption during peak hours by offering time-varying prices[12] that charge higher electricity rates during peak load hours and lower rates during other times.
2. **Incentive-based DR**—consumers are encouraged to reduce their consumption during peak hours by offering incentives such as discounted retail rates or separate payments[13] for pre-contracted and/or measured load reductions.

DR programmes can also be classified as manual or automated depending on whether consumers must manually control the loads. While the conventional appliances require consumers to turn them off/on or down/up, there have been initiatives such as OpenADR [28] and DRED [29] to standardise so that DR messages can be directly processed by communication-enabled electrical loads themselves.

In the aforementioned programmes, the balancing authorities centrally decide and execute DR events. Researchers have also proposed decentralised DR systems such as nPlug [30] that autonomously identify grid load conditions and appropriately schedule the loads to avoid peak grid load hours.

---

[12] A few illustrative tariffs are time-of-use (TOU) pricing, real-time pricing (RTP) and critical peak pricing (CPP).

[13] A few illustrative ones are direct load control (DLC), interruptible/curtailable (I/C) service, demand bidding/buyback (DB), emergency demand response programmes (EDRP), capacity market programmes (CMP) and ancillary services market programmes (ASMP).

Thus, demand in microgrids can be altered, either automatically (while respecting user preferences) or manually, in response to the available supply by increasing/decreasing the consumption levels and/or by advancing/postponing the use of electrical loads.

5.1.2 Charging Energy Storage Systems

Energy storage systems (abbreviated as storage) such as batteries form a special class of loads because of the following reasons:

- The difference is that electricity stored in batteries can be reused latter. However, some of the energy may be lost due to round-trip losses associated with charging and discharging of batteries, and through self-discharge [31].
- Batteries must be charged using prescribed charging methods. So their current consumption can follow complex patterns. For instance, charging a lithium-ion battery must follow the sequence of constant-current charging, saturation charging, and finally trickle or top-up charging. Within the prescribed method, the aforementioned demand management methods can be employed [32].

Batteries can be charged depending on the availability of electricity and future energy requirements while respecting the charging methods (rate and sequence) prescribed by the manufacturers.

5.2 Producer

A participant can supply electricity to fellow participants in a CM by sharing either the energy generated by solar PV and/or the energy stored in the batteries in their premises.

5.2.1 Local Energy Generation

As explained before, solar PV systems are intermittent energy sources that cannot be dispatched. That is, the amount of electricity they produce cannot be turned up or down to match the demand. However, it can be adjusted (by controlling the output current): decreased (decreasing is known as curtailment) all the way to zero or increased to a level constrained only by available DC power output of the solar array[14] at that point in time.

In fact, multiple utility companies across the world curtail solar energy to different levels. The report [33] from NREL provides a comprehensive survey of such

---

[14]A set of solar panels connected in series is called a string, and a set of strings connected in parallel is called an array.

curtailment practices and experiences in the USA. Similarly, in CMs, the output from solar PV installations can be adjusted to match the load and storage requirements.

5.2.2 Discharging Energy Storage Systems

When there is a shortfall in energy generation, batteries can be discharged to service the load requirements. However, as discussed above, batteries can only support a limited number of charge and discharge cycles. The total number of such cycles a battery can withstand depends on how deeply the battery is discharged each time and the rate at which it is discharged. Manufacturers commonly rate lead-acid batteries at a very low 0.05 C, or a 20-hour discharge.

Batteries are seldom fully discharged, and manufacturers often use the 80% DoD and 50% DoD to rate a lithium-ion battery and a lead-acid battery, respectively. This means that only 80% of the available energy is delivered and 20% remains in reserve. As discussed before, cycling a battery at less than full discharge increases service life [34].

In addition to conventional battery storage systems, batteries in electric vehicles (EVs) can also serve as energy storage systems. For instance, Mahindra e-Verito [35], an electric car manufactured in India includes 21.2 kWh lithium-ion battery. An EV, connected to a bi-directional converter, can control the flow of energy from and to the grid depending upon the requirement. Hence, vehicle-to-grid (V2G) or vehicle-to-home (V2H) policies could also be deployed to address the energy shortfall in community microgrids.

As described, community microgrids can continuously balance supply and demand with contributions from producers, consumers and prosumers.

6 Case Studies: Community Microgrids

In this section, we describe two case studies that demonstrate the feasibility and value of community microgrids. First one is from Bangladesh and the second is from Australia.

6.1 SOLshare—Community Microgrids in Bangladesh

SOLshare [36] offers peer-to-peer community microgrids which deliver solar power to households and businesses, and enable them to trade their (excess) electricity for profit. In rural areas of Bangladesh, approximately 3 million homes that are cut off from central electric grid mainly rely on energy from domestic solar generation units called solar home system (SHS). SOLshare installs micronetworks, similar to the one illustrated in Fig. 8, that allow individual homes to share the electricity generated in

Fig. 8 SOLshare community microgrid in Bangladesh

their homes with roughly a dozen other homes, of which some are equipped with solar panels and others not. Every home in a community microgrid would choose to be a prosumer (seller) or a consumer (buyer). The company has built a peer-to-peer energy trading platform based on distributed ledger technology. By creating a community microgrid of interconnected SHS, SOLshare enables a consistent, inexpensive and reliable energy network for an entire village. Additionally, the homes equipped with SHS get a new source of income by selling excess energy to the community. SOLshare envisions setting up 10,000 community microgrids by 2030, reaching a total of over one million users and expanding outside Bangladesh as well.

6.2 Totally Renewable Yackandandah—Community Microgrids in Australia

Totally renewable Yackandandah (TRY) [37] is a 100% volunteer run community group, with the goal of powering a small Victorian town, Yackandandah, with 100% renewable energy and achieving energy sovereignty by 2022. The primary objectives for TRY are reducing energy consumption and adopting renewable energy sources, storage and energy sharing.

In December 2017, the community microgrid shown in Fig. 9, a first of its kind in Australia, was launched at Yackandandah with a partnership among TRY, Mondo Power and AusNet Services. The first phase of deployment included a community microgrid involving 169 households with a cumulative battery capacity of 110 kWh installed in 14 households and solar panels with a total generation capacity of 550 kWp installed in 106 households. In February 2019, Yackandandah achieved the

Fig. 9 Yackandandah community microgrid in Victoria, Australia

important milestone of generating 1 GWh of their own renewable energy. The same model is being considered for deploying community microgrids at other townships in Australia.

7 Summary

From the previous sections, it can be seen solar PV-based community microgrids (CMs) enable circular economy through the following means:

– **Minimise land usage**—as solar panels can be installed on roofs, no separate land is required for power plants.
– **Minimise wastage**—as energy is distributed locally, network losses are minimised.
– **Minimise impact on the environment**—as no fuel is used for generating electricity, neither greenhouse gases are emitted nor waste products such as ashes are produced.
– **Optimise resource usage**—since generated energy is shared among the participants and demand is made to follow supply, resources get optimally used.

More importantly, as described in the case studies, community microgrids can function as a reliable alternative and/or an augmentation to conventional power grids. In essence, community microgrids are emerging as an effective solution to address, arguably, the most important problem—access to clean and cheap energy [38]—currently faced by humanity.

Exercises

1. Find out the name of the distribution company and/or retailer that supplies power to your household/school. Find out what proportion of electricity supplied is from renewable energy resources.
2. Use Global Solar Atlas [26] to determine solar energy generation potential for your location during different seasons of a year.
3. Check your monthly electricity bills for the last 3 months and try to identify major sources of energy consumption at different times of day on different days of the week.
4. Based on solar energy generation potential for your location and based on energy demand of your household/school, verify how much of the demand is coincidental with solar energy generation, and if a solar PV system can offset some of the energy demands directly. If so, how much solar capacity installation would be required to completely fulfil the demand?
5. If most (say, more than 75%) of the energy demand from your household/school does not coincide with solar generation times, determine if battery storage can be added to store electricity generated by solar PV to offset the demand during evenings/nights. Compute the size of battery required to implement this system.
6. Gather information about your local government policies and subsidies for solar PV systems and about total costs from your local solar PV installers. Combine those details with the answers for the previous two questions to compute the total capital investment required for installing solar PV in your household/school.
7. Perform a cost-benefit analysis of installing solar PV considering the capital investment, interest rate, potential increase in electricity rates, reduced electricity bills and expected life of solar PV equipment, etc.
   Hint: Bijli Bachao [39] has published a number of articles that could help in completing these exercises.

Acknowledgements We thank the editors for inviting us to write this chapter and for providing invaluable guidance and feedback. We are grateful to Sawaros Thongkaew of the STEAM Platform [40] for drawing the architecture block diagram (Fig. 1).

References

1. Wikipedia, Prosumer. https://en.wikipedia.org/w/index.php?title=Prosumer.
2. IEA. (2019). World energy outlook 2019—electricity. https://www.iea.org/reports/world-energy-outlook-2019/electricity.
3. Ollagnier, J. -M. (2020). View from davos: What's the role of electricity in the 'circular' economy?'. https://www.forbes.com/sites/jeanmarcollagnier/2020/01/21/whats-the-role-of-electricity-in-the-circular-economy/#e1c5fe20d7e3.
4. Enel, Futur-e. https://corporate.enel.it/en/futur-e.
5. Froese, M. (2017). Decommissioning canada's oldest wind farm. https://www.windpowerengineering.com/decommissioning-canadas-oldest-wind-farm/.
6. EIA, How much electricity does a nuclear power plant generate. https://www.eia.gov/tools/faqs/.

7. Nuclear explained. Retrieved from https://www.eia.gov/energyexplained/nuclear/us-nuclear-industry.php.
8. IEC. (2017). Efficient electrical energy transmission and distribution. IEC Technical Document.
9. Electric power transmission and distribution losses (percentage of output). https://data.worldbank.org/indicator/EG.ELC.LOSS.ZS.
10. Spees, K. (2008). Meeting electric peak on the demand side: Wholesale and retail market impacts of real-time pricing and peak load management policy, technical report, Carnegie Mellon University.
11. EPA, Centralized generation of electricity and its impacts on the environment. https://www.epa.gov/energy/centralized-generation-electricity-and-its-impacts-environment.
12. NREL. (2009). Using distributed energy resources—a how-to guide for federal facility managers, technical report, U.S. Department of Energy by the National Renewable Energy Laboratory (NREL).
13. Wikipedia, Swanson's law. https://en.wikipedia.org/wiki/Swanson%27s_law.
14. Team, M. E. (2008). A guide to understanding battery specifications. https://web.mit.edu/evt/summary_battery_specifications.pdf.
15. Austin, R. (2018). What are the best batteries for solar off grid?. https://understandsolar.com/best-batteries-for-solar-off-grid/.
16. Goldie-Scot, L. (2019). A behind the scenes take on lithium-ion battery prices. https://about.bnef.com/blog/behind-scenes-take-lithium-ion-battery-prices/.
17. Rapier, R. (2020). Environmental implications of lead-acid and lithium-ion batteries. https://www.forbes.com/sites/rrapier/2020/01/19/environmental-implications-of-lead-acid-and-lithium-ion-batteries/#253c70fb7bf5.
18. Ton, D. T., & Smith, M. A. (2012). The U.S. department of energy's microgrid initiative. *The Electricity Journal, 25*(8), 84–94.
19. Chin, N. C. Microgrids are powering remote communities—and helping southeast asia's eco resorts live up to their name. https://tinyurl.com/r4ct7d2.
20. Southeast asian eco-resorts drawn to microgrids by favorable economics. https://microgridknowledge.com/eco-resorts-microgrids/.
21. Doing good with microgrids in Africa and India. https://microgridknowledge.com/microgrids-africa-india-solar/.
22. Panasonic and Indian utility team up to roll out urban microgrids. https://microgridknowledge.com/urban-microgrids-india-panasonic/.
23. Wikipedia, Duck curve. https://en.wikipedia.org/wiki/Duck_curve.
24. Wikipedia, Power density. https://en.wikipedia.org/wiki/Power_density.
25. Smil, V. (2010). Power density primer: Understanding the spatial dimension of the unfolding transition to renewable electricity generation (part iv—new renewables electricity generation).
26. Global solar atlas by world bank. https://globalsolaratlas.info/map.
27. Zhang, Q., & Li, J. (2012). Demand response in electricity markets: A review. In *9th international conference on the European energy market*, vol. 05, pp. 1–8.
28. Herberg, U., Mashima, J. G., Jetcheva, S., & Mirzazad-barijough, S. Openadr 2.0 deployment architectures: Options and implications.
29. Demand response framework and requirements for demand response enabling devices (dreds). (2017). https://shop.standards.govt.nz/catalog/4755.1:2017(AS%7CNZS)/scope?.
30. Ganu, T., Seetharam, D. P., Arya, V., Hazra, J., Sinha, D., Kunnath, R., et al. (2013). Nplug: An autonomous peak load controller. *IEEE Journal on Selected Areas in Communications, 31*(7), 1205–1218.
31. Wikipedia, Self-discharge. https://en.wikipedia.org/wiki/Self-discharge.
32. Battery-University, Charging lithium-ion batteries. https://batteryuniversity.com/learn/article/charging_lithium_ion_batteries.
33. Bird, L., Cochran, J., & Wang, X. (2014). Wind and solar energy curtailment: Experience and practices in the united states, technical report, National Renewable Energy Laboratory (NREL).
34. Basics of battery discharging. https://batteryuniversity.com/learn/article/discharge_methods.
35. Mahindra e-verito. https://www.mahindraelectric.com/vehicles/eVerito/.

36. Me solshare. https://www.me-solshare.com/.
37. Totally renewable yackandandah. https://totallyrenewableyack.org.au/.
38. Smalley, R. (2005). Future global energy prosperity: The terawatt challenge. *MRS Bulletin, 30,* 06.
39. Jain, A., Srivastava, A., & Jain, M. K. Bijli bachao. https://www.bijlibachao.com/using-renewables/solar.
40. Steam platform. https://www.steamplatform.org/events?id=43.

Deva P. Seetharam is an independent researcher working on systems that are based on IoT and AI. He has been working on data-driven energy systems for more than a decade. First, he founded and led the Smarter Energy Systems group for IBM Research India. After leaving IBM, he cofounded and served as the CEO of DataGlen, a company focused on enabling Distributed Energy Resources through IoT and Machine Learning Technologies. He holds a bachelors degree in electronics and communication engineering from Bharathiar University and a Masters degree from the Media Laboratory at the Massachusetts Institute of Technology.

Harshad Khadilkar is a scientist with Tata Consultancy Services Ltd where he leads the Planning & Control team, and is also a visiting associate professor at the department of aerospace engineering, IIT Bombay. He holds a bachelors in technology from IIT Bombay, and a Masters and PhD from the department of aeronautics and astronautics at the Massachusetts Institute of Technology.

Tanuja Ganu is a Principal Research SDE Manager at Microsoft Research, India. She works at MSR India's center for Societal impact through Cloud and Artificial Intelligence (SCAI) that focuses on creating, nurturing and deploying technologies that will have large scale impact on society. Prior to this, Tanuja was the Co-Founder and CTO of DataGlen Technologies, a B2B startup that focuses on achieving decarbonization and rapid adaption of distributed and renewable energy resources using IoT & Machine Learning technologies. Earlier, she worked as a Research Engineer at IBM Research, India. Tanuja holds an MS in CS from Indian Institute of Science (IISc).

Renewable Energy and Circular Economy: Application of Life Cycle Costing to Building Integrated Solar Energy Systems in Singapore

Rashmi Anoop Patil, Veronika Shabunko, and Seeram Ramakrishna

Abstract This chapter seeks to provide a representative example of Life Cycle Costing (LCC) for building-integrated solar energy systems in Singapore. First, renewable energy is introduced from the circular economy perspective, to better understand its significance in promoting sustainability. Solar energy among all renewable energy sources is the most promising for resource-stressed tropical cities such as Singapore. For such densely populated built-environments, innovative energy systems such as the building-integrated photovoltaic (BIPV) systems serve as good options to capture and use renewable energy incident on large facade areas. To estimate the financial feasibility of implementing BIPV systems and the cost-competitiveness in comparison with conventional building materials, the LCC serves as an enabling tool to evaluate the viability of these energy systems as per the market need. A step-by-step illustration of the LCC analysis, based on the tool developed by the BIPV Centre of Excellence at the Solar Energy Research Institute of Singapore (SERIS), is provided as a case study. This highlights the significance of LCC in evaluating the economic benefits of a practical application of a BIPV system for a representative high-rise building in Singapore. The benefits of transiting toward such sustainable and clean energy systems from a consumer and environmental perspective are summarized as the conclusion of the chapter.

Keywords Life cycle costing · Renewable energy · Building-integrated PV · Life cycle thinking · Circular economy

R. A. Patil (✉) · S. Ramakrishna
Circular Economy Task Force, National University of Singapore, Singapore, Singapore
e-mail: rashmi.anoop33@gmail.com

S. Ramakrishna
e-mail: seeram@nus.edu.sg

V. Shabunko (✉)
Solar Energy Research Institute of Singapore (SERIS), National University of Singapore, Singapore 117574, Singapore
e-mail: veronika.shabunko@nus.edu.sg

S. Ramakrishna
Department of Mechanical Engineering, National University of Singapore, Singapore 117576, Singapore

Learning Objectives

- To understand and quantify the impact of renewable energy sources such as building-integrated solar energy systems for resource-stressed countries such as Singapore.
- To understand the application of LCC calculation for BIPV systems in the Singapore building sector.
- To understand the significance of LCC in transition to renewable energy sources.

1 Renewable Energy and Circular Economy

The circular economy (CE) is often understood by novices as a concept related to materials circularity and zero-waste ecosystems, with somewhat subdued focus on energy sources [1]. However, CE also emphasizes optimizing the performance of an economic system (conceptually understood as a system involving resource allocation, and production, distribution, and consumption of products and services within a given area) by consciously transiting to renewable energy (RE) such as solar, wind, hydropower, tidal, and geothermal energy, and decoupling fossil fuel utilization for energy generation [1]. The optimization of energy performance in an economic system and transition to renewable energy sources is generally a gradual process and can be realized by embracing the following modifications:

1. reducing the production and consumption of non-renewable energy sources such as crude oil, coal, and natural gas
2. improving the energy efficiency of industrial processes and electrical/electronic products
3. reducing energy wastage (due to transmission losses and/or overuse) and recycling energy generating systems (after end-of-life).

At the core of such a transition, two objectives influence the reduction in the use of fossil fuels as energy sources: (i) separating economic growth from the use of scarce resources such as fossil fuels and (ii) reducing the carbon footprint or in other words decarbonizing the current energy production and consumption. Advances in technology for energy generation and storage, evolving environmental and social consciousness, and governance initiatives such as the carbon tax are providing impetus to this transition to renewable energy sources. This development is a game-changer for the energy industry and the market, and the consumers as well.

To appreciate the role and the importance of RE in the context of the CE, its imperative to first understand the concept of RE. RE can be defined as the energy produced from sources that do not deplete over time or can be replenished within one's lifetime [2]. Various forms of RE exist that can be harnessed using established technologies. For example, the energy in the wind currents (driven by the heat from solar energy [3]) can be captured using wind turbines. Hydropower that depends on the flow of water (another form of solar energy, as the sun powers the hydrologic cycle [4]) can also be harvested using turbine technology. The bioenergy stored in the plant biomass is eventually derived from solar energy. Innovative technologies such as photovoltaic (PV) cells that make up solar panels can capture the incident solar energy to generate power. When all these derived energy sources are collectively considered, it is evident from a common viewpoint, that the incident solar energy is the main source of most forms of renewable energy [2].

This chapter presents a discussion on how harnessing renewable energy such as solar energy can be beneficial both financially and environmentally in the long run (Fig. 1). Such a study provides key insights into the transition dynamics involved in incorporating renewable energy systems. For the purpose of illustration, we present a real-world example of the PV technology, that has seen tremendous progress in recent years. This technology is based on the photovoltaic effect—a well-known method for generating electricity by using solar cells to convert incident sunlight into a flow of electrons (Fig. 2). The electricity generated by solar PV cells can be connected to the grid or stored using a battery. As the PV technology advanced in the recent decade, it has been possible to integrate the solar energy systems with the building facades, making the technology more feasible for the built environment.

Currently, the deployment of solar PV systems is growing globally due to the significant reduction in the production costs of PV panels. These PV deployments include both grid-tied and off-grid installations. According to the International Renewable Energy Agency (IRENA), the globally installed solar capacity had reached nearly 480 GW by the end of 2018 [5]. IRENA also estimates that solar PV power installations could grow almost six-fold over the next 10 years and reach a cumulative capacity of ~2,840 GW globally by 2030 and rise to ~8,520 GW by 2050 [5]. These projections show that in the coming decades, solar energy harnessing systems will be a major contributor to the global energy output. This transformation

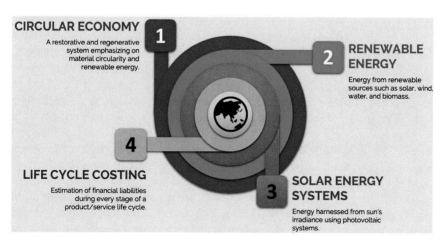

Fig. 1 Schematic illustration of the link between circular economy, renewable energy, and LCC analysis for BIPV systems. Renewable energy as an integral part of the circular economy is first discussed followed by a case study on the capture and use of solar energy using BIPV systems and its LCC. Design adapted from a template; Copyright PresentationGO.com

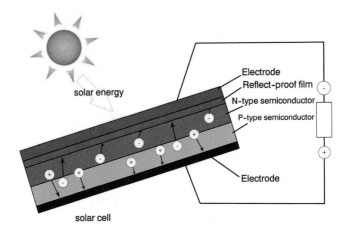

Fig. 2 Schematic illustration of the photovoltaic effect in solar cells that converts incident sunlight into electricity. Reproduced with permission from [6]; Copyright 2018, Elsevier

will facilitate in contributing to a circular economy[1] as well as the Paris Agreement requirements[2] for mitigating climate change.

The discussion in the next section focuses on the transition dynamics of incorporating a solar energy system such as the building-integrated photovoltaic (BIPV) system suitable for urban built environments. The economic analysis and overall competitiveness of solar energy system are usually expressed in form of the levelized cost of electricity (LCOE). It is a well-established method to evaluate various power generation options [7]. The LCOE approach considers the entire lifecycle cost of a solar energy system over the duration of the project and the associated electricity generated. However, for the case of a BIPV system, the economic viability is evaluated differently, as it typically replaces conventional construction materials of the building envelope. Therefore, the cost competitiveness is evaluated against the conventional construction materials per square meter rather than the cost of conventional electricity generated, expressed in form of life cycle costing (LCC). LCC can be understood as an evaluatory study that succinctly captures the cost associated with the project (in this case, the incorporation of BIPV system) from the beginning of the project to the end of its useful life (generally, the useful lifetime of the solar panels). The LCC study includes the initial costs that include realizing the project (e.g., acquiring the solar panels and cost of installation and integration with the existing building-related energy network/grid) and operating the renewable energy system till the end of its useful life. The LCC provides a clear cost picture of the project to compare with other alternatives with respect to the long-term benefits and the impact on running costs. It can thus be understood that LCC assists in decision

---

[1] https://www.ellenmacarthurfoundation.org/circular-economy/what-is-the-circular-economy.
[2] https://unfccc.int/process-and-meetings/the-paris-agreement/the-paris-agreement.

making regarding the necessity of the project by providing critical economic analysis that compares initial investment options and assists in identifying the least cost alternatives, and economic benefits in the long run.

First, we briefly provide a background on the relevance of building-integrated solar energy systems to the urban settlements in the tropics such as Singapore and the need for its LCC for the Singapore market. An illustrative application of LCC to the BIPV system in Singapore then follows as a representative real-world case study on incorporating renewable energy systems in the built environment. The takeaway message from such a study would be a basic understanding of the decision process involved in transitioning to renewable energy systems that can contribute to circularity in energy in the years to come.

1.1 Building-Integrated Solar Energy Systems in Singapore

Building-integrated solar energy system refers to the deployment of photovoltaic systems as a structural and/or esthetic component of the exteriors of buildings such as the facade curtain walls, cladding, windowpanes, and rooftops to cater to the energy requirements of the building without utilizing extra space [8]. Here, it should be noted that the BIPV systems installation differs from the BAPV (Building Applied Photovoltaics) type, a traditional approach of fitting modules to the existing surfaces by superimposing (such as installations on rooftops), once the construction has been completed. BIPV systems are very suitable for densely built-up cities with limited land and often competing for roof usage (e.g., with mechanical and electrical equipment such as air conditioners), but abundant solar irradiance. One such example is Singapore, which is packed with high-rise buildings and where BIPV technology could help to greatly expand the possible areas for deployment of solar PV modules (see sample installation in Fig. 3).

Currently, around 95% of Singapore's energy requirement is sourced from imported natural gas[3] and, hence, the country is moving towards embracing sustainable energy. In this context, transiting to a BIPV system comes with an initial financial investment but also offers environmental benefits in the long run. To facilitate such a transition, the LCC for BIPV systems provides a practical estimate of the cost-competitiveness and financial viability of BIPV technology in the Singapore market.

---

[3] Singapore's Energy Story, Energy Market Authority. https://www.ema.gov.sg/ourenergystory

Fig. 3 Photographs of sample BIPV installations on the facade at the School of Design and Environment, National University of Singapore (NUS). (From left) Non-transparent colored BIPV panels, semi-transparent modules, and all-black modules. Reproduced from the BIPV brochure; Copyright 2014, SERIS

2   Case Study: Applying LCC to Singapore's BIPV Systems

This case study[4] will focus on illustrating the LCC tool application for BIPV systems considering the Singapore market and climatic conditions for evaluating the economic dimensions of this sustainable energy option [9]. A step-by-step illustration of the LCC analysis is provided below and such an analysis is suitable for consumers/real estate developers investing in BIPV systems to arrive at an economyically sound decision.

2.1   LCC Parameters

Before moving to the actual calculation, there is a need to understand the different technical and financial parameters [10] involved in/influencing the LCC calculation and know their practical value range. These parameters are briefly explained and relevant values for each of them are given below.

1.   **Discount Rate:** In general, the discount rate is defined as the rate of return used to discount future cash flows back to their present value. In this case, it is a part of the overall general financing cost for the real estate project. Here, the cost of BIPV facades should be included in the total cost of the real estate investment

---

[4]Building Integrated Photovoltaics (BIPV) Life-Cycle Cost (LCC) Calculator by Solar Energy Research Institute of Singapore (SERIS), NUS (program administrator for National Solar Repository (NSR)). Copyright 2014, SERIS. https://www.solar-repository.sg/lcc-calculator/index.cfm

as the BIPV installation replaces the conventional construction material for the façade.

The BIPV LCC tool recommends setting a discount rate of 5% of the total real estate investment for the BIPV installation as a default number, which may be modified based on the country [9].

2. **Power/Unit area:** The power generated per unit area of a BIPV module varies with the module technology, type, transparency, and color (as the energy absorbed by the color varies). It is computed by dividing the total power (in Watts) output of the BIPV installation by the total module area (in m^2) of the BIPV installation.

   The BIPV LCC tool recommends the following values for different types of BIPV technologies: non-transparent black BIPV's output is approximately 165 W/m^2 and other non-transparent color modules such as gray-blue, light blue, and golden provide a power output of nearly 135 W/m^2. The semitransparent modules provide the least power output of around 50 W/m^2 [9].

3. **System Price:** This includes the cost of each unit system, for instance, the BIPV module price, inverter price, cabling cost, installation fee, and framing cost in case of curtain walls.

   In Singapore, a colored BIPV system per unit costs around 500 SGD (at May 2020 module cost) [11].

4. **Operating & Maintenance Cost:** Operating and maintenance cost for activities such as facade cleaning depends on the total area of the system installed and is slightly higher than the maintenance cost for conventional rooftop photovoltaic systems.

   The BIPV LCC tool recommends a conservative operating and maintenance cost of 4–6 SGD/m^2 [11].

5. **Annual Reserve for Inverter Replacement:** This is computed based on a warranty extension cost every fifth year with escalating premiums (i.e., 25%, 40%, 60%, and so on). It is assumed that inverter cost per BIPV module power output (W_p) will reduce to around 0.07 SGD/W_p within the next 13 years and remain stable thereafter.

   The BIPV LCC tool recommends to utilize a value of 5.5 SGD/kWp for annual reserve charge for inverter replacement based on assumptions in the case of Singapore: 4.8 SGD/kW_p for 20 years, 5.5 SGD/kW_p for 25 years, and 6.0 SGD/kW_p for 30 years [9].

6. **Inflation:** The inflation rate is used to escalate operating and maintenance costs, inverter replacement cost, and residual value cost.

   It is recommended to set the annual inflation rate in Singapore to be 2% for calculation purposes [9].

7. **System Lifetime:** A value for the system lifetime can be approximated based on the building's life expectations or in-line with the output warranty of the BIPV module manufacturer.

   The BIPV LCC tool recommends to set 25–30 years as the system lifetime [9].

8. **System Performance Rate:** The performance rate is an indicator of how well a solar PV system converts available sunlight into electricity. In practical applications, there are losses due to DC to AC conversion, cabling or mismatch issues, temperature effects, shading caused by nearby buildings, soiling, reflections, all of which will decrease the system performance.

   A well-designed system for Singapore's weather conditions can achieve a performance rate of > 80%. The BIPV LCC tool recommends to consider the system performance rate to be 75% [9].

9. **System Degradation Rate:** After the first year's specific energy yield (i.e., available irradiance multiplied by the system's performance rate), the system degrades by a certain percentage from year to year. For temperate climates, the degradation rate is usually assumed at 0.5% and for tropical climates (such as that of Singapore), the BIPV LCC tool recommends to use a degradation rate of ~1% [9].

10. **Residual Value:** This refers to the amount of value (in SGD) that can be recovered at the end of the system's operational life after subtracting the dismantling and recycling costs.

    The BIPV LCC tool recommends using 0 SGD for the residual value in calculations [9].

11. **Reduced Solar Heat Gain Coefficient (SHGC):** The SHGC is a fraction of solar radiation entering through a window, door, or exterior glass walls/skylight, either transmitted directly and/or absorbed, and subsequently released as heat into the building interior. Replacement of exterior glass panels by BIPV modules that absorbs some parts of the solar radiation leads to potential energy savings as a result of cooling load reduction in air-conditioned environments which is expressed as reduced SHGC.

    The SHGC reduction can be as high as 0.4 by using glass–glass BIPV modules, or 0.2 by using see-through BIPV modules (calculation is based on a common air-conditioning coefficient of performance of 3.7).

    The reduced SHGC is considered to be 0 in conservative calculations [9].

12. **Electricity Grid Emission Factor:** The grid emission factor measures the average CO_2 emission per kWh of electricity. It is calculated using an average operating margin (OM) method which is a generation-weighted average CO_2 emission per unit net electricity generation of all generating power plants serving the system.

    Singapore's average operating margin-based grid emission factor (OM-GEF). is 0.4188 kg CO_2/kWh in 2018 (EMA 2018). When the multi-silicon PV technology's emission rate of 28 g is subtracted from the GEF, the updated value is 390.8 gCO_2/kWh.

13. **Carbon Tax Savings:** Carbon Tax refers to the carbon pricing of per tonne of CO_2 emissions levied by the government. Carbon Tax Savings of the BIPV system coupled with avoided CO_2 over the system's lifetime will be estimated for LCC.

    Carbon Tax Charge (levied on consumers) = Electricity Consumed based on. metered values x OM-GEF x Carbon Tax Price.

Table 1 List of parameters' values considered for LCC calculation

Parameter	Value
Discount Rate	5%
Power/Unit area	135 W/m^2
System Price	500 SGD/unit
Operating & Maintenance Cost	5 SGD/m^2 (mean)
System Lifetime	25 years (lower end of the range)
Annual Reserve for Inverter Replacement	~5.5 SGD/kW_p
Inflation	2%
System Performance Rate	75%
System Degradation Rate	1%
Electricity Grid Emission Factor	390.8 g CO_2/kWh (EMA 2018)
Electricity Arrangement	NCC
Carbon Tax Price	5 SGD/tCO_2

The BIPV LCC tool recommends to set carbon tax price to 5 SGD/tCO_2 as per Singapore regulations in 2019 [9].

14. **Energy Economics:** The electricity price payable to the utility provider (in SGD cents/kWh) for either the non-contestable client (NCC) or contestable client (CC) is decided based on the average values of the respective customer group with future electricity price progression. Hence, opting to be CC, buying from a sustainable source and adopting BIPV options is more economical and environmentally friendly in Singapore.

The parameter values considered for the step-by-step LCC calculation in the following section are listed in Table 1.

2.2 Step-by-Step LCC Calculation for BIPV System

Let us consider a building in Singapore with BIPV system deployed on the east exterior wall/facade, spread over an area of 500 m^2 as shown in Fig. 4.

1. **Calculating the total initial investment**
   The total area of the BIPV system installed (east facade) = 500 m^2 (i).
   System price for unit area (per m^2) = 500 SGD (ii).
   Therefore, the total initial investment is given by
   = (i) × (ii).
   = 500 × 500.
   = 250,000 SGD.

Fig. 4 A schematic illustration of a representative building with BIPV installation on its east facade spread over an area of 500 m^2. Reproduced from the BIPV LCC Calculator available at National Solar Repository of Singapore website; Copyright 2014, SERIS

2. **Calculating the total operating and maintenance (O&M) cost for the BIPV system's lifetime**

   System lifetime (N) = 25 years.
   The annual average operating and maintenance cost per unit area (m^2) of the system = 5 SGD (iii).
   Discount rate (r) = 5% (or 0.05).
   Inflation rate (in Singapore) (i) = 2% (or 0.02).

   The annual average operating and maintenance cost of the BIPV system is given by
   = (i) × (iii).
   = 500 × 5.
   = 2500 SGD$(O\&M)_{initial}$.

   The total operating and maintenance cost of the BIPV system over its lifetime is given by
   $(O\&M)Total = (O\&M)initial * \sum_{n=1}^{N} \left(\frac{1+i}{1+r}\right)^n$
   where N is the system's lifetime,
   i is the inflation rate, and.
   r is the discount rate.
   Substituting the corresponding values in the above equation,
   $(O\&M)_{Total} = 2500 * \sum_{n=1}^{25} \left(\frac{1+0.02}{1+0.05}\right)^n$
   $(O\&M)_{Total} = 2500 * \sum_{n=1}^{25} (0.9714)^n$

$$= 2500 \text{ x } (0.9714 + (0.9714)^2 + \cdots + (0.9714)^{25}).$$
$$= 2381 + 2313 + \cdots + 1187.$$
$$= 42{,}960 \text{ SGD}.$$

3. **Calculating the total inverter replacement reserve $((IRR)_{Total})$ for the BIPV system's lifetime**

Calculating Installed Capacity:

The power generated per unit area of the BIPV system (non-transparent colored module) $= 135$ W/m^2 (iv).
The installed capacity of the BIPV system for an area of 500 m^2 is given by
$$= \text{(iv)} \times 500.$$
$$= 135 \times 500.$$
$$= 67.5 \text{ } kW_p \text{ (v)}.$$

Calculating the Inverter Replacement Reserve:

The annual inverter replacement reserve per $kW_p = 5.5$ SGD.
The annual inverter replacement reserve for the installed capacity is given by
$$= 5.5 \text{ x (v)}.$$
$$= 5.5 \times 67.5$$
$$= 371.25 \text{ SGD} (IRR)_{initial}.$$
The annual discounted inverter replacement reserve for the installed capacity, $(IRR)_{discounted}$, is given by
$$\frac{(IRR)_{initial}}{1+r}$$
$$\frac{371.25}{1+0.05}$$
$$= 354 \text{ SGD}$$
The inverter replacement reserve for the BIPV system over its lifetime is given by
$(IRR)Total = (IRR)discounted * \sum^{N}$.
where N is the system's lifetime,
i is the inflation rate, and.
r is the discount rate.
Substituting the corresponding values in the above equation,

$$(IRR)_{Total} = 354 * \sum_{n=1}^{25} \left(\frac{1+0.02}{1+0.05} \right)^{n-1}$$

$$(IRR)_{Total} = 354 * \sum_{n=1}^{25} \left(\frac{1+0.02}{1+0.05} \right)^{n-1}$$

$$= 354 \text{ x } (1 + 0.9714 + (0.9714)^2 + + (0.9714)^{24}).$$
$$= 354 + 343 + 334 + \ldots + 176.$$
$$= 6{,}380 \text{ SGD}.$$

4. **Calculating the benefit of electricity production over the BIPV system's lifetime**

Calculating the total green energy produced by the BIPV system

The BIPV system degradation rate (SDR) = 1% (or 0.01).
Energy generated by the system in the first year is calculated to be 41,272 kWh ($E_{initial}$).
Total green energy produced over the system's lifetime is given by
$E_{Total} = E_{initial} * \sum_{n=1}^{N}(1 - SDR)^{n-1}$
Substituting the corresponding values in the above equation,
$E_{Total} = 41,272 * \sum_{n=1}^{25}(1 - 0.01)^{n-1}$
$= 41,272 \times (1 + 0.99 + (0.99)^2 + + (0.99)^{24})$.
$= 41,272 + 40,859 + 40,451 + ... + 32,427$.
$= 916,976$ kWh(vi).

Calculating the total electricity produced by the BIPV system

The BIPV system performance rate = 75% (or 0.75).
So, the total electricity produced over the system's lifetime is given by
= (vi) x 0.75.
$= 916,976 \times 0.75$.
$= 687,732$ kWh(vii).

Calculating the economic benefit due to the total electricity output

The per unit of electricity tariff in July 2019 was 24.2 cents in Singapore and is considered here for evaluation.
Therefore, the electricity benefit is given by,
$= (vii) \times 0.242$.
$= 687,732 \times 0.242$.
$= 166,431$ SGD.

5. **Calculating the total LCC for the BIPV system**

As shown in Table 2, the total LCC for the BIPV system is calculated by adding up the expenditures (inclusive of the total initial investment, the total operating and maintenance costs and the total inverter replacement reserve), and subtracting the economic benefit obtained by the total power produced by the installed modules.

6. **Calculating the LCC for the BIPV system per unit area**
Total LCC = 132,909 SGD.
Therefore, the LCC for the BIPV system per unit area is given by
= 132,909 / (i).
= 132,909/500.
$= 266$ SGD/m^2.

Table 2 Total LCC calculation in SGD

Item description	Cost (SGD)
Total initial investment	250,000
Total Operating and Maintenance Cost	42,960
Inverter Replacement Reserve	6,380
Benefit from electricity production	−166,431
Total Life Cycle Cost	132,909

7. **Environmental benefit—Calculating the Carbon Tax savings over the BIPV system's lifetime**

Electricity Grid Emission Factor $= 390.8$ g CO_2/kWh.
CO_2 savings for the system's lifetime is given by
$= $ (vi) x 390.8
$= 916,976 \times 390.8$
$= 358.35$ tonnes.

Carbon Tax Price in Singapore $= 5$ SGD/tCO_2.
The Carbon Tax savings over the BIPV system's lifetime is given by
$= 358.35 \times 5$.
$= 1791.75$ SGD.

It should be noted that this representative example of LCC is provided here for a basic understanding of the underlying cost calculations that generally determine the economic feasibility of a BIPV system over its lifetime. A similar set of calculations is modeled as a three-step process where the user can input custom values for the parameters to assess the economic viability of the selected BIPV system. This model is presented as a user-friendly tool by SERIS on the National Solar Repository (NSR) website at https://www.solar-repository.sg/lcc-calculator and is useful to further explore the effect of various parameters on the LCC values.

2.3 Decision Making Using LCC Analysis

Decisions about transiting to new energy systems typically involve a certain degree of uncertainty about their initial and running costs, and potential savings and energy benefits over long run. LCC greatly increases the likelihood of initiating a new project or integrating a new system that is economically feasible in the long run. Yet, there are hidden challenges associated with the LCC model and results. LCC studies are usually performed for a new project in the initial stages when only financial estimates are available and extrapolations are generally made to fill up the gap in the estimates that may be due in the years to come. This uncertainty in critical input values such as costs and savings means that the LCC outcomes may differ from the actual outcome and occasionally this difference may be significant. There are, in general, two approaches to address this concern.

(i) Sensitivity analysis: A technique generally recommended for energy and water projects, sensitivity analysis is useful for identifying which of several uncertain input values has the most significant impact on the economic evaluation of the project. It also provides an estimate of how variability in the input values affects the economic evaluation of the project.

(ii) Break-even analysis: Decision-makers such as project builders, consumers, or real estate developers, many a time, want to know what minimum benefit a project can yield over the long run, say few years or a decade, and still, cover the cost of the investment.

A detailed description of these analyses is beyond the scope of this chapter. However, the reader should appreciate the uncertainties associated with any LCC model and its application and note that these gaps can be addressed using sensitivity and/or break-even analysis.

Having discussed the sensitive dependencies of the LCC study on timing and economic uncertainties, it is also important to consider the financial benefits of LCC in the long run. To illustrate this point, we consider a real-world scenario in which vertical facade applications were considered for adoption based on their economic viability. A representative example for such a study is the one conducted by SERIS on the colored BIPV facades installed at the School of Design and Environment (SDE), NUS campus, as a part of the first real-world test-bedding project in Singapore [11].

Based on empirical evidence as reported in [11], we note the following key results that are representative of the general benefits regarding economic viability and longterm financial performance, an LCC study can provide for an energy project.

1. Even though the initial investment for a BIPV facade is higher than the conventional options such as cladding, the market value of the energy generated offsets the difference in the initial investment over 7 years. On the contrary, the expenditure for the maintenance of conventional facades as compared with those for the BIPV systems keeps accumulating, as a result of which conventional facades become an expensive investment.

2. The implementation of BIPV systems (colored type) over the long-term, say 30 years, works out to be more economical compared to the conventional cladding facades. Over this period, the LCC value of the BIPV facade can eventually drop to almost half of the LCC value of the conventional facade which would have increased significantly over time due to maintenance costs.

3. If the benefits of carbon emissions savings are considered, say in the case of Singapore, the LCC value for BIPV systems can further decrease significantly.

3 Concluding Opinion

With rising consciousness about resource depletion and sustainable energy, innovative energy systems are required to capture and use incident solar energy on buildings. This transition to new technologies such as the BIPV discussed in this chapter should be preceded by a thorough understanding of the costs involved and the feasibility factors that determine the effectiveness of integrating these systems with the existing buildings in the long run. For example, measuring the impact of transitions to BIPV systems, involves parameters such as energy savings, reducing carbon emission, and breaking even financially. LCC provides an evaluatory model to largely understand and assist these transitions by critically examining the economic benefits and challenges of incorporating new technologies or methodologies into existing systems.

The LCC application illustrated in this chapter is relevant to tropical ecosystems wherein renewable energy sources such as solar energy is abundantly available. From a user perspective, this analysis is most suitable for architects, real estate developers, and property owners as they can tailor the integration depending on the cost-effectiveness of the BIPV system for various placement orientations and installation areas. LCC, in this case, provides a detailed breakdown of initial investments and recurring costs with the total amount of electricity generated over the life cycle of the system installed. It is also helpful in estimating carbon emissions avoidance and energy savings (due to its cooling effect) in the long run which signifies how greener BIPV systems are in comparison with the conventional energy sources.

In the context of CE, adopting innovative renewable energy sources such as the BIPV systems into urban ecosystems becomes important. It reduces the burden on traditional power grids that are mainly dependent on fossil fuels and consequently lowers the environmental impact of energy production and consumption.

Questions

1. Explain conceptually how the transition towards a circular economy in the energy sector can be realised.
2. Define LCC and LCOE.
3. Explain the need for solar energy systems such as BIPV modules in resource-stressed countries such as Singapore.
4. Consider a building that uses BIPV systems on its east and west exterior walls. The modules are spread over 250 m^2 area on the upper half of each of the sides to capture the solar energy during the day. Using the online LCC tool posted by SERIS on the National Solar Repository (NSR) website (link below), estimate the LCC for such an implementation. Also, highlight with reason, any differences in LCC between this implementation and the case in which the BIPV module is installed on only one side of the building (area: 500 m^2), say east as illustrated in the chapter. The NSR tool can be accessed at https://www.solar-repository.sg/lcc-calculator.

Suggested Reading

1. Wang, J., Yu, C., Pan, W. (2020). Relationship between operational energy and life cycle cost performance of high-rise office buildings. *Journal of Cleaner Production*, 121300.
2. Ruoping, Y., Xiaohui, Y., Fuwei, L., et al. (2020). Study of operation performance for a solar photovoltaic system assisted cooling by ground heat exchangers in arid climate, China. *Renewable Energy*, 102–110.

Acknowledgments The authors would like to acknowledge the Solar Energy Research Institute of Singapore (SERIS), National University of Singapore (NUS), (program administrator for National Solar Repository (NSR)) for the case study analysis presented in this chapter (Sect. 2). The LCC analysis tool is available on the NSR website and can be accessed at https://www.solar-repository. sg/lcc-calculator. The authors wish to acknowledge Monika Bieri, solar economics specialist, for her valuable contributions in developing the BIPV LCC calculator and Dr. Anoop C. Patil, N.1 Institute for Health, NUS, for his valuable suggestions to improve this chapter.

References

1. Korhonen, J., Nuur, C., Feldmann, A., et al. (2018). Circular economy as an essentially contested concept. *Journal of Cleaner Production, 175,* 544–552.
2. Twidell, J., Weir, T. (2015). Renewable energy resources. Routledge.
3. Bhatia, S. (2014a). Wind energy. In A. Renewable (Ed.), *Bhatia S* (pp. 184–222). Energy Systems: Woodhead Publishing India.
4. Bhatia, S. (2014b). Hydroelectric power. In A. Renewable (Ed.), *Bhatia S* (pp. 240–269). Energy Systems: Woodhead Publishing India.
5. IRENA. (2019). *Future of solar photovoltaic: Deployment, investment, technology, grid integration and socio-economic aspects* (pp. 1–73). Abu Dhabi: International Renewable Energy Agency.
6. Husain, A. A., Hasan, W. Z. W., Shafie, S., et al. (2018). A review of transparent solar photovoltaic technologies. *Renewable and Sustainable Energy Reviews, 94,* 779–791.
7. Congedo, P., Malvoni, M., Mele, M., et al. (2013). Performance measurements of monocrystalline silicon PV modules in South-eastern Italy. *Energy Conversion and Management, 68,* 1–10.
8. Jelle, B. P., Breivik, C., & Røkenes, H. D. (2012). Building integrated photovoltaic products: A state- of-the-art review and future research opportunities. *Solar Energy Materials and Solar Cells, 100,* 69–96.
9. Shabunko, V., Bieri, M., Reindl, T. (2018). Building integrated photovoltaic facades in Singapore: Online BIPV LCC Calculator. In *2018 IEEE 7th world conference on photovoltaic energy conversion (WCPEC)* (pp. 1231–1233). IEEE.
10. Sozer, H., & Elnimeiri, M. (2007). Critical factors in reducing the cost of building integrated photovoltaic (BIPV) systems. *Architectural Science Review, 50*(2), 115–121.
11. Shabunko, V., Reindl, T. (2019). High-level financial assessment of colored BIPV faades: Case study in Singapore. In *46th IEEE photovoltaic specialists conference (PVSC)*. Chicago, IL: IEEE.

Rashmi Anoop Patil is a circular economy enthusiast and an engineer by profession with an undergraduate degree in Electronic Engineering from Visveswaraiah Technological University, India. As a member of the Circular Economy Task Force at the National University of Singapore (NUS) (led by Prof. Seeram Ramakrishna, Chair, Circular Economy Task Force, NUS), she is currently researching circular economy concepts. She is passionate about sustainability and ecofriendly businesses.

Dr. Veronika Shabunko is Head of the BIPV Centre of Excellence at the Solar Energy Research Institute of Singapore (SERIS) and Senior Research Fellow at the National University of Singapore (NUS). She is specialized in interdisciplinary research focusing on Smart and Sustainable cities, by combining energy efficiency and conservation measures, innovative high-performance building integrated photovoltaic (BIPV) systems, and reliable integration of renewable energies into the infrastructure on an urban scale. Her research aims to develop techno-economic solutions for the sustainable development of densely populated cities and to overcome the challenges of climate change. She is a scientific lead of the working group for the establishment of relevant national regulations and standards for innovative BIPV technologies in Singapore with governmental and regulatory bodies, and an active contributor to the IEA PVPS (International Energy Agency, Photovoltaic Power Systems Programme), Task 15 on "Enabling framework for the acceleration of BIPV".

Professor Seeram Ramakrishna *FREng, Everest Chair* (https://www.eng.nus.edu.sg/me/staff/ramakrishna-seeram/), is among the top three impactful authors at the National University of Singapore, NUS (https://academic.microsoft.com/institution/165932596). NUS is ranked among the top five best global universities for engineering in the world (https://www.usnews.com/education/best-global-universities/engineering). He is the Chair of Circular Economy Taskforce. He is a member of Enterprise Singapore's and ISO's Committees on ISO/TC323 Circular Economy and WG3 on Circularity. He also the Chair of Sustainable Manufacturing TC at the Institution of Engineers Singapore and a member of standards committee of Singapore Manufacturing Federation (http://www.smfederation.org.sg). He is an advisor to the Ministry of Sustainability & Environment—National Environmental Agency's CESS events, (https://www.cleanenvirosummit.sg/programme/speakers/professor-seeram-ramakrishna; https://bit.ly/catalyst2019video; https://youtube.com/watch?v=ptSh_1Bgl1g). European Commission Director-General for Environment, Excellency *Daniel Calleja*

Crespo, said, "Professor Seeram Ramakrishna should be praised for his personal engagement leading the reflections on how to develop a more sustainable future for all", in his foreword for the Springer Nature book on Circular Economy (ISBN: 978-981-15-8509-8). He is a member of UNESCO's Global Independent Expert Group on Universities and the 2030 Agenda (EGU2030). He is the Editor-in-Chief of the Springer NATURE Journal Materials Circular Economy—Sustainability (https://www.springer.com/journal/42824). He is an Associate Editor of eScience journal (http://www.keaipublishing.com/en/journals/escience/editorial-board/). He is an opinion contributor to the Springer Nature Sustainability Community (https://sustainabilitycommunity.springernature.com/users/98825-seeram-ramakrishna/posts/looking-through-covid-19-lens-for-a-sustainable-new-modern-society). He teaches ME6501 Materials and Sustainability course (https://www.europeanbusinessreview.com/circular-economy-sustainability-and-business-opportunties/). He also mentors Integrated Sustainable Design ISD5102 project students. Microsoft Academic ranked him among the top 25 authors out of three million materials researchers worldwide based on H-index (https://academic.microsoft.com/authors/192562407). He is named among the World's Most Influential Minds (Thomson Reuters) and World's Highly Cited Researchers (Clarivate Analytics). Listed among the top three scientists of the world as per the Stanford University researcher study on career-long impact of researchers or c-score (https://drive.google.com/file/d/1bUJrvurVVBbxSl9eFZRSHFif7tt30-5U/view). He is an Impact Speaker at the University of Toronto, Canada Low Carbon Renewable Materials Center (https://www.lcrmc.com/). He is a judge for the Mohammed Bin Rashid Initiative for the Global Prosperity (https://www.facebook.com/Make4Prosperity/videos/innovation-inclusive-trade/479503539339143/). He advises technology companies with sustainability vision such as TRIA (www.triabio24.com), CeEntek (https://ceentek.com/), Green Li-Ion (www.Greenli-ion.com) and InfraPrime (https://www.infra-prime.com/vision-leadership). He is a Vice-President of Asian Polymer Association (https://www.asianpolymer.org/committee.html). He is a Founding Member of Plastics Recycling Association of Singapore (PRAS). His senior academic leadership roles include University Vice-President (Research Strategy), Dean of Faculty of Engineering; Director of NUS Enterprise and Founding Chairman of Solar Energy Institute of Singapore (http://www.seris.nus.edu.sg/). He is an elected Fellow of UK Royal Academy of Engineering (FREng), Singapore Academy of Engineering and Indian National Academy of Engineering. He received PhD from the University of Cambridge, UK, and The TGMP from the Harvard University, USA.

Circular Economy in a Water-Energy-Food Security Nexus Associate to an SDGs Framework: Understanding Complexities

Alex Godoy-Faúndez, Diego Rivera, Douglas Aitken, Mauricio Herrera, and Lahcen El Youssfi

Abstract The world currently faces significant challenges in its adaptation to and mitigation of climate change in order to provide goods for the growing demand for food, water, and energy—key inputs into a modern society. Today, there is hardy industrial competition for resources that includes agriculture, heavy industry (mining, manufacturing, etc.), forestry, and light industries (electrical component manufacturing, textiles, etc.). Agriculture currently accounts for 70% of global water withdrawals; 30% of total global primary energy consumption, via production and distribution; and 51% of aggregate global energy use. Together, they produce considerable volumes of wastes, increase pressure on ecosystems, and impact local communities' water, energy, and food security. These three sectors form the water, energy, and food security nexus and are integral to achieve sustainable development goals. In extractive economies such as that of Chile, it is critical to understand how the water–energy–food nexus is linked to circular economy models of short production chains of agriculture and mining to transit to the circular agrofood system and green mining. As a country whose primary industries are agriculture and mining, Chile would be

A. Godoy-Faúndez (✉) · D. Rivera · M. Herrera
Centro de Investigación en Sustentabilidad y Gestión Estratégica de Recursos (CiSGER), Facultad de Ingeniería, Universidad del Desarrollo, Avenida Plaza 680, Las Condes, Santiago, Chile
e-mail: alexgodoy@udd.cl

D. Rivera
e-mail: diegorivera@udd.cl

M. Herrera
e-mail: mherrera@ingenieros.udd.cl

A. Godoy-Faúndez · D. Rivera
ANID/FONDAP-15130015, Water Research Center for Agriculture and Mining (CRHIAM), Chillán 3812120, Chile

D. Aitken
Sustainable Minerals Institute International Centre of Excellence Chile, SMI-ICE Chile, The University of Queensland, Av. Apoquindo 2929 Piso 3, Las Condes, Santiago, Chile

L. El Youssfi
Khenifra Superior School of Technology, Sultan Moulay Slimane University, Beni Mellal, Morocco
e-mail: l.elyoussfi@usms.ma

© Springer Nature Singapore Pte Ltd. 2021
L. Liu and S. Ramakrishna (eds.), *An Introduction to Circular Economy*,
https://doi.org/10.1007/978-981-15-8510-4_12

one of the world's five countries suffering from the highest water stress in 2040, and the World Resource Institute is forecasting Chile as having the worst distribution of water resources. Therefore, it is imperative to understand the complexities and linkage of both industrial activities, agriculture and mining, in determining the cross-spatial scale fluxes between water, energy, and waste. With this understanding, one may recognize the patterns and relationships between each of the variables and how they contribute to resource-based conflicts.

Keywords Chile · Mining · Agriculture · Water-energy-food nexus · Sustainable development goals · Circular economy

Learning Objectives

- Through Chilean cases, learn about the adaptation and mitigation challenges of the agricultural and mining sectors.
- Through Chilean cases, learn of the overlapping environmental impacts of industrial activities in the territory.
- Understand the mining processes and pathways to achieve green mining.
- Understand the agriculture activities and transit to accomplish a circular agrofood system
- Understand the environmental impacts of industrial activities overlapped in the territory.

1 Introduction

According to the Global Risks Report [1], our civilization faces multiple risks on natural and human systems derived from climate-related issues. It dominates all the top-five long-term risks. Climate change impacts have already been observed due to an increase in extreme weather events threatening our civilization with unprecedented global challenges on water-energy-food security. These risks depend on the magnitude and rate of warming, geographic location, levels of development, and vulnerability, forcing our choices to implement climate-change mitigation and adaptation strategies and approaches by governments and businesses [2].

The climate hazards of global warming of 1.5 °C will cause damage to critical infrastructure and significant biodiversity loss between 1,000 and 10,000 times higher than the natural extinction rate. Those impacts would drive the collapse of vulnerable ecosystems in convergence with major natural disasters (e.g., earthquakes, tsunamis, volcanic eruptions) and human-made environmental damage by depleting natural resources and environmental crimes striking harder and more rapidly than expected. Scientists' evidence about depletion and overconsumption of natural resources has been provided as an alert signal of we are exceeding "planetary boundaries" in many áreas [3]. In 2019 was the "Earth Overshoot Day", defining when humanity's demand for ecological resources and services in a given year exceeded what Earth can regenerate in that year reached alarming 1.7 planets are required to deliver humanity's demand on Earth's ecosystems. That overshoot would collapse ecosystems and reduce world stocks of natural resources—natural capital—essential to human well-being by the destruction of ecosystems services by breaking the production systems to provide goods and services [4].

The drivers of global changes are linked to human population dynamics and economic development associated with production and consumption patterns. Both need to decouple growth to raw material extraction [5] required to sustain the ecosystems by technological innovation and transformational change for achieving the Sustainable Development Goals [6]. But, although governments and businesses are working on in a transition toward a low-carbon economy by decarbonization introducing renewable energies and environmental-friendly technologies, these efforts are not enough to accomplish with Paris Agreement goals [7]. About the human population, the UN's estimated around 8.5bn people will live on the planet by 2030, over half (5.3bn) will be 'middle class'. Despite raising out of poverty could be successful in being achieved, it will come with a high pressure on the natural environment if we do not change the resource-intense drive toward the material comforts of middle-class lifestyles. These socio-economical forces will accelerate the rise in greenhouse gas (GHG) emissions leading to global warming, ecosystem destruction, and pollution with dangerous levels of plastics, nitrogen, and phosphorous that will lead to significant and, in some cases, irreversible changes to the Earth's environment.

According to World Bank *"Chile is highly exposed and vulnerable to multiple hazards with such as earthquakes, volcanic activity, and tsunamis as well as hazards which can change due to climate change such as wildfires, floods and landslides, and droughts. Chile is suffering a megadrought between 2008-today that affected much of the southern and central areas. Climate change is expected to change the frequency, intensity, exposure, and magnitude of multiple hazards that have historically affected Chile, namely, wildfires, floods and landslides, droughts, and impacts of sea-level rise. The accumulation of risks, exposure, and multiple hazards can have important implications for economic* growth and development in regions, particularly for electricity generation, agriculture, and public health" [8]. The economic impact of climate change in Chile [9] depends on the location of industrial activities based on ecosystem services. Chile expects a reduction of rainfall increasing shortages at far and near north and rising of temperature as well. These changes will impact mining and agriculture activities throughout the country, both critical sectors of the Chilean economy (Fig. 1).

The water shortage expected could be relieved by circularizing the processes which depend on water and energy consumption as a consequence of the linear economy. In Chile, production and manufacturing follow a take-make-waste model, which has enabled unprecedented levels of material comfort at affordable prices but high environmental impacts. Therefore, businesses need a transformative change beyond transition by an integrated approach to fix the pathways required for sustainable development. The International Resource Panel has provided insights about pathways to achieve SDGs by adopting material efficiency strategies for a low-carbon economy to 1.5 °C goal set by the Paris agreement. The panel proposes the reduction of GHG emissions throughout the adoption of the circular economy, 3R perspective

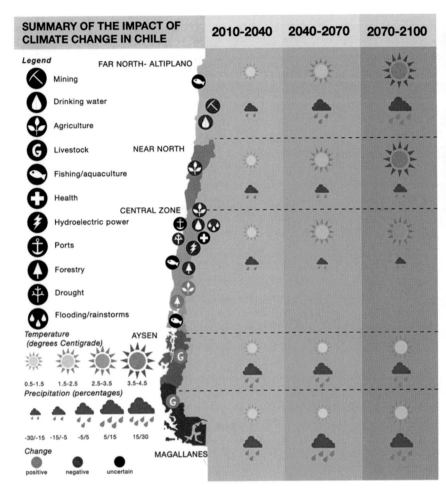

Fig. 1 Summary of the impact of climate change in chile

(reduce, reuse, recycle), and sustainable materials management as materials-climate change nexus [10]. Therefore, how we manage resources, make and use products, and what we do with the materials creates both opportunities and challenges afterward. If we want to understand how to manage our natural resources, as the first step, we need to understand the tradeoff between water-energy-food nexus into production systems to provide solutions regarding material and energy flows through the human systems.

2 Water-Energy-Food/Waste Nexus

Water and energy are critical inputs into a modern society with waste being an almost unavoidable consequence; all three are inextricably linked despite often being managed in isolation [11]. All industrial activities show and interdependence of water-energy consumption, putting on the risk the provision of food. The diagram in Fig. 2 shows the interdependencies of three national securities for welfare, named as nexus [12]. Therefore, the Energy, Food, and Water Security depend on energy, land, and water managing used by human activities to provide goods and services for our well-being. In the extractive's economies, the technology used in processes and the contained pollution by treatment plants as crucial factors to reduce their environmental impact. However, these sectors are intensive in water and energy consumption by converting raw materials into goods or supplies for the supply chain that can be used at downstream manufacturing.

The nexus is vulnerable to climate change too (Fig. 3). Climate and land-use change need to be managing to achieve goals such as fertilizer reduction into corps production reduction by using by-products; continual treatment of water discharges to reuse, recycling or transfer to other industrial activities; or include renewable energy into operations. As a consequence, the extractive economies will reduce their environmental impact [13].

Agriculture is one industry strongly dependent upon cheap and readily available sources of both water and energy. The agricultural industry accounts for around 70% of global water withdrawals [14] and 30% of total global primary energy consumption via production and distribution [15]; a key sector for food security. The industrial competition for resources includes heavy industry (mining, manufacturing etc.), forestry and light industries (electrical component manufacturing, textiles etc.).

Fig. 2 Diagram water-energy-food security [12]

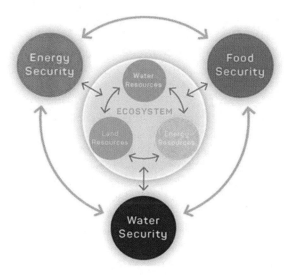

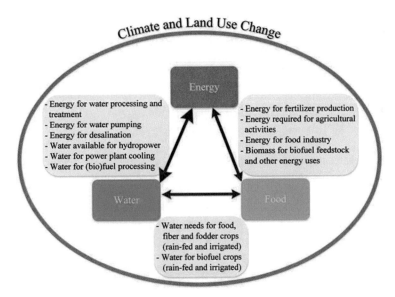

Fig. 3 Climate and land-use change

These industries are generally high consumers of energy and produce considerable volumes of waste (a valuable source of resources in scarcity). Many newly developed countries are now producing huge flows of waste materials as a by-product of their rapid development, which has led to widespread environmental degradation. However, these materials in the context of a circular economy (the industrial economy framework that promote recycling, preventing the loss of valuable materials towards zero-waste system) represent a significant source of resources to be recovered as energy and materials. According to the Energy Information Administration, industry accounts for 51% of total global energy use [16]. Heavy industry accounts for the highest volume of waste generation, in the European Union, the construction industry was responsible for 32.6% of waste, mining, and quarrying for 29.2% and manufacturing for 10.7% [17].

The competition between different sectors for energy and water alongside the generation of high volumes of waste leads to conflict between users, particularly in areas where water availability and energy are limited (water and energy scarcity). Global warming, increased populations, high consumer demands, and poor environmental management are leading to increased conflict over key resources [18]. It is important to understand the patterns and relationships between each of the variables contributing to resource-based conflicts. With this knowledge, strategies can be implemented to alleviate conflicts on a local level because the impact of extraction have adverse impacts into Sustainable Development Goals (Fig. 4) than could be managing to achieve positive a contribution [19].

Chile is an ideal case study for this issue due to the existence of industrial sectors overlapping in the same locations by competing by energy and water resources. Given

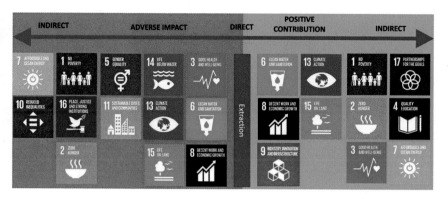

Fig. 4 Raw materials extraction and impacts on sustainable development goals. European Union Office

the co-location between human activities and population, the industry's differential intensities could be the underlying cause of the continuous deterioration of the ecosystems, reflecting a loss in the quality of life. They may partially explain water, energy, and food insecurities. For Chile, it is necessary to understand the existing relationships between sociodemographic, industrial, and environmental variables and the Water-Energy-Waste/Food nexus. These are key to water, energy, and food security in a country facing severe water scarcity and where its two main industrial activities, agriculture, and mining, overlap. Therefore, their understanding would allow the link to be linked to Sustainable Development indicators and ecosystems' quality.

Water availability in Chile: Water availability in Chile varies greatly depending upon altitude and latitudinal gradients and the influence of substantial interannual variability of rainfall and snow. Indeed, the streamflow's partition from rainfall or snowpack depends strongly on the temperature regime according to both phenomena. Thus, there is a strong dependency of water availability—timing and volume—upon precipitation. Water availability is lowest in the northern regions (18°–33° S) due to the presence of a high-pressure center—the Pacific Anticyclone—leading to arid and semi-arid conditions [20] (Fig. 5).

In the region of Antofagasta, the historic mean annual rainfall is 44.5 mm compared with 650 mm in the central metropolitan region. The central regions of Chile have a Mediterranean climate, streamflow is mainly pluvial in the winter season (85% of annual rainfall) but driven by snowmelt in early spring. During summer, rainfall is less than 10% of the annual mean, and groundwater systems, i.e., baseflow, provide streamflow. The Southern region (45°–55° S) presents low population densities and a high density of streams, making it attractive for hydropower generation but under intense competition with agriculture and the forestry industry. Thus, Central and Northern region has the largest population concentration, but water availability is less than half of the threshold considered necessary for sustainable development, while Southern region exhibits twelve times more available water than of North and

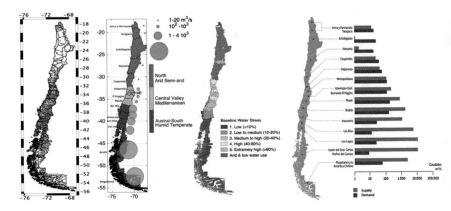

Fig. 5 Water balance and Water Stress in Chile. The figure shows from left to right the waterseheds, water availability, water stress, and water balance between supply and demand

Central regions [21]. Figure 6 shows an integrated view of precipitation; river flow, aquifer recharge; dam storage; water availability, and water balance by region [22].

Water Consumption: Water in Chile is unequally distributed due to its length, varied geography, and climatic conditions. Although the country's highest water consumer is agriculture with 85% of consumption, mining industry consumption with only 7% is relevant, particularly due to the location of operations and around 11% of the industry. The high water consumption in the mining industry in the north, alongside deficient availability, has led to high water scarcity. In the central regions, the high demand for agriculture and industrial and domestic use has also created water scarcity despite relatively high availability. With growing demand and decreasing water availability (quality and quantity), we expect more conflicts between water

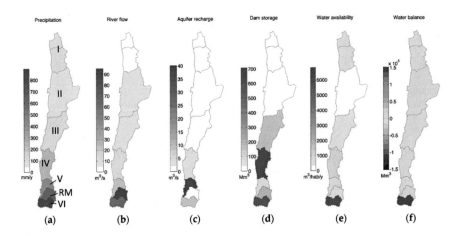

Fig. 6 Maps displaying the values of **a** precipitation; **b** river flow; **c** aquifer recharge; **d** dam storage; **e** water availability and **f** water balance by region

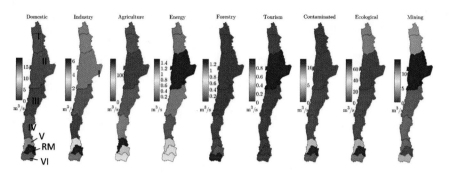

Fig. 7 The water demand of each sector by region (Aitken et al.)

users, stakeholders, and decision-makers. Some regions do not have sufficient water to support the current water demands, and the situation worsens when extended to the future (Fig. 7).

Additional water resources and management strategies are thus required to meet both current and future demands. The reduction in water availability, the increase in the use of groundwater reserves, and the current water demands cause widespread problems and make drinking water inadequate due to the occurrence of pollutants from industrial activities. In the Central Zone of Chile, water availability and quality is some of the main limitations to economic growth and development. Most of the Central Area (from Coquimbo to O´Higgins Regions) is considered an arid area, and the rest of the central Zone (Maule and Biobío) is mostly a semi-arid zone. Given that 78% of the Chilean population lives in that area (13.5 million inhabitants), there is a great need to develop advanced research investigating how society will interact with ever-scarcer water resources.

Energy Consumption: According to the Chilean electricity grid, hydroelectricity provides the most significant energy (42%), although this is primarily generated in the south of the country. The remaining generation methods are via thermoelectric sources, mainly coal (25%), gas, and diesel (17%). Chile's thermoelectric plants are mainly located from the central south to the north, where there are limited water sources for hydroelectric generation. Chile's vast copper mines also have an energy problem. Consumption currently accounts for 39% of the country's electricity consumption in an area with limited hydroelectric resources.

The current electricity generation mix is highly vulnerable to reduced water availability due to climate change and the importation of fossil fuels from other countries. Competition for the consumption of electricity is also very high with the greatest demand from the industrial sector. Demand is expected to increase at about 6% per annum [23]. This situation has led to the cost of electricity in Chile being one of the highest in Latin America, with the cost increasing. Nevertheless, the Chilean government is encouraging investment in renewable energies such as solar and Biomass in the north and south of Chile. Indeed, the minister of energy provides information by energy potential maps according to natural resources throughout Chile to promote

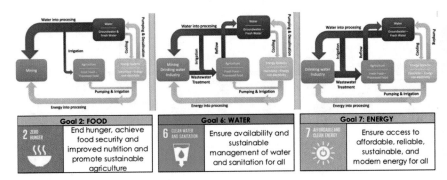

Fig. 8 Water-energy-food flows. The figure shows the flows at north, central, and Southern Chile linked to sustainable development goals

new investment in the electricity sector. However, energy production by industrial and urban wastes has not yet been studied, providing big opportunities to materials and energy recovery beyond solar or wind technologies.

According to Chilean geography, we can draw three kinds of Water-Energy-Food nexus (Fig. 8). At north of Chile, the main flows of water and energy are driven by mining; at central Chile the agriculture activities drive both consumptions meanwhile at south the flows are driven by industrial and drinking water activities. In all of them we can propose solutions based on 3Rs of water resources, energy efficiency approaches and improving the technology.

In conclusion, the Water-Energy-Food/Waste nexus depends on water and energy resources and the industries placed throughout the country [24]. In Chile, the nexus is driven by Agriculture and Mining activities, changing the flows of the north, central, and south regions negatively impacting the environment. All of those interactions directly impact three Sustainable Development Goals. A wrong balance will be reducing the availability of fresh food, increasing the prices, and impacting the nutritional balance of the population. Despite the circular economy being a fundamental framework for developing, this is a very complicated pathway in extractive economies because both activities have short supply chains [25]. Therefore, a circular economy mindset could modify the extraction activities encouraging the use of by-products and recycling, reuse, and reduction of water and energy consumption. Figure 9 shows that both activities are at the beginning of the supply chain putting a big challenge to circularize both their processes.

3 Circular Economy and Water-Energy-Food/Waste Nexus: An Opportunity for Agriculture and Mining.

Aforementioned show a clear competition by water and energy resources increasing the pressure on the territory. How can we make a safe transition to support a circular economy thinking in the nexus? [26]. The agricultural industry is a smaller contributor

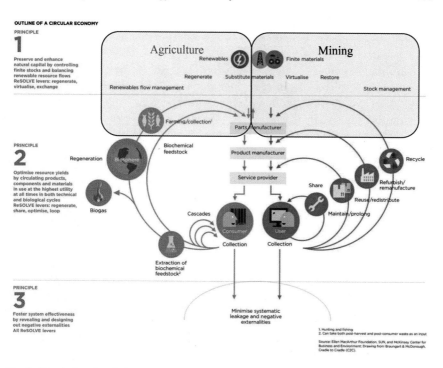

Fig. 9 MacArthur foundation report, growth within: a circular economy vision for a competitive Europe

to GDP at around 3.5% but is a large employer, employing around 10% of the Chilean workforce [27] and providing Chile with strong food security. However, the impacts of the scarcity of water have a high uncertainness. In the Global Food Security Index, Chile is ranked 27 out of the 109 countries assessed, but according to World Resources Institute, the forecasting to 2040, Chile is ranked as a water-stressed country [28], 24 out of 167 countries. Agriculture activities are concentrated within the Central Zone, where more than 270,000 farmers exist, developing 30.5 million hectares, around 40% of the continental surface of the country. Despite these challenges, it is expected that by 2030, the food processing industry will account for one-third of the country's economy. However, the water demand of the industry is extremely high, it has been estimated that consumption from agricultural irrigation accounts for 85% of total water consumption in Chile [29]. Chile has declared its intention to strengthen its agricultural sector, increasing its competitiveness in the country's economy and becoming an international food provider. However, this ambition will cause increasing pressure on already stressed land use and water resources; shared resources with another industrial activity.

What model could follow Chilean agriculture? According to circular food model proposed by Wageningen University, the "*Crop production in a circular agrofood system is designed to 'lock in' minerals and organic material, so that they can be used to their fullest potential. For instance, crops will utilise nutrients from the soil more effectively than now. In the future, crops such as potatoes or rice may be able to bind atmospheric nitrogen for their food. Precision agriculture offers a very targeted way of providing plants with the necessary nutrients. A circular agrofood system uses residual flows from agriculture and the food industry to produce animal feed. These flows may be the parts of a plant that we now think of as having no use, such as straw and foliage. By using insects, worms or fungi, we can convert this matter into nutrient-rich raw materials for animal feed. Cattle and sheep in a circular agrofood system would consume grass and herbs in pastures that are unsuitable for growing food…. This allows the animals to convert residual flows and crops that are unsuitable for human consumption into high-quality, protein-rich food for people. This could be milk, eggs, or meat.the animals' manure is also a valuable source of organic material that replenishes the soil and completes the circular agrofood system"* (Fig. 10). The Chilean farmers are starting to transit to circular agrofood system by introducing basic and straightforward approaches by understanding the nexus.

Chilean farmers have started increasing the water reflow to increase water availability at groundwater. Also, the government provides funding to implement Precision agriculture techniques to reduce water and fertilizer consumption. Besides, Chilean farmers have started understanding the tradeoff between water and energy and several irrigation techniques. Figure 11 shows a comparison between furrow and pressurized irrigation. Despite different approaches, the choice of one depends on local conditions about water and energy flows. In the north of Chile, the pressurized

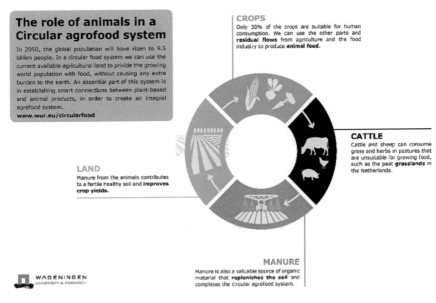

Fig. 10 Circular agrofood system

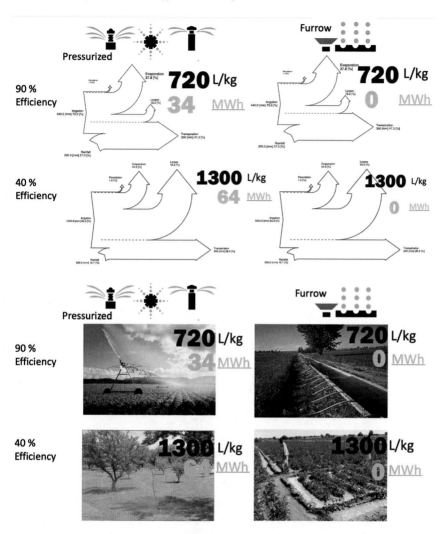

Fig. 11 Water and energy tradeoff of irrigation systems

irrigation could be better than furrows; meanwhile, the furrows could be better at the south of Chile. Further, the furrows could be the good one when the goal is to increase the reflow. Therefore, before choosing one, the analysis needs to include the characterization of local conditions to understand how different technologies can contribute or not to water-energy-food security.

A good example is the Wine industry, the most famous Chilean product produced entire in Chile. They are converting vineyards to less intensive crop production and organic systems by agroecology or other innovations. They are including composting and recovering by-products and energy by anaerobic digestion as well. This kind of waste treatment will be reinforced by a new law promoting organic recycling.

Further, the retailers by working with local suppliers and farmers will reduce food losses and food waste by encouraging consumers to consume local and seasonal. In Chile, waste management is a significant problem due to the high volumes of waste generated and the inadequate management system. In 2009, 16.9 million tonnes of waste were generated in Chile, 10.4 million tonnes from industry, and a total of 6.5 million tonnes from municipal sources (REF). Effectively all waste in Chile is landfilled, most going to sanitary landfills (69%), some going to old landfills with low regulations (22%), and a smaller fraction going to dumps with no regulations (9%). The landfilling of waste presents numerous issues, the cost of disposal, the use of land, the contamination of soil and groundwater, and the generation of emissions from the landfill and vehicles. However, the waste stream in Chile presents an enormous opportunity for the reclamation of materials and the generation of energy through various conversion techniques. It is expected that if energy conversion techniques were to be implemented, around 6% of Chile's electrical energy demand could be met through waste to energy, thus reducing energy resource scarcity [30]. Finally, the law could increase the composted materials used by farmers and green public spaces into cities.

Mining Sector: Our economy is heavily dependent upon the mining industry, and recent contractions within the industry have led to a slowing of Chile's economic growth rate. The mining industry in Chile accounts for around 13% of GDP, the majority of which is from copper recovery, but a high number of the services sector and suppliers depend on this kind of activity. A recent decline in copper prices, reduction in mining investments, and lower private consumption have led to a lower GDP growth rate of 1.9% and an increase in unemployment. However, the mining industry in Chile is expected to grow at an annual rate of 1.6% annually from 2015 to 2019. This large capacity for ore processing has a considerable impact on the environment [31] resulting from the release of large solid waste deposits, sterile piles, and lixiviation piles, as well as high consumption of energy and water. In 2015, water consumption reached 7%, and it is essential to note that almost all of this consumption occurs in the scarce water north. Chile's vast copper mines also have an energy problem, and consumption currently accounts for 39% of the country's overall electricity consumption, the majority of which is generated in the hydro rich regions of the South. Due to the impacts of global warming, Chile has been suffering from a drought for the last 15 years, which has affected electricity generating capacities. Chile's mines pay twice as much for their energy as their peers in neighboring Peru.

Chilean mines are now actively seeking new energy solutions for energy security and to reduce costs for grid-connected and off-grid operations. With energy accounting for 20–40% of operating costs, reducing electricity expenditure is now a significant operational and strategic goal for Chile's mining leaders. The use of renewable energy is getting to play a significant role in meeting this aim for both remote and grid-tied mines. The north of Chile lacks access to water; however, there is an abundance of solar energy and wind energy potential, both of which mining operations are beginning to take advantage. On August 26th, 2014 Chile launched a joint venture between Pattern Energy, a US firm, and Antofagasta Minerals, its biggest wind farm to date on a coastal hilltop 400 km north of Santiago with an installed capacity of 115 megawatts, which will provide around 20% of the electricity demand of Los Pelambres mine [32]. Antofagasta Minerals is also developing a solar plant with SunEdison in the Atacama region to supply the same mine. Various other renewable energy projects are also in the planning stage, the development of a wind farm by the state-owned Codelco, a geothermal energy project for BHP Billiton's Escondida mine, and a solar plant planned for the Collahuasi mine. The contribution of renewable energy to Chile's industries and the national grid is increasing, indeed, between January and July of 2015, Chile added 600 MW of renewable energy capacity to its grid, more than twice as much as in the whole of 2013, nearly 9% of Chile's installed capacity. A result of this is that many mining operations are also now investigating and developing the use of renewable energy sources, thus cutting reliance upon the national grid and making economic savings in the longer term. Examples of such developments are Antofagasta Minerals' wind farm development providing 115 MW to the Los Pelambres mine, and Collahuasi's planned 25 MW photovoltaic plant.

In terms of water consumption, primary mining operations monitor their annual water consumption and include results in sustainability reports. The more proactive companies are implementing sustainable water management strategies where water recycling is maximized, and losses minimized. A considerable fraction of water consumption is now provided via seawater pumped from the coastline in some operations. However, such strategies have considerably increased the electricity consumption of operations, particularly concerning pumping seawater and desalinization, putting yet further strain on the electricity grid. However, more and more companies are getting energy by solar and wind proposing deploy swapping water.

As the mining operations are located at Andeans, they provide water to farmers located near-coastal zone, leaving the water available upstream and reducing competition by water rights. Finally, they implement water recycling strategies and new wastewater treatment plants, leaving the water discharges available to irrigation downstream. The goal is to reduce toxic substances such as arsenic, zinc, cyanide, and mercury can build up and leak in local groundwater sources contaminating water supplies and impacting farming communities and local residents. In the central regions, the mining industry's impact is less, although conflict remains as population sizes are closer to mining sites and the agricultural industry is more extensive than in the north. Indeed, as climate change reduces water availability, water scarcity is increasing, meaning hydroelectric electricity generation is reduced, and there is greater competition between all users of water.

About mining wastes, the industry has started recovering minerals from tailing dams [33] due to a reduction in ore grade found into mining pits. This strategy is looking to reduce the energy in water consumption this is could be the kickoff to start a complementary recovery material from wastes and mineral recycling industry as proposed Spooren et al. in 2020 [34] in the Fig. 12

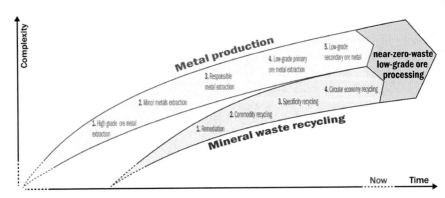

Fig. 12 Pathway to mineral waste recycling

4 Final Remarks

The Chilean economy is an extractive economy based mainly on agriculture and mining activities intensive in water and energy consumption. Although both activities do not have long supply chains except the wine industry, some simple circular strategies can be deployed to reduce water and energy demand. These reductions contribute to the strength of the water-energy-food/wastes nexus and reduce the energy-water-food insecurities. In agriculture activities shorten the supply chain by working with local farmers can contribute to reducing the food losses. Reusing the food waste to recover energy and producing compost will allow the replacement of fertilizers to be used as amended. The understanding of water-energy tradeoff in irrigation systems could improve decision-making to adopt one of them and deploy it into the territory. For the mining activities, a pathway to recover minerals from tailing damns and start the mineral waste and scrap recycling could be a pivotal step to reduce their impact on the near environment. Finally, the implementation of different strategies in both sectors will change the flows into the nexus that need to be assessed according to the previous characterization of local and geographical conditions.

Questions

Q1 Will circular economy ensure sustainability into extractive economies?

Q2 Why is relevant understand the relationships between water-energy-food nexus?

Q3 Why is relevant understand the location and geographical conditions into nexus?

Q4 How can we circularize extractive operations?

Q5 List three most important lessons learnt based on Chile's experience and efforts towards circular economy.

Answers

A1

Circular economy emphasizes resource efficiency in the extractive economies as well as the capacity to use by products to reduce raw materials extraction.

A2

If we understand the nexus, we can understand the flows of the materials through local systems and their tradeoff between nexus's components.

A3

The geographical locations define the ecosystems and their ecosystems services as source of natural resources. These resources define the intensity of industry.

A4

We can circularize by increasing the efficiency of production, recycling material and energy flows as well as replacing raw materials by using by-products of downstream supply chain.

A5

Open.

5 Acknowledgments

Authors thank the project ANID/FONDAP/15130015: Water Research Center for Agriculture. Special thanks to the professor Lerwen Liu, Ph.D. for all her time provding me insights and advice on the conceptualization of Circular Economy applied to Water-Energy-Food Security.

References

1. World Economic Forum. (2020). The global risks report 2018 (15th ed.). Retrieved June 29, 2020, from https://reports.weforum.org/global-risks-report-2020/
2. IPCC. (2018). Summary for Policymakers. In Masson-Delmotte, V., P. Zhai, H.-O. Pörtner, D. Roberts, J. Skea, P.R. Shukla, A. Pirani, W. Moufouma-Okia, C. Péan, R. Pidcock, S. Connors, J.B.R. Matthews, Y. Chen, X. Zhou, M.I. Gomis, E. Lonnoy, T. Maycock, M. Tignor, & T. Waterfield (Eds.), *Global Warming of 1.5°C. An IPCC Special Report on the impacts of global warming of 1.5°C above pre-industrial levels and related global greenhouse gas emission pathways, in the context of strengthening the global response to the threat of climate change, sustainable development, and efforts to eradicate poverty* (32 p.). Switzerland, Geneva: World Meteorological Organization.

3. Steffen, et al. (2015). Planetary boundaries: Guiding human development on a changing planet. *Science, 347*(6223), 1259855. Retrieved February 13, 2015.
4. McKinsey Global Institute. (2020). Reduced dividends on natural capital?. Retrieved June 29, 2020 from https://mck.co/2Vs0RwQ.
5. UN Environment (2019) Global environment outlook—GEO-6: Summary for policymakers *Nairobi*https://doi.org/10.1017/9781108639217.
6. Sachs, J., Schmidt-Traub, G., Kroll, C., Lafortune, G., & Fuller, G. (2019). Sustainable development report 2019. New York: Bertelsmann Stiftung and Sustainable Development Solutions Network (SDSN). This work is licensed under a Creative Commons Attribution 4.0 International License.
7. Paris Agreement to the United Nations Framework Convention on Climate Change, T.I.A.S. No. 16-1104. Retrieved December 12, 2015
8. Climate Change Knowledge Portal. https://climateknowledgeportal.worldbank.org/country/chile/vulnerability
9. CEPAL. (2012). La economía del cambio climático en Chile. Naciones Unidas, Santiago, Chile, 134 p. https://www.cepal.org/es/publicaciones/35372-la-economia-cambio-climatico-chile
10. IRP. (2020). Resource efficiency and climate change: Material efficiency strategies for a low-carbon future. In E. Hertwich, R. Lifset, S. Pauliuk, N. Heeren, (Eds.), A report of the International Resource Panel. United Nations Environment Programme, Nairobi, Kenya.
11. Simpson, G. B., & Jewitt, G. P. W. (2019). The development of the water-energy-food nexus as a framework for achieving resource security: A review. *Frontiers of Environmental Science, 7*, 8. https://doi.org/10.3389/fenvs.2019.00008
12. GIZ. (2016). Water, energy & food nexus in a nutshell. http://www.water-energy-food.org/fileadmin/user_upload/files/2016/documents/nexus-secretariat/nexus-dialogues/Water-Energy-Food_Nexus-Dialogue-Programme_Phase1_2016-18.pdf.
13. Liu, J., Yang, H., Cudennec, C., Gain, A. K., Hoff, H., Lawford, R., et al. (2017). Challenges in operationalizing the water–energy–food nexus. *Hydrological Sciences Journal, 62*(11), 1714–1720. https://doi.org/10.1080/02626667.2017.1353695
14. The United Nations World Water Development Report (2015). https://www.unesco.org/new/fileadmin/MULTIMEDIA/HQ/SC/images/WWDR2015Facts_Figures_ENG_web.pdf.
15. FAO. (2011a). Energy smart food for people and climate. Rome: Food and Agriculture Organization of the United Nations. https://www.fao.org/docrep/014/i2454e/i2454e00.pdf.
16. Annual Energy Outlook presents an assessment by the U.S. Energy Information Administration of the outlook for energy markets through 2050. https://www.eia.gov/outlooks/aeo/.
17. Europe 2020 headline indicators. https://ec.europa.eu/eurostat/statistics-explained/index.php/Europe_2020_headline_indicators.
18. UNDP Annual Report. https://annualreport.undp.org.
19. Mancini, L., Vidal Legaz, B., Vizzarri, M., Wittmer, D., Grassi, G., & Pennington, D. (2019). Mapping the role of raw materials in sustainable development goals. A preliminary analysis of links, monitoring indicators, and related policy initiatives. EUR 29595 EN, Publications Office of the European Union, Luxembourg, 2019 ISBN 978-92-76-08385-6, https://doi.org/10.2760/026725, JRC112892.
20. Rutllant, J., Fuenzalida, H., & Aceituno, P. (2003). Climate dynamics along the arid northern coast of Chile: The 1997–1998 Dinámica del Clima de la Region de Antofagasta (DICLIMA) experiment. *Journal of Geophysical Research, 108*, 4538.
21. Valdés-Pineda, R., Pizarro, R., García-Chevesich, P., Valdés, J.B., Olivares, C., Vera, M., Balocchi, F., Pérez, F., Vallejos, C., Fuentes, R., Abarza, A., & Helwig, B. (2014). Water governance in Chile: Availability, management and climate change. *Journal of Hydrology, 519*(PC), 2538–2567. https://doi.org/10.1016/j.jhydrol.2014.04.016.
22. Aitken, D., Rivera, D., Godoy-Faúndez, A., & Holzapfel, E. (2016). Water scarcity and the impact of the mining and agricultural sectors in Chile. *Sustainability, 8*, 128.
23. Chile: International Energy Agency. https://www.iea.org/countries/chile
24. Meza, F. J., Vicuna, S., Gironás, J., Poblete, D., Suárez, F., & Oertel, M. (2015). Water–food–energy nexus in Chile: The challenges due to global change in different regional contexts. *Water International*. https://doi.org/10.1080/02508060.2015.1087797

25. MacArthur Foundation Report, Growth Within: A Circular Economy Vision for a Competitive Europe. https://www.ellenmacarthurfoundation.org/publications/growth-within-a-circular-economy-vision-for-a-competitive-europe.
26. Del Borghi, A., Moreschi, L., & Gallo, M. (2020). Circular economy approach to reduce water–energy–food nexus. *Environmental Monitoring Assessment: Water-Energy-Food Nexus, 13,* 23–28.
27. Employment in agriculture (% of total employment) (modeled ILO estimate). The World Bank. https://data.worldbank.org/indicator/SL.AGR.EMPL.ZS.
28. Ranking the World's Most Water-Stressed Countries in 2040 https://www.wri.org/blog/2015/08/ranking-worlds-most-water-stressed-countries-2040.
29. Jorge, J., López, M. A., Martín, S., Álvaro, S., & Luis, & Melo, Ovidio. . (2009). Administration and management of irrigation water in 24 user organizations in Chile. *Chilean Journal of Agricultural Research, 69*(2), 224–234. https://doi.org/10.4067/S0718-58392009000200012
30. González Martínez, T., Bräutigam, K., & Seifert, H. (2012). The potential of a sustainable municipal waste management system for Santiago de Chile, including energy production from waste. *Energy, Sustainability and Society, 2,* 24. https://doi.org/10.1186/2192-0567-2-24
31. Oyarzún, J., & y Oyarzún, R. . (2011). *Minería sostenible: Principios y prácticas.* España: Ediciones gemm.
32. Winds of change. https://www.economist.com/blogs/americasview/2014/09/energy-chile.
33. Lèbre, E., Corder, G., & Golev, A. (2017). The role of the mining industry in a circular economy: A framework for resource management at the mine site level. *Journal of Industrial Ecology, 21*(3), 662–672. https://doi.org/10.1111/jiec.12596.
34. Spooren, J., Binnemans, K., Björkmalm, J., Breemersch, K., Dams, Y., Folens, K., et al. (2020). Near-zero-waste processing of low-grade, complex primary ores and secondary raw materials in Europe: Technology development trends. *Resources, Conservation and Recycling, 160,* 104919.

Alex Godoy-Faúndez I am Director of Sustainability Research Center & Strategic Resource Management and Academic Director of Master's program in Sustainability Management at School of Engineering at Universidad del Desarrollo & Research Associate to the Water Research Center for Agriculture and Mining awarded in Fifth National Competition for Research Centers in Priority Areas of National Commission for Scientific & Technological Research (CONICYT in Spanish). Further, I am Head of Waste to Energy Research & Technology—Chile at WTERT Council and Research Associate at Earth Engineering Center at Columbia University.

My research interest is the Sustainability Science to understand the relationships Engineering, Society, and Sustainability. My main goal is to develop frameworks of analysis for understanding the causes and effects of current events to contribute to sustainable development. I apply system modeling to understand the local systems—characterized by its natural resources, climate, and infrastructure needs engineering, societal and political frameworks—for the service of decision-making techniques to the public policy level, linking science, technology, society, and education. Today, my research projects are focused on the Water-Energy-Food Security nexus.

In 2006, I was recognized as one of the 100 young leaders by the Center of Leadership Universidad Adolfo Ibañez. Awarded in Academic Excellence and Best Paper Award in The International Conference on Environmental Science and Technology 2006-2014, sponsored by American Academy of Science; Young Researcher Award at the 4th European Bioremediation Conference 2008, Greece and one of the winners in the "Innovation Challenge Week" and Prize "Innovation & Entrepreneurship 2008". In 2011, I was a Visiting Scholar and Scholar in Residence 2012 at the University of Wisconsin-Green Bay. Coauthor, First Environmental Chilean Survey about "Environment and Climate Change: Attitudes and Perceptions" (2009–2011).

Throughout my career, I have had a strong involvement as a member of advisory boards into governmental institutions such as CONICYT, National Commission in Innovation and Development (CNID), Ministry of Finance, Ministry of Environment as a member of the Advisory Council in three administrations. I have been a member of the presidential commissions for innovation in water resources (CNID) and long-term planning of the electricity system (Ministry of Energy). In 2015, I was a member of the Chilean delegation for COP21, advising to Ministry of Finance for Paris Agreement. In 2019, I was a member of the scientific committee on Climate Change in an advisory body of the Ministry of Science, Technology, Knowledge, and Innovation organizing the COP25.

Also, I have been a referee in peer-reviewed journals and environmental projects in institutions such as CONICYT and CORFO (National Developing Bank). My career has provided me with the opportunity to combine my multidisciplinary experience on environmental, chemical, and social-environmental management and innovation. As a result, I have become very interested in how to use technological management applied to environmental management and green technologies in connection with public policies, environmental economics, society, and education. My contributions are summarized in more than 90 presentations to national and international conferences, on 40 publications indexed as author or co-author, one book, and six chapters, being also associate editor and reviewer in various journals indexed.

At the international level, I am very committed to Science Diplomacy. I am an alumnus of the Global Young Academy. The GYA is an international organization aiming to empower and mobilize young scientists to address issues of importance to early-career scientists. My focus was SDGs and WEF nexus. Further, I am a member of Nexus KAN of Future Earth organization, an international research program for global sustainability working on Water-Energy-Food Security Nexus. In Chile, he has contributed with the Proposal for Science Diplomacy about STIC Chile-United States as a member of the Chile-United States Council for Science, Technology, and Innovation (STIC) in the US Embassy.

Between 2017–2019, He was member of the High-Level Panel of Experts Steering Committee on Food Security and Nutrition (HLPE), Committee on World Food Security (CFS),

Food and Agriculture Organization (FAO), United Nations (UN), coordinating two global reports.

Since 2019, I am Review Editor of Chap. 13 titled "National and sub-national policies and institutions) for the Working Group III to the IPCC Sixth Assessment Report (AR6) and lead author of Chapter "Moving from linear to a circular economy: What this means for business? For Global Environment Outlook (GEO) for Business at Science Division, United Nations Environment Programme. I have served as an expert reviewer of the 2019 Global Sustainable Development Report (GSDR) prepared by the independent group of scientists appointed by the United Nations Secretary-Genebra and for "10 New Insights in Climate Science 2019" report of Future Earth & The Earth League, Stockholm, 2019 for COP25. Also, I served as Panel Member of Water Quality Task Force (TF) for the International Water Resources Association (IWRA) working on the Water Quality Project: "Developing a Global Compendium on Water Quality Guidelines".

Finally, in 2019, I was awarded as Doctor Honoris Causa—Doctor of Laws at the University of Wisconsin—Green Bay for his contribution to the creation of public policies and innovations in the private sector on sustainability issues. Today, I am a current mentor for the HBA Sustainability Certificate for Centre for Builiding Sustainable Value at Ivey Business School, Canada.

Recycling of Waste Electric and Electronic Products in China

Keli Yu, Heran Zhang, and Yunong Liu

Abstract This chapter mainly focuses on the development of waste electrical appliances and electronic products/waste electric and electronic products (WEEP) management in China and its key effects to circular economy. China has experienced 5 stages since the 1990s when the collection and disposal of WEEP started. By trying different ways to deal with WEEP, China has found the possible route to manage it. The article discusses the progress, achievements, and the status of WEEP management in China (including the categories, dismantling capacity & technologies, operational mode, reuse, environmental, and social influences). We provide case studies on Aihuishou (China's largest electronic products collection platform) and Guiyu (transformation of a well-known e-waste recycling town). China now is the country with the world's largest WEEP dismantling and disposal capacity. The recycling of WEEP is also a microcosm of China's circular economy.

Keywords Waste electric and electronic products (WEEP) · Policy · Management · Collection · Dismantling technique · Reuse · Effects

Learning Objectives

- National policies on WEEP.
- China has experienced 5 stages on WEEP management.
- The annual dismantling capacity of the qualified national WEEP dismantling enterprises is about 152 million units.

K. Yu (✉) · H. Zhang · Y. Liu
China National Resources Recycling Association, Room 8321, No. 13 Yue Tan Bei Xiao Jie, Xicheng District, Beijing 100037, China
e-mail: ykl@crra.com.cn

H. Zhang
e-mail: zhr@crra.com.cn

Y. Liu
e-mail: lyn@crra.com.cn

© Springer Nature Singapore Pte Ltd. 2021
L. Liu and S. Ramakrishna (eds.), *An Introduction to Circular Economy*,
https://doi.org/10.1007/978-981-15-8510-4_13

- The total dismantling quantity of WEEP of 109 qualified enterprises from 2012 to 2018 was 441,735,909 units.
- China's standard collection rate of WEEP has reached the international leading level.
- New collection modes of WEEP are emerging.

1 Introduction

China is a major producer, exporter, and consumer of electrical and electronic products. With the rapid development of the economy, the pollution of the environment problem has become increasingly severe, therefore, the Chinese government places great importance to the recycling and disposal of WEEP.

Definition

Waste Electrical and Electronic Equipment (WEEE): The Regulations on Administration of Collection and Disposal of Waste Electrical Appliances and Electronic Products provides that "the Regulation shall apply to the recovery, disposal, and other relevant activities of waste electrical and electronic products listed in the Catalogue". This means that a wide range of electrical and electronic products will have to be recycled and disposed of in accordance with the Regulation. These products listed in the Catalogue are TV sets, refrigerators, washing machines, air conditioners, and microcomputers.

"Home appliances old for new" Policy: Refers to the Chinese government has used government funds to encourage consumers to trade in old ones.

Urban Mining: The recyclable iron and steel, non-ferrous metals, produced and stored in the process of industrialization and urbanization in waste electrical and mechanical equipment, wires and cables, communication tools, automobiles, home appliances, electronic products, metal and plastic packaging and waste. Resources of rare and precious metals, plastics, rubber, etc., are used in an amount equivalent to that of primary mineral resources.

2 Development Process of Waste Electric and Electronic Products in China

(1) Primary Stage

Since the 1990s, informal recycling of WEEP started in China. It was collected by small vendors and shipped to domestic electronic waste distribution centers, mostly to Guangdong and Zhejiang Province. At that time, illegal importing of waste electrical and electronic products was rampant, which lead to serious negative social and environmental impact [1–3].

(2) The emerging stage of formal dismantling enterprise

After 2005, formal recycling companies began to appear in China, however, in general, the development of formal recycling companies was very slow due to the lack of collection channel, and some big companies also tried to build physical collection channels spontaneously. Companies like TES-AMM started their operation and provide recycling service for some electrical and electronic producers.

(3) "Home appliances old for new" policy

From June 2009 to December 2011, the Chinese government implemented a policy named "Home appliances old for new" nationwide, which not only promoted the consumption of home appliances, but also incentivized the formal recycling of WEEE. The successful implementation of the policy has also improved the energy efficiency of household appliances, reduced environmental pollution, and recycled resources such as steel, non-ferrous metals, plastics, rubber, and other resources available for recycling, which is very important to developing a circular economy. During the period of "home appliances old for new" (June 2009–November 2011), more than 40 formal enterprises were officially established, and the amount of recycled waste home appliances added up to 83.733 million.

At present, Guangxi Province has implemented the "Home appliances old for new" policy again since last year. At the same time, some e-commerce platform, such as Tmall, JD.com, Pinduoduo, also have launched many "Old for new" promotion to stimulate consumers to trade in their used electrical and electronic products and buy new ones.

(4) Construction of urban mining industrial parks

In 2010, in order to implement the "Circular Economy Promotion Law", and promote the development of recycling industry, China's National Development and Reform Commission and the Ministry of Finance initiated the construction of "urban mining industrial parks". Seven regional recycling parks were selected for the construction of "urban mining" as the first batch for demonstration.

In these parks, waste electrical appliance recycling enterprises carry out the treatment and disposal of waste household appliances, and have achieved good treatment results. Up to now, a total of 49 "urban mining" industrial parks have been established nationwide.

(5) "Regulations on waste electrical and electronic equipment Recycling and Treatment" and the implementation of financial subsidy scheme

From January 1, 2011, the "Regulations on Administration of Collection and Disposal of Waste Electrical Appliances and Electronic Products" came into effect. After the promulgation of the regulation, China gradually established a waste electric and electronic products recycling system, including planning, a funding scheme (funds collected from electrical and electronic product manufacturers including both domestic and foreign, and distributed to the dismantling plants) for the disposal of WEEP and an audit management. Since the implementation of the Regulation, a total of 16 complementary regulations, regulatory documents and standards have been issued to govern the various stages of collection, treatment, and disposal of waste electrical and electronic equipment.

Since 2012, China has implemented a financial subsidy scheme for the disposal of WEEP. Funds are collected by the government from the producers/importers of electric and electronic products. WEEP are collected by producer or collectors and are dismantled by qualified dismantling plants (5 batches of 109 waste electrical and electronic products dismantling and processing enterprises were selected and listed as the qualified WEEP recyclers that can receive financial subsidy). Dismantling plants have to report the dismantling amount of WEEP per quarter, and the provincial department of ecological environment audits the report to confirm the data of dismantling outcome. After that, the Ministry of Ecological and Environment (MEE) has to further review the data and submitted the result to the Ministry of Finance (MoF). Finally, the Ministry of Finance will approve and appropriate funds to the audited dismantling plants (Fig. 1).

The dismantling volume of WEEP of 109 qualified enterprises from 2012 to 2018 is shown in Table 1. The implementation of the financial subsidy scheme has effectively standardized the disposal of WEEP and reduced their potential environmental risks (Fig. 2).

Since the implementation of the financial subsidy policy for WEEE, from 2012 to 2018, the dismantling products generated by the formal dismantling facilities have reached 7,552,600 tons, including iron and steel, recycled copper, recycled aluminum, and recycled plastics.

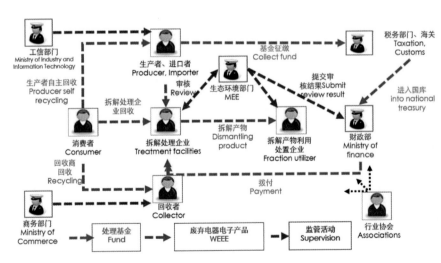

Fig. 1 Implementation of the financial subsidy scheme for WEEP disposal funds. MOC stands for Ministry of Commerce

Table 1 The first batch of urban mining demonstration base

NO.	Name	Province/municipality
1	Tianjin Ziya Circular Economy Industrial Park	Tianjin
2	Ningbo Jintian Industrial Park	Jiangsu province
3	Hunan Miluo Circular Economy Industrial Park	Hunan province
4	Guangdong Qingyuan Huaqing Circular Economy Park	Guandong province
5	Anhui Shoutianying Circular Economy Industrial Park	Anhui province
6	Qingdao Xintiandi Vein Industry Park	Shandong province
7	Sichuan Southwestern Renewable Resources Industrial Park	Sichuan province

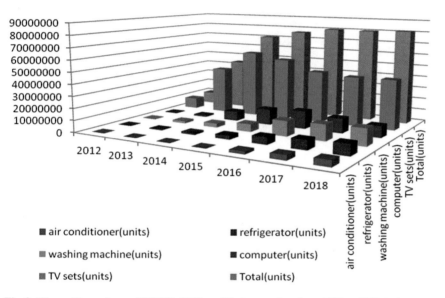

Fig. 2 Dismantling volume of WEEE of 109 qualified enterprises from 2012 to 2018 (units)

Estimated by Xianlai Zeng [4], in 2018, the generation of WEEE (estimation only includes five kinds of WEEP including TV sets, refrigerators, washing machines, air conditioners, and computers) was about 9.981 million tons in theory, while the weight of WEEE treated by qualified facilities in China was nearly 2.006 million tons, the formal recycling rate of waste electrical appliances can be calculated as 20.1%.

The formal recycling rate = the weight of WEEE treated by qualified facilities/the weight of WEEE generation in theory * 100%.

The implementation of the "Regulations on the Management of the Recycling and Disposal of Waste Electric and Electronic Products" and supporting policies have formulated detailed technical and management requirements for dismantling and disposing of waste electrical appliances, guiding them to carry out production

Fig. 3 WEEP collection in community (Picture is from the following website. https://new.qq.com/ omn/20180619/20180619A1THLK.html)

management in accordance with the relevant requirements for financial subsidies, and environmentally sound operations.

Tip

Waste electrical and electronic products management in China has experienced 5 stages.

The total dismantling volume of WEEE in 109 qualified enterprises from 2012 to 2018 reached 441,735,909 units (Fig. 3).

Case Study

The collection system of WEEE in Beijing

The construction of collection system is considered to be very important in promoting the recycling of WEEE in China. In 2017, Beijing has launched a pilot project of new WEEE collection system and several designated collection channels were selected as demonstration cases. In the first batch, there are five pilot types including.

1. sanitation enterprises rely on domestic waste separation and collection network recycling,
2. electrical and electronic product manufacturing enterprises rely on sales network recycling,
3. Electrical and electronic product sales enterprises "old for new" recycling,
4. recycling companies to expand the scope of services, and

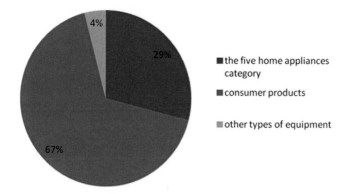

Fig. 4 Ratio of different kinds of WEEP in 1.09 million units

5. Internet companies "Internet + collection".

13 companies were selected as demonstration cases by the government. According to preliminary statistics, in 2018, a total of 1.09 million units of various WEEP were collected by 13 pilot units, exceeding 16% of the planned volume.

Figure 4 shows the breakdown of different kinds of WEEP in the 1.09 million units. The five home appliances category include TV, air conditioner, washing machine, refrigerators, computers), consumer products include mobile phones, laptops, tablets, cameras, etc. and other types of equipment include printers, telephones, network equipment, etc.

In April 2018, these selected companies initiated the "Beijing Waste Electrical and Electronic Products Recycling Industry Alliance" to share information and resources, exchange experiences, and coordinate to solve common problems. With the active participation of industrial associations and pilot companies, the research and development work aimed at regulating WEEE recycling and forming relevant industrial standards has also started as planned.

Key Points

Beijing City, the first pilot city in China to launch the construction of formal WEEE collection system, conducted demonstration activities for the collection of waste electrical and electronic products and achieved good results.

3 Status of Dismantling and Treatment of WEEE in China

(1) Dismantling capacity

At present, the annual dismantling capacity of the major home appliances (TV sets, refrigerator, washing machine, air conditioner, computer) of the 109 qualified facilities is about 152 million units per year (Fig. 5).

The dismantling capacity by region is shown in Fig. 6, unit (million units).

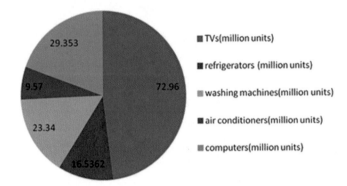

Fig. 5 The annual dismantling capacity of the 109 qualified facilities

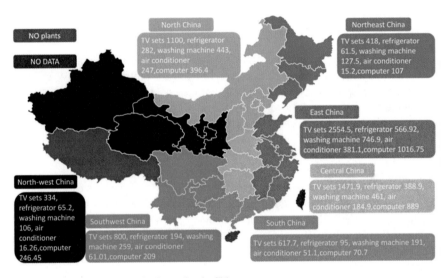

Fig. 6 The dismantling capacity by region in China

According to the annual dismantling volume of WEEP in the qualified facilities published by the Ministry of Ecology and Environment (MEE), relevant information on industrial development can be shown in Fig. 7.

In the past two years, WEEP dismantling enterprises show the phenomenon of polarization. The dismantling volume of large dismantling companies (mostly listed companies) with strong financial support and dismantling capabilities has continued to increase, while the dismantling volume of some small and medium-sized dismantling enterprises has dropped significantly.

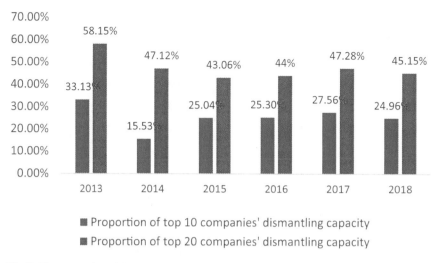

■ Proportion of top 10 companies' dismantling capacity
■ Proportion of top 20 companies' dismantling capacity

Fig. 7 The proportion of the top 10/20 companies in total dismantling in 2013–2018

(2) Operational model

The basic model of China's WEEE dismantling is shown as follows (Fig. 8):

(3) The categories and technologies of dismantling

At present, the environmental-sound treatment and disposal of waste electrical and electronic products in China mainly aims at major home appliances. The dismantling process is shown as follows (Fig. 9):

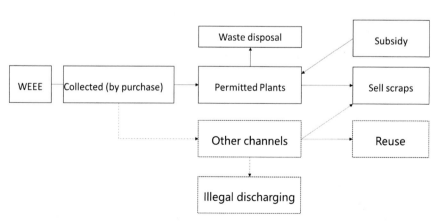

Fig. 8 Basic model of China's WEEP dismantling

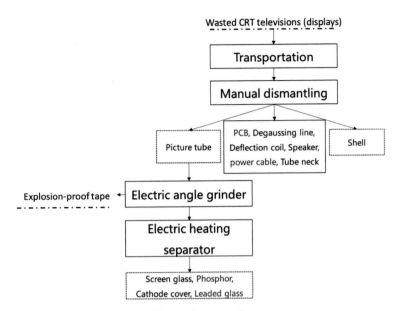

(a) Wasted CRT televisions (displays) dismantling

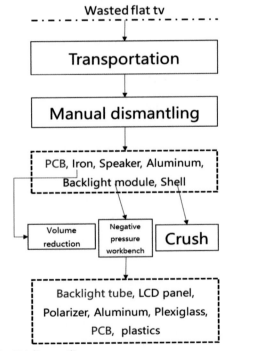

(b) Wasted flat TV dismantling

Fig. 9 Dismantling process of wasted CRT televisions (displays), wasted flat TV, wasted refrigerators, wasted room air conditioners and wasted washing machines and dismantled products

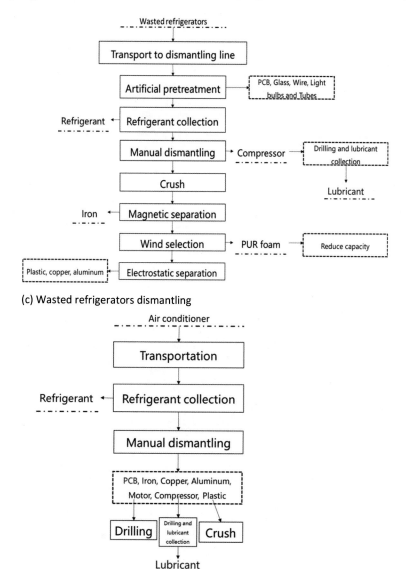

(c) Wasted refrigerators dismantling

(d) Wasted room air conditioners dismantling

Fig. 9 (continued)

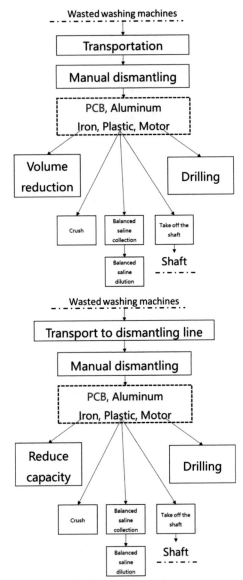

(e) Wasted washing machines dismantling

Fig. 9 (continued)

Waste printed circuit boards (WPCB) are one of the key products in the dismantling process. The disposal of waste circuit boards is an important part of environmental protection. At present, the 109 qualified dismantling plants transport most of the WPCB and components to companies with qualifications for scrap circuit board disposal. The waste circuit board disposal technologies used by these qualified companies are mainly physical, thermal, and hydronic methods. In addition, some copper smelting companies are also developing WPCB disposal business.

In addition to the five categories of home appliances, according to the "Waste Electrical and Electronic Products Disposal Catalog (2014 Edition)", there are nine new types of WEEP (printers, copiers, fax machines, monitors, range hoods, Electric water heaters, gas water heaters, mobile communication handsets, telephone) are also included in the treatment list, but there is no relevant policy to guild the disposal of nine new types of WEEP at present. In addition, for small household electrical appliances other than the 14 categories, some of WEEE dismantling facilities use the overall crushing and sorting process.

(4) Reuse of the used electric and electronic equipment (Table 2).

Regarding the reuse of electrical and electronic products, China has developed rapidly, in recent years, an active second-hand electronic product market. Some second-hand product trading Internet platforms have been in operation, such as Paipai, Xianyu, Aihuishou, etc. It mainly focuses on second-hand electronic product, especially mobile phone, due to the high update rate of mobile phone, the trading volume on the platform is huge. On the other side, because of the market for second-hand reuse of large appliances is relatively inactive, large household appliances (such as refrigerators, washing machines, etc.) are in low-trading volumes. For example, second-hand refrigerators may have obvious noise, vibration, and poor cooling effects, high risk in safe refrigeration of food; second-hand washing machines may have rust in the tub, aging of the transmission belt, and health risk. In addition, the power consumption of home appliances that have exceeded their working life may increase.

According to the survey conducted by China National Resources Recycling Association (CRRA), the reuse of used waste mobile phone screens is about 50%, and keyboards, shells and other components, is relatively low, about 10%.

Although the second-hand electronic product market has grown at a certain scale with sorting and classification and other preliminary operating specification and processing formulating, regulations and industry standards for refurbishment and reuse of components are yet to be developed.

(5) Environmental and social impact

The use of secondary materials generated from the dismantling of waste electrical and electronic products has obvious energy saving and emission reduction effects. According to the estimates of CRRA, recycling 1 ton of WEEP can save 1.97 tons of standard coal, and reduce wastewater, solid waste, carbon dioxide, and sulfur dioxide by 24.23, 13.61, 4.73, and 0.046 tons, respectively, according to the statistics from

Table 2 List of main policies and regulations related to WEEE recycling in China

No.	Name	File document number
1	Catalogue of Waste Electrical and Electronic Products for Disposal (2014 Edition)	No. 5, 2015
2	Measures for the Administration of Licensing of Waste Electrical and Electronic Products	No. 13
3	Administrative Measures for the Restricted Use of Hazardous Substances in Electrical and Electronic Products	No. 32, 2016
4	Measures for the Administration of Collection and Use of Waste Electrical and Electronic Products Processing Fund	No. 34, 2012
5	Guidelines for Qualification Examination and Licensing of Waste Electrical and Electronic Product Disposal Enterprises	No. 90, 2010
6	Guidelines for the preparation of development plans for the disposal of waste electrical and electronic products	No. 82, 2010
7	Guidelines for establishing a data information management system and reporting information for waste electrical and electronic product processing enterprises	No. 84 of 2010
8	Guidelines for Dismantling and Disposing of Waste Electrical and Electronic Products and Production Management Guide (2015 Edition)	No. 82, 2014
9	Guidelines for Review of Disassembly and Disposal of Waste Electrical and Electronic Products (2015 Edition)	No. 33, 2015
10	Waste electrical and electronic product processing fund subsidy standards	No. 91, 2015
11	Notice on further clarifying the scope of products collected by the waste electrical and electronic products processing fund	No. 80, 2012
12	Notice on improving policies for disposal of waste electrical and electronic products	No. 110, 2013
13	Notice on adjusting matters related to the declaration and review of waste electrical and electronic products	No. 117, 2016
14	Technical policies for the prevention and control of pollution from waste household appliances and electronic products	No. 115, 2006
15	Technical specifications for pollution control of waste electrical and electronic products	No. 1, 2010
16	Guidelines for Dismantling and Disposing of Waste Electrical and Electronic Products and Production Management Guide (2019 Edition)	2019

MEE, as of 2018, China's WEEP recycling facilities have recycled 8.142 million tons of WEEP. Based on this estimate, as of 2018, China's WEEP facilities have accumulated energy saving and emission reductions as follows (Table 3).

According to this calculation,[1] as of 2018, the emission reduction benefit from the formal dismantling and disposal of WEEP was 6.549 billion yuan.

---

[1] http://www.dlyj.ac.cn/article/2014/1000-0585/14709.

Table 3 Energy saving and emission reduction effect of WEEP (million tons)

Type	Save standard coal	Reduce wastewater	Reduce solid waste	Reduce CO_2	Reduce SO_2
Effects of energy saving and emission reduction of WEEP	16.0397	197.2807	110.8126	38.5117	0.3745

The recycling of waste electrical and electronic products has significant social benefits. According to the survey conducted by CRRA, in 2018, the average number of employees in China's waste electrical and electronic product processing enterprises was 191, it is estimated that 18,763 jobs were directly created by WEEP facilities nationwide. In addition, in the recycling process, there are also a large number of scattered WEEP collectors who are engaged in related work.

By the end of 2018, a total of 5 WEEE recycling companies were named as China's national circular economy education demonstration bases. In addition, many enterprises have been rated as local circular economy education demonstration enterprises by different provinces and cities, which have strongly promoted the development of China's circular economy publicity and education.

Case Study

Case 1
Aihuishou

Aihuishou is China's largest electronic products collection platform, the first "Internet + environment" type company, which is mainly engaged in professional and safe collection of electronic products, sale of second-hand products, mobile phone rental. It focuses on recycling of mobile phones, laptops, digital cameras, and other 3C (Computers/Communications/Consumers) products, with the second-hand mobile phone trade accounting for 80–90% of its business. By the end of 2017, Aihuishou had opened more than 200 direct sales stores in China's major cities and more than 30,000 cooperative stores in other cities with over 1500 employees. In 2016, Aihuishou processed 5.2 million mobile phones with a turnover of about RMB1.5 billion (Fig. 10).

It provides a three-step collection service, which is convenient and fast:

1. The customer fills in the equipment information according to the equipment list template. Agreed on-site survey time.
2. On-site survey and verification.
3. The two parties sign the contract; then collect the equipment, and pay the customer.

Key Points

Internet + collection have become a new form of WEEP collection.

Fig. 10 offline collection of Aihuishou (Picture is from the following website. http://dy.163.com/v2/article/detail/EGP575VB0514R9KM.html)

Case Study

Case 2

Guiyu Town

Since the 1970s, Guiyu has gradually developed into a center for recycling and dismantling waste household appliances and waste plastics. The town is mainly engaged in collection, dismantling, processing, and utilization of waste electronic appliances and plastics. More than 5,500 farmers were engaged in the WEEE recycling business across 21 villages. There were more than 300 private enterprises/home workshops with more than 60,000 employees in 2005 [5]. The annual dismantling and processing of waste electronic appliances and plastics added up to 1.55 million tons. However, due to outdated recycling technology and the lack of proper management, the scattered WEEE recycling in Guiyu caused serious pollution to the environment, and Guiyu became one of the most notorious centers for e-waste illegal importing and informal recycling in the world.

On account of the serious environmental problems in Guiyu, under the guidance of relevant national ministries and commissions, as well as provincial, municipal, and district governments, Guiyu launched a comprehensive environmental pollution control project which was proposed to complete at the end of December 2015. At present, the overall restoration work and the construction of the circular economy

Fig. 11 Water quality in Guiyu a few years ago (Picture is from the following website. http://www1.xcar.com.cn/bbs/viewthread.php?tid=21671949&extra=page=1&page=2 (2004 year))

industrial park are progressing smoothly, and heavy pollution and illegal dismantling in Guiyu has been significantly curbed (Figs. 11 and 12).

In order to promote Guiyu's comprehensive improvement of environmental pollution and the transformation and upgrading of the electronic waste dismantling and processing industry, the Guiyu Circular Economy Industrial Park started its construction in 2010. The planned construction period of the park is 10 years divided into two phases (2010–2015, and 2016–2020). The first phase of the park was completed in 2015.

A total of 411.84 acre of land is acquired in the park, and the current use area is 156.5 acre, which are 82.37 acre of actual land in Huamei District and 74.13 acre of Nanyang District. The park is divided into four areas, including the trading area, dismantling area, further treatment area, and environmental protection infrastructure. The trading area has two parts: a centralized trading center and a product trading market. The dismantling area has a general dismantling plant (a total of four phases), TCL Deqing Environmental Protection Company. Further treatment area is divided into three types: smelting project, hydronic project, and physical method treatment project. The environmental protection facilities include: industrial sewage treatment plant, domestic sewage treatment plant, hazardous waste transfer station, garbage compression transfer Stations, domestic waste landfills, and flue gas treatment facilities (Fig. 13).

Prior to 2012, Guiyu's annual electronic waste treatment capacity could reach the level of one million tons, and the current treatment capacity is around 400,000–500,000 tons per year, and the overall scale has dropped significantly. According to the statistics of the industrial park, the total volume of electronic waste transactions

Fig. 12 Guiyu town water quality status (2015 year)

in the second, third, and fourth quarters of the park in 2015 was 110,000 tons and the first quarter of 2016 was about 110,000 tons (excluding the TCL Deqing processing volume). At present, about 90% of the circuit boards processed in the park were domestic origin. Based on the information from all parties in the survey, it is estimated that the current PCB processing capacity in the park is about 45,000t/year, of which foreign sources should be less than 5000t/year. Since a large number of home workshops have entered the park, the informal recycling in Guiyu has changed

Fig. 13 Park flue gas treatment equipment (The first picture is from the following website https://www.sohu.com/a/164860752_99908715. The second picture is from the following website. http://sn.people.com.cn/big5/n2/2016/0113/c340887-27533674.html)

significantly. Most of the home workshops have moved into the park for operation. Guiyu e-waste dismantling has improved significantly in terms of pollution control.

Guiyu has been engaged in the dismantling and processing of electronic waste for decades. It has rich experience in fine manual disassembly, high recycling rate, and resource efficiency. Valuable materials are accurately and meticulously identified. Guiyu has created positive impact in efficient recycling of valuable materials and improved resource utilization (Fig. 14).

Key Points

In Guiyu Town, a famous e-waste recycling center, great changes have taken place in pollution control, and the environment has been greatly improved. A centralized industrial park is already operating meanwhile, illegal and informal recycling is diminishing.

4 Conclusion

China advocates the development of circular economy. In terms of recycling WEEP, various attempts have been made since the 1990s. China now is the country with the world's largest WEEP dismantling and disposal capacity. At the same time, the management level and treatment technology are also improving. Although the

Fig. 14 Components of waste electrical and electronic products (The picture is from the following website. http://news.bjsyqw.com/2017/0208/96507.shtml)

qualified facilities in China mainly recycle big home appliances, it provides valuable experience for the subsequent treatment and disposal of other kinds of WEEP.

Questions

1. What stages did China go through in recycling of WEEP, and the characteristics of the Chinese model?
2. How many enterprises in China formally dispose waste electrical and electronic products, what is the dismantling capacity?
3. What's the new mode of collecting waste electrical and electronic products in China?
4. What is the status of Guiyu Town, a world-renowned e-waste disposal area?

Answers

1. China has gone through 5 stages in management of waste electrical and electronic products since 1990s, they are primary stage, the budding stage of formal dismantling enterprise, home appliances "old for new", Urban Mining, the implementation of "Regulations on waste electrical and electronic equipment Recycling and Treatment" and financial subsidy scheme phase.
   China's success provides a new management mode of WEEP, it can be called the new extended producer responsibility system, which may make a lot of sense for developing countries.

2. In China, a total of 109 dismantling and processing electronics companies have been included in the fund subsidy. The annual dismantling of "four machines and one brain" is 152 million units.
3. Internet + collection has become a new way of collecting WEEP.
4. Guiyu Town has begun comprehensive environmental control, the dismantling of electronic waste and the transformation and upgrading of processing industry. Most demolished households have moved into the park for operations as required. The dismantling of e-waste in Guiyu has significantly improved pollution control, and the image of water, air, and roads has also improved markedly.

Patient information and guidelines

Suggested Reading
Keli Yu, Heran Zhang, Jinfeng Qiu, Yunong Liu. Waste Electrical and Electronic Products Recycling Industry Development Report (2019).

References

1. Bigum, M., Brogaard, L., & Christensen, T. H. (2012) Metal recovery from high-grade WEEE: A life cycle assessment. *Journal of Hazardous Materials, 207–208*(2012), 8–14.
2. Cui, Jirang, & Forssberg, Eric. (2003). Mechanical recycling of waste electric and electronic equipment: A review. *Journal of Hazardous Materials, 99*(3), 243–263.
3. Luo, Y., Luo, X. J., Lin, Z., et al. (2009). Polybrominated diphenyl ethers in road and farmland soils from an e-waste recycling region in Southern China: Concentrations, source profiles, and potential dispersion and deposition. *Science of the Total Environment, 407*(3), 1105–1113.
4. Zeng, X., Gong, R., Chen, W. Q., & Li, J. (2016) Uncovering the recycling potential of 'new' waste electrical and electronic products in China. *Environmental Science & Technology, 50*(3), 1347–1358.
5. https://new.qq.com/omn/20180619/20180619A1THLK.html.

Keli Yu male, 39, He started to study e-waste in 2007 and has accumulated more than 10 years of research and industrial experience in this field. He worked at the Basel Convention Regional Centre for Asia and the Pacific for 5 years. From 2013 to 2015, he worked for the Asia-Pacific Regional Office of the United Nations Environment Programme. From 2015 to present, he has worked for China National Resource Recycling Association (CRRA). He mainly engaged in research on WEEE management, industry situation survey. As the project leader, he completed a number of research reports. Through the influence of CRRA in the WEEE industry, he led to finish China waste electrical and electronic products recycling industry development report (2019). He graduated from Tsinghua University in 2007, Master degree.

Heran Zhang male, 32, worked for China National Resource Recycling Association (CRRA) since 2017, mainly carried on waste electric and electronic products (WEEP) industry research. During the period, he wrote several reports and articles on China's WEEP, and organized a few meetings about the WEEP treatment and disposal. He graduated from Peking University in 2015, Master degree.

Yunong Liu female, 28 years old. In 2017, she worked for China National Resource Recycling Association (CRRA), and was on loan to the Solid Waste and Chemical Management Center MEE for two years, mainly engaged in waste electrical and electronic waste (WEEE) management related research, review of Chinese WEEE fund, supervision of WEEE processing enterprises, organized and held a few WEEE dismantling training sessions. Master degree in environmental engineering.

Transforming e-Waste to Eco Art by Upcycling

Vishwanath Davangere Mallabadi

Abstract The global volume of electronic waste is expected to reach 52.2 million tones or 6.8 kg per person by 2021. A recent UN report—A New Circular Vision for Electronics—highlights that the world produces as much as 50 million tons of electronic and electrical waste (e-waste) a year. However, only 20% of this is formally recycled, remainder 80% either ending up in landfills or being informally recycled. It is unreasonable to find an ethical recycler and it has been a major challenge to pledge to eradicate e-waste from planet. Apparently, upcycling e-waste could be exploited as compelling solution. The article discusses about upcycling as an alternate solution to e-waste crisis. There is an Inspiration Gallery in the end, which would motivate readers to start creating eco art.

Keywords Circular economy · Sustainable production and consumption · Upcycle · Recycle · e-waste · Waste management · Waste transformation · Environment and health

Learning Objectives

- Help environment and climate change, mitigation and adaptation in context with e-waste management.
- Exploiting and leverage on potential of Upcycling using e-waste.
- Build awareness and promoting Upcycling.
- Stimulate artistic creativity.

V. D. Mallabadi (✉)
21/1, Vaishnavi Nilaya, 1st Main, Maruthi Extension, Srirampuram, Bangalore 560021, India
e-mail: ewasteart@gmail.com

© Springer Nature Singapore Pte Ltd. 2021
L. Liu and S. Ramakrishna (eds.), *An Introduction to Circular Economy*,
https://doi.org/10.1007/978-981-15-8510-4_14

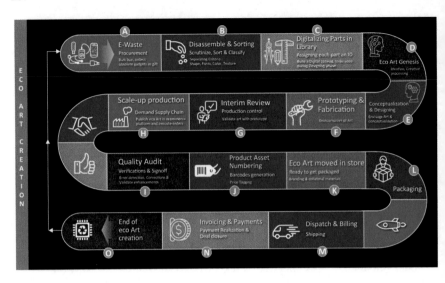

1   Introduction

If the world goes on with business as usual with their high tech lives and polluting lifestyles. Eventually we will have to face adverse consequences. In fact, e-waste recycling in terms of equipment and processing is very expensive and time-consuming, for example, extracting precious metals, like gold from PCB. Whereas same gold plated PCB could be upcycled and transformed into beautiful jewelries.

In next 10–20 years, the world may have to face catastrophic results wherein on our planet there would not be any place left to dump e-waste. Eventually human race will be surrounded by massive landfills. Hence, empowering young generation to participate in sustainability mission to save our planet is very important. To make a phenomenal difference, appreciate young generation to get motivated to work on this crucial initiate with empathy. They should imbibe and become a part of upcycle revolution. Induct youth to acquire knowledge on how upcycling is done. Induce curiosity, and activate desire for learning unique things. In process eager to take it as an opportunity for coaching and mentoring youth.

My motto is to create something qualitatively unique. I always get fascinated with experimenting and searching for new and abstract ways to create unique things.

Waste Management is becoming strenuous day by day. Considering the current scenario, we are generating more e-waste than we can recycle. So the unused e-waste needs to be recycled. There has to be an innovative way to deal with the pile-up of e-waste. With the human touch, e-waste can come back to us in restored form. Apparently, Upcycling will be the way forward for environmental sustainability and reducing landfills. I give higher priority to Upcycling e-waste rather than Recycling.

Upcycling e-waste can play a major role as Planet Healer, by practicing consistently our earth become greener. There is an unseen beauty in the e-waste and it can be transformed into new lease of life. Moreover, foreseeable eco art benefits, the concept of upcycling has been practiced by entrepreneur, a typical example is company Deko Eko, an Upcycling platform to promote unique sustainability concept. The case study projected here will demonstrate how discarded electronic gadgets could be converted into something interesting and artistic. None of the e-waste is useless for an artist. No e-waste is waste; it can be turned into amazing art pieces. As a matter of choice, Upcycling will be superior and have major benefits over recycling. A second life to discarded electronics waste by turning them into something worthwhile. As fashion forecaster, WGSN trend-spotters in Jewelry category, ensures insight and trends are accurate and actionable. Because new generations are demanding techy/geeky Jewelries.

Bringing creative vision to e-waste that increases its value tremendously. I care for the environment and concerned about the future generation.

1. Giving a new life to e-waste is surprisingly a gorgeous functional works of art. Upcycling is an eco-friendly alternative way of getting rid of landfill and discarded electronic waste in turn it willingly acquire second life. Upcycling will also help in leading healthy living for future generation.
   E-waste Collection, Dissembling, and Sorting (A and B of eco art creation)

Procurement—Bulk purchase, Identify e-Waste vendors, recyclers of scrapped electronic material, for example, E-Parisar Private Limited. Collect obsolete gadgets as gift from friends. Dissembling starts with dismantling, segregating, and scrutinizing. Post placing parts in different Bins. Auditing and separation criteria is based on Shape, Forms, Color, and Texture.

Digital Inventory starts by updating each part in a Library by assigning each part with ID. Finally, a Digital catalog is build, which will be used during Designing phase.

I had upcycled several hundred parts of phones and computer parts and created beautiful jewelry like earrings, cufflinks, pendants, etc. This has been a perfect example of how wisely one can manage waste.

2. **Eco Art Genesis** (C, D, E, F and G of eco art creation)

Introspecting e-waste will initiate the visualization process. Thinking process begin by evaluating and visualizing e-waste as a prospective potential to get transformation. Conceptualization starts after scrutinization process, by inspecting e-waste texture, shape, color, and so on. Post working on POC (proof of concept) finally comes out Art with something qualitatively unique.

I am obsessed by e-waste, and use them to make an interesting artifact. The context behind each creation is to reincarnate a second life out of discarded objects. In other world dead becomes alive again. I do not call e-waste as objects, they are the amazing base materials to get rescued or transformed into beautiful sculptures. Any e-waste material is an open immense possibility for me to be repurposed. Be it a PCB, electronic components or a dead instrument. I try to bring in some meaning to the art.

Creating artwork out of electronic junk is a creative process. One has to be eco-minded, applying artistic skills and method for upcycling with e-waste. Indeed it is an eco-friendly alternative to the traditional disposal of tech products and gives green consumers with a passion.

3. **Eco Art Scale-up production** (H, I, J, K, L, M, N and O of eco art creation)

After Prototyping, demand Supply Chain process begins.

eco Art works will be published in ecommerce platform and execute orders.

4. **Eco Art Education**

Art is an aesthetic pleasure. It fall predominantly under luxury and some products are usable too. Post initiating e-Art projects from last 6 years, I have been received lot of appreciations. But recognition is not enough, eco Art should be acknowledged worldwide.

(a)   Showcasing eco Art in seminars/conference.
(b)   Workshops and campaigns.
(c)   Solo Art Show—Transforming e-Waste into e-Art in Wipro Technologies.
(d)   Promote eco Art via Social Media, Facebook and Instagram etc.
(e)   eco Art Foundation.

For example, a pilot campaign launched by Nokia "Planet Ke Rakhwale". https://www.downtoearth.org.in/news/ewaste-management-nokia-sets-exa mple--41799

5. **Eco Art Entrepreneurship**

Each individual has a God gifted potential skill to render service to the humanity. One has to recognize talent, which he/she is good at and passionately work on it. To exploit talent one has to consistently hard work to achieve results. Be a creative entrepreneur, this passionate intuition gives me energy even when I am tired.

1. Exhibition sales.
2. Online ecommerce shop.
3. Crowdfunding.
4. Sponsored/funded Projects.
5. e-Waste Alliances worldwide.
6. Collaborating with International Artist.
7. Collaborating with Fashion Industry.

Source https://dekoeko.com/success-stories/Orange
Some of the major companies implementing Upcycling in context with Fashion Industry.

2 Upcycling: The New Wave of Sustainable Fashion

In a world still churning out trendy throw-away fashion pieces at breakneck speed, the idea of upcycled or refashioned apparel can be an anomaly. But it is a continuously growing trend and is one of the most sustainable things people can do in fashion. As upcycling makes use of already existing pieces, it often uses few resources in its creation and actually keeps 'unwanted' items out of the waste stream.

Upcycling stops adding stuff to a world that is already overwhelmed with material things. It also reuses materials that may otherwise end up in the landfill in creative and innovative ways—producing original often one-of-a-kind items from what many consider to be waste. The world of upcycling has exploded in the past few years [1–20].

https://www.triplepundit.com/story/2014/upcycling-new-wave-sustainable-fashion/58691.

Upcycled Fashion Explores Designer Imagination Fashion labels, like From Somewhere, are upcycling discarded textile materials to turn waste into beautiful, unique garments.

http://ecosalon.com/upcycled-fashion-explores-designer-imagination/.

Upcycling is no new concept, it has been a part of our Indian culture since 1930s and 40s when our families used to reuse almost everything—but now the old is new again, with some improvements. Making something new and artistic with something you once believed is useless is true art.

Source: https://economictimes.indiatimes.com.

Example 1 Process of recreating Van Gogh's master painting "The Starry Night" with resistors.

1. Simplifying complex details.
2. Retained contours, shapes, and form of each landscape elements.
3. Similar color resistor is replacing each brush stroke.

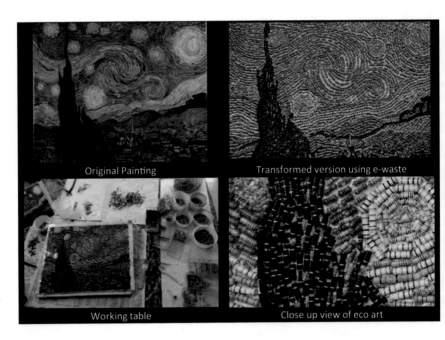

Example 2 Process of recreating collage Portrait.

1. Identifying appropriate snap to be used as base portrait reference.
2. Posterising, a process of simplifying an image or tone separation, highlights, mid-tones, and shadows.
3. Identifying best PCB match of tonal values, to be used for the portrait.

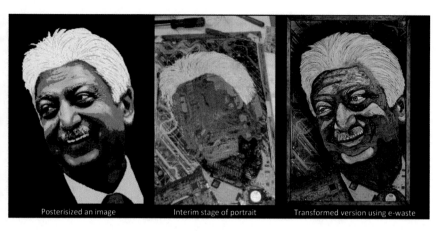

Posterisized an image | Interim stage of portrait | Transformed version using e-waste

Example 3 Process of recreating Future City using "Heat sink".

Heat sink has a typical character of having thin slits, the core functional purpose of it is to desperate heat efficiently, and I visualize and see these symmetrical shapes as a single building structure.

Originally Heat sink is designed in different shapes, depending on integrated chip size, it could be circular, square, or rectangular.

Proportions and scale have to be kept in mind while building structure.

Each tower has its unique characteristics, keeping real functional form in consideration. Helipad structure is also built to show modern structural engineering.

Arial birds eye view | Closeup of tower

Each tower placed in one row

Perspective view to show realism   View to highlight helipad

Used CPU chip's as the pathway, connecting different towers | Demonstrating GPS navigation scenario

3 Case Study

Best example of Upcycling is Jewelry designs using e-waste. It could be earrings, Necklace, Bracelets, and Pendants.

Typical recycling e-waste starts with huge quantity of scrap and post-processing very less quantity of metals would be extracted. Whereas same quantity of scrap could be transformed via, Upcycling by adding more value.

4   Jewelry Created from e-Waste

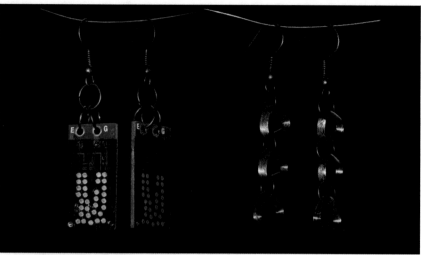

Typical example of effective usage of e-waste in making eco art

Flower bouquet created from e-Waste   Dazzle Bird

Some unique examples of wearable art and Jewelries created by me from e-Waste. Eco art is an expression, which is undoubtedly eye catching and makes people comprehend potential of transforming e-waste to amazing art. Some of my works are very intimate.

Upcycling brings a plethora of benefits:

1. Helps in preserving natural resources.
2. Helps in reducing landfills and pollution.
3. Reducing carbon footprint.
4. Eco art has a better environmental value.
5. A typical example is upcycling in garments industry.
6. A new movement has major impact on fashion-scape. It boosts creativity, construct community, and acts as a catalyst when demonstrating others
7. Companies can take up upcycling industry and turn e-waste to useful products

Discontent may be the reason manufacturers produce new gadgets. Apparently, obsolete technology has to be discarded and treated as e-waste. Recycling may not be the ideal solution to reduce landfill, one of the reasons could be it uses expensive equipment and not every kind of e-waste can be recycled. Whereas upcycle turn unwanted, low-value goods into something more useful, appraised with higher value. We should try to pay homage to these obsolete products. I believe they can be reincarnated by attaining an implicit character and look and feel which make them deserve a better destiny than a dump.

I wanted to emphasize on upcycling and we should buy less and use creativity to transform existing obsolete equipment. We cannot stop buying things, throwing outdated products and buying advanced model, relegating objects to an anticipated death that, most of the times, has no reason to be. Obsolete objects cannot speak for themselves. They can be transformed into an amazing art. Adding greater value and also giving new life.

Everyone should become militant to make a difference in fixing irreversible damage done to earth by reducing e-waste landfills. We should inculcate our new generations; rather upcycling should have been innate character. My goal is to spread knowledge on upcycling and try to change user behavior. Eco art creation is a process of ingenious inventiveness. Zero-waste has been an innate part of nature. Think twice what can you do with an e-waste before throwing it away. Eco art is a way to preserve the wisdom.

1. **Can anyone create eco art?**

    1. Yes, each individual has to develop strong visualization to work on eco art.
    2. One has to undergo practical training.
    3. As prerequisite, one has to have knowledge of all electronic parts used in the devices.
    4. Visual identity recognition is most important to work on different project.
    5. One has to know the basics of material science, mechanical engineering, fabrication techniques.
    6. Appropriate power tools, used to achieve expected results.

2. **Challenges faced during mass productionizing eco art works**

    1. Getting bulk number of similar parts, at times it has may not be feasible.
    2. Each eco art produced will be unique, because its hand crafted. Therefore, there will be minor differences. 100% replica is not possible.
    3. Some of the e-waste items are hazardous; hence, precaution has to be taken while working on it. Add on safety while working.
    4. Training individual may be a big challenge, because each person conceptualization power would be different.

3. **Why Upcycling is important?**

    1. Upcycling is an important alternative measure to deal with waste management.
    2. Sustainable eco art designs bring extra value to the trash.
    3. Upcycling does help in Zeroing Waste and also reducing Landfill.
    4. Upcycling conserve and save limited resources, by reducing the need for new raw materials to create new products.
    5. It brings second life to discarded/dead products or e-Waste, surpassing what they were originally created for.
    6. An Innovative way to use scrap, also stimulates the creativity and handmade work.
    7. Upcycling is reworking, reinventing, trying to see new forms. In other words, it is a kind of tribute.
    8. Upcycling plays a major role in dealing with threat to the global environment for the future generation.
    9. It helps in sustainability and creates a new wave of Sustainable Fashion Example—Making attractive jewelry for the fashionista or techy geek.
    10. Upcycling stops adding more stuff to a world that is already overwhelmed with material things.

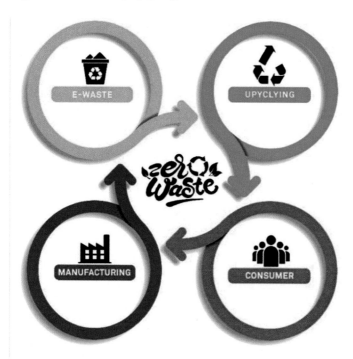

Mumbai is being buried under a mountain of its own trash; e-Waste is threatening to bury our lives
https://www.livemint.com/Politics/xZdqBRJhyDqYhx1UAmOvjK/Mumbai-is-being-buried-under-a-mountain-of-its-own-trash.html.

Each day, more than 500 trucks line up along a two-lane dirt road in an eastern suburb, waiting to add to a mountain of refuse tall enough to submerge the White House twice over.

1. **Potential of Transforming e-Waste**

   (a) Usable Products.
   (b) Abstract Art.
   (c) Wall Art.
   (d) Sculptures.
   (e) Creatures.
   (f) Robots.
   (g) Jewelry Designs.

Actionable Thoughts…

1. Take Upcycling e-Waste as an industry.
2. To promote awareness by conducting workshops.
3. Entrepreneurs could take up as a business, in lieu of treating e-Waste conventionally.
4. We should live with a lot less stuff.
5. Contribute to Circular Economy.
6. Promoting eco art by campaigns, workshops create a better future for the Young Generation.
7. Collaborating with current Recycling companies, and advocate to upcycle e-waste and transform into amazing art.
8. Start a major innovative crowdfunding campaign, to increase awareness and changing the lifestyle and behavior. People could contribute by pledging. For example, a pledge not to dispose e-waste illegally and transform into Art. Finally, a total global change can be made. I am appealing for partnership organizations, environmentalist, government bodies, schools, universities, passionate individuals, influencers, and celebrities to contribute in this mission. I request all of you to join this movement. Please connect with me to build an effective movement.
9. Propose Government, Upcycling e-Waste as a major alternative solution.

5 Conclusion

Upcycling has a great opportunity to tackle gigantic landfill. In context with waste management, I believe upcycle could be a major differentiator as an alternate solution.

I hope the work I do will help young generation. The thinking my initiative is, if each person stops discarding products and upcycles or reuses them instead, it will make a major difference in healthy living.

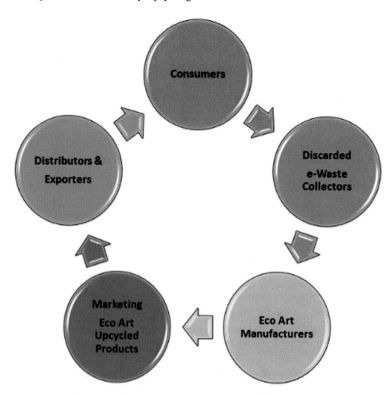

Upcycling could be an effective way to deal with environment sustainability and to arrest the problem of reducing landfills. I urge Government to support me and help me in building the institutions.

Be part of the Upcycling Movement...
Helping the World to become more Greener...

Questions

1. Can non-skilled person do Upcycle?
2. Do Upcycled products have endless life?
3. How to price the Upcycled products?
4. Can any e-waste product could be Upcycled?

Answers

1. Yes, with proper training one can learn how to upcycle.
2. Since transformed product is no more has a functional, it has reincarnated as artefact. Hence, it has no end of life.
3. Since raw material cost would be very less, but all Upcycled products would have value as par with other conventional products.
4. Yes, there is possibility. One can transform.

Exercises

Make use of any e-waste device and create some interesting art.

Raw Materials: Use any form of e-waste, for example, obsolete mobile phone data card, Hard disk, CD/DVD, mouse, keyboard, etc.

Instructions: One could mix more than one material. Use power tools and industrial glue gun, in case of cutting/joining different materials, respectively.

Procedure: Visualize the forms/art to be created from selected raw material. Create a rough sketch and start developing in stages. Review outcome and do enhancements if required.

Inspiration Gallery

Suggested Reading

1. Upcycling for Sustainable Living.pdf.
2. Vishwanath an eco Artist—Aspiring Environmentalist.pdf.

References

1. eco-Art works, published by HINDU News on 18 Feb, 2020 https://youtu.be/v8JJCbfIlws
2. https://www.behance.net/eco-art
3. Art during Corona Virus Outbreak https://youtu.be/fTiIslPACT0
4. https://www.2tout2rien.fr/le-recyclage-delectronique-en-sculpture-de-vishwanath/ 2tout2rien is a French blog/webzine on the unusual, art.... all the bizzareries quirks of the world....
5. "Upcycling e-waste for eco-art".
6. https://buyequip.com.au/ewaste-fashion-wear-old-electronic-gadgets/ Australian Portal—"don't waste IT".

7. https://www.recyclart.org/bust-sculptures-made-from-e-waste/ "e-Waste Upcycling Idea— Ewaste Fashion: Amazing Ways to Wear Your Old Electronic Gadgets". These self-described 'unique geek wearable earrings' by Indian sculptor and jewelry designer Vishwanath Davangere Mallabadi deliver exactly what they promise. Definitely one for the forward thinking, these green dangles are made from mini transceivers and are designed to let the world know that wearable e-waste is here to stay.
8. https://in.pinterest.com/dmvishu/pins/
9. https://ewasteart.wordpress.com/
10. Indian Talent Magazine - September 2019 (The latest Indian talent news and events) http://www.indiantalentmagazine.com/archives/september_19/#p=52
11. Astonishing Future City made with discarded heat sink https://www.behance.net/gallery/771 08263/Smart-City-built-with-e-waste-Heat-sink-other-parts https://www.behance.net/gallery/69855899/
12. Article published in Telangana Today—Hyderabad—Tabloid—Features on 18th November 2019 https://epaper.telanganatoday.com/Home/ShareArticle?OrgId=ba14182c&imageview=0
13. Article published in The New Indian Express. "From Waste to Best" http://www.newindian express.com/cities/hyderabad/2019/nov/05/from-waste-to-best-2057542.html
14. Article published in THE HINDU (Hyderabad & Bangalore). Vishwanath Mallabadi bats for eco art with e-waste. In section, "PLANET HEALERS ENVIRONMENT" https://www.the hindu.com/sci-tech/energy-and-environment/vishwanath-mallabadi-e-waste-wipro-ecoartist/ article29944218.ece "From waste to worthy" https://www.thehindu.com/sci-tech/energy-and-environment/vishwanath-mallabadi-e-waste-wipro-ecoartist/article29944218.ece
15. Article published in "The Pioneer" in Page 9/12 https://www.dailypioneer.com/uploads/2019/ epaper/november/hyderabad-english-edition-2019-11-06.pdf
16. Article published in "Telangana Today". "Where e-waste is transformed into a thing of a beauty". Vishwanath Mallabadi takes unused gadgets and turns them into quirky pieces of art. https://telanganatoday.com/where-e-waste-is-transformed-into-a-thing-of-a-beauty
17. "From waste to worthy". Bengaluru based Vishwanath Mallabadi, sees beauty in discarded circuit boards and recreates Van Gogh's Starry Nights with resistor on wood https://epaper.the hindu.com/Home/ShareArticle?OrgId=GSH6MRDME.1&imageview=0
18. Eco Friendly Passionate IIT Vishwanath Making 200 Products Of Computer Waste Material I ABN TV https://youtu.be/5P37jJeZjT0
19. News Media coverage by ABN—TV Channel, Telugu. Eco Artist Vishwanath Making Products With Computer Waste Materials https://www.youtube.com/watch?v=CaEOsPWxbJ8
20. News Media coverage by V6 TV, Telugu News Channel. Hyderabad Host Indo Data Week 2019 https://youtu.be/eao84Ws_M7Y

Vishwanath Davangere Mallabadi was born in Chitradurga, Karnataka, India on 22 October, 1962. I grew up in an atmosphere where art was respected and encouraged and still is. I thus inherited a priceless legacy. My father, Late D. M. Shambhu was a famous sculptor and painter and dedicated his life in restoring and maintaining valued artifacts under the aegis of the Government of India. In fact my father's uncanny eye for detail and finesse has had a profound impact on me throughout my professional career.

My Father and Mother have been my prime motivator's. I married my lifelong companion, Dr. Nirmala and have a resourceful daughter. My father wanted me to be a Doctor, but I was not medically waste inclined. However the traits of a Medical Practitioner were imbibed by me during the time when I assisted my Father in his endeavours in various artistic forms and mediums.

I decided to pursue my professional career in Art & Design. My self-validation skills with constant encouragement from my Father developed my creative talent. By age 18, I had developed obvious skills as an artist and reinforced my skills on completing my professional degree in B.F.A. in Applied Art. I studied at the College of Art (1983–1987) in New Delhi, India.

At the age of 23, I used to explore second hand trash objects and try to convert them into worthy objects. I started exploring new art forms within e-waste. Fortunately, I was part of HAM radio, wherein I developed my technical skills, learned fundamentals of technology, electronics and fabrication. Subsequently it inspired me to know more about electronic product design. My philosophy and rationale towards creating something extraordinary is by doing fusion between creativity and technology. I always enforced my acquired technical knowledge and blend with my artistic skills.

I thus blended Creativity with Modernity and forms. I always wanted to challenge myself by doing something unique. Thoroughly inspired, I started exploring my passion by creating new art forms. I chose e-waste as a medium to express my ideas.

I am passionate and dedicated about the work I do and I thoroughly enjoy it. The effort involves a lot of patience, love and determination. For me, no waste is waste, I see interesting forms in e-waste. I upcycle them and try to give a second life to inanimate objects. I scrutinize e-waste and restore potential components, finally transforming the e-waste into unimaginable amazing masterpieces. The creation process seeks serendipity and aesthetics. I try to bring in the creative vision to discarded e-waste and add higher environmental value than it had in its original state.

Upcycling is a key concept to add value to junk e-waste. I have created many unconventional and iconic designs out of e-waste. I try to interconnect distinctive e-waste pieces into a coherent composition improvise with different forms and keep working until I satisfied with the results. I enjoy discovering what I am creating while I am doing it and I have an irresistible urge. I practice keeping an open mind to the subconscious and like to be surprised by the outcome of my work. I get inspired and fascinated by inner forms of electronic equipment, it's intricate unnoticed textures and vibrant colors and intrinsic characters.

For me e-waste has a life of its own, and at times, it provokes me to reincarnate. I am consistently aiming for new and unusual compositions, with visually arresting expressions. My remarkable eco art works do offer inspiration to young artists.

My talent has been recognized by India Talent Magazine and exhibited my works at the International conference for Sustainable Development Goals (SDG) and got a lot of recognition by News and Television Media.

I am keenly interested in conducting workshops, lecturing in conferences and seminars to promote e-waste awareness. My intention is to bring an emotional connect between people and e-waste. Eventually I am anticipating large eco art installation works.

Currently I am working on a collage portrait of Shri Narendra Modi India's Prime Minister. My future plans are to create contemporary abstract art series on "An Indian Icon". I strongly believe that art from e-waste could be an alternative method for recycling and preventing landfills. However, this unique spin on upcycling e-waste doesn't just give birth to stimulating art, it also generates revenue from the sale of usable products. My never ending investigation on e-waste and creating beautiful, spectacular art will last forever.

Being a designer and environmentalist, I am concerned about the future generation and trying to do my bit for the betterment of the coming generation. I encourage everyone to leverage their hidden potential and be creative.

A unique eco warrior in true Gandhian spirit.

Waste Management Practices: Innovation, Waste to Energy and e-EPR

Stephen Peters and Keshan Samarasinghe

Abstract Solid waste is the unseen side of the circular economy. Governments need innovative policy to create functional markets to keep waste generated within the supply chain. Transitioning to minimal or no net waste requires innovation to avoid compromising quality of life. This means transfer from the current linear thinking model with large expensive centralized infrastructure to discrete distributed chains of infrastructure and using the circular economy principles. Most developed countries have already initiated prevention of waste. The waste management hierarchy (reuse, recovery, recycling, thermal recycling and disposal) can be optimized by innovation. By creating smaller circular waste pathways closer to the source of waste generation, more expensive end of life solutions can be rightsized due to higher resource recovery rates (from 10% up to 80%). The best-case scenario for National level waste resource waste management is 67% material recycling, 25% thermal recycling, and 8% to sanitary landfills. Through further innovation in product design and business delivery models, the thermal destruction and landfill can be further reduced. Extended producer responsibility (EPR) schemes create higher resource recovery efficiency and can be revenue positive to Governments. Through public-private partnership models, these EPR schemes can be digitized to improve the efficiency of circular economy.

Keywords Recycling · Waste to energy · Extended producer responsibility · Public-Private partnership

S. Peters (✉)
Sustainable Development and Climate Change Department, Asian Development Bank, Manila, Philippines
e-mail: speters@adb.org

K. Samarasinghe
Consultant, Asian Development Bank, Manila, Philippines
e-mail: keshsamarasinghe@gmail.com

© Springer Nature Singapore Pte Ltd. 2021
L. Liu and S. Ramakrishna (eds.), *An Introduction to Circular Economy*,
https://doi.org/10.1007/978-981-15-8510-4_15

Key Concepts

- **Waste Transformation**: Ability to recycle or upcycle waste into value-added materials.
- **Functioning Supply Chains**: Enhance collection processes and recovering activities to maximize value creation on waste materials.
- **Digitization**: Information technology processes used to improve the efficiency of the functions of supply chain.
- **e-Extended Producer Responsibility**: Materials management philosophy embedded in digitization to strengthen circular economy.

Objectives

- To introduce the opportunities and barriers to material recycling-based society and the role of the waste to energy.
- To introduce the benefits of digitized extended producer responsibility through public-private partnership to increase contribution and improve the efficiency of circular economy and where waste to energy fits in.

1 Introduction

"**Waste is the Government's Problem**." This simple statement is at the heart of the global waste crisis. Governments have limited budgets to support transformation in the waste industry, let alone deriving benefits from the waste. Interventions via urban infrastructure and large waste to energy plants are often over-designed as magic bullet solutions. Private and informal sectors are often considered as a hindrance to achieving sanitation and excluded from planning scenarios.

It is critical to understanding what the characteristics of your waste are, how much of it you create and how it is collected from the starting points. Some densely populated communities with higher waste generation per capita will opt for incineration to ensure public sanitation. This model is typical linear thinking. Other communities who have limited resources, seek to reduce the impact from waste by recovering valuable material from waste, creating jobs, and enhancing economic development. However, the performance of recycling efforts is dependent upon the community attitude about waste and waste management options. Whilst segregating waste at source is the best practice for recycling, the behaviour is difficult to incentivize. In this chapter, you will see how innovation in waste management can enhance circular economy adoption and improve the material recycling sector.

2 What Is My Waste?

In 2016, the world has generated 2.01 billion tonnes of municipal solid waste (MSW) [5]. Asia has a population of more than 4.45 billion. It is estimated that Asian cities will generate 1.8 billion tonnes of MSW in the year 2025 and continue to grow at approximately 15% annually [4].

Every day, we produce household waste, industrial waste, medical waste, agricultural waste, wastewater, and importantly hazardous and toxic waste. The rising population, the growth of economic activity and rapid urbanization are the reasons for changing consumption patterns and increasing per capita solid waste generation around the world. The characteristics of our daily waste generated indicate where we live, how affluent we are and how conscience we are. Also, it links to how likely we are to separate our waste to allow for recycling or upcycling to create value-added materials from waste.

The character of our waste can be described by the relative weight of various items such as food waste, plastics, paper, textiles, rubber, metal, aluminium, ash, soil and other constituents. In Rural Thailand, the poorest communities will produce 0.3 kg day^{-1} person^{-1} [2]. This is due to the high resource recovery of food waste and well-developed waste recycling supply chain through companies like Wongpanit. In nearby towns the generation rate is up to three times higher. In Bangkok, some areas produce 3.8 kg day^{-1} person^{-1}. Figure 1 shows the estimated average composition of general municipal solid waste in world.

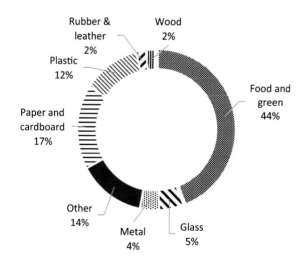

Fig. 1 Composition of municipal solid waste *Source* World Bank 2018. What a Waste 2.0: A Global Snapshot of Solid Waste Management to 2050

Rubber & leather 2%

Wood 2%

Plastic 12%

Food and green 44%

Paper and cardboard 17%

Other 14%

Metal 4%

Glass 5%

3   What Are My Technical Options?

How the waste is presented for collection is the first challenge in determining technical options. The contaminant level within waste materials will determine suitability for sorting by mechanical treatment technologies, localized process (biogas/composting), upcycling (refuse-derived fuels/plastic to fuels), reuse (handicrafts/artworks), recycling or centralized thermal treatment. The waste is sorted into discrete types of material. One of the key barriers to have only material recycling-based society is that recyclability of waste. Where waste has low recyclability index, thermal treatment (incineration/pyrolysis/depolymerization) can be more appropriate to treat non-recyclable waste instead of bulk landfilling. Where air pollution control systems are included in the plant, these facilities are commonly referred to waste to energy plants or energy from waste plants.

Dumpsites, not sanitary landfills, are the prevailing solution for developing countries, whilst many developed countries have used incineration to reduce the volume of waste before disposing in sanitary landfills. Many cities look to waste to energy solutions to reduce the volume going to landfills, optimize land use in densely populated cities and derive some energy benefits. 172 such plants are established in South Korea and suppling heat for district heating system and energy for domestic consumption. In 2010, 3.1 million tonnes of waste were treated in 35 waste to energy plants [1]. In Denmark, 29 waste to energy facilities treated 3.5 million tonnes of waste which corresponds to about 26% of total waste generated in the year 2005. The waste to energy facilities contributed about 3% of the total Danish electricity production and about 18% of the total district heating production [3].

Whilst these waste to energy technologies make urban sanitation easier, they crowd out the wealth and job-creating opportunities of recycling and upcycling. Also, the management of heavy metals, dioxins and fly ash requires advanced air pollution control system to meet emissions standards. The management of these persistent organic pollutants in fly ash presents challenges. The minimum capacity threshold for commercially operated waste to energy plant is 600 tonnes per day with the optimal size being twice that capacity, according to industry sources.

Zero waste communities are springing up all over the world and make use of composting or biogas technology for digestible waste, recycling, upcycling. Also local communities are encouraged to follow waste management hierarchy within a circular economy context. However, quantity is a matter of selection of appropriate waste treatment method under the hierarchy. The efficiency of mechanical recycling is lagging since these communities, however, don't consider the hazardous and toxic waste inherent in the production of many commonly used products. Therefore, high volume of residual waste is generated from recycling plants and the quality of the recycled products deteriorates.

A strong supply chain which maximizes resource recovery whilst ensuring public sanitation should be developed to ensure future sustainability. At the heart of this approach, Fig. 2 is presented to understand the transformative processes in the waste supply chain.

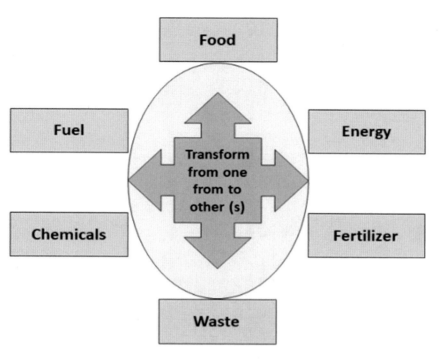

Fig. 2 The waste supply chain and competing use. *Source* Author, Stephen Peters

By using other forms of waste such as wastewater, this transformative process can be improved and minimize impact on overall environmental and commercial viability. Often horticultural waste management approaches are economically viable.

Under the transformative processes, it is important to understand the transformation options. Table 1 shows some of the commonly transformed materials.

4 Case Study of Material Recovery Facility—Saahas Zero Waste, India

Saahas Zero Waste (SZW) is a social-environmental company works on circular economy by recovering valuable resources from MSW. Through its 18 years of experience of working with MSW, the company has shown that 90% of the resources can be recovered if a wholistic approach is taken in waste management by trained personnel and if the management process is in close proximity to waste generation.

The company is conducting a program called Zero Waste in India. The process flow of the Zero Waste program is presented in Fig. 3. Under this program, 26 tonnes of bulk waste is collected from industrial zone and residential complexes daily. A portion of wet-waste is treated in a compost site and the remaining portion is used

Table 1 Transformation options

Materials Transformability

Waste	Food	Fertilizer	Fuels	Energy	Chemicals
Solid—rice husk, food scraps, EFB, fiber, MSW, offal, spent grain, ash Liquid—POME, process waste, sewage, sludge Gas—waste gases, waste heat, emissions, fly ash, radiant heat	**Crops**—corn, cassava, palm, sweet sorghum, sugar, wheat, rice, edible oils, fruits, algae, grasses, trees **Livestock**—chicken, beef cattle, diary, duck, sheep, deer, fish, seafood	NPK UREA Silica Phosphate Soil Conditioners Biochar Ash	**Solid**—briquettes, pellets, biochar **Liquid**—biofuels, bioethanol, DME, biodiesel, FAME, LPG, LNG **Gas**—NG, CNG, BioCNG, hydrogen, syngas	Thermal Electrical Stored Transportable Distributed Microgrid Centralized Grid Emerging DC Nanotech	Recycled plastic resins Plates Cutlery Biochemical Industry

Source Author, Stephen Peters

BioCNG = compressed purified biogas, CNG = compressed natural gas, DC = direct current, DME = dimethyl ether, EFB = empty fruit bunch, FAME = fatty acid methyl ester, LNG = liquefied natural gas, LPG = liquefied petroleum gas, MSW = municipal solid waste, NG = natural gas, NPK = nitrogen phosphorous & potassium fertilizers, POME = palm oil mill effluent

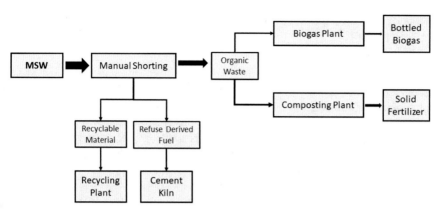

Fig. 3 Process flow diagram of zero waste program. *Source* SAAHAS Zero Waste (SZW) communications materials, 2018

for biogas production. The biogas is bottled and used for commercial purposes and the compost packets are sent to the local market.

Dry waste is sent to the material recovery facility (MRF) for further segregation, aggregation, compaction, and dispatch to recycling facilities and co-fueling units such as cement kilns. Most of the waste management activities are carried out manually. It is worth noting that of 250 employees, approximately 80% are women.

SZW provides professional management of the local supply chains process including customer acquisition feedback, awareness on reducing single use products, and enforcing segregation at source. A Management Information System is used to analyze data. Therefore, it has high efficiency and accountability of collection, transportation, and treatment processes. The project directly supports climate mitigation by reducing GHG emission when burning waste is considered as the third-highest cause of GHG emission in India. It is estimated that annual CO_2 emission reduction by the project is about 12,300 tonnes.

Because of the Zero Waste program, the regularity of income has improved the employees' quality of life. Investors are attracted due to the high level of financial viability and high level of visibility of the project as one supportive of mainstreaming the waste management industry.

SZW's innovation success is due to engaging community around the source of waste and attracting community members to interact on a commercial basis. This approach is considered best practice for localized waste management.

5 The Role of the Governments, General Public, Business and CSOs

Communities require assistance in overcoming the policy hurdles, financial capacity, technical options, process risks and breaking the knowledge barriers. National level thinking lags community-led models. In 2016, South and Central Asia have disposed 74% of their MSW in landfills. These proportions are also as high as 59 and 55% in South-East Asia and Eastern Asia [4]. There is a huge gap between required infrastructure and available facilities in waste management sector as a result of infrastructure investment shot fall in public sector. Private Public Partnership (PPP) can fill the investment gap.

In many developing countries, PPP can create a waste to energy over-capacity trap to ensure commercial viability (the Lock-In Effect). Often risk is allocated poorly, not well researched or documented. Through well-developed PPPs investments, the waste management sector can integrate new waste treatment technologies and business models. Table 2 presents *"Twelve pathways to a circular economy: Using waste to energy interventions to reduce their budget on our ecosystem"*. SZW is an excellent example of pathway 5—Decentralized waste sorting and recycling.

6 Case Study of Resource Recycling System of Municipal Solid Waste

The community in Taipei, China places high importance on resource conservation, recycling and sustainable development. Municipal solid waste generation rapidly increased as a result of the rapid urbanization and economic development. Consequently, initiatives were introduced to improve the waste management sector over recent decades. Figure 4 shows the summary of initiatives which have been taken from 1987 to 2005 and the performance of the recycling industry. Initially, only market-driven recycling industry was initiated to manage a small portion of recyclable waste. In 1987, polluter-pay principle was introduced which made producers were responsible for recycling physically and financially. Based on the producers pay principle, eight recycling funds were created to collect recycling fees and the funds are administered by the Environmental Protection Administration (EPAT). Based on the statistics, the recyclables collection rate increases 47% in the past 20 years. In recycling industry, annual revenue is greater than USD2 billion and 10% of income is generated from the recycling fund (Recycling Fund Management Board 2019).

Resource recycling management regulations were put in place to strengthen the waste management sector. Firstly, management measures for manufacturers and importers were introduced to calculate amount of **regulated recyclable waste** (RRW) products which are sent to the local market. Secondly, proper standards were implemented for waste collection, transportation, storage, and recycling of RRW. Attractive subsidies were given for RRW collection and recycling with an application

Table 2 Twelve pathways approach

No.	Pathway	Description of critically important issues
1	End of life—waste destruction (Provincial)	The use of advanced flue gas cleanup is critical in these waste to energy plants. Uncontrolled flue gas is highly toxic. It MUST BE scrubbed of fly ash prior to emission. Emissions must be monitored in real time and shared with stakeholders.
2	End of life residues (Immobilization)	The capture, proper handling and immobilization of fly ash from flue gas are critical. Locking the fly ash up in geotechnical materials with rigorous testing is the current best practice.
3	Centralized sorting—eco-industrial park model	Siting an end of life facility in a larger Industrial park allows for sharing of energy and heat from the process but also allows for processes to extract higher value or to recover waste into usable products.
4	Centralized recycling and upcycling	Siting these value addition and material recovery technologies with an Eco-Industrial park leads to simpler environmental management. It also allows for tracking of materials and sharing with industry involved in recycling and upcycling.
5	Decentralized sorting and upcycling	Cities will have existing material recovery facilities and transfer stations as part of their existing waste supply chain. These sites are excellent locations to add recovery or localized solutions to reduce waste quantities for subsequent transport to centralized locations.
6	Digitization at source	Various APPs existing for collection of waste including household truck collection, centralized community collection and opportunistic collection of higher value items by the informal sector. The capture of higher value materials at source creates more secure feedstocks for decentralized recycling and upcycling technologies. The informal sector can be matched to householders and businesses using apps. The degree of digitization will reduce transport costs for cities and increase participation and efficiency of the circular economy. It does need to be well regulated.
7	Landfill, soil and river clean up	Most landfills in developing countries are not sanitary. Over the coming decade, these landfills will be mined and remediated to limit groundwater pollution. Digitization of this cleanup will allow for more efficient "not for profit" interventions through direct payment.

(continued)

Table 2 (continued)

No.	Pathway	Description of critically important issues
8	Regional eco industrial parks	The eco-industrial park model can be expanded to support those Small Island Developing States and Archipelagic States with extended marine or river supply chains. Creating the economy of scale for recycling and upcycling of higher values items can work hand in hand with the destruction of harmful waste which cannot be treated in country.
9	Digitized extended producer responsibility schemes	The success of extended producer responsibility schemes is based on their ability to add the embedded waste management cost to materials prior to sale of product. These costs collected by EPR funds should reflect the impact of a product. These funds cover the cost of the Government underwriting environmental management and should be revenue positive.
10	Strengthening recycled output supply chains	The challenge for many recyclers is that they are unable to sell their recycled product. Creation of supply chains, price discovery, product quality verification and buyer identification are required to support recyclers/upcyclers.
11	Supporting FMCGs in product redesign—recycled %	Large consumer goods manufacturers (or Fast-Moving Consumer Goods—FMCG) are aware that the days of single use plastics are numbered. Creating recognition and support for these groups in product redesign is a crucial as FMCG companies have such a large impact.
12	Strengthening governance and enforcement	The above pathways need consistent direction in policy and also strong enforcement. Strengthening enforcement capacity is critical.

Source Author, Stephen Peters

review process. Finally, transparent registration, monitoring, certification of RRW, auditing, and cancellation of registration were carried out to encourage enterprise rigour and investor confidence.

MSW and Recycling Management Structure and Function

EPAT uses three sub-agencies—the Department of Waste Management, the Bureau of Environmental Inspection and the Recycling Funds Management Board. All regulations for the waste management sector, all regulation of recycling industry, incinerators, landfill management, and recycling fund management are controlled and operated by these three sub-agencies of EPAT.

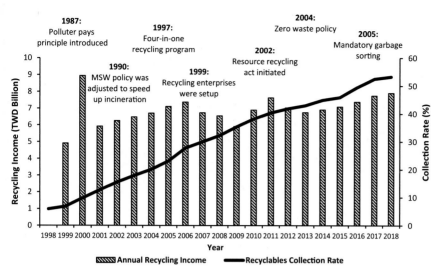

Fig. 4 Resource recycling management policy

Four-in-One Recycling Program

The Four-in-One recycling program consists of the government, manufacturers, consumers, and recycling enterprises. Community-based recycling organizations promote waste segregation at source and recovering recyclable materials from MSW. In addition, recycling enterprises regularly collect recyclable materials from households, commercial places, and municipalities. Local authorities collect and sort recyclable material from waste and they were able to generate an income by selling recyclable materials. A recycling fund is financed by contributions from waste generators and manufacturers under the EPR program. As a result of having sufficient funds to incentivize recycling, a thriving recycling sector has developed. Figure 5 shows the recycling management fund function and operation system (Environmental Protection Administration (EPAT), 2017).

RRW can be categorised into two parts such as containers and objects. The containers are iron, aluminium, glass, paper, plastic, and pesticide containers, and the objectives are battery, automobile, tire, IT equipment, home appliance, and light bulbs. All industries which manufacture RRW product, are responsible for the end of life waste treatment. Thus, the industries provide financial support to recyclers through the recycling fund. Because of the EPR initiative, segregation efficiency of MSW is higher than rates in developing countries.

This change took the community 30 years to achieve. Internationally, policymakers are struggling to reconcile how to integrate the circular economy into waste management and are often seduced to accept the magic bullet of oversized waste to energy plants. Realigning Policymakers interests requires incentives for Governments. Generation of revenue is one such incentive.

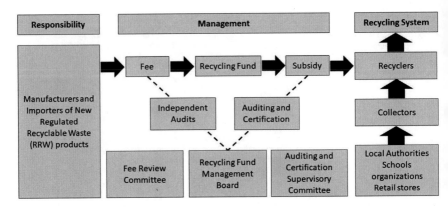

Fig. 5 Recycling management fund function and operation. *Source* EPAT Communications Materials, 2019

EPAT returns about 20% of its revenues to Government whilst deploying the remainder to the operations of waste management supply chain, subsidies to recyclers and operations costs.

e-EPR

Digitization of extended producer responsibilities offers a faster route to implementation. Many countries have start-ups promoting waste trading and second-hand sales APPs. These start-up offer access to consumers willing to engage in the circular economy. The development of the back-end function for materials pricing, revenue collection, allocation of waste resource and trend analysis requires a centralized software provider. Linking APPs with back-end software providers will allow for faster rollout of EPR schemes.

In discussions with the authors, industry sources highlighted other benefits.

1. If the supply can be tracked, material provenance can be tracked. This is especially important in developing countries where FMCG cannot use recycle plastics without assurance of "No Child Labour" from multi-stakeholder supplies chains.
2. Data created by such systems can provide hard data which can be used by financial institutions to substantiate loans to project developers in the supply chain, especially those using distributed technologies business models.
3. FMCG companies will come under increased consumer sales pressure from organizations whose products [6] create less waste and hence will have a commercial advantage under EPR schemes.

A number of software providers are experimenting in this space. The success of such an innovation is not to secure dominance of a supply chain but delivery of utility to consumers, revenue to Government and sanitation/environmental/social benefits to the community. There is no digitized model at this time which provides these three benefits explicitly across the whole waste supply chain.

7 Conclusion

In Asia Pacific region, solid waste generation and volume of mismanaged waste have increased tremendously during the past decades as a result of the rapid urbanization and poor waste management practices. Innovative thinking is required to respond to these challenges. The twelve pathways approach using waste to energy provides context under which this innovation can be achieved.

The two case studies presented show localized innovation in action. Policymakers can create a space for the private sectors provides revenue opportunities to Government, jobs, increased awareness of public sanitation and inbound investment. PPP's are a useful tool when well researched with appropriate sizing for infrastructure.

Digitizing existing schemes to link the existing supply chains participants speeds the implementation of circular economy principles and aids in the transition from a linear model to an increasingly circular model.

8 Exercises

1. Identify the circular economy business models in your area and list four innovative business delivery models to strengthen the circular economy?

2. List five advantages and five disadvantage of extended producer responsibility schemes.

3. Assuming that you are the director of the waste management department of your country, what are the strategies and policies introduced to promote extended producer responsibility for the government?

Suggested Reading
The Ecology of Commerce, Hawken.

References

1. Bourtsalas, A. C. T., Seo, Y., Alam, T., & Seo, Y. (2019). The status of waste management and waste to energy for district heating in South Korea. *Waste Management, 85,* 304–316. https://doi.org/10.1016/j.wasman.2019.01.001.
2. Hiramatsu, A., Hara, Y., Sekiyama, M., & Honda, R. (2009). Municipal solid waste flow and waste generation characteristics in an urban-rural fringe area in Thailand, (May 2014). https://doi.org/10.1177/0734242X09103819.
3. Ramboll, (2006). *Waste to energy in Denmark.* Denmark: Virum.
4. United Nations Environment Programme (UNEP). (2017). *Asia Waste Management Outlook.* Retrieved from http://www.rrcap.ait.asia/Publications/AsiaWasteManagementOutlook.pdf.
5. World Bank Group (WBG). (2018). *What a waste 2.0: A global snapshot of solid waste management to 2050.* Washington, DC.
6. https://saahaszerowaste.com/.

Stephen Peters Steve Peters is a Senior Energy Specialist (Waste-to-Energy) in the Energy Sector Group of the Sustainable Development and Climate Change Department, Asian Development Bank. He is responsible for developing the knowledge base, project development and implementation in waste to energy and supporting projects across waste, the circular economy, and ocean impacts. Prior to joining ADB, Mr. Peters was the Director at Stratcon Singapore Pte Ltd where he consulted on waste to energy, waste to fuels, biomass, biogas biofuels, and innovation in clean technology. He was a Founder and Technical Director for the Technology and Development Network which is a network of professionals in clean technology in Asia. He was the Managing Director at Waste to Energy Pte Ltd, a Stratcon JV biogas developer in ASEAN from 2004 to 2010; Deputy Project Manager at Kumagai Gumi Co Ltd., in Singapore in 1997–1999; Director at Stratcon Pty Ltd. in Australia in 1993–1999; and a Director at Foster Sturrock Pty Ltd in Australia from 1988 to 1993. He is the immediate past Chairman of the Asia Pacific Biogas Alliance. Mr. Peters obtained his Master's and Bachelor's degrees in Civil Engineering both from University of Melbourne, Australia in 1990 and 1987 respectively.

Keshan Samarasinghe Keshan Samarasinghe is a Clean Energy Technology Specialist (Consultant) in Asian Development Bank. He obtained a postgraduate degree from the Asian Institute of Technology, Thailand on Environmental Engineering and Management, and a bachelor's degree from University of Moratuwa, Sri Lanka on Chemical and Process Engineering. Possesses practical as well as research-based experience on environmental engineering and chemical engineering in regional context covering countries Sri Lanka, Maldives, Thailand and Philippines.

Agricultural and Municipal Waste Management in Thailand

Suneerat Fukuda

Abstract Circular economy is simply the economic system in which wastes are minimized and resources are best utilized. This approach is aligned with "Bioeconomy, Circular economy and Green economy (BCG)", one of Thai government's flagships for national social and economic development plan. In agricultural sector, circular economy concept has long been successfully adopted through waste minimization and renewable concept. Thailand is the ASEAN leader in bioenergy production. One important factor is the government's long term renewable energy plan which has supported the implementation of bioenergy projects. Recovery as food, energy, fertilizer and other value added products from agricultural wastes are already in commercial practice in agro-processing industry, i.e. sugarcane and palm oil industry as good examples. These provide an excellent foundation to another step of circular economy where more value addition can be extracted along the value chain. On the other hand, achieving circular economy is still far for municipal solid waste (MSW), the more difficult one to manage. Although some fractions are recycled and recovered for energy production and other purposes, large amount of MSW is still not properly disposed. Incineration, which seems to be the best short term solution, often encounters local unacceptance due to environmental concern and the difficult operation due to poor quality unsorted wastes. For sustainable waste management under circular economy concept, considering and planning based on the whole life cycle of MSW as well as raising people awareness to change public behavior on waste generation will be needed for better and easier management.

Keywords Circular economy · Agricultural wastes · Municipal solid wastes · Sustainable management · Energy

S. Fukuda (✉)
The Joint Graduate School of Energy and Environment (JGSEE), Center of Excellence on Energy Technology and Environment (CEE), King Mongkut's University of Technology Thonburi, Bangkok, Thailand
e-mail: suneerat@jgsee.kmutt.ac.th; suneerat.fukuda@gmail.com

© Springer Nature Singapore Pte Ltd. 2021
L. Liu and S. Ramakrishna (eds.), *An Introduction to Circular Economy*,
https://doi.org/10.1007/978-981-15-8510-4_16

List of Nomenclatures

TJ   Terajoule
Ktoe   Kilo ton of oil equivalence
GWh   Gigawatt hour
MWe   Megawatt of electricity
MWh   Megawatt hour
PM2.5   Particulate matter with particle size below 2.5 μm

1 Introduction

The term "wastes" covers agricultural and agro-processing wastes, municipal solid wastes and other wastes including industrial wastes and hazardous wastes. The management and treatment of wastes are different based on many factors including the amount and their characteristics and the process from which they are generated. Only agricultural/agro-processing wastes and municipal solid wastes will be focused in this chapter. Industrial wastes and hazardous wastes will not be included since these wastes need special treatment and management.

With the concept of waste minimization and recycling, agricultural and agro-processing wastes have been utilized for energy and non-energy purposes. Less success has been for municipal waste management. This chapter shares some insights about the supply chain and current management of agricultural wastes and municipal wastes and management in Thailand within the concept of circular economy. Some initiatives as the way forward are also presented as information for future better management and sustainability.

2 Agricultural Wastes

Thailand is an agricultural country. A lot of wastes are generated annually from agricultural plantation and processing of its economic crops including rice, sugar cane, cassava, oil palm and so on. These wastes are present in solid form or known as "biomass" and in form of organic wastewater. Table 1 illustrates major biomass types and potential. Wastes are utilized for both energy and non-energy (e.g. animal food, nutrient recycle into soil as fertilizer) purposes and a large portion of some wastes is still not utilized at their full potential, the top three of which are rice straw, cane tops and leaves and palm fronds and leaves [1].

Table 1 Major biomass types and potential [1]

Type of biomass	Generation (ton)	Current utilization (ton)	Remaining potential (ton)	Heating value (TJ)	Crude oil equivalence (ktoe)	Power generation potential (GWh)	Installed capacity (MWe)
1. Rice straw	19,005,628.14	8,112,801.26	10,892,826.89	134,308.56	3,188.71	7,461,586.42	942.12
2. Rice husk	8,145,269.20	8,006,283.36	138,985.84	1,879.09	44.61	104,393.81	13.18
3. Cane tops and leaves	17,016,248.08	1,845,487.74	15,170,760.34	234,843.37	5,575.58	13,046,853.89	1,647.34
4. Cane bagasse	28,026,761.54	28,026,761.54	0	0	0	0	0
5. Corn leaves and trashes	9,315,603.52	465,780.18	8,849,823.34	86,993.76	2,065.38	4,832,986.86	610.23
6. Corn cob	1,215,078.72	1,094,081.58	120,997.14	1,163.99	27.64	64,666.25	8.16
7. Cassava rhizome	6,045,508.40	164,196.52	5,881,311.88	32,288.40	766.58	1,793,800.12	226.49
8. Cassava cake	1,813,652.52	1,813,652.52	0	0	0	0	0
9. Cassava pulp	8,463,711.76	8,463,711.76	0	0	0	0	0
10. Palm trunk	1,957,280.00	0	1,957,280.00	14,757.89	350.38	819,882.84	103.52
11. Palm fronds and leaves	18,065,006.01	1,707,454.87	16,357,551.14	28,789.29	683.51	1,599,405.00	201.95
12. Palm empty fruit bunch	4,099,859.52	1,891,985.90	2,207,873.62	15,985.00	379.51	888,055.83	112.13
13. Palm fiber	2,434,291.59	2,434,291.59	0	0	0	0	0
14. Palm shell	512,482.44	512,482.44	0	0	0	0	0
15. Bean leaves and stems	65,017.48	3,250.87	61,766.61	1,002.47	23.8	55,692.89	7.03
16. Rubber tree roots and twigs	1,094,365.00	218,873.00	875,492.00	5,751.98	136.56	319,554.58	40.35

(continued)

Table 1 (continued)

Type of biomass	Generation (ton)	Current utilization (ton)	Remaining potential (ton)	Heating value (TJ)	Crude oil equivalence (ktoe)	Power generation potential (GWh)	Installed capacity (MWe)
17. Rubber wood wastes	2,626,476.00	2,626,476.00	0	0	0	0	0
18. Rubber wood slab	2,626,476.00	2,626,476.00	0	0	0	0	0
19. Rubber wood sawdust	656,619.00	656,619.00	0	0	0	0	0
20. Coconut empty bunch	292,909.57	56,824.46	236,085.11	3,635.71	86.32	201,983.93	25.5
21. Coconut carb	333,310.89	329,976.78	3,334.11	54.11	1.28	3,006.26	0.38
22. Coconut shell	252,508.25	230,540.03	21,968.22	393.89	9.35	21,882.79	2.76
23. Cashew nut shell	70,038.56	1,674.28	68,364.29	375.32	8.91	20,851.11	2.63
TOTAL	134,134,102.21	71,289,681.68	62,844,420.53	562,222.85	13,348.12	31,234,602.58	3,943.76

2.1 Current Situation of Agricultural Waste Management

Thailand is the ASEAN leader in bioenergy production. One important factor is the government's long term renewable energy plan which has supported the implementation of bioenergy projects over the past few decades. Earlier attempt for waste minimization at downstream of the industrial process was initially for environmental purpose and/or lowering cost of disposal. Later, more advanced technology adoption and the promotion of government policy have enabled the energy recovery from wastes. There are a number of success cases in industry. For example, the use of husk from rice processing as combustion fuel for steam production or biogas production from cassava processing wastewater for power generation. The recovered energy is used for own process consumption to reduce the net energy import.

2.1.1 Sugarcane

An excellent and commercially proven example showcasing the circular economy concept is in the sugar mill where the wastes along the supply chain can be utilized for value creation (Fig. 1). In Thailand, sugarcane is grown once a year and harvested at around the end of the year. Cane is transported to sugar mills for sugar production for around 4 months. In the crop year 2017/2018, total cane production in Thailand was about 111 million tons [2].

In the sugar production process, there is a major waste after extracting juice from cane, which is bagasse. All bagasse, i.e. around 30% by weight of fresh cane, is used as fuel for steam and power, which are then used for consumption in sugar production processes. In case where the high energy efficiency technology is in

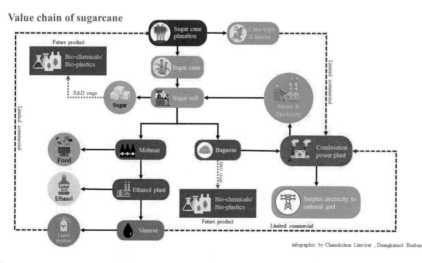

Value chain of sugarcane

infographic by Chanokchon Limvirut , Duangkamol Buaban

Fig. 1 Value chain of sugarcane

(a)   (b)

Fig. 2 **a** Typical scene of cane opening burning to remove tops and leaves. **b** Manual harvesting after burning. *Source* (https://taibann.com)

place, the sugar mill can become energy self-sufficient or even with some surplus. Sugar cane molasse, i.e. the syrup left over after crystallization, is another waste stream from sugar production process. In Thailand, molasse is the main feedstock for the production of bio-ethanol which is used to blend with gasoline at 10–85% for use as gasohol in transportation. At the end of ethanol process, a wastewater stream so called "vinasse" is generated. Full of organic components, vinasse is sent to cane field as liquid fertilizer. When exceeding the need for fertilizer, vinasse is utilized as fuel in two ways. First, it provides organic substrates for biogas production after which biogas is used in form of gaseous fuel. Second, vinasse is concentrated and burned in slurry form. Apart from energy applications, small fraction of bagasse can be used for others like manufacturing particle boards and food containers. All wastes are minimized along the process and values can be created.

However, there are still rooms for improvement. Unfortunately, open burning to get rid of cane tops and leaves before manual cane harvesting has been seen as normal practice in many areas of cane plantation in Thailand (Fig. 2). The limitation of machine harvesting in some areas and to increase productivity of manual harvesting, 60% of the total cane production are "burnt" cane [3]. This open burning not only reduces the value of cane, kills the opportunity for recovery of energy or other benefits like nutrient recovery as fertilizer, but also creates smog (PM2.5), one of life threatening problem nowadays. Logistics management due to their bulky nature and necessary treatment for efficient end-use are also the factors.

2.1.2 Oil Palm

A lot of similar good practices can also be seen for other biomass industries such as oil palm industry (Fig. 3). Compared to other oil producing crops, oil palm has the highest oil yield giving it the most productive yield of oil production per plantation area rai, i.e. 6–10 times more than those of other oil-bearing crops grown globally. Thailand is the third most crude palm oil production in the world, behind Malaysia

Value chain of oil palm

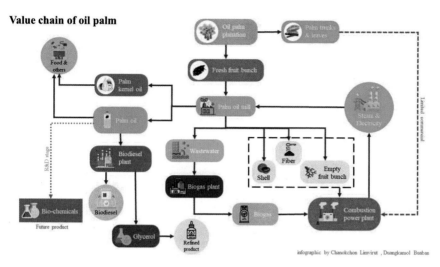

infographic by Chanokchon Linnvirut , Duangkamol Buaban

Fig. 3 Value chain of oil palm

and Indonesia, with an annual output of approximately 2 million tons/year. 85% of Thai oil palm plantations and crude palm oil extraction mills are in the South of the country [4]. The crude palm oil is refined into cooking oil or used as feedstock for bio-diesel. Attempts have been made on developing oleo-chemical industry to diversify the use applications of palm oil which will help make the price of palm less vulnerable, but the commercial applications especially for the high value products are still limited.

The palm fresh fruit bunch (FFB) is sent to palm oil mill. The FFB is processed with steam and the fruits are pressed to get the oil as main product. The oil yield is around 20% by weight of FFB. There are a number of waste streams from the process. Wastewater is used to produce biogas, which is returned to boiler in palm oil mill for energy co-production. Shells can be used as fuels or activated carbon production, while fiber and empty bunches are used as boiler fuels. Palm oil mills can reduce the import of electricity from the national grid and at the same time minimize the wastes. For biodiesel production, waste glycerol is considered as by-product which can be refined and processed into valued added products like soap, food additives.

Every 20–25 years, palm trees are cut down since the fruit production stops. At plantation areas, palm leaves & trunks still remain in numerable quantity as shown in Table 1. However, efficient and cost-effective logistics management between distributed plantation areas and palm oil mills needs to be developed.

2.1.3 Biogas and Biomethane

Conventional systems of biogas production accept low solid content, i.e. 5–10%. Organic wastewaters such as those from palm oil factories, cassava factories, animal

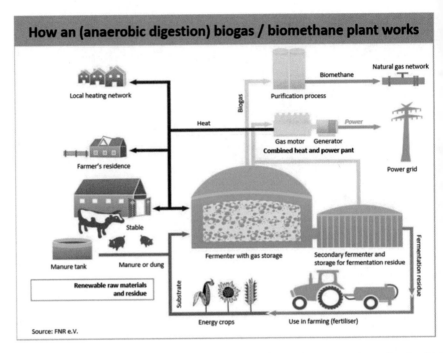

Fig. 4 Biogas/biomethane production line (*Source* https://www.cleanenergywire.org/dossiers/bio energy-germany)

farms are typical feedstocks. Nowadays, lignocellulosic biomass or agricultural wastes in form of solid is pretreated and mixed with wastewater to increase biogas production potential. Biogas is a methane (CH_4) rich and combustible gas. It can be used instead of liquefied petroleum gas (LPG) for cooking or other fuels for heating and power generation. The digested solid can be used as high quality bio-fertilizer.

Biogas can be upgraded into biomethane, which has the thermal and chemical properties similar to natural gas. The upgrading process is done by removing carbon dioxide (CO_2) to enrich CH_4 concentration to the level of natural gas. Moisture and hydrogen sulfide (H_2S and also known as rotten egg gas) are also removed. Biomethane can be compressed and used to substitute natural gas in transportation sector. Figure 4 illustrates the biogas/biomethane production line. Figure 5 presents some CBG fueled vehicles in Thailand (developed by EDRI/CMU[1]) as well as others.

---

[1]Energy Research and Development Institute-Nakornping, Chiang Mai University, Thailand.

Fig. 5 Compressed Biomethane Gas (CBG) fueled vehicles in Thailand and UK (*Source* https://www.bangkokbiznews.com/news/detail/852023; http://www.ngvjournal.com/s1-news/c1-markets/new-british-study-highlights-biogas-potential-for-heavy-duty-trucks/)

2.2   The Way Forward

Bio-based industry is the strength of Thailand for future economic development. The Thai government recently announced the "Bioeconomy, Green Economy and Circular Economy" or BCG as a central development strategy, serving as the foundation of our economic development platform. And under the framework of our newly established Ministry of Higher Education Science, Research and Innovation, "BCG in Action" is one of the four research and innovation flagships to enable sustainable development for green economy. The aim is to create a sustainable supply chain and benefit to all stakeholders along the supply chain.

2.2.1 Sustainability Issues

Open burning to get rid of agricultural wastes including cane tops and leaves before manual cane harvesting, rice straw or corn trash before next cropping is still in current practice in some areas. Government policies and mandates as well as raising farmers' awareness are needed. However, solving the problems have to consider a holistic view of the supply/value chain.

For the case of cane open burning, several recommendations are proposed where a combination of these may be used.

- **Farm mechanization**

Plantation planning to physically allow harvesting machines that can collect canes and leaves at the same time. A function to instantly bale the leaves should be included to increase density, ease storage and reduce the transportation cost. Some portion of leaves may be returned to the soil for nutrient recycling. Examples are in Figs. 6 and 7.

- **Utilization pathways for energy and non-energy**

Currently sugar mill boilers use bagasse as the main fuel with small ratio of cane leaves mainly due to the poor chemical properties of leaves compared to bagasse. Energy RD&D is needed to ensure the high efficiency of boiler when using high ratio

Fig. 6 Mechanized harvesting *Source* (https://northcoastcourier.co.za)

Fig. 7 Baler for cane leaves (*Source* http://www.mitrpholmodernfarm.com)

of cane leaves such as making fuel pellets or torrefied pellets (Fig. 8). Alternatively, research for advanced technology to convert cane leaves into high value products will help attract farmers not to burn away the leaves.

• **High efficiency, low emission**

At the factory or power plant where biomass is used, dust from transportation and storage must be minimized. Adopting high efficiency energy conversion system not only reduces the amount of fuel used, but also the air pollution including carbon monoxide (CO) and particulate matter. Finally, air pollution control devices should be in place if necessary.

• **Government policy**

Fig. 8 Biomass fuel pellet from sugarcane tops and leaves (*Source* TRF Newsletter No. 129)

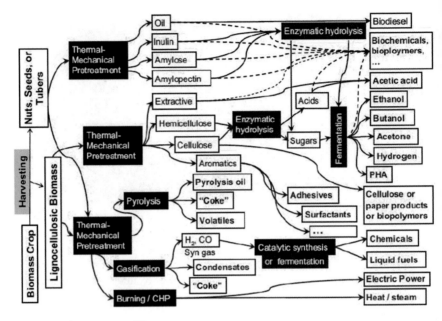

Fig. 9 Biorefinery concept [9]

To accelerate the "no-burning" scheme, strong measures and attractive incentives are needed in processes along the supply chain. These are soft loan for investment in harvesting machine, (short-term) special feed-in-tariff for heat and power production from cane leaves.

2.2.2 High Value Addition

Under the biorefinery concept (Fig. 9), high value products derived from biomass can be obtained. In addition to energy, high value fractions can be extracted for the production of food, pharmaceuticals, biomaterials and so on. Apart from the speed of technology advancement/breakthrough, promoting policy and market will be the key determining factors for their commercialization. There are some advanced biorefinery companies such as Borregaard, Norway (Fig. 10), while many are also under development in Thailand and the world. One example of biorefinery research initiatives, as shown in Fig. 11, is Integrated Biorefinery Laboratory (IBL), the collaborative laboratory co-established by JGSEE/KMUTT[2] and BIOTEC/NSTDA.[3]

---

[2]The Joint Graduate School of Energy and Environment (JGSEE) and Center of Excellence on Energy Technology and Environment (CEE) at King Mongkut's University of Technology Thonburi, Thailand.

[3]The National Center for Genetic Engineering and Biotechnology (BIOTEC) at the National Science and Technology Development Agency (NSTDA).

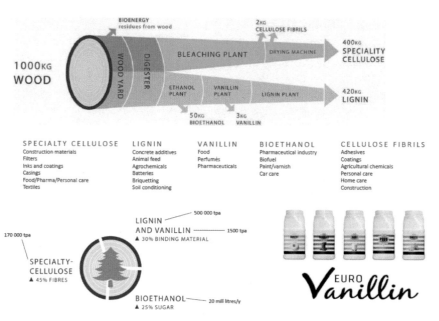

Fig. 10 Borregaard's integrated biorefinery [10]

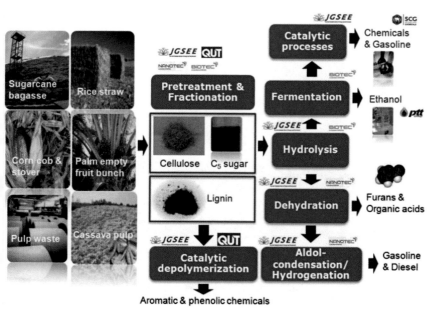

Fig. 11 Integrated biorefinery laboratory

3 Municipal Solid Waste (MSW)

Municipal solid waste (MSW) is a heterogeneous mixture of matters generated from human daily activities. Generally, the amount of wastes increases with the growth of economy and population as well as lifestyle changes. As shown in Fig. 12, the amount of waste generated in Thailand in 2018 was 27.8 million tons, an increase of around 1.6% from 2017 [5]. The amount of waste generated in the capital city, Bangkok, is also increasing steadily as shown in Fig. 13. The MSW generation rate is slightly more than 1 kg/day/person, which is higher than other Asian cities like Tokyo, Seoul, Singapore [6] but in the same range with other ASEAN big cities like Kuala Lumpur (as shown in Table 2). Even higher waste generation rate, i.e >2 kg/person/day, is found for a tourist city like Phuket.

MSW in Bangkok consists mainly of plastics, paper, food waste and bio-wastes (Fig. 14), which is similar to most of other Asian cities (Table 2). Although some efforts for waste separation (Fig. 15) have been made, only those for recycling can benefit. Without organic wastes sorted out at source, MSW comes in a form difficult for disposal e.g. containing too high moisture content or toxic materials. In 2016, waste generation in Thailand was around 27 million tons but only 35% was properly disposed [5].

3.1 Current Municipal Waste Management

Out of 27.8 million tons of waste generated in 2018, around 34% were separated and re-utilized for recycling and natural fertilizer [5]. The rest went to proper and improper disposal as shown in Fig. 12 and Table 3. Among waste generated, around

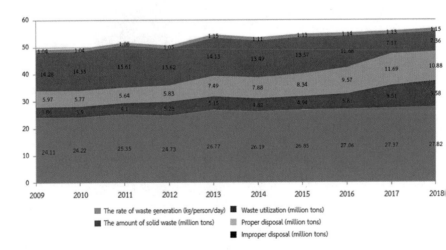

Fig. 12 Solid waste generated in Thailand in 2009–2018 [5]

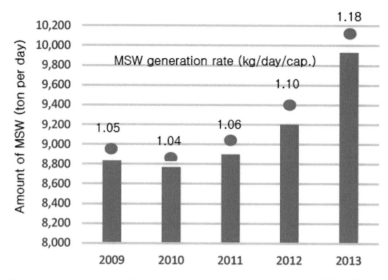

Fig. 13 Amount of waste generation and generation rate in Bangkok during 2009–2013 [6]

2 million tons were municipal plastic waste. Around one fourth was recycled, which were mainly plastic bottles. The rest, non-recycle, were mainly plastic bags and others. Improper disposal includes open dumping, opening burning, throwing away to water sources. These activities have led to contamination in the sea, as we have already seen the negative impact on seas and marine lives.

In some big and tourist citie with large capacity of wastes (i.e. >100 ton/day) like Bangkok or Phuket, MSW is simply incinerated with energy recovery or so-called waste to energy (WTE) plant. There are a few other small incinerators and sanitary landfills for landfill (CH_4 rich) gas recovery. However, a significant portion of MSW nationwide is just dumped in landfill sites, most of which do not meet sanitary regulation and cause further problem of leakage. Some wastes are even dumped in open air. Although incinerator seems to be the best short term rescue before wastes will overflow the city, **Not In My BackYard** phenomena have stopped or delayed many projects. One of the reasons is that people do not trust that incinerators will not create pollution which will affect their livings. Some examples case studies of waste management are given below.

- **Waste to energy plant in Bangkok**

  Bangkok produces 8,700 tons of garbage per day. Not until May 2016 when the first incineration plant was built and in operation, MSW from Bangkok was collected and sent to landfill in other nearby cities. This waste to energy (WTE) plant, located in Nong Khaem District in the western area of Bangkok, has a capacity to burn 300–500 tons of city wastes per day to produce electricity up to 9.8 MW. It is the third large incineration plant in Thailand after those in Hat Yai and Phuket. The lack of proper waste separation and contamination of other

S. Fukuda

Table 2 MSW generation and composition in cities of Asian developing countries [12]

City (Country)	Waste generation		Composition (in percentage)									
	Tons/day	Kg/cap.day	Decomposable organic	Paper	Plastics	Textile	Glass	Metals	Rubber/leather	Wood	Ash	Others
Surabaya (Indonesia)	2160	0.8	72.41	7.26	10.09	2.68	1.7	1.41	0.46	2.39	1.48	0.12
Jakarta (Indonesia)	6000	0.65	68.12	10.11	11.08	2.45	1.63	1.9	0.55	NA	NA	4.12
Allahabad (India)	500	0.4	45.3	3.6	2.86	2.22	0.73	2.54	41.66			
Puducherry (India)	370	0.59	42	30	10.4	4.5	5	4.1	2.5	1.5	NA	NA
Kathmandu (Nepal)	523.8	0.66	71	7.5	12	0.9	1.3	0.5	0.3	NA	NA	6.7c
Bangkok (Thailand)	8778	1.54	42.68	12.09	10.88	4.68	6.63	3.54	2.57	6.9	NA	10.04
Phuket (Thailand)	364	2.17	49.39	14.74	15.08	2.07	9.67	3.44	2.28	NA	NA	3.33
Yala (Thailand)	80	1.049	49.3	14.5	19.9	–	10.08	0.4	–	5.1	NA	NA
Kuala Lumpur (Malaysia)	3798.9	1.62	61.5	16.5	15.3	1.3	1.2	0.25	0.6	0.4	0.7	NA
Rasht (Iran)	420	0.8	80.2	8.7	9	0.4	0.2	0.7	–	0.4	NA	0.4
Dhaka (Bangladesh)	5340	0.485	68.3	10.7	4.3	2.2	0.7	2	1.4	–	NA	10.4

*NA Not available

Fig. 14 Average composition of municipal solid waste (MSW) in Bangkok between 2007 and 2011 [11]

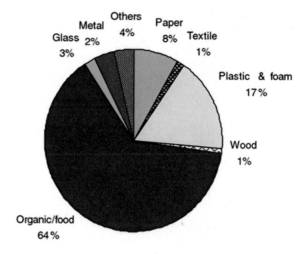

Fig. 15 Waste separation and recycle (*Source* (1) www.bkkkids.com/blog/recycling-in-bangkok/; 2) www.prachachat.net/economy/news-283137)

kinds of wastes especially hazardous waste are still the problem for burning and operators must take extra cautions.

- **Kamphaeng Saen landfill waste to energy project**
  With the aim to reduce fossil-based electricity production, this landfill waste to energy project is the first in Thailand to extract landfill gas (LFG) and convert it into electricity, which is then fed to the National Grid (as shown in Fig. 16). The project facility is located at the main landfill site in Kamphaeng Saen District, Nakhon Pathom Province, around 100 km west to Bangkok, where approximately 5,000 tons of waste per day from Bangkok is sent. Each year this CDM registered project reduces 550,000 tons of emissions, and generates approximately 25,000 MWh of renewable electricity [7], delivering towards United Nations Sustainable Development Goals (SDGs), especially SDG13 Climate action.
- **Waste sorting at source and RDF production**
  In the past, initiatives for waste sorting at source and pretreatment like processing into Refuse Derived Fuel (RDF) were on demonstrated in communities but most of the projects could not continue successfully due to both technical failure and the

Table 3 Status of municipal waste disposal and transfer sites [5]

Total of 3,205 municipal solid waste disposal and transfer sites			
2,786 in operation		419 are closed	
Public sites	Private sites	Public sites	Private sites
2,398	388	371	48

Total of 2,764 municipal solid waste disposal and total of 22 transfer sites
Total of 647 proper municipal solid waste disposal and transfer sites

Type	Number (site)	
	Public	Private
Sanitary landfills/Engineered landfills/Semi-aerobic landfills	90	19
Controlled dumps with a capacity of less than 50 tons/day	386	87
Incinerators with air pollution control system	16	11
Incinerators for energy production	0	6
Compost system	6	3
Mechanical-biological treatment system/Refuse derived fuel production	18	5
Total	516	131

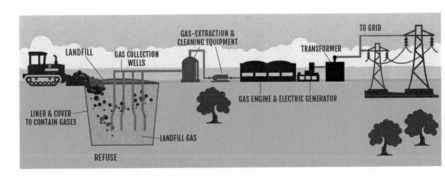

Fig. 16 Landfill gas to energy (*Source* http://www.advanceddisposal.com)

lack of local perception and willingness. Partnership between private sector and local administrative office together with the involvement of local people seems to help to success. Currently, some of MSW and RDF are continuously used as fuel/raw material for cement manufacture as well as for WTE plants.

3.2 Government Policy on Waste Management

Government policies on waste management do not seem to align with each other. There have been campaigns for waste reduction and recycling, for example, producer participation in eco-friendly products and package designs and more recently the limit and ban of single use plastic products. However, there is no formal recycling scheme or plan to prevent the increasing levels of solid waste. Separation or sorting at source is also not a common practice.

At the same time, waste to energy especially for power generation is the target of the Alternative Energy Development Plan (AEDP) (2015–2036). With a challenging target of 900 MWe installed capacity (100 MWe at the end of 2019) [8], there is a great concern on the lack of waste feed to the WTE plants. While the waste management including the 3R (reuse, reduce, recycle), waste sorting and waste pretreatment is implemented, focusing waste on energy production seems to move forward with about 10 new WTE plants to be commissioned in a next few years with total installed capacity of around 100 MWe. In long term (20-year span), the city development and changing behavior will influence tremendously the characteristics of wastes and the future scenario for WTE plant operation. Therefore, a clear and consistent policy integrating related agencies is needed for sustainable waste management.

3.3 Master Plan of Sustainable Waste Management

The Pollution Control Department (PCD) under the Ministry of Natural Resources and Environment, as the implementation agency, has developed a 20-year master plan for sustainable management of solid and hazardous waste (2018–2037) with the framework that integrates 3R, Circular Economy/Waste to Resources and Polluter Pays Principle (PPP).[4] Strategies with some activities already implemented under this master plan are summarized below.

Although sustainable waste management in Thailand seems to be moving towards the right direction, it can be clearly foreseen that contribution will be only on voluntary basis or driven by a limited number of big private companies. Polluter Pays Principle is not likely be adopted since there is currently no legal duty to the environment in Thailand and there are a lot of arguments on its socio-economic impacts. Careful design of action is needed to guarantee efficiency or cost effectiveness in environmental protection.

---

[4]Polluter Pays Principle (PPP) states that whoever is responsible for damage to the environment should bear the costs associated with it. (Taking Action, The United Nations Environmental Programme.).

Strategies of municipal solid waste management under this master plan include [5, 6].

- Promote waste reduction and separation at sources by creating awareness and asking participation from people and private sectors, e.g. contribution to limit and ban single use plastic products.

- Clustering for centralized solid waste management by encouraging and supporting local administrative organizations to adopt appropriate waste disposal technologies and practices for different types of separated wastes. Private sectors are also encouraged to participate in joint investment (Public Private Partnership) through financial incentives, relaxation of related laws and regulations and so on, together with local administrative organizations. This scheme could evidently be seen to attract lots of interests for investment.

Clustering waste management		
Model L	>300 Tons/day	• Waste sorting at waste to energy (WTE) plant • Waste to Energy plant by RDF/Incineration/Biogas/Composting • Rejected to landfill
Model M	50–300 Tons/day	Same as Model L
Model S	<50 Tons/day	• Source separation • Integrated disposal technology

- Recommend solutions and follow up the improvement of land fill sites by site cleaning (land reclaiming) and producing RDF from old landfill mined wastes for sending to be co-fired at cement plants.[5]
- Launch guideline and criteria regarding solid waste management along the whole waste supply chain and all available treatment technologies.
- Plastic waste management including
  a. draft a plastic waste management roadmap (2019–2027) with the aims of (1) 100% re-utilization of all plastic wastes by 2017, and (2) reduce

or stop using plastics products and packaging, or shift to alternative eco-friendly materials, e.g. stop using bottle cap seal in 2019 or stop using foam meal boxes and plastic straw by 2022.

b. reduce import of plastic scraps.

- Marine debris management by reducing amount of solid wastes thrown away into the sea by target groups.

4 Exercises

(1) Identify a potential agricultural waste and how to utilize it based on circular economy concept.

Answer:

(2) Draw the whole life cycle of MSW and identify where improvements can be made by 3R concept.

Answer:

References

1. DEDE, Biomass potential database for crop plantation year 2013 (in Thai).
2. OCSB (Office of the Cane and Sugar Board), Cane production report 2017/2018 (in Thai).
3. Bank of Thailand, Analysis Report "Why burning cane?" 2019 (in Thai).
4. Krungsri Research, THAILAND INDUSTRY OUTLOOK 2017–19: OIL PALM INDUSTRY, June 2017.
5. Pollution Control Department (PCD), Booklet on Thailand State of Pollution 2018.
6. Kerdsuwan, K. (2017). *Status of waste management and future policy directions for renewable energy from waste and biomass in Thailand, Expert Group Meeting on Sustainable Application of Waste-to-Energy in Asian Region.* Korea: United Nation Office For Sustainable Development.
7. www.naturalcapitalpartners.com/projects/project/kamphaeng-saen-landfill-gas-to-energy.
8. Department of Alternative Energy Development and Efficiency (DEDE), Alternative Energy Development Plan (AEDP) 2015–2036.
9. Amidon, T. E., Bujanovic, B., Liu, S., & Howard, J. R. (2011). Commercializing biorefinery technology: A case for the multi-product pathway to a viable biorefinery. *Forests, 2,* 929–947.

---

[5]Co-firing RDF (mainly consisting of plastic) at cement plants is so far the best existing technology in Thailand regarding environmental concern due to the high temperature (i.e. >1500°C) and long residence time (i.e. >2 s) during combustion inside cement kiln

10. www.borregaard.com.
11. Thitanuwat, B., Polprasert, C., & Englande A. J. (2016). Quantification of phosphorus flows throughout the consumption system of Bangkok Metropolis, Thailand. *Science of the Total Environment, 542,* 1106–1116.
12. Dhokhikah, Y., & Trihadiningrum, Y. (2012). Solid waste management in asian developing countries: Challenges and opportunities. *Journal of Applied Environmental and Biological Sciences, 2*(7), 329–335.

Suneerat Fukuda is the Director of the Joint Graduate School of Energy and Environment (JGSEE) and Centre of Excellence in Energy Technology and Environment, King Mongkut's University of Technology Thonburi, Thailand. Dr. Fukuda received BSc in Chemical Technology from Chulalongkorn University, Thailand, in 1996. Then, she received M.Sc. and Ph.D. in Chemical Engineering from Imperial College London, UK, in 1998 and 2002, respectively.

Dr. Fukuda's major expertise is solid fuel utilisation for energy with focus on biomass/waste conversion for heat and power, coal combustion, co-firing of coal and biomass in circulating fluidised bed and bio-oil production and upgrading. Together with fundamental scale studies, she has established JGSEE pilot plant for the purpose of solid fuel combustion and gasification research and industrial services. Large-scale fluidised-bed facilities, high-temperature drop-tube facility as well as related analytical methods were developed. Over 15 years of study has made her research laboratory one of the best in this field in Thailand and ASEAN. Dr. Fukuda is the leader of research projects on biomass/waste supported by government agencies and industries and has published a number of papers in peer-reviewed journals as well as technology guidebooks. She has also been involved in a number of projects/collaborations supported by international organisations e.g. BMBF, French Embassy, JST/JICA.

New Paradigm for R&D and Business Model of Textile Circularity

Edwin Keh

"Everything Old is New Again"—Recycling of Used Apparel.

Abstract Recycling of garments is difficult due to the complexity of blended materials used during manufacturing. There is also the logistics challenge of processing these used materials in cities where there are no factories. Added to all this is the sheer size of this growing problem. Environmental sustainability is an urgent global challenge. We not only need to come up with innovative technologies but also to scale up these innovations rapidly into viable businesses. New research paradigms and new business models are all necessary to materialize technology and innovation into viable and impactful solutions. We share our story of the accelerated development of innovative and scalable solutions in recycling used apparel through a public-private partnership. Our key innovations include a resource-efficient and low-cost hydrothermal materials separation system and an automated, intelligent, and chemical-free mechanical recycling system that enables us to process used garments into high-value raw materials for new clothes as well as new business model for scaling up the garments recycling.

Keywords Post-consumer apparel recycling · Recycling methods · Hydrothermal recycling system · Textile circularity · Apparel circularity · R&D · Innovation · Public-private partnership

Learning Objectives

1. Circularity of Textile
2. Technology and innovation for circularity

E. Keh (✉)
The Hong Kong Research Institute Research of Textiles and Apparel, Hong Kong, China
e-mail: edwinkeh@hkrita.com

The Wharton School, University of Pennsylvania, Philadelphia, USA

© Springer Nature Singapore Pte Ltd. 2021
L. Liu and S. Ramakrishna (eds.), *An Introduction to Circular Economy*,
https://doi.org/10.1007/978-981-15-8510-4_17

3. Developing the urgency mindset for circularity and focus on solution development
4. Multidisciplinary research and development for circularity R&D
5. Accelerate circularity through Industry partnership.

1 Introduction and the Problem

A small city like Hong Kong (population 7.4 M in 2019) throws away more than 300 tons of apparel everyday.[1] Some of this material can be reused, resold, or given away to charity. The rest of the used materials are either too damaged or too soiled to be used as is.

Up till 2018 the easy solution is to export this, along with other recyclable waste to China.[2] Then in January 2018 China closed its doors to imported waste. All of the sudden Hong Kong, along with many other cities in the world, was left with few options to deal with their old apparel and other waste. In Hong Kong these became landfill material. In other parts of the world they are landfilled or incinerated (Fig. 1) Neither of these are sustainable solutions for our growing global cities.

Recycling and the reuse of old apparel material is a much talked about solution.

While we have the options to use our clothes longer, buy more second hand apparel, or give more of our old apparel to others, we are still left with some unusable damaged, and soiled garments. These need to be reprocessed or they are destined to be landfilled or incinerated.

Today they can at best be down-cycled to be used in a lesser value application Possible uses include as insulation, carpeting, or rags.

The other problem we face with any type of reprocessing is the logistics challenge Used materials are now mostly found in the consuming economies and not in the manufacturing economies. The 300 plus tons of daily discarded clothes from Hong Kong, for example, are in a city that no longer has yarn mills, fabric processing plants, or apparel factories. The same is true across most Europe and the Western economie (Fig. 2). For the last 50 years we have built a highly effective linear apparel supply chain that makes apparel in the East for consumption of apparel in the West.

To complicate matters, most of our clothes today are made from blended materials. To reduce cost, improve comfort, and enhance performance, most of our clothes use a blend of synthetic petroleum-based materials (the most common is polyester), cellulosic materials (most common is cotton), and protein materials (like various wools and silk). Blended materials are great in use however these materials are so well constructed as fibers that they are very difficult to separate. Unfortunately, if used blended materials are not deconstructed, they are very difficult to reprocess re-dyed, and reuse in any high-value application (Fig. 3).

---

[1] https://www.scmp.com/news/hong-kong/health-environment/article/2179680/all-dressed-and-nowhere-go-except-landfills-fast.

[2] https://www.nationalgeographic.com/environment/2018/11/china-ban-plastic-trash-imports-shi fts-waste-crisis-southeast-asia-malaysia/.

Fig. 1 How much clothing are we throwing away

Reprocessing, separation, and deconstruction of used apparel are usually done in the following three general ways:

1. Mechanical Recycling—the most mature and common: Here clothes are cut shredded and pulled apart down to yarn and then to fiber before the material is reconstituted and spun back into fabric. The advantage of mechanical recycling is that this can use existing machinery for the processing of virgin material. The process is quite straightforward and well understood. The major disadvantage is that the resultant materials are usually of shorter fiber lengths and have inferior performance and coarser hand feel.
2. Chemically breakdown used apparel and to recover useful materials: This usually involves the use of various acids or ionic fluids in the process. The advantage of this process is its efficiency in the recovery of synthetic materials. The disadvantages include the relatively high costs of the chemicals used and the resultant waste stream of used processing chemicals.
3. Biological methods using various enzyme and fermentation processes to separate and breakdown materials. The advantage of this method is the relative energy efficiency of the processes. The disadvantage is these processes are slow and inefficient.

A dash in the table means that data is not available.

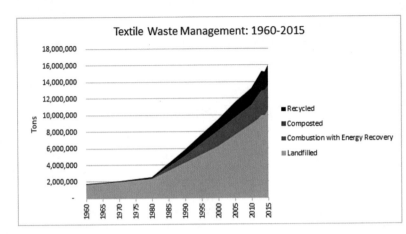

1960-2015 Data on Textiles in MSW by Weight (in thousands of U.S. tons)

Management Pathway	1960	1970	1980	1990	2000	2005	2010	2014	2015
Generation	1,760	2,040	2,530	5,810	9,480	11,510	13,220	15,240	16,030
Recycled	50	60	160	660	1,320	1,830	2,050	2,260	2,450
Composted	-	-	-	-	-	-	-	-	-
Combustion with Energy Recovery	-	10	50	880	1,880	2,110	2,270	3,020	3,050
Landfilled	1,710	1,970	2,320	4,270	6,280	7,570	8,900	9,960	10,530

Sources: American Apparel and Footwear Association, International Trade Commission, and the Secondary Materials and Recycled Textiles Association

*MSW=Municipal Solid Waste

Fig. 2 Municipal waste and recycling volume growth in the USA in the last half-century

What happens to the clothes, shoes, and linens you give?

5%
Unusable – if textiles are wet, moldy, or contaminated with hazardous materials, they cannot be reused or recycled

20%
Reprocessed into fiber used in new products, such as carpet padding and insulation

30%
Recycled and converted into items like commercial wiping rags

45%
Reused as secondhand clothing, in the US or other countries

Source: Secondary Materials and Recycled Textiles Association Media Kit, www.smartasn.org/about/SMART_PressKitOnline.pdf

Fig. 3 What happens to clothes

2 Partnership of HKRITA, H&M Foundation, and Novetex

2.1 About HKRITA

The Hong Kong Research Institute of Textile & Apparel (HKRITA) was established in 2006 by the HK Government to conduct applied research. It is funded by the Innovation and Technology Fund (ITF). HKRITA's mission is to engage in useful applied research to support the textile, apparel, and fashion industries.[3]

The operating leadership of HKRITA consists of a CEO, the Head of Research, the Head of Project Management, and the Head of Business Development. They are supported by a small administrative team that takes care of the project proposals, funding paperwork, the management of intellectual properties developed by the center. The in-house research team has about 50 research scientists. They are complemented by research work that is funded by HKRITA in various universities locally and more and more globally.

Research work began in earnest at about 2009. The center starts 20–30 new projects a year, and there are at any time about 50–60 ongoing projects. Project direction is driven by industrial needs in regular discussions with stakeholders. The ITF funds the operations of the center and is a major funding source of all projects.

In 2016, HKRITA embarked on a series of experimental projects to solve some of the major challenges with circularity and the recycling of used or excess apparel materials. In addition to the ITF funding, these projects are supported by the H&M Foundation and Novetex.

---

[3] See www.hkrita.com.

2.2 About H&M Foundation

The H&M Foundation is a non-profit global foundation. The Stefan Persson family privately funds it. They are the founders and main owners of the H&M Group. The Foundation's mission is to drive long-lasting positive change and improve living conditions by investing in people, communities, and innovative ideas. The Foundation has a goal of accelerating the progress needed to reach the UN Sustainable Development Goals by 2030.

Since 2013, the family has donated 1.5 billion Swedish krona ($200 million/€163 million) to the H&M Foundation.

The Global Change Award (GCA) is the Foundation's annual challenge to startups and entrepreneurs to come up with new ideas around circularity and sustainability. GCA has given out $1 M Euro prize money each year to the winners.

In 2016 the Foundation announced a four-year research agreement with HKRITA to work on solutions to circularity and sustainability for the fashion industry.

2.3 About Novetex

Novetex was founded in 1976 by industrialist Mr. CHAO Kuang-piu (who also founded Hong Kong's Dragon Air Airlines). The Chao family still owns Novetex. It produces about nine million kilograms of wool annually from its main factory in Zhuhai (China). Head quartered in Hong Kong with over 1000 employees, Novetex has expanded its operations over the last four decades to become one of the world's largest single-site spinners. The factory in Zhuhai (China) has four spinning mills and one dyeing mill that can produce high-quality yarn of different blends and count from different types of fibers such as Yak hair, organic cotton, Merino wool, and other types of wool products.

In 2015 Novetex approached HKRITA to work on the challenges of improving the efficiency of yarn recycling and exploring new business models to make recycling a sustainable and viable business. This system was subsequently named the "Billie System" by Novetex.[4] Billy or Billie was ChAO Kuang-pui's nickname.

As the interests of Novetex and the H&M Foundation are similar they began to collaborate on various HKRITA research projects together.

---

[4]https://thebillieupcycling.com/interview-with-chairman-ronna-chao/.

3 Solving the Problem

HKRITA developed an innovative "Hydrothermal Recycling System" during the first series of experimental research project with the H&M Foundation funded by ITF.[5] This set of projects set about to look for practical ways to separate mixed synthetic and cellulosic fabrics.

The hydrothermal system is very energy efficient and low cost in processing used materials using heat and pressure as the mechanism for separation of synthetic and cellulosic fabrics. The recycled synthetic materials are recovered in fiber form. This makes their reuse as valuable material for apparel commercially viable.

Initially we explored at the laboratory scale the various chemical-recycling methods. The drawbacks of the harsh chemical waste streams meant that new problems were being created. To deal with this, the research team challenged themselves to try to eliminate or reduce the various chemicals to see if the process can be simplified. As the use of the various harsh chemicals was reduced, the team was excited to find methods to effectively do the work of the material separation with just heat and pressure aided by a small amount of benign green chemicals. This breakthrough in developing a scalable hydrothermal recycling technology in the HKRITA was achieved in the spring of 2017 and announced in the fall of 2017.

In Parallel in 2016, HKRITA and Novotex started to work on automation systems for mechanical recycling of yarns and fibers during manufacturing processes. In addition, Novotex saw an opportunity and growing demand in the market place for recycled materials for apparel manufacturing. The challenge was to create effective processing methods and to reduce production costs through automation.[6] The "Billie System" was developed to effectively and efficiently recycle from garments to yarns and fibers with automation. It is an intelligent and highly automated mechanical recycling system. The goal of building this system in Hong Kong is to provide a local processing solution to all the available recyclable materials in HK.

4 Scaling the Research—Disrupting the Disruption

The usual path of applied research is to systematically go from laboratory scale work to bench scale work to refine the results. The work is then either scaled up from there or additional experiments are performed. While this work is going on, there are the tasks of filing patents and the registration of IPs. Then the work goes onto the stage of publicizing the results and the experiments involving academic publications and conference presentations. Following this process, it usually takes a long time for research in the lab to impact industry and solve real-world problems.

The hydrothermal experiments are known as ITP/103/15TP and ITP/025/17TP. For both these projects the ITF was the majority contributor to the research funding.
The Novotex recycling automation project is known as ITP/002/16TI. Novotex is the majority funding contributor for this project.

The research team in HKRITA and its partners felt that the subject matter of sustainability is not only time sensitive, it is also an existential threat, and critical While the recycling of apparel may not be the silver bullet to save our planet, the apparel, fashion, and textile industries are all significant contributors to the pollution of our planet, they also produce significant greenhouse gases during production and transportation. These industries consume significant amount of resources, energy and water. The sooner various recycling technologies are scaled up to industria scale, the better. There is an environmental crisis with conventional manufacturing and the use of limited raw materials and resources in these industries is unsustainable

5   Integrating the Solutions and Making an Impact Through an Accelerated and Circular R&D Model

In 2017 when the lab work showed significant promise on several of our projects there was general frustration and concern that the path to commercialization will be too long. The concern is that we will arrive at an industrial solution too little and too late for the environment. In discussions with the Innovation and Technology Fund (ITF), and with the H&M Foundation, we found a shared desire to find a faster path to scale and industrial use.

The alternative path to scale is to abandon the conventional path and use a product development model favored by the PC and software industries. As the technology industry favors fast product launches, new products are launched as soon as these reach a level of acceptable functionality. Enhancements and improvements are subse quently released as upgrades. New products are released as version 1.0, with new generations and fixes released in subsequent versions and generations.

Instead of taking the usual predictable and safe path to scale, we conduct research and experiments to continuously enhance and improve in industrial scale. Our path of research projects is from lab to the development of industrial-scale solutions as soon as possible through concurrently solving system engineering problems, while working out the necessary "science at scale" solutions. This could pose significant financial and reputational risk.

A lot of multi-discipline challenges will have to be taken on by our small research team concurrently. These include operational ergonomics considerations, system integration issues, materials logistics challenges, and new business model design All of these, as well as other yet to be considered issues will have to be resolved in an accelerated time frame. The complexity of managing the project is significant One of the ways HKRITA in mitigating the complexity is to begin with the desired market outcomes and work backwards from there to create the research projects By understanding the desired solutions, we can more effectively frame research questions to address them.

Another approach we have adopted is the "open platform", a research model that we have been using for a few years. Here we build a collaborative research platform (forming a public-private partnership) with participation and engagement for multiple stakeholders along the supply chain. The idea is to invite strategic partners to work together. These could include system engineers, raw material suppliers, manufacturers, importers, and retailers. HKRITA would pull all the parties together as co-sponsors of these "open platform projects". HKRITA would also contribute project funding and help design and manage the project. All sponsors contribute to and direct the course of the project, and share know-how and learning. When the project is completed all sponsors share the benefits of the outcomes. There are several advantages to this arrangement. These include immediate ownership of the outcomes, better awareness of all stakeholders concerns, and a much faster road to adoption. Often times this engagement not only accelerate the R&D processes, but a lot of issues are resolved during the execution of projects and resulting in accelerated adoption of the outcome.

5.1   The Fateful Decision and the Long Hot Summer

In the Spring of 2017, both the hydrothermal industrial system and Billie system were both funded and everyone involved in the research team was busy in designing and building these industrial-scale systems.

In May 2018, as work on many fronts was in progress, there was a meeting of the HKRITA research leadership, the Novotex team, and the H&M Foundation teams in Copenhagen. The question to consider was to commit to system launch dates. To draw the most attention, an opportunity was to launch during HK Fashion week when many of the industry are gathered in one place. It would be even more impactful to simultaneously launch both the Hydrothermal system and the Billie System, since they are being designed to be in the factory building in Hong Kong. Together they make a powerful statement about the viability of recycling, and the reality of a circular apparel business.

Hong Kong Fashion week is the first week of every September. In May 2018 the big decision on timing is whether the launch date is September 2018, when these projects are fresh, new, exciting systems, or be conservative and aim for September 2019 when there has been ample time for delays, tests, and tweaks.

So much is on the line if either fails to launch on time, or launch but experiences significant issues in public. The alternative is to launch later but risk the loss of momentum.

After a series of discussions, and anxious calls to confirm work schedules, and soul searching, the teams came to the conclusion that just as we embarked on disrupting the research model, we should in the same spirit of risk taking challenge ourselves to launch to an aggressive timeline. Collectively we worked to mitigate the risks of failure and delays. Everyone had to work together to make the 4 months to launch time line possible.

This decision in May was the beginning of the long hot summer of work, suspense, excitement, and ultimately a successful launch. Aggressive engineering work schedules were planned and a large-scale international launch event was planned.

On Sept 4th 2018, both the Billie system and the Hydrothermal industrial systems were successfully launched attracting significant global attention. Since the launch, we have been working on building bigger systems based on our existing designs requested by partners and customers

The Billie system has started shipping the yarns produced. These are currently being sent to China where products have been made and sold to brands. The first products (sweaters for the "Conscious Collection" in H&M Stores) are already in the market.

An enhancement to the hydrothermal system is in work to remove dye while the separation is in process. The addition of dye filters will ensure the outputs to be immediately ready for production.

Work is at hand to design and build significantly larger systems in other locations.

6 Technology Innovation Enabling New Business Model for Circular Economy

Post-consumer materials are hard to recycle because of the complex logistics involved. Mixed materials are also hard to separate and reprocess.

From day one of our R&D, we engage stakeholders along the entire supply chain of apparel who provide us technical and business guidance. The outcome of our scalable technology innovation, hydrothermal industrial system, and Billie system enables the On-site recycling of garments at the retail stores in the first world cities This has solved the problems of both materials separation and logistics complexity enabling scalable and low-cost recycling of garments.

Currently, the Billie System at capacity is able to produce yarns from recycled materials that cost less than that from virgin materials at comparable performance. The Garment to Garment system will be more profitable with increased output. Our industry partners are also gaining first mover advantage through practicing circularity in the apparel industry.

Our accelerated R&D efforts in solving problems in the circular economy have provided our industry partners competitive business model and profitable business outcome.

7 Challenges Ahead

Protein materials separation is still an unsolved issue. Protein-based materials such as wools and silks are expensive and widely used. The hydrothermal system destroys proteins in process, the Billie System blends these with other less valuable materials. There is still a need to separate these before putting materials into either system so as to get more value out.

Application for the cellulosic materials is still in the works and so there are still unknowns about how valuable these outputs can be. Ideally a better process can be developed to recover these as fiber as well so that they can be reused without future processing.

The ergonomics of these as large industrial-scale systems will still need more research and experiments. We have yet to understand what modifications are needed when these systems are scaled to ten or one hundred times their current outputs.

System Integration was the biggest challenge for both these systems in development. Getting smaller systems to operate as parts of one large system is complicated. Operational problems are hard to diagnose and debug. Experts and researchers from different disciplines need to be willing to come out of their domain comfort zones which could be intimating and confusing at times. Working in a diverse team could be frustrating sometimes. However, by managing expectations of the early users, the version 1.0 adopters, these problems eventually get resolved.

8 Conclusions

There are different paths to useful research application. A multi-discipline team willing to take on some risks can accelerate progress and rapidly introduce solutions for industrial use. Circularity is an urgent challenge and requires disruptive thinking at all levels. This includes the most fundamental levels of ideation and experimentation. Proceeding at speed requires collaboration of researchers from different disciplines, the willingness of stakeholders to take on new risks, and for the teams involved to be focused on the same outcome.

In a situation where research work and scaling are done in near simultaneously manner, a tolerance for imperfection is required. Things will not all work and some level of compromises maybe required for the sake of rolling out a faster solution.

At the same time communication is critical. There should be general buy-in from researchers, sponsors, and end users that there are lots of moving parts and enhancements are ongoing. The willingness to quickly adapt solutions is necessary as required by the urgency of our environmental challenges.

Finally, the transition to circular apparel can be accelerated through public-private partnership using the Open Platform solution development model involving stakeholders along the entire supply chain.

Appendix

About the Hydrothermal System

The hydrothermal recycling system. *Photos* Edwin Keh

The first Hydrothermal Separation System can process about 150 kg of material a day.

How does it work.

The system consists of 2 pressurized tanks that treat and separate materials. These tanks then discharge the materials into filtration tanks where the synthetic materials are separated from the cellulosic materials. This 2 tank system allows the treatment of materials in tandem by moving pressure and heat back and forth between the tanks.

The hydrothermal system developed at HKRITA is very energy efficient and is a low-cost method to process used materials. The use of heat and pressure as the mechanism for separation is a significant breakthrough.

The recycled synthetic materials are recovered in fiber form. This makes their reuse as valuable material for apparel commercially viable. Using the conventional methods for recycling of synthetic materials, there is a lot of reprocessing involved. Materials are cleaned, melted, and reprocessed back into fiber form by extrusion. These recycling pathways are usually energy and chemical intensive, consuming a lot of water, and weakening the materials in the process. Our hydrothermal system keeps material in fiber form so it is a very short and straightforward way to reuse. It is a very fast process and the pressure, heat, and water use can be reused with minimal loss. What is more, materials separated and recovered in this manner do not seem to exhibit any weakening of fiber strength or performance. The recycled materials can be used as they are for the making of yarns and can be knitted into garments with simple blending of a percentage of virgin material to strengthen the tensile strength f the final yarn. Early trials of the yarn in production are very positive. More than 0% of the resultant yarn and fabrics can be made up of old materials. There is ongoing work to increase the percentage of old material.

The cotton in the processed materials is separated and is recovered in a powdered orm. This cellulosic material is too short and powdery to be turned into fiber as is. The material absorbs many times its weight in moisture. So there is ongoing work o use this material as a performance coating material to other fibers, to use this as n aid to irrigation in farming, or as a raw material to make new cellulose-based ynthetic fibers. Early work with this versatile cellulose material shows promise.

The Billie System

The Highly Automated Recycling Factory, the Billie System in Tai Po, Hong Kong.

This is an intelligent and highly automated mechanical recycling system. The goal of building this system in Hong Kong is to provide a local processing solution to all the available recyclable materials in HK.

The system uses a combination of multiple engineering sub-systems integrated with machine vision tools in a manufacturing and processing plant that clean, separate, and reprocess old materials into new usable yarns. The process is almost entirely automated, completely dry, and can be completed in a very compact area. The Tai Po system can process 3 tons of materials a day.

Collected old garments are first introduced into the system via a series of ozone processing tanks. These tanks kill off over 99% of bacteria, germs, and other harmful microorganisms in the collected material. The advantage of using ozone as the treatment method is that it is a dry process and does not use harsh or harmful chemicals. The cleaned materials are then transported via conveyor belts to operators who remove any hardware on the used garments. These are mostly buttons and zippers. This is the only manual process of the current system. The materials are then sorted and send along in trays. The trays are scanned and separated by robotic arms by material composition and color. The now sorted and cleaned materials are then put in separate bins. Full bins are carried away by Automated Guided Vehicles (AGVs) to a system library. When an order is received for a type of yarn, the system selects the appropriate material composition in the appropriate colors to be mechanically processed into fibers and twisted into yarns. A percentage (20–50%) of virgin fiber is introduced at this point to improve the performance of the new yarn, and also to improve the new color and appearance. New yarns go through a secondary cleaning and sterilizing using UV lights. The produced yarns are comparable to new yarn in appearance and functionality and usually cost less to produce. Since no water or chemicals is used in this production process, the new yarn has a much reduced carbon footprint.

A video of the operations of the Billie System can be found here https://thebillie upcycling.com.

The Garment to Garment Retail Shop

The garment to garment (G2G) shop in the mills, Tsuen Wan, Hong Kong

The G2G retail shop is a retail recycling experiment that is being conducted inside the Mills Shopping Mall. Here customers are asked to bring their old garments and watch as workers turn their old garments into fiber and then spin this back into yarns and 3D knit new garments. The idea is to use raw materials provided by customers to make new clothes.

Conventional Recycling Methods

Mechanical Recycling—the mechanical shredding and reconstitution of used materials. Used garments and fabrics are fed into shredding machines that turn these back into loose fibers. The fibers are then either reconstituted into nonwoven materials (for example, various insulation materials), or these are then combed, carded, and spun back into yarns for woven or knitted applications. The disadvantage of mechanical recycling is that materials are damaged and fibers are shortened when being pulled apart so the value of processed material is lessened. Various performance characteristics are also weakened, and in general the material will feel coarser.

I took these pictures in a recycling factory in Zhuhai China in 2018 showing the processes of used apparel recycling:

1. Cut up pieces of used apparel are fed into a shredding machine

2. Materials are then pulled apart by progressively finer shredding machines til
    they are totally disentangled and are reduced to fiber form

3. When they are in this loose fiber form they are then either used for nonwoven applications. This particular factory makes yarns for knitted garments, so in this case the fiber is combed and carded so that the fibers are oriented in the same direction.

4. The fiber from the used material is twisted and blended together with a percent of virgin material, to improve hand feel and tensile strength. The material is then twisted together into yarns.

5. Yarns are then twisted together and pulled to finer and finer as tension is tighter to improve yarn strength. Fine yarn is then finally put on cones and these are now ready to be knitted or woven into new fabrics and apparel.

See also this interesting video https://www.youtube.com/watch?v=AjHVMl AOZa4.

Chemical and Biological Recycling—Chemical and Biological recycling is no as interesting visually. Materials are processed in vats of liquids (usually large stee tanks), various chemicals or enzymes are used to separate and breakdown materials These are then dried and reprocessed

Chemical Recycling

https://www.chemistryworld.com/features/recycling-clothing-the-chemical-way/4010988.article.

https://greenblue.org/work/chemical-recycling/.

http://greenblueorg.s3.amazonaws.com/smm/wp-content/uploads/2018/05/Che mical-Recycling-Making-Fiber-to-Fiber-Recycling-a-Reality-for-Polyester-Tex tiles-1.pdf.

Biological Recycling

https://www.youtube.com/watch?v=PfLQdC4VqWg.

This is a video of a biological recycling method we developed at HKRITA

Question

1. How can research teams breakdown research silos and work closer with each other?
2. Are there other circular solutions to consider?

Future Readings

About the H&M Foundation. https://hmfoundation.com.

About the Novetex Billie System and Novetex. https://thebillieupcycling.com.

About HKRITA. http://www.hkrita.com.

About HKSAR's ITF. https://www.itf.gov.hk.

Edwin Keh is the CEO of the Hong Kong Research Institute of Textile and Apparel (HKRITA). He is also on Faculty at the Wharton School, University of Pennsylvania and at the Hong Kong University of Science & Technology. He teaches supply chain operations.

Until April 2010 Edwin was the SVP COO of Walmart Global Procurement.

Prior to Walmart Edwin managed a consulting group that worked on supply chain, manufacturing, and product design for NGOs, charities, as well as commercial clients.

Edwin managed sourcing at Payless Shoes, Donna Karan International, Country Road Australia, Abercrombie & Fitch and Structure stores. Prior to graduate school he worked for the United Nations High Commission for Refugees.

Edwin serves on the Board of Whittier College, CA, and is on the advisory board of multiple social enterprises, and NGOs.

Edwin holds multiple IPs for inventions at HKRITA. His projects have won several Global Invention Awards.

2019 Edwin was the "Innovative Retailer of the Year" Inside Asia.

2018 & 2019 Edwin was on the Expert Panel of H&M Foundation's Global Change Award.

Edwin is in Debrett's "100 Most Influential People in Hong Kong".

Edwin was the 2011 recipient of the Production and Operations Management Society's *Excellence in Production and Operations Management Practice Award*.

Industry 4.0 and Circular Economy Digitization and Applied Data Analytics

Parvathy K. Krishnakumari, Hari Dilip Kumar, Shruti Kulkarni, and Elke M. Sauter

Abstract In this chapter, we aim to inspire questions and discussion on why the circular economy is relevant to industries for ensuring sustainable growth and continuous improvement. We also will discuss some of the challenges faced by manufacturing industries in adopting circular economy practices, and how applied data analytics in the framework of Industry 4.0 can help in overcoming these.

Keywords Applied data science · IIoT · Big data · Digital twins · Circular economy · Industry 4.0

Learning Objectives

Understand the basics of Industry 4.0 and its relevance to the circular economy.
Get familiar with digital technologies for Industry 4.0: Internet of Things (IoT), big data analytics, digital twins.
• Become familiar with a Data Analytics Framework applicable to industrial or business processes in the circular economy.
• Learn about applications of Data Analytics for the circular economy.

P. K. Krishnakumari
Jheronimus Academy of Data Science, 's-Hertogenbosch, The Netherlands
e-mail: p.k.krishnakumari@tue.nl

H. D. Kumar
Independent Researcher, Bengaluru, India

S. Kulkarni (✉)
Indian Institute of Science, Bengaluru, India
e-mail: shrutik@iisc.ac.in

E. M. Sauter
Gumac Tech, Den Haag, The Netherlands

© Springer Nature Singapore Pte Ltd. 2021
L. Liu and S. Ramakrishna (eds.), *An Introduction to Circular Economy*,
https://doi.org/10.1007/978-981-15-8510-4_18

1 Introduction

The impact of industrial processes on the environment increased price volatilities complexities, and risk in supply chains, and even impacts on the health of people have alerted business leaders and policymakers to the necessity of rethinking materials and energy use. There is a call to be more innovative and competitive and to make a transition from a linear to the circular economy. This shift is further accelerated by regulatory frameworks being introduced in many countries, impacting citizens, local governments, companies, and industries in many parts of the world.

In an industrial system, a circular economy can be defined as a economy of closed loops that is restorative or regenerative by design, using systems thinking at its core, and which shifts toward the use of renewable energy sources. On the other hand, Industry 4.0 is termed as the "era of cyber-physical systems" and incorporates computing, digitization, networking, and physical processes into traditional manu-facturing and industrial platforms [1]. Both Industry 4.0 and the circular economy seek to improve products and processes and optimize resource usage and costs. While the circular economy can drive the transition of manufacturing industries toward systemic sustainability, Industry 4.0 can drive innovation and spur the digital transformation toward *smart* and *resilient* manufacturing enterprises.

2 Industry 4.0

Before the end of the eighteenth century, most commodities being used by humans such as tools, food, and clothing, were made by hand or using non-standardized processes. Since then, recurring technological innovations have transformed the economy, starting with the first industrial revolution, where water- and steam powered tools and techniques were used to optimize production. With the introduc-tion of electricity, and later steel into production processes came the advent of the second industrial revolution in the early twentieth century. This augmented the ability of these processes to be scaled and improved and thus concepts like mass produc-tion were introduced. Distinct from Industry 3.0, which involved the automation of single machines and processes, Industry 4.0 encompasses **end-to-end digitization** and data integration of the value chain (Fig. 1). Digital products and services and connected physical and virtual assets allow for transforming and integrating opera-tions and internal activities, building partnerships and supply chain, and optimizing customer-facing activities [2].

The impact of Industry 4.0 is finding its way into both horizontal and vertical value chains. The digitization of horizontal value chains integrates and optimizes the flow of information and goods from the customer through their own company to the supplier and back. This includes proactive controlling of internal processes and procedures within departments of the company (such as planning, manufacturing

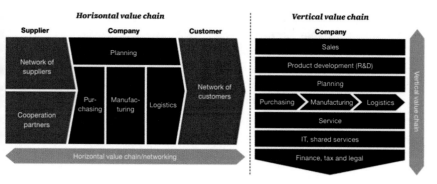

Fig. 1 Industry 4.0 requires digitization of horizontal and vertical value chains [2]

selling/purchasing, servicing, and logistics). Horizontal digitization also encompasses external value chain partners needed to ensure efficient service delivery to customers. Vertical digitization, on the other hand, deals with securing a consistent flow of information and data from product design and development, to manufacturing, to sales and logistics. By optimally connecting manufacturing systems, enabling preventive and predictive maintenance of systems, and better analytical capabilities in the verticals, digitization improves quality and flexibility at reduced costs [2].

3 Technology Suite for Industry 4.0

Smart technologies such as the Internet of Things (IoT), artificial intelligence (AI), big data analytics, robotics, digital twins, cybersecurity, and 3D printing are part of the set of technologies available within the Industry 4.0 ecosystem [1]. All of these technologies involve the use, generation, collection, or output of data at some point in the workflow.

In a survey of 225 German industrial companies by PricewaterhouseCoopers (2014) [2], more than half the respondents stressed that the analysis and use of data were highly important. A total of 90% of the companies (surveyed from five different industry sectors) were convinced that the ability to efficiently analyze and effectively use large volumes of data was of vital importance for the success of their business models.

In the next section, three of these technologies will be explored (IoT, big data analytics, digital twins) in relation to the applied data analytics framework.

a) IoT-Powered Digitization

To ensure adequate monitoring, evaluation, and control, the number of sensors, embedded systems, and connected devices being deployed in Industry 4.0 systems is increasing rapidly. These interconnected systems provide an infrastructure for continuous data influx, forming part of the Internet of Things (IoT). The application of IoT developments to create value for industrial processes, supply chains, products, and services is termed as the **Industrial Internet of Things** (IIoT) [3]. Merely being able to collect data will not provide value to industries and this requires scalable infrastructure, with timely analysis pipelines set up in place to facilitate data-informed and data-driven decisions.

(b) Big Data Analytics

A smart manufacturing enterprise is therefore made up of smart machines, plants and operations, all of which have high levels of embedded intelligence. These linked systems are based on open and standard Internet and cloud technologies that enable secure access to devices and information. This allows for the generation of "big data" streams about the manufacturing, operations, processing, and logistics information from industries. Big data is thus a new ecosystem, characterized by an exponential generation of data from different sources, which cannot be processed with traditional software tools.

Big data can be described by the 5 V's features [4]:

- Volume: a large amount of data from multiple sources including Internet of Things (IoT) devices, satellites, and mobile phones.
- Velocity: the speed of data transactions during collection, processing, and analysis
- Variety: the heterogeneity of data (different data formats, different contents).
- Value: the process of using big data analytics to extract valuable knowledge from the data.
- Veracity: the protocols that ensure accuracy and correctness in relation to data quality, data governance, metadata, privacy, and legal concerns.

This data is to be processed with new, advanced analytics tools, driving greater business value by embedding data-informed and data-driven decision-making at the core of industries. The variety of data sources to be analyzed to make the right decision is always challenging. To deal with such diverse and **high volume data,** tools and techniques ranging from business intelligence to machine learning are used to extract maximum value. These tools and techniques can be used for pattern/process mining performance metric tracking, analyzing user behaviors, predictive maintenance, near real-time monitoring of processes, and optimizing value chains.

Apart from deriving value from *internal* data, these techniques hold the capability to perform **Big Data Fusion** with *external* data sources, such as environmental and economic indicators, and their proxy datasets. Data fusion is "the process of integrating information from multiple sources to produce specific, comprehensive unified data about an entity" [5]. Data fusion enhances the value of data, allowing correlations and comparisons to be made that ultimately grant better decision making power as information is contextualized. This in turn enables improvement to efficiency and profitability, innovation and cybersecurity, and better safety and performance at reduced CO_2 emissions impact [6].

c) Digital Twins

The concept of a **virtual** representation of a **physical** product is one of the techniques that can bring the manufacturing industry to a "smarter" state. Termed as a digital twin, product avatar, or cyber-physical equivalence, this is a "holistic model-based description of a product for current and future life cycle stages" [7]. The core concept of the digital twin of enterprises is traced back to the Information Mirroring Model published in 2005 [8]. This primitive model is refined multiple times to support complex systems and termed as digital twin in Virtually Perfect: Driving Innovative and Lean Products through Product Life cycle Management [9].

Digital twins are mainly utilized within automotive, aerospace and marine manufacturing, and the building industry (as Building Information Models or BIMs). They are useful for the pre-building phase, where a physical object is designed as a digital model. However, with an already existing physical object, one can digitize it, capture required data, and then add to the model [10]. Digital twins are capable of providing a distributed approach to simulate, interact, and manage a product's life cycle information among different domains and stakeholders [7]. This can be done to high levels of granularity, with detailed information on products, attributes, and services at the item level.

Material passports (or product biographies) are now being implemented in a wide range of industries, as they act as the equivalent of digital twins. Passports aggregate qualities and characteristics of a product or material, such as origin, location, composition, size, owner, prize, and steps along the supply chain [11]. As passports can provide a high-level overview of the state and nature of a product, they are not only being adopted by the construction industry as a complement to BIMs, but also by the fashion, furniture, and food industries as well. Knowing information about the material or product not only adds economic value, but can facilitate reuse/remanufacture, proactive maintenance, and more rapid identification of risks [11].

Material passports could have a considerable impact in the circular economy, as they can grant different parties traceability and first-hand information about an object's trajectory and constitution. However, material passports fail to secure their full potential as standalone documents and do not fully address issues of material circulation in practice. Research proposes converting these material passports into a **Linked Data ecosystem** by enriching them with semantic web technologies [12]. When data is brought into this smart interlinked environment, circular economy actors can interact with it, and pose queries about materials, people, and products. In this manner, data and technology provide a first mechanism for smart connection of inputs and outputs between relevant stakeholders, enabling further material exchanges between them.

4 A Circular Industry 4.0

This section provides an indication of how the principles of CE would work in practice in conjunction with the technology and data of Industry 4.0.

(a) Closed-Loop Manufacturing

Closing the loop with regard to material flows is an important step toward establishing a circular economy. In a **closed-loop manufacturing** setting, an industry's own waste materials are reused for other production processes. Alternatively, recycled materials are used for creating new products, thus avoiding the consumption of virgin material. For this to be operational at the material level, closed-loop interactions should be intentional by design, strategically planned at the start of a product's life cycle, and further implemented and monitored.

Having an **Industrial Internet of Things (IIoT)** setup across companies and businesses can allow for the implementation of such closed-loop principles, particularly those involving close monitoring and frequent communication between parties. The concepts of **extended life** and **asset maintenance** take a higher relevance under this setting, given that the IIoT devices can emit early warning signals when a product's component is flawed, or when its service life is approaching its end. In this way, with the aid of digital devices, the value-added component of the product can be identified and recaptured in time to undergo upcycling, instead of lower value options such as recycling or landfilling.

(b) Reverse Logistics

Once a flawed product is identified, the responsible company would then undertake **reverse logistics**, where products are collected from their "final destinations" and taken back to factories for repair. Here **IoT devices** continue to play a role, with their geolocation capabilities enabling routing options for the collector. Reverse logistics should be made as efficient as possible, for example, by picking up products after completing "forward logistics" deliveries, which would avoid having a truck return with empty space. Attention should be paid to **Big Data Fusion** at this stage, as data regarding waste streams can be fused with spatiotemporal transportation data, and recycling industry data. This will contribute to successfully directing the product into its next use cycle. Used products will then arrive at factories to undergo the refurbishment or remanufacturing needed to return the product to operating order. Alternatively, collected products could be sent to the appropriate disposal process or downcycling along the materials chain.

(c) Life Cycle Management

Life cycle management is the process of tracing and managing all the activities and information flows of a product, from its design or development stages, all the way through its use/service and maintenance stages, and sometimes including disposal [13]. The use of **digital twins** (such as material passports), together with IoT devices, permits almost continuous tracking of individual products as they travel along different networks of people, processes, and business systems [13, 14]. In the beginning-of-life [BOL] stages, digital twins can help in simulation and modeling to conceptualize the product. In the middle of life [MOL] stages, digital twins are geared toward optimizing product use, as they provide indications of required service and maintenance activities for manufacturers and customers [13]. The circular economy draws greater attention to the end-of-life [EOL] stages, as it presents the consumer with more options than the traditional linear disposal approach. Having information about the product is especially deemed necessary at this EOL point, as it allows for different interventions to be made before the product's next use cycle.

(d) Servitization

The different use cycles, refurbishment, and dismantling processes characteristic to circular economy challenge the identity and stability of products, as this is constantly shifting and revalued [13]. This offers an opportunity for industries to adopt innovative business models that can generate multiple revenue streams. **Servitization** is a utility-driven approach that adds value to products by providing services [15]. Parties engage in an intangible transaction, a service, which might involve physical goods, but where ownership rights are removed from the product [15]. In some cases, ownership of the product is kept with the manufacturer and offered as a lease or rental to consumers.

Manufacturing companies should clearly offer disassembly and repair services. Consequently, they should offer incentives to consumers, encouraging them to return used items to factories for remanufacturing. Further, these new business models can be embedded with data-driven processes, such as **RFID tags** and **QR codes**. These IoT devices can then interact with dashboards and web applications, allowing for constant monitoring of products, and adjusting their relationships with actors as they move along use cycles [13]. This enables data-informed decision-making between consumers and producers/service providers. As more services move from product ownership to leasing, closed-loop supply chains have to be flexible and designed for collaboration, resilience, and scalability. When successfully implemented, this creates a long-term relationship with the clients, with an opportunity for industries to grow and adapt based on closed-loop feedback as well.

The market potential for servitization models is very promising. Pay-per-use services are gaining popularity as they allow customers to access products matching their fluctuating needs. According to Richard Girling's book Rubbish! published in 2005, 80% of products made get thrown away within the first 6 months of their

life [16]. Pay-per-use models could help stabilize some of these issues by decou
pling economic growth from resource consumption [16] as well as encouraging the
efficient and responsible use of resources.

(e) Renewable Energy Transition

A global energy transition is critical to meet the objectives of the circular economy
This energy transition will be accelerated by policies that enable technological inno
vation, notably in the field of renewable energy and energy efficiency. As such, the
life cycle of materials that make up a product or process cannot be viewed separately
from the energy sector.

The current discourse on circularity is mainly focused on material cycles
neglecting that these will require large-scale development of **renewable energy (RE**
resources to be powered sustainably [17]. Industries use energy predominantly fo
production, apart from lighting and other business uses. Industrial production make
up just over half of all global energy use and is expected to grow by about 1.5%
globally each year through 2035. Methods for improving **energy efficiency** shoul
therefore be identified, and using data can help transition to renewable energy in th
best way.

5 Applied Data Analytics for Circular Economy

This section provides a framework that can be used to analyze any industrial o
business process from the perspective of data. Readers may implement the framewor
for a sector of interest, in order to get a roadmap of the possibilities data could create

Data Science is the ability to combine statistical methods, algorithms, and model
to extract valuable information from structured and unstructured data. The outcome
possible with the tools and techniques of data science can be demonstrated usin
Gartner's model of data ascendancy (Fig. 2).

Fig. 2 Gartner's model on
data ascendancy

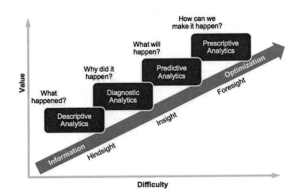

Let us take as an example of the manufacturing unit of an industry aiming to improve energy efficiency. **Descriptive analytics** of the energy consumption data (historic and near-real time) can give valuable insights into the statistics of energy use by different processes. This analysis can help identify areas that need to be improved. The next step to improving energy efficiency is to identify root causes for higher-than-expected energy consumption by certain processes. This is termed as **diagnostic analytics**. Descriptive and diagnostic analysis can couple exploratory data analysis with pattern- and data-mining techniques, to identify hidden patterns, bottlenecks, and interdependencies within the energy consumption profile of the manufacturing unit. The next step in the process is to predict the energy consumption of the unit. **Predictive analytics** leverage the insights and features identified in the previous steps, and model short- and long-term energy consumption based on historic data in conjunction with assumptions about certain features. **Prescriptive analytics**, which is considered the most difficult, will add the most value to the unit, by prescribing different consumption profiles leading to optimized energy and cost savings.

This modeling framework is one of the important stages in a **Data Science Project Life cycle** as shown in Fig. 3 [18].

The life cycle consists of multiple stages which are all iterative and interconnected. **Business understanding** is the stage where the end users of the project work collaboratively with data scientists, analysts, and engineers. In this stage, the potential hidden within the data is explored, as well as definition of the research questions and hypotheses that bring the most value to the organization. The success of a data science project in most cases is heavily reliant on the business case defined; hence,

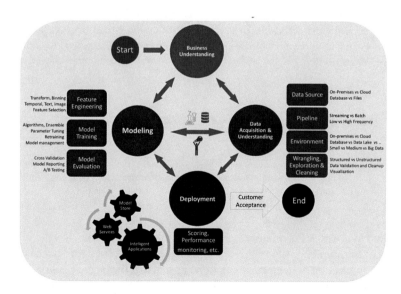

Fig. 3 Data science life cycle [18]

this step is very important. The **data acquisition and understanding** phase identi fies the data assets and performs exploratory data analysis. The data pipelines and environment are also decided and experimented within this step, in alignment with the business goals previously defined. This step involves multiple decision-making steps, an example being the choice between using a cloud-based or on-premise solu tion. These decisions will be taken based on inputs from business units; one of the criteria will be an analysis of cost-benefits. The **modeling** phase applies business intelligence tools, programming languages, algorithms, and simulation models to leverage data to describe, diagnose, predict, or prescribe based on the business use case defined. This stage involves deciding the evaluation model to be used to keep track of performance. The **deployment** phase involves packaging the engineering and analysis into a data product. This data product can be a dashboard, developed using business intelligence tools such as Power BI or Tableau, or it can be a customized web application developed using frameworks such as Django or packages such a Shiny. The data product could be an application program interface (API) or a backend application. There are many possible definitions, but in the context of this chapter we define a **data product** as a tool that helps a user interact with, understand, and convert insights into data-driven or data-informed actions.

The **Deployment** phase requires more than just data science skills, because the user interface/experience is a vital part of the product that must be designed for usability and scalability. It, therefore, also requires a solid understanding of the end users and business goals to make the most impact. A curated collection of blog post published by [19] discusses some important principles and concepts of designing data products. Four sections—audience, data, design, and delivery—are highlighted in the framework for a good data product, as shown in Fig. 4. To ensure that user are engaged while using the data product, the publication also stresses certain desig principles, such as solve a problem, enable casual use, tell a story, lead to action, and encourage exploration.

6 Systems Thinking and Systems Connectivity

The world is a complex, interconnected, finite, ecological-social-psychological-economic system. We treat it as if it were not, as if it were divisible, separable, simple, and infinite. Our persistent, intractable, global problems arise directly from this mismatch.

–Donella Meadows

The Ellen MacArthur Foundation highlights "**thinking in systems**" as integral to the process of transitioning to the circular economy [20, 21]. A "system" can be though of as any connection of interacting components that together serve a "purpose" [22 for example, a steam engine, whose purpose is to locomote, or the education system of a country, whose purpose might be to skill citizens to serve society. The fundamenta principle of "systems" thinking is that everything is interconnected or dependent o something else in the system for survival [23].

	Actions	What we've learned	Resources to learn more	Completion
AUDIENCE	Target users vs. target buyers	Look beyond the analysts who use the data to find the people who see the value and have the budget to back up their interest.	7 Companies That Totally 'Get' Their Buyer Personas (HubSpot) bit.ly/Buyer_Personas	☐
	Understand users' data fluency	You'll find a lot of variation in the users' ability to work with data. How can you thrill data junkies while supporting novices?	Data Is Useless Without the Skills to Analyze It (HBR) bit.ly/data-without-skills	☐
	Fit users' workflow	An engaging solution fits into users' existing processes, delivers data in the right format, and is timely for decision-making.	Form and function in data delivery (Juice) bit.ly/data-form-function	☐
	Inform user actions	Many solutions provide data that is interesting but doesn't connect to decisions. What information can people act on?	Choosing the Right Metric (Juice) bit.ly/30Days-Metric	☐
	Balance flexibility vs. guided analysis	Too much flexibility puts the burden on users to find answers. Too little flexibility is frustrating and won't serve a broad audience.	How to Tell Stories with Data (E. Segel) bit.ly/stories-with-data	☐
DATA	Get the right data	All data is not created equal. Find the best data with the most valuable insight.	More Meaningful Big Data bit.ly/meaningful-data	☐
	Choose your metrics	Apply unique or specialized metrics that make the data highly valued in your industry.	Franken-measures (Juice) bit.ly/franken-measures	☐
	Make your data special	Spend time with your data. Do the analysis and understand its value.	Big Data + Intimate Stories (Juice) bit.ly/data-intimate-stories	☐
	Know what's fair game	Be aware of the laws and constraints around sharing your data internally and externally.	Privacy requirements (Teach Privacy) bit.ly/data-privacy-security	☐
DESIGN	Define product architecture	Content modules, visualizations, and features are woven together into a structure to make it easy for users to accomplish their goals.	The Dribbblisation of Design bit.ly/dribbblisation	☐
	Plan for data variation & outliers	The design needs to anticipate the worst-case scenarios for data. E.g. sparse data, null values, odd distributions, and long labels.	Random Data Generator bit.ly/random-data-gen	☐
	Enable sharing	Users will want to share your data, collaborate with colleagues, and distribute their findings. Consider features for capturing and exporting user parameters, configurations, and annotations.	InVision's prototyping app has beautiful ways for sharing content bit.ly/invision-app	☐
	Create logical flow	Provide a guided path for users to explore and find insights. Help your audience by giving them an obvious place to start and the visual indicators to deepen their analysis.	Guide to Dashboard Design, see Pt. 2 (Juice) bit.ly/guide-dash-design	☐
	Write clear labels and description	Use plain, jargon-free language to explain the data. Craft titles, labels, and legends that explain the meaning of the data and descriptions of actions to be taken with the results.	8 Rules for Great Writing (K. Vonnegut) bit.ly/write-with-style	☐
DELIVERY	Focus on customer development	Start small by thinking about discrete data applications that solve real customer problems and answer their questions.	Four Steps to the Epiphany (S. Blank) stanford.io/clo4lu	☐
	Prioritize features	Novel and fun features can outpace their need. Make sure features line up with company's goals.	60 second business case (J. Brett) slidesha.re/S8YnvQ	☐
	Find the right price	Ask users how they plan to use the product and assess the value they will derive.	KISS Metrics on Pricing bit.ly/kissmetrics-pricing	☐
	Run experiments	When you're not sure (and you rarely will be), create an experiment, learn and move on.	Run Experiments to Tackle Risks (A. Maurya) bit.ly/run-experiment	☐
	Reach customers	How do you get marketing budget without proven sales? Find early adopters, excel at service, and ask for referrals.	Why I Hate Funnels (B. Chestnet) bit.ly/hate-funnels	☐

Fig. 4 Data product checklist [19]

Systems thinking enables an understanding of how the parts of a system interact to produce the behavior of the whole. The economy, society, and environment are interdependent and intimately weaved among each other. A true understanding of "systemic" challenges is therefore possible only with an understanding of how the distinct elements interact to produce dynamics, behaviors, and emergent properties [23] of the whole.

The systems approach replaces the traditional picture of the world and the (linear) economy as a controllable, predictable, and understandable "**machine**," with a picture of the (circular) economy as a "**living ecosystem**" with delicately balanced interdependencies that can lead to effectiveness of scale at the wide economy level.

As such, systems thinking enables the identification of **root causes** leading to implementation of better solutions and it also provides the **conceptual framework** required to understand the functioning of the economy as a whole. Finally, systems thinking can identify **opportunities** for "closing loops" in biological and technical cycles with products, components, and materials recirculated into the market at the highest quality possible and for as long as possible.

Industry 4.0 is of particular relevance to systems thinking for the circular economy. The Industry 4.0 technology suite provides a mechanism of connectivity—both to couple agents (humans, devices, machines, living beings) within the system, and to provide a platform for interaction between them in the form of apps, dashboards, and information flows.

Data and information flows are therefore important constituents of systems in the circular economy. The highest value of data is reached through adopting the systemic view—incorporating data sources, but also considering technological elements, users, organizations, and other actors in the environment. This perspective can incorporate, for example, the **digital interconnected devices** and "urban pulses" of smart cities [24]. The **data analytics framework** discussed in the previous section therefore takes on new meaning when embedded in the systems thinking approach for circular economy.

7 Case Studies

In the next section, we showcase different case studies that have used parts of the applied data framework in the context of the circular economy. Table 1 provides a high-level overview of three case studies: Bundles, Rendicity, and Move About. Then a closer look is taken on the Gogoro use case which delves on the Industry 4.0 technologies put in place to deploy scooters powered by shared batteries.

As evidenced by these case studies, Industry 4.0 and adoption of data analytics are at an early stage in the context of circular economy. The authors therefore challenge students to apply data science in this context—to find new business models, ways to reach new pay-per-use consumers, demonstrate best practices, and optimize current processes.

7.1 Gogoro: Scooters Powered by Shared Batteries

With more than 13 million scooters among 23 million citizens, Taiwan has the highest per-capita density of scooters by far. Given environmental concerns, regulatory authorities in Taiwan had made previous attempts to incentivize the shift to electric scooters, including through a substantial subsidy to the consumer. These had however, failed to create a widespread impact on the public, in part due to challenges like longer battery recharge time compared to refill time of fossil oil, battery-charging stations not easily or widely accessible, etc. [27].

In recent years, Gogoro, a Taiwanese electric scooter manufacturer, has adopted an innovative approach to accelerating the adoption of electric vehicles, through a model of electric scooter ownership mixed with battery sharing. The Gogoro Smartscooter

Company information	Business features
Bundles **Origin**: NL **Founder**: Marcel Peters **Awards**: It has received approximately EUR 800,000 in financing from Rockstart Accelerator and crowdfunding initiatives **Targets**: Bundle aims for resource effectiveness through the use of products exclusively produced by the company "Miele[a]" (which is believed to be the only remaining manufacturer to use reused or recycled material) **Innovation**: The usage of "IoT" within the Bundles start-up enables the company to keep constant track of their machines and perform resource effective improvements **Closed-loop manufacturing**: None. Bundles can not be considered as a closed-loop manufacturing company as there is no remanufacturing, reuse, or recycling of the products or components. Therefore, a partnership with a responsible manufacturer would be helpful to close the loop	"Bundles" offer a service to produce clean laundry in homes on a pay-per-wash basis and charges customers a monthly subscription fee while the company retains ownership of the appliances • It uses the Internet of Things (IoT) to lease washing machines to customers in the form of a "pay-per-use" system. Bundles aims for a customer-controlled relationship where the customer is able to end their contract at any point as well as specifying their costs based on how much they use their product • The customers pay per month with the addition of deposit depending on the consumers' requirements and frequency of use. The tariff can be adjusted constantly as a consequence of changes in usage
Rendicity **Origin**: NL **Founders**: Dr. James Kennedy and Dr. Philip Healy **Awards**: It secured a EUR 20,000 cash prize and was named the Best Early Stage company at the Munster final of the 2015 Inter Trade Ireland all island Seedcorn Investor Readiness competition in November 2015[b] **Targets**: Rendicity aims to reduce companies' carbon footprints and electricity bills **Innovation**: The usage of cloud-based services within the rendicity start-up enables the users to access high-performance computing power resources effectively **Closed-loop manufacturing**: None	"Rendicity" allows customers to create cloud render farms on demand • It provides end users of applications requiring intensive computer power with access to high-performance computing power via the cloud on a pay-per-use basis and with a user-friendly interface • It is estimated that the world's 1.5 billion computers consume about 90,000 MW of electric power, comprising 10% of global consumption [25]. Subscription to public clouds via platforms like Rendicity enables organizations to spend less on electricity for powering and cooling their computing hardware, and allows them to save on space needed to house IT infrastructures and resources

(continued)

Table 1 (continued)

Company information	Business features
Move About **Origin**: Oslo, Norway **Founder**: Ulf Jacobsson **Awards**: KPMG, Microsoft, Statkraft, Chalmers, ABB, SP, Entra Eiendom, and various municipalities are among its customers. It has received investment capital several times and is now growing organically without need for additional capital **Targets**: To be a part of a sustainable urban mobility movement that compliments public transportation, bike share programs, and other initiatives to provide an alternative to private vehicle ownership **Innovation**: It uses organizational/business model innovation as well as service/process innovation **Closed-loop manufacturing**: None	"Move About" is the world's first and largest car sharing fleet exclusively using electrical vehicles According to [26] Move about meets following success factors: • Economic—reduced costs of the system through optimized fleet management; zero risk for the customer through a service offer; low prices of the service • Political—favorable policies in place in Norway for the business start-up • Social—improved convenience for the customers (high level of satisfaction reported in regular customer surveys), improved public image • Ethical—a strategic choice of green PR as a selling point in Sweden: targeting organizations with a will to be environmental pioneers or early adopters • Spatial/Urban—collaboration with other private and public actors: customers who take it upon them to provide the infrastructure; energy companies that provide the electricity necessary for charging the vehicles; proactive municipalities that are customers; etc.

a "Home-Miele." https://www.miele.com/en/com/index.htm. Accessed 6 Jul. 2020

b "Rendicity secures €20,000 cash prize and goes forward to…." https://intertradeireland.com/news/rendicity-secures-e20000-cash-prize-and-goes-forward-to-national-final/. Accessed 6 Jul. 2020

cannot be charged from a power outlet. Instead, the user must visit a nearby "battery-swapping platform" in order to exchange depleted batteries for charged ones. This approach addresses the needs of users in developing countries to have the experience of "ownership" of a vehicle (important due to cultural reasons) while having a "sharing" experience when it comes to batteries [3]. The battery-swapping approach lowers the cost of the scooters, reduces the number of batteries in circulation, and allows for optimization of charging costs [3]. Gogoro has seen a tenfold increase in adoption over the years, with over 100,000 electric scooters being sold in 2019 alone.

Rosa et al. [27] examine this approach in the context of the acceleration of shared economy aspects of the circular economy. The technologies of Industry 4.0—including IoT, **predictive data analytics,** and cloud computing—enable real-time monitoring of battery performance, the cloud computing-enabled battery-swapping platforms, **predictive maintenance** to improve battery quality, and **data-driven allocation** of battery-charging station locations and associated battery capacity.

Based on expert opinion, 13 critical factors for the success of this model—from the specifically designed liquid-cooled permanent magnet synchronous motor system to the elements of safety and impact on the environment—were identified. Using interpretive structural modeling (ISM) to determine the linkage between factors, the authors find that Industry 4.0 can indeed accelerate the sharing economy within the circular economy, provided that it starts from the collection and analysis of data related to user behavior and needs.

While Gogoro's model does have some challenges, it enables the detection of batteries that are approaching the end of their "first lifetime" as electric scooter batteries, at about 70% of new capacity. With built-in reverse logistics, the company envisions an entire "second life" for these hundreds of thousands of batteries, including energy storage in buildings, powering data centers, and homes [6]. With 94% of electric scooter market share in Taiwan, Gogoro is expanding in international markets, and views itself as a platform enabler for electric mobility, and "a cost-effective, worry-free solution" for municipalities and entrepreneurs [28].

5 Conclusions: Technology, an Aid to the Circular Transition

Worldwide, the implementation of the Circular Economy is at an early stage and not without its challenges. Further, technology is advancing at unprecedented rates, while leaving unanswered many technological questions of relevance.

Industries are currently facing **difficulties** transitioning to circular models, including financial barriers, lack of awareness about the benefits, lack of technical skills, deficiencies in data infrastructure, poor government support and legislation, and a lack of support from the marketplace. Further, **global value chains** are faced with integration conflicts (both on the policy and the technology side), pointing at the need for the establishment of collaborative partnerships and platforms across all levels to ease migration to the internationalized circular economy [29].

The technologies of Industry 4.0, in particular, IoT, data analytics, and AI bring several **benefits** to the picture, enabling the circular economy innovation across industries in several ways. They can aid in the design of circular products, components, and materials, through iterative machine-learning-assisted design processes allowing for rapid prototyping and testing. They can enable combining real-time and historical data from products and users to enable pricing and demand prediction, predictive maintenance, and smart inventory management, resulting in increased product circulation and asset utilization. Finally, they can optimize infrastructure, improving the reverse logistics required to "close the loop" on products and materials, by enabling processes to sort and disassemble products, remanufactured components and recycle materials [30].

However, the transition to a circular economy is a **systems-change** and will require much more than just technological innovation. New business models, the involvement of research institutes, a stimulating policy environment, access to finance, and a long-term strategy accommodating all these are important. Any envisioned development pathway toward a circular economy must also be coherent with prevailing paradigms like decarbonization, climate agreements, and the sustainable development goal (SDG) 2030 agenda. Finally, a well-implemented circular economy will directly support socio-economic inclusion and the drivers of social and environmental justice, by reducing resource insecurity and environmental degradation.

Questions

1. Sketch your vision for an IoT system of devices within your house and how these IoT things might interact to help you make circular decisions about your material possessions.
2. Outline the composition of a material passport (digital twin) of any product of your choice. Include attributes that could be relevant for other use cycles or processes such as repair, refurbishing, and disassembly.
3. Apply the data analytics framework to a process from your profession or domain area. What are the questions you might answer with data that can take your organization to an anticipatory state rather than a reactive state?

References

1. Piccarozzi, M., Aquilani, B., & Gatti, C. (2018). Industry 4.0 in management studies: A systematic literature review. *Sustainability* (Switzerland), *10*(10), 1–24. https://doi.org/10.3390/su10103821.
2. Pwc industrie 4 0. (2014). Industry 4.0: Building the digital enterprise—PwC. Retrieved July 6, 2020, from https://www.pwc.com/gx/en/industries/industries-4.0/landing-page/industry-40-building-your-digital-enterprise-april-2016.pdf.

3. Mcfarlane, D. (n.d.). *Industrial internet of things applying IoT in the industrial context.* Retrieved July 5, 2020, from https://www.ifm.eng.cam.ac.uk/uploads/DIAL/industrial-int ernet-of-things-report.pdf.

4. Bello-Orgaz, G., Jung, J. J., & Camacho, D. (2016). Social big data: Recent achievements and new challenges. *Information Fusion, 28,* 45–59. https://doi.org/10.1016/j.inffus.2015.08.005.

5. Cheng, B. (2012). *Data fusion by using machine learning and computational intelligence techniques for medical image analysis and classification.* Retrieved from https://scholarsmine. mst.edu/doctoral_dissertations/1974.

6. Conway, J. (2015). *The industrial internet of things: An evolution to a smart manufacturing enterprise.* Retrieved from https://www.mhi.org/media/members/15373/131111777451 441650.pdf.

7. Holler, M., Uebernickel, F., & Brenner, W. (2016). Digital twin concepts in manufacturing industries—A literature review and avenues for further research. In *18th International Conference on Industrial Engineering (IJIE)* (pp. 1–9).

8. Grieves, M. (2005). Product Lifecycle Management: The new paradigm for enterprises. *International Journal Product Development, 2*(1/2), 71–84.

9. Grieves, M. (2011). *Virtually perfect: Driving innovative and lean products through product lifecycle management.* Cocoa Beach, FL: Space Coast Press.

0. Digital, G. E. (n.d.). *Digital twins and IIoT.* GE Digital. Retrieved June 14, 2020, from https:// www.ge.com/digital/blog/digital-twins-bridge-between-industrial-assets-and-digital-world.

1. Bosch, P. (2020). *Is the "Digital Twin" the new "Circular Win"? Yes, it is!: Madaster.* Retrieved June 14, 2020, from https://www.madaster.com/en/newsroom/blog/digital-twin-new-circular-win-yes-it.

2. Sauter, E., Lemmens, R., & Pauwels, P. (2018). CEO & CAMO ontologies: A circulation medium for materials in the construction industry. In R. Caspeele, L. Taerwe, & D. Frangopol (Eds.), *Life-cycle analysis and assessment in civil engineering: towards an integrated vision* (pp. 1645–1652). London: Taylor & Francis.

3. Corallo, A., Latino, M. E., Lazoi, M., Lettera, S., Marra, M., Verardi, S., et al. (2013). Defining product lifecycle management: A journey across features. *definitions, and concepts.* https:// doi.org/10.1155/2013/170812.

4. Spring, M., & Araujo, L. (2017). Product biographies in servitization and the circular economy. *Industrial Marketing Management, 60,* 126–137. https://doi.org/10.1016/j.indmarman.2016. 07.001.

5. Tauqeer, M. A., & Bang, K. E. (2018). *Servitization: A model for the transformation of products into services through a utility-driven approach..* https://doi.org/10.3390/joitmc4040060.

6. Gould, H. (2014). 10 things we learned in recent circular economy live chat. *The Guardian.* Retrieved from https://www.theguardian.com/sustainable-business/2014/oct/06/10-things-we-learned-in-recent-circular-economy-live-chat.

7. Desing, H., Widmer, R., Beloin-Saint-Pierre, D., Hischier, R., & Wäger, P. (2019). Powering a sustainable and circular economy—An engineering approach to estimating renewable energy potentials within earth system boundaries. *Energies, 12*(24), 4723. https://doi.org/10.3390/en1 2244723.

8. Microsoft Team Data Science Lifecycle. (2017). Retrieved July 07, 2020, from https://docs. microsoft.com/en-us/azure/machine-learning/team-data-science-process/overview.

9. Gemignani, Z., Gemignani, C., Galentino, R., & Schuermann, P. (Eds.). (2015). *Data fluency.* Wiley. https://doi.org/10.1002/9781119182368.

0. Ellen MacArthur Foundation. (2013). *Towards the circular economy* (vol. 1). Retrieved from https://www.ellenmacarthurfoundation.org/assets/downloads/publications/Ellen-MacArt hur-Foundation-Towards-the-Circular-Economy-vol.1.pdf.

1. Ellen MacArthur Foundation. (2017). *Systems and the circular economy.* Retrieved July 5, 2020, from Ellenmacarthurfoundation.org website: https://www.ellenmacarthurfoundation.org/exp lore/systems-and-the-circular-economy.

2. Waldegrave, L. (2017). *What is systems thinking?* Retrieved June 29, 2020, from https://cir cular-impacts.eu/blog/2017/09/30/what-systems-thinking.

23. Acaroglu, L. (2017). *Tools for systems thinkers: The 6 fundamental concepts of system. thinking.* Retrieved June 29, 2020, from https://medium.com/disruptive-design/tools-for-sy: tems-thinkers-the-6-fundamental-concepts-of-systems-thinking-379cdac3dc6a.
24. Verma, P. (2019). *Urban pulse: The state of smart cities today (and tomorrow).* Retrieved Jul 2, 2020 from https://www.delltechnologies.com/en-us/perspectives/urban-pulse-the-state-of smart-cities-today-and-tomorrow/.
25. Van Heddeghem, W., Lambert, S., Lannoo, B., & Demeester, P. (2014, September). *Trend in worldwide ICT electricity consumption from 2007 to 2012.* Retrieved July 5, 2020, from ResearchGate website: https://www.researchgate.net/publication/260438995_Trends_in_wo ldwide_ICT_electricity_consumption_from_2007_to_2012.
26. CASI Project. (2020). *Car sharing of electric vehicles Move About AB, Scandinavia.* Retrieve July 5, 2020, from Futuresdiamond.com website: http://www.futuresdiamond.com/casi2020 casipedia/cases/car-sharing-of-electric-vehicles-move-about-ab-scandinavia/.
27. Rosa, P., Sassanelli, C., Urbinati, A., Chiaroni, D., & Terzi, S. (2019) Assessing relation between Circular Economy and Industry 4.0: A systematic literature review. *Internationa Journal of Production Research.* https://doi.org/10.1080/00207543.2019.1680896.
28. Anbumozhi, V., & Kimura, F. (2018). *Industry 4.0: Empowering ASEAN for the circula economy.* © Economic Research Institute for ASEAN and East Asia. http://hdl.handle.ne 11540/9381.
29. Ziegler, C. (2015, January 5). *Meet Gogoro, the outrageous electric scooter of the future* Retrieved July 5, 2020, from The Verge website: https://www.theverge.com/2015/1/5/7484171 gogoro-smartscooter-electric-scooter-removeable-battery.
30. Green Action Plan for SMEs. (2016, July 5). Retrieved July 5, 2020, from European Commis sion website: https://ec.europa.eu/growth/smes/business-friendly-environment/green-action plan_en.
31. Maoz, M. (2013). *How IT should deepen big data analysis to support customer-centricit* Gartner G00248980. Retrieved from https://www.gartner.com/en/documents/2531116/how-i should-deepen-big-data-analysis-to-support-custom.

Parvathy K. Krishnakumari is a data scientist passiona about leveraging data science tools and techniques to acce erate the achievement of Sustainable Development Goals 203(She has a Bachelor of Technology in Electrical and Electronic Engineering and a Master of Technology in Energy Manage ment. She has worked extensively on Applied Data Science i the domain of Renewable Energy at premier research institute in India and Singapore. She then went on to complete a Profe: sional Doctorate in Data Science from Eindhoven University i the Netherlands.

Currently based in the Netherlands, Parvathy is a Da Science Consultant at the World Bank Group and is a Da Science Researcher at the Jheronimus Academy of Da Science. She also has been appointed as a 'Mentor of Chang(under Niti Aayog's Atal Innovation Mission in India. Throug her data science startup DAV Data Solutions in India, sh provides Data Entrepreneurship training, mentoring, an internship to selected students in India.

Hari Dilip Kumar An engineer by training, Hari is a generalist problem-solver with about a decade of rich and varied experience in sustainability and social impact. He has worked on clean energy product development & research, water in agriculture, off-grid energy access, food security, and community-based sustainability transitions. He is passionate about science & technology for social benefit, and keenly interested in innovation, human-centered design, and social entrepreneurship.

Hari has contributed significantly to groundbreaking projects that have been recognized globally for innovation and impact, including twice by the United Nations. Based in Bangalore, India, he currently works as an independent consultant and researcher with startups & non-profits on a range of sustainable development challenges. He writes about his experiences and interests in the field at The Sustainability Problemsolver.

Dr. Shruti Kulkarni Data Science & Analytics evangelist and advocate of using data science for sustainability. Shruti is currently a data science manager by profession with passion for sustainability research. Shruti had a brief stint with AI pedagogy and with one of the oldest software R&Ds in India. She had also held research exchange positions at the Japan Advanced Institute of Science and Technology (JAIST). Shruti has finished her master's and Ph.D. from the department of management studies, IISc, Bangalore. Her research interests include computational sustainability, climate change-energy nexus and beyond.

Elke M. Saute is a Geographic Information Specialist and open innovation enthusiast that is passionate about building creative technological solutions for the betterment of the environment. She has a Bachelor's degree in Civil Engineering from Texas A&M University and a Master's Degree in Geographical Information Management Applications (GIMA), which is a joint collaboration between the universities of Delft, Utrecht, Twente, and Wageningen (WUR) in The Netherlands. Her master's thesis topic investigated Linked Spatial Data as a Collaboration enabler across the Circular Economy.

Elke works with her Netherlands-based startup, Sumac Tech, contributing to companies such as Space4Good, FarmHackNL and the Mothership Missions by performing geo-spatial analyses and organizing tech communities to develop solutions using AI and satellite technologies in support of agriculture and vulnerable landscapes.

Innovation for Circular Economy

Jovan Tan and Virginia Cha

Abstract Innovation can be magical. It has the potential to reduce the unprecedented resource stress on our planet while creating vast new economic opportunities for businesses to capitalize and prosper. With this promising proposition, business leaders are encouraged to design innovations that contributes to the betterment of society. This includes designing innovations for the circular economy, which is the centerpiece of discussion for this chapter. This chapter explores the dynamics of a successful innovation and discusses the current state of innovation for the circular economy. It further introduces the concept of Restorative Innovation—an innovation economic model that explains a pattern of innovation-driven growth for innovative solutions designed to restore our health, humanity, and environment. By the end of this chapter, readers will have a baseline understanding of innovation and the importance of designing innovation for the circular economy. Above all, readers will also appreciate the possibilities of creating and capturing positive value for both our economy and our society through Restorative Innovation.

Keywords Restorative Innovation · Innovation · Circular Economy · Impact · Outcome

Learning Objectives

To understand the fundamentals of innovation.
To learn about the current state of innovation designed for the circular economy, its challenges, and examples.
To be introduced to the Restorative Innovation framework.

J. Tan (✉)
National University of Singapore, 21 Lower Kent Ridge Road, Singapore 119077, Singapore
e-mail: jovan@u.nus.edu

V. Cha
INSEAD, 1 Ayer Rajah Avenue, Singapore 138676, Singapore
e-mail: virginia.cha@insead.edu

© Springer Nature Singapore Pte Ltd. 2021
L. Liu and S. Ramakrishna (eds.), *An Introduction to Circular Economy*,
https://doi.org/10.1007/978-981-15-8510-4_19

1 Introduction

The advent of technology has ushered in a state of society that is constantly evolving at an unprecedented rate. This state of flux has introduced a lot of uncertainty in the global business climate. As a result, many businesses—including industry incumbents, are finding it increasingly harder to keep pace with the rate of change and to compete effectively in their respective domains. The competitive advantages that once gave these businesses a defensible position is also no longer as impregnable as they were. In addition, businesses are also starting to realize that they simply cannot cost cut their way to profitability.

Therefore, for businesses to continue thriving in this fast-changing environment and to guard against potential disruptors, businesses must be forward-looking and be different. They must commit to strengthening their existing capabilities, while relentlessly identifying new growth opportunities and developing them into key strategic levers of growth. That is the intrinsic motivation of why business leaders innovate, and why innovation—the process of commercializing and exploiting inventions [1]—is imperative in our contemporary business environment.

Innovation revolves around value creation—where the value created is what consumers are willing to use and pay for. Hence, the precursor to innovation entails upon businesses to accurately spot the shifts in consumer behaviors and demands. In general, these shifts are gradual and are either geographic or industry specific. However, every once in a while, there will be a new eye-catching catastrophe that grabs our attention and consciousness. In recent years, it was the stark wake-up call from a special report released by the Intergovernmental Panel on Climate Change (IPCC). This special report denotes the impacts of global warming of 1.5 °C and stresses on the limited time we have on taking action to minimize extreme weather events, species loss, water scarcity and many other climate impacts that endanger lives, economies, and livelihoods [2].

This incident drove a monumental shift in global consumer behavior and demand towards green and responsible consumerism. Consumers are starting to feel most responsible for the future of the planet and are willing to play their part by rejecting goods that are detrimental to either health, humanity or the environment. Instead, conscientious consumers are now seeking alternatives to live a greener, cleaner and more equitable lifestyle. Therefore, for businesses to continue providing value and establishing a long-term relationship with their consumers, they will have to re-look at their current product offerings and introduce improvements or new solutions that are more sustainable.

As the rise of green and responsible consumerism is becoming ubiquitous, the question arises of—"How should businesses ride on this wave?" and taking a step further, "How does one innovate responsibly and introduce innovations that contribute to the betterment of society?"

There is no singular answer to this question as each approach to create and deliver value through innovation is unique. As such, this chapter sets out to guide and introduce the latest thinking on how we can innovate for circularity. To achieve our objective, we have divided this chapter into 2 parts.

Part One (Sects. 2, 3, 4, 5) aims to establish a baseline understanding of innovation through a short case study from one of Apple's most successful innovation story. Building on the learnings from Apple's case study, we will extend our discussion and apply it towards understanding the current state of innovation for the circular economy.

Part Two (Sects. 6, 7, 8, 9, 10) introduces the Restorative Innovation framework that explains a pattern of growth for innovative solutions that are designed to do good for our health, humanity, and environment. In this part of the chapter, we will establish the theoretical depth of Restorative Innovation while encapsulating the breadth of its usefulness and applicability through a short case study on TRIA, a Singapore-based company that is trying to "close-the-loop" for the global food services industry.

Case Study: Lessons from Apple's iPod Success

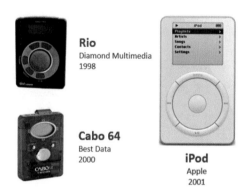

Examples of music players before the first iPod was introduced in 2001

Even though Apple's iPod may be the most iconic and successful digital music player in the world today, it was not the first to be introduced to the market. By the launch of the first iPod, there were at least 50 other portable music players for sale 3].

Yet, none of iPod's existing competitors were able to effectively drive product adoption and dominate the market. The primary reason was with the user experience and in particular, the downloading and transferring of digital music. Consumers were turned off by the process of doing so—legally or not—as it was extremely time-consuming and tedious.

Hence, Apple knew that introducing another digital music player by itself would not work. For the iPod to appeal to the masses, Apple would have to create a platform that facilitates a fast, intuitive, and seamless process for iPod users to purchase and manage their music. Apple also took an ecosystem approach towards delivering the benefits of digital music on the go, with a staged approach towards building this ecosystem.

The first release of the iconic digital MP3 player solved the problem of fast download of music with the integration of Apple's FireWire and Toshiba's ultra-slim 1.8", 5 GB hard disk, coupled with the superior UI/UX and design. Apple was able to release the first iPod with a value proposition of "1000 Songs in Your Pocket." The iPod became a smashing success 3 years later, when the iTunes and the ability for users to purchase individual music scores enables the delivery of exponential benefit to music lovers. The strategic iPod and iTunes combination would later be acknowledged as the pivotal moment in Apple's successful attempt to revolutionize the portable entertainment market.

Apple took a step beyond mere product innovation and significantly innovated on the iPod's business model too. By allowing their users to effortlessly purchase music directly from the iTunes music store, Apple was able to take a commission off each piece of purchased music and this approach brought in additional service revenue for Apple. This model is an improvement to Gillette's famous blades-and-razor model. In this instance, Apple created the ecosystem where they were the only ones offering low cost "blades" (low-margin iTunes music) to lock in purchase of the "razor" (the high-margin iPod) [4].

"RAZOR-AND-BLADES" BUSINESS MODEL EXPLAINED

The "Razor-and-blades" is a business model in which one item is sold at a low price in order to increase sales of a complementary good, such as consumable supplies. An example would be to offer high-margin razor below cost to increase volume sales of low-margin razor blades.

Other examples include:

Coffee Machines
(Razor)

Coffee Capsules
(Blades)

Printer
(Razor)

Ink Cartridges
(Blades)

To achieve this feat, Apple had to develop a synergistic business model that incorporates its hardware (iPod), software and services (iTunes). By perfectly synchronizing the various elements, Apple was able to tap into new revenue sources while addressing the users' biggest pain point in using a digital music player. In just three years, the iPod became a near $10 billion product and contributed to almost half of Apple's revenue.

In a nutshell, iPod's success is largely attributed to the fact that Apple was able to accurately diagnose their users' pain point in using a digital music player and applied the right treatment, which is also commercially viable. Instead of introducing another iPod that is better and faster (product innovation), Apple introduced a business ecosystem around digital music consumption (business model innovation) that seamlessly integrates and uplifted the entire product experience to a point where consumers are willing to use and pay for.

Apple changed the economics of purchasing digital music for the consumers, and in doing so, has created a business fortune for the company. To effect this change, Apple had to convince music publishers that they would make more revenue from a small unit sale (each individual score), but with an exponentially growth, versus the then-existing paradigm of charging a large sum for an entire (CD) album.

To effect the fundamental economics of how consumers pay for music, Apple had to (1) create a ready consumer base with the initial releases of iPod, (2) change the supply chain economics of selling music so it is efficient and low-cost to distribute music, and (3) change the availability of music choices to meet the demand of consumers. We will return to these three fundamental and inter-related economic actions later in this chapter.

3 Understanding Innovation

So, what have we learned from this case study? Firstly, we learnt that innovation can be perceived as an outcome that an organization seeks to achieve [5], and good innovations solve problems that currently only had poor solutions or none. In Apple's case, the desired outcome was to eradicate the time-consuming hassle and complexity of downloading and transferring digital music to the iPod. The exponential benefit of digital music cannot be fully realized without a full ecosystem approach, with breaking apart the economic constraints for wide-spread music consumption.

A good innovation is capable of addressing a persons' need and it is critical to understand that people do not just buy a product or service, they "hire" a product to do a job [6]—or in the words of the renowned Harvard Business School marketing professor Theodore Levitt, "People don't want to buy a quarter-inch drill. They want a quarter-inch hole!". This perspective places strong emphasis on the need for innovators to deeply understand the "human needs" they are trying to fulfil and recognize that these "human needs" changes across the time horizon.

As organizations and individuals possess different needs at different points in time, there is no universal answer or a singular approach to innovation. Innovation, by nature, is contextual to the problem these organizations are trying to solve. Tc innovate effectively, one must have a clear and complete understanding of the problem and situation at hand before devising the treatment.

A great technique to obtain clarity of the higher purpose for which customers "hire" a product or service is to understand what's the "Job to Be Done". iPod served their users with a clear "Job to Be Done". Users were seeking a fast, intuitive and seamless process to purchase and manage their music library. Hence, they "hired" the iTunes music store to get the job done.

WHAT IS THE "JOB-TO-BE-DONE"?
(More Examples)

To keep your teeth clean   To tell the time   To hold your water

Exercise:
What do you think is the "Job-to-be-Done" of your mobile phone?

As we shall see, there are a diversity of innovation types and methods to help innovators formulate a strategy to get the "Job" done. We have detailed a few common types of innovation in Table 1 below. The list in non-exhaustive and the example provided can overlap with other types of innovations too.

As we can see from this table, innovations can be applied across multiple dimensions (Product, Business Model, Market Demand, Process), and can be Radical or Incremental. We also must remember that not all innovations have to be ground breaking and radical. Many innovators and technologies dismiss the potential market impact from minor incremental innovations.

As with Apple's case, Apple did not introduce the first-generation iPod with iTunes. It was an incrementally better MP3 player than others in the market place—with increased storage capacity to store up to 1,000 CD-quality songs, ultra portability at only a fifth of the volume of then-hard drive-based players, and faster transfer speed with FireWire® [7]. The iPod entered the market on the single focused benefit, combined with an easier user-interface, for which Apple users have come to expect as hygiene factor. By the third-generation iPod, each successive model had mere incremental hardware improvements from its predecessors, and it was not th key enabler for its massive breakthrough in 2003. The breakthrough came from th minor yet significant software changes to the iPod operating system to enable it to work flawlessly with the iTunes music store.

The third-generation iPod was not an outlier. It is natural for successful organizations to pursue incremental innovation more frequently than radical innovation. Th problem with radical innovation is that it often involves translation of breakthrough

Table 1 Types of innovation

Types of Innovation		Definition and characteristics	Example(s)
Product; service; technology	Incremental	Slight Improvements and/or minor differences from existing competitors in the market (i.e. additional 1 or 2 features) with a little or no change in consumer behavior and habits	Each successive iterations of smart phones
	Radical	Profound/Breakthrough changes and approach to the product, its overall experience and consumer usage behavior, without altering the value proposition & problem it intends to solve	3D Printers; LASIK; CRISPR
Business model	Incremental	Known and/or proven business model but applied to a different solution and/or industry	Selling solar energy to state-own utilities
	Radical	Untested and Unproven business model (how customer purchases and/or access the product)	Rolls Royce's 'Power-by-the-Hour'; power purchase agreements as financial products
Market demand	Incremental	Proven market demand from same (or more) target group(s) of customers as the competitors.	E-payments platforms; Online games
	Radical	Targeting an untested market demand with no known direct or close proximity competitors with similar value proposition	Cryptocurrency; second life
Process	Incremental	Incremental changes and/or tweaks to existing methodology or process to improve and achieve greater overall efficiency of a process	Adding robotics to an automated factory line
	Radical	Implementation of a new or significantly improved methodology or process to improve and achieve greater overall efficiency through a change in user's behaviors and habits	Toyota production system (lean manufacturing) or ford's assembly line process

Source Cha et al. [8]. Innovation risk cube

research that comes with heavy resource commitments—largely capital and labor—and use of special resources which are not easily attainable. Besides, there are also considerable regulatory hurdles to cross, substantial risks in the ability to deliver the expected product with stable performance and the unpredictable market recep tivity to the innovation. Above all, a separate empirical research by the authors have suggested that the pursuit of radical innovations does not guarantee nor necessarily increase the chances of a venture success [8].

Successful innovations are built on other innovations. It is not a zero-sum game as successful innovations are often found by recombining ideas across boundaries [9]. Likewise, in the iPod case study, we discovered that improving the product's performance alone was not the panacea that cemented iPod's success. It was the fusion of the incremental product improvements and an integral business model—with a deeply-embedded and robust profit formula—that really propelled the iPod to be widely accepted and adopted by the mass consumers.

To innovate effectively, innovators should follow a layered and structured formula of introducing the specific innovation dimension into the marketplace, gauging and learning from the customer reactions, and subsequently integrating the market insights into iterations of innovations. This serves to eliminate uncertainty and mini mize risks while naturally building in evidence-based decision making in the process. Another successful formula is for organizations to pursue incremental innovation alongside radical innovation, to balance their innovation effort by allowing small wins in pursuit of big wins [5].

4 State of Innovations Designed for Circular Economy

While innovation brings growth, the motivation behind this pursuit of growth has always been couched in an economic model. This economic model follows a well defined set of performance indicators that can be optimized. In individual businesses typical performance indicators are sales revenue, profit generation and growth in market capitalization. At face value, these performance indicators may seem harm less. However, organizations across the past few decades are fanatically obsessed in achieving stellar performance across these indicators. So much so that they are relentlessly innovating for the sake of producing and delivering products and service that are faster, cheaper and with little to no consideration to the ecological aspect of their actions.

Here we cite the example of China's now defunct bike sharing companies, because you can see the paradigms magnified and amplified due to the sheer size and speed of the Chinese market. Bike-sharing started as an innovative business concept with good intentions. The idea of crowd-sharing assets can lead to a reduction of mate rial usage while fulfilling the needs of consumers. However, with the execution of the innovation, the exact reverse happened. These companies were blinded by the desire to achieve unicorn status and the need to seize market share as fast as they could. Consequently, these companies invested huge amounts of capital into the ector regardless of the actual demand [10], which led to excessive overcapacity and lamaged bikes, forcing many of the companies to declare bankruptcy and leaving nassive number of bicycles being strewn haphazardly.

A worker rides a shared bicycle past a huge pile of unused shared bikes in a vacant lot in Xiamen, Fujian province, China, on December 13, 2017. (Source: Reuters and The Atlantic)

This is a recent prime example of a hazard that arise from our relentless pursuit f economic growth. While it is still fundamentally important for organizations to ursue economic growth, we need to also remember that organizational growth must e ecologically sustainable for our ecosystem too.

In the same vein, we must also not forget that the world's population is continuing o grow, and it is projected to reach 8.5 billion in 2030 and 9.7 billion in 2050 [11]. A rowing population will also proportionately increase the use of resources to satisfy ur basic needs. However, as we reiterate, Earth's resources are limited. Without regenerative timescale, the continued depletion of Earth's resource will further hreaten the survivability of our planetary ecosystem. Therefore, circular economy as become such an appealing concept because it advocates for an innovative process f rethinking and redesigning products to a state where the materials required to onstruct it can be recycled indefinitely without degradation of its properties.

The transition to a circular economy is a logical proposition. It will do more than aving the planet. It will also create vast new economic opportunities [12]. Despite ie potential upsides, we have a strong sentiment that innovation for the circular conomy today are still largely driven by the rising pressure for organizations to ehave more sustainably and responsibly. As a result, organizations tend to react rith haste by delivering quick-fix solutions as opposed to developing one that is olistic and can maximize the impact translation to society.

Many organizations also relegate innovations around sustainability to be a Corpo- ate Social Responsibility (CSR) initiative, and do not invest into creating a long- erm, economically robust model for these projects. Not surprisingly, the projects ould subsequently be side lined due to conflicting, higher priority, and/or short- erm profit-driven demands. Only when circular economy innovations can generate ue economic impact to the project sponsor, and can be measured, that we will e sustained efforts and outcomes to the betterment of our environment. Simi- rly, perhaps due to the lack of corporate attention, many models of innovations (as summarized in Table 1 above) do not emphasize the unique characteristics of circula economy needs.

Therefore, as we extend our discussion towards understanding the current stat of innovation designed for the circular economy today, we would also want to take this opportunity to inspire more innovators to introduce holistic innovations in ou society. We aim to do that by highlighting some commendable initiatives and sharing our thoughts on how we feel they can take a step further with their circularity efforts

5 Examples of Innovations Designed for the Circular Economy

A remarkable product innovation designed for the circular economy is BASF ecovio®. It is a high-quality and versatile bioplastic that is certified compostable an bio-based. As a material, it has a wide range of applications. It can be used for injec tion molding to produce hard plastic goods or used as flexible plastics for shoppin; bags, waste bags and food packaging. With reference to its technical specifications ecovio® possess superior attributes and is undeniably an eco-friendlier alternative t petroleum-based plastics [13]. With its compostable properties, ecovio® can also b converted into compost in specially designed and operated facilities.

BASF ecovio® Organic Waste Bag (Source: BASF)

Despite its product superiority and capabilities, BASF does not provide any end of-life options for their customers to properly dispose and reap the benefits of usin ecovio®. As such, for customers who lacks access to a composting facility, they wi presumably dispose their waste in a general waste stream, which will likely end u being incinerated or landfilled, and defeats the good intentions of adopting ecovio®

Though it may be beyond BASF's scope and position as an organization, in ou opinion, BASF can take a step further to ensure that their customers' waste loo is truly circular. They can form key partnerships with organizations that has a established waste collection system for compostable products. This will reduce th leakage of post-consumption ecovio® products into the environment on land, int waterways, and the ocean. It will also ensure that these post-consumption product re sent to the appropriate facilities and returned to Earth as compost and hence, erving its purpose.

Another notable example is Philips' commercial lighting services. For commer-ial premises, Philips installs, maintains, and manages the lighting throughout its fecycle. This makes it possible for their customers to purchase light as a service ather than invest in new hardware upfront [14]. By innovating on the business model nd modelling it after a managed service business instead of direct sale, Philips was ble to unlock a new revenue stream.

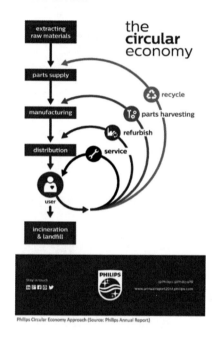

Philips Circular Economy Approach (Source: Philips Annual Report)

Coupled with a strong and grounded profit formula, this business model innovation lows Philips to generate recurring revenue while being able to collect its lighting quipment back for reuse or recycling. This approach has enabled Philips to achieve ircularity in their commercial lighting business.

Furthermore, Philips has also placed considerable thoughts on the ecological spects of their products as they innovate to improve its performance. Some of eir efforts include modularizing the connectors of their new lighting product and aking them more energy-efficient and more durable. Taking a step further, Philips an explore ways to modify this successful commercial offering and cater it towards e consumer retail market.

Another interesting example is a young Singapore-based startup Insectta. Insectta an urban insect farm that breeds black soldier fly larvae by feeding it with discarded od waste. It then takes these black soldier fly larvae and convert it into biomaterials, cluding one known as chitosan—a biodegradable polymer typically derived from ustacean shells [15]. Insectta claims that these biomaterials can be repurposed as semiconductors in devices such as phones and computers, or as protein and probiotic in animal feed additives.

Insectta's black soldier fly larvae feeding on spent brewery grains and soybean pulp.
(Photo source: Teng Yong Ping/Yahoo Lifestyle Singapore)

Insectta has also innovated on their business model. While their competitor harvest and sell their insects whole to customers, Insectta created a mechanism that allows it to separate their black soldier fly larvae into different parts through biorefinery processes to produce its chitosan, organic semiconductors, and protein and probiotic products. This differentiation allows Insectta to triple the larvae's final product value.

In summary, to effectively design innovations for the circular economy, it would be beneficial for both organizations and innovators to have a holistic understanding of innovation, especially the dynamics of a successful innovation. As a brief recap and summary, we learnt that good innovations could create and capture new value. Further, we learnt that these innovations can either be incremental or radical, and they come in all shapes and sizes. Above all, we learnt that innovations build on other innovations. For organizations to affect a real positive impact to the stakeholders they serve, they would first need to have a clear understanding of what's the "Job-to-be Done". After which, they must also be bold in experimenting with numerous types of innovations until they are able to develop and introduce an all-rounded solution that customers are willing to "hire" to get the job done.

Lastly, we must not forget that the fundamental reason for a monumental shift in global consumer behavior and demand towards green and responsible consumerism is climate change. To combat climate change, we need substantial innovations to be built, deployed, and scaled at a massively fast pace.

Therefore, in the following part of this chapter, we will be exploring what happens if we attempt to build, deploy and scale innovations that are designed to do good for our health, humanity and the environment at a massively fast pace. How would they grow? and what are the implications? For that, we turn towards understanding the concept of Restorative Innovation, which is co-created by the authors of this chapter

Introduction to Restorative Innovation

Restorative Innovation is an innovation model that aims to accelerate the adoption of innovative solutions designed to restore our health, humanity, and environment, with interventions aimed to drive down the economic costs. With more organizations and individuals starting to understand that making a profit and doing good do not have to be mutually exclusive, we are starting to see a surge of nascent entrepreneurs launching new ventures aimed at trying to make the world a better place. However, it is empirically observed that these impact-driven enterprises would have very diminished impact or become social enterprises or charitable organizations. We believe this is due to the lack of a robust and predictable economic model to justify and anticipate returns from continued investments into the innovation development. It is universally acknowledged that introducing innovative products and services is an expensive undertaking, an undertaking that is fraught with very high risks and uncertainty.

Typically, the beginning of the innovation adoption curve is flat or erratic. Hence, a model to project the growth curve under certain conditions and parameters can offer a guide to the potential future growth if the hypothesized conditions are met. Disruptive Innovation, a model developed by the late Clayton Christensen, enjoys widespread support and adoption by innovators and entrepreneurs because it offers a predictable model for the switchover effect.

THE DISRUPTIVE INNOVATION MODEL

This diagram contrasts product performance trajectories (the red line showing how products or services improve over time) with customer demand trajectories (the blue lines showing customers' willingness to pay for performance).

As incumbent companies introduce higher-quality products or services (upper red line) to satisfy the high end of the market (where profitability is highest), they overshoot the needs of low-end customers and many mainstream customers.

This leaves an opening for entrants to find footholds in the less-profitable segments that incumbents are neglecting. Entrants on a disruptive trajectory (lower red line) improve the performance of their offerings and move upmarket (where profitability is highest for them, too) and challenge the dominance of the incumbents.

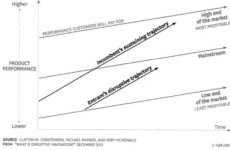

SOURCE CLAYTON M. CHRISTENSEN, MICHAEL RAYNOR, AND RORY MCDONALD
FROM "WHAT IS DISRUPTIVE INNOVATION?" DECEMBER 2015   © HBR.ORG

Graphics adapted wholly and copyright belongs to Harvard Business Review.

In a nutshell, Disruptive Innovation offers innovators a predictive model of how innovations from the low-end, typically with an initial sub-par performance but satis fies a niche market demand, can replace its mainstream counterpart, in part due to the mainstream technology overshooting its performance to customers' needs [16].

Such a model does not exist for innovations designed to restore our health, humanity, and the environment. Therefore, even with all the best intentions, the lack of a good profit formula or visibility on expected returns would produce an enterprise that is either economically unsustainable or it will lack the ability to scale beyond its initial, evangelists-led market.

We also believe that innovations can be the key to solve the critical issues facing our earth. It is well established that innovations, when applied correctly, have the ability to enable organizations to achieve continued economic growth. However there is minimal research and evidence to support the fact that innovations can also empower these impact-driven enterprises to simultaneously do good for the society and thrive economically.

We also observe that innovations which are restorative or for the betterment for our health, humanity and the environment tend to be priced at a substantially higher level than its mass market alternatives at launch—at a price point that prohibit widespread adoption—and remains at a niche level, often relying on corporate social responsibility (CSR) or other social innovation initiatives to drive adoption. Thus how do we offer a model to enable a predictable path towards a business-sustainabl growth? The Disruptive Innovation paradigm addresses the switch-over effect from new entrants when the performance of the Disruptive Innovation reaches on-par with the mainstream needs [16].

Therefore, it is imperative to develop a new innovation model and framework that can accurately expound on the growth pattern of innovations that are restorative and contribute to the betterment of society. The authors thus embarked on a research project which culminated into the Restorative Innovation framework, a framework to illustrate how innovators can develop solutions for the betterment of society, amass profits, and achieve scalable growth, all at the same time.

We offer this framework to pave the way for a whole new set of possibilities as Restorative Innovation changes the global narrative of innovation for the good. This new narrative offers entrepreneurs and innovators the way to explain how their innovative solution can bring profits and returns to investors, paving the way for capital injection to develop the restorative solution. Moreover, Restorative Innovation can also be used as the innovation model to explain how we can use innovation to accelerate our societal progress towards achieving the United Nations Sustainable Development Goals (SDGs), where the goals are largely connected with positively improving our health, humanity and the environment.

7 The Theory of the Latent Conscientious Consumer

To explain Restorative Innovation, let us first establish that we all want to be a conscientious consumer. We all have an inner desire to live a healthier lifestyle, be more humane, and to protect the environment. These needs, we shall label them as latent needs as they tend to be subconscious, omnipresent, and intrinsic in all of us. Even though we possess these latent needs, our decisions to pursue and satisfy them are often overridden by explicit counter signals from our fast-moving consumptive environment.

Let us use grocery shopping as an example to illustrate such counter signals. When we think about healthy eating, the first thought that naturally comes to our mind is to consume organic products. However, when we start shopping for these organic products, we would discover that there is a staggering price difference between the organics and non-organic products. Using cow's milk as an example, a study by Nielsen in 2018 [17] states that the average unit price of organic cow's milk is 34% higher than the average unit price of its non-organic alternative, and moreover, depending on where you purchase groceries, the premium you pay for organic produce can even go as high as 300% more [18]. This exceeds the upper-bound estimate of 50% more price premium that consumers are willing to pay for a commodity organic food item [19], which correlates to our own research by way of qualitative interviews that a switch-over effect occurs between 30 and 50% premium over mass goods. Unless the conscientious good is within this premium range, many consumers end up suppressing the desire to satisfy their latent needs and instead, anchor to what is mass produced and widely available.

Here we can clearly see the psychology of price anchoring as having an effect to suppress our latent needs. Due to our highly optimized supply chain for mass production, the parallel, not-yet-optimized supply chain is used only to serve niche market segments like organic produce. While we inherently know that organic produce is healthier for ourselves and our families, we are unwilling to pay the higher price, thus creating an ever-increasing price gaps between the two supply chains. Note that we use the word 'unwilling', and not 'unable' because we submit to you that for many of our households in developed countries, we can absolutely afford the healthier option. However, due to the price anchoring, our latent needs are suppressed from the choice dilemma, often not even entering our conscious decision-making. Thus, how do we break this cycle of ever-increasing price gaps?

8 How Restorative Innovation Works

Given that deep within us, we all desire to be a conscientious consumer and are willing to pay a slight premium for goods that cater to our latent needs. Our proposition is, we can close the price gap and eliminate the inherent complication of price anchoring with innovation and technology.

To do so, we need impact-driven leaders to not only introduce more innovative solutions that cater to our latent needs, we also need them to actively problem solve on the following dimensions: (1) increase the production availability and capacity, (2) improve the efficiency of their supply chain, and (3) drive the adoption of the innovation with targeted go-to-market strategies.

By improving on these dimensions, either singly or collectively, the price premium will be reduced. This drives the innovation into a downward price cycle due to the cost savings. The savings can then be passed on to increase the convenience and accessibility for consumers to adopt these new innovative solutions. More importantly, the savings can also be used to reduce the overall cost of production and the selling price. This will greatly strengthen the product's appeal and attract more mainstream consumers to adopt it over existing mass-market alternatives.

The increased product adoption will further accelerate the reduction of its respective production cost. This will pave the way for these innovative solutions to gradually enter the acceptable price zone. As a result, mainstream consumers will make the switch and adopt these innovative solutions that cater to their latent needs and allow the restorative effect to proliferate.

The illustration on the Restorative Innovation Model below will help reader visualize the growth pattern of these innovative solutions that are designed for our latent needs.

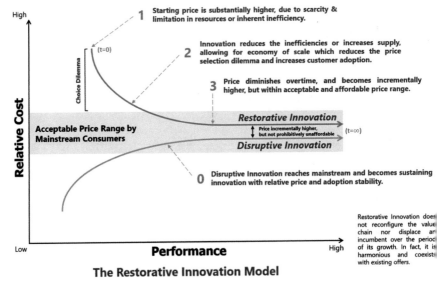

The Restorative Innovation Model

The Principles of Restorative Innovation

Impact-driven leaders play an instrumental role in enabling the above growth pattern for their innovative solutions. These leaders are passionate and intrinsically driven by their altruistic vision. They prioritize the potential impact translation of their innovative solution more than the scale of potential earnings. Therefore, these leaders possessed a radically different decision-making mindset as opposed to the profit-driven leaders.

It is empirically observed that these leaders want the best for their consumers and will not compromise on the product quality. They will take a solution-based approach to ensure that their innovative solution is holistic and is able to maximize the impact translation to society. In addition, they will also carefully consider all aspects of the product and ensure that only the best and most appropriate materials are incorporated into the finished product. This ensures that their newly introduced innovation achieves its functional purpose without harming our health, humanity, or the environment.

These innovative solutions would often possess superior product attributes and capabilities as compared to their mass market alternatives, which would then result in a higher initial true cost of production. In most instances, the higher true cost of producing the innovative solution will be directly translated to a higher starting price point. This explains our natural interactions with the explicit counter signals against our desire to satisfy our latent needs. An example of a product with superior attribute and a higher starting price point would be ecovio®, the bioplastic material that BASF has created. As we have discussed earlier, unlike its petroleum-based plastics alternatives, ecovio® is able to be composted and return to Earth as compost.

Nevertheless, some impact-driven leaders may choose to introduce their innovative solutions at a lower price point in efforts to hasten consumer adoption. In most cases, these leaders will use their resources to artificially lower the selling price and make it as competitive as the existing mass market alternatives. Gradually, with greater consumer adoption, the efficiency of the production and supply chain will improve and that would translate to a lower true production cost.

Below is a table summary to encapsulate what we have learned on Restorative Innovation.

Table Summary of Restorative Innovation

Pre-conditions	1. We possess a latent need to be a conscientious consumer and we want the best for their health, humanity and the environment
	2. Products with impact to these 3 dimensions are multi-stratified. They can be catered for the low-end, mass or high-end market
	3. Restorative Innovation are always enabled and championed by impact-driven leaders, who often possess a decision-making mindset that is more inclined towards impact translation than profit generation

(continued)

(continued)

Principles and characteristics	1. As compared to its existing mass market alternative, Restorative Innovation solutions often possess superior attributes, especially on the latent need attributes that customers value
	2. Restorative Innovation always come with a higher initial true cost even though it contains new value propositions to attract a new (and often) niche customer segment who is willing to pay a price-premium for the attributes
	3. Restorative Innovation inherently spreads socially and can influence and co-opt more adopters as the choice dilemma reduces (degree of latent needs vs. willingness to pay)
	4. Restorative Innovation can co-exist with mainstream products for an extended (even indefinite) period and create a parallel and duo consumption pattern

10   Case Study: Closing the Loop with TRIA Bio24

TRIA is a Restorative Innovation champion and an aspiring leader in designing and implementing innovation for the circular economy. TRIA is a sustainable food packaging company based in Singapore.

As a food packaging company, TRIA is not a new kid on the block. Today, the are manufacturing and supplying disposable food packaging and food service ware (collectively known as foodware) to some of the largest food services player in Asia

Founded by impact-driven leaders Ng Pei Kang and Tan Meng Chong, TRIA' mission is to enable and empower the global food services industry to move toward more sustainable patterns of consumption and eventually emerge as a truly zero wast service provider.

To accomplish this feat, TRIA has invested heavily to innovate on multipl fronts and has translated its Research and Development efforts to create and launc

Bio24, a breakthrough zero waste system that is perfectly aligned with the 12th Goal of the United Nations Sustainable Development Goals (UN SDGs)—Responsible Consumption and Production. Tout as the world's first holistic "table-to-farm" system, the Bio24 system is a marvel to behold. The entire system is capable of converting foodware, together with food waste, into good-quality compost within 24 h. This integrated approach provides an elegant solution to recycle single-use food-ware & food waste without the need for segregation nor any changes to the client's existing operations.

For this closed-loop system to work, it requires several components to work in unison. Firstly, TRIA's foodware must be constructed using their proprietary material, NEUTRIA®—a bioactive polyester derived from plant-based sources—which allows it to be fully compostable. In every aspect, NEUTRIA® is superior in product performance over its petroleum-based plastic counterpart and it does not compromise on its functionalities.

When thrown into TRIA's Bio24 digester—which is the second component to this equation—the NEUTRIA® material ages and breaks down rapidly. Fed with natural enzyme, NEUTRIA®, together with the food waste will undergo a chemical-free catalytic degradation process and be converted to compost within 24 hours. TRIA will retrieve and enrich the quality of compost to make it suitable for commercial agricultural use. Ideally, TRIA aims to sell the commercial organic compost back to their own clientele base for use in their own farms, and hence, truly closing the loop.

To ensure this outcome is achieved, TRIA has also orchestrated and established a landmark cross-industry partnership, known as the Bio24 Alliance. The Bio24 Alliance includes a waste management company, who is responsible for collecting and transporting the waste to the digester plant, and an impact-driven outcome-based certification body who will independently quantify, analyze and publish the performance of the Bio24 ecosystem and their respective participating clients in achieving zero waste. These key partners will contribute their expertise and pool their resources to enable and ensure that TRIA is able to consistently close the loop for their clients and contribute to the betterment of society.

(TRIA's Digester Plant in Singapore)

Bio24 is TRIA's answer to the commonly faced challenges in the recycling industry. Before the inception of Bio24 as a closed-loop system, there was no sustain able solution (from both environment and economic aspects) in managing single-us foodware. Even if the foodware is constructed with eco-friendlier options like renew able, biodegradable, or compostable materials, most of it will still be burnt down in the incinerator or disposed in the landfills creating negative environmental impact.

Among all the eco-friendlier options, composting offers the most attractive propo sition as it has the potential to turn waste materials into resource. Even so, a repo by the Food and Agriculture Organization (FAO) of the United Nations found the depending on the methods used to carry out the composting process, the activ composting period ranges from 3 weeks to 2 years [20] in an industrial compostin site. If left to compost in an open environment, it can take decades for it to degrade As the active composting process is carried out in an industrial composting site, the entire composting process would require storage space and resources to manage, thu making it economically non-viable to operate and consequently, it creates a situatio where most composting sites are rejecting bioplastics. Existing food waste digestor and infrastructures are also a contributing factor. As they are incapable of simultane ously managing both foodware and food waste, food waste recyclers are required t segregate and remove inorganic contents from their collected waste before sendin it to be composted. This introduces additional operational costs and complicate the entire composting process. Similarly, for recycling bioplastics, there are solver solutions that can chemically degrade the bioplastic material. While effective, it als relies heavily on the purity of the feedstock and thus, waste will still need to b separated before the bioplastics can be recycled.

Holistically, the Bio24 system is a state-of-the-art ecosystem approach to simulta neously and sustainably manage and treat both single use foodware and food wast By synergistically adding and blending layers of innovation together, TRIA was abl to create a breakthrough experience that alleviates the most faced challenges in th recycling industry. Following the footsteps of Apple's successful iPod and iTune combination, TRIA tailored its NEUTRIA® material composition to work sean lessly with its complementary catalytic digester. This deliberate optimization allow for the Bio24 to achieve peak degradation performance of at least 20 times faster than the current top of the line composting method (24 hours vs 3 weeks). On top of this product innovation, TRIA has also implemented process innovation. TRIA has streamlined Bio24's operations and deployment process to allow for their participating clients to conduct their "business as usual". With this current deployment process, TRIA is confident in onboarding and enabling their clients to go zero waste within a month or two. TRIA will work closely with their client to design and manufacture their foodware according to their functional requirements. Afterwards, all that is left is for the participating clients to replace their existing foodware with the NEUTRIA® foodware. Everything else stays the same and this means there are no further changes to their current modus operandi. The participating clients will not need to sort and separate their waste before recycling, neither do they need to be trained on any new processes and procedures before utilizing Bio24. This enables an effortless recycling process and it further eliminates the need for incurring additional time and manpower cost.

In addition, TRIA has also placed considerable effort on innovating and extending its business model. Apart from generating revenue through the sale of its NEUTRIA® foodware, TRIA has also paved the way for a new revenue source by ingeniously conceiving a cost-effective method to enrich and sell the by-product of the Bio24 system—which are organic compost—to members of the agricultural community.

Bio24 may seem to be the perfect formula to accelerate societal progress towards a greener, cleaner, and more equitable world. However, the implementation of the Bio24 system comes with its own set of challenges and resistances too. Evidently, TRIA's Bio24 has met all the conditions to be classified as a Restorative Innovation. It is led by passionate impact-driven leaders and its innovation has superior performance and attributes, especially on the latent need attributes that customers value—which in this case is environmental consciousness. As a result, TRIA's Bio24 system follows the Restorative Innovation growth pattern and started off with a higher initial true cost due to scarcity and limitations in resources and/or inherent inefficiencies.

For a 12oz Cold Cup		
Material	**Degradation**	**Cost**
PP Plastic	Incineration or Landfilling (up to 1000 years)	≈SGD$0.07
Styrofoam		
Paper (Lined)		≈SGD$0.08
PLA Plastic	Biodegradation (3-6 months)	≈SGD$0.12
NEUTRIA®	Composting (24 Hours)	≈SGD$0.13

Adopting Bio24 creates a choice dilemma. In TRIA's case, this choice is made on the consumer's behalf by TRIA's clients. In the food services business, each unit of sale yield a tiny profit margin. In most cases, a percentage of this profit has been re-allocated to cover the cost of foodware. Hence, the additional premium to adopt NEUTRIA® proves to be the greatest resistance as these food services businesses will opt for foodware that fulfils their functional requirement and within their budget.

Exceeding their budget will correspondingly reduce their profit. As such, even if they are environmentally conscious, it is not easy for them to readily make the switch.

Nonetheless, TRIA believes that the sustainability uptake for the food service providers can be a zero-sum game and they have creatively experimented with various strategies to encourage adoption. With their vast experiences in the food packaging sector, TRIA has observed that their brand-savvy clients have the strongest inclination towards sustainability. As these brand-savvy clients leverage heavily on the use of marketing and branding as their competitive advantage, they tend to be constantly on a lookout to stay relevant with the latest trends by re-positioning and re-aligning themselves to it.

Fortunately, today, there is a monumental shift in consumer's behavior towards green and responsible consumerism. As such, TRIA capitalized on this opportunity and offered these prospective clients an attractive proposition where TRIA will render their award-winning design expertise to help them design and develop foodware that are appealing, distinctive, and can be incorporated as part of their overall brand experience, should they choose to adopt NEUTRIA® foodware and be part of the Bio24 ecosystem.

The strategy to bundle in award-winning designs as part of the NEUTRIA® food ware is the critical first step towards increasing availability of the Restorative Innovation goods. The early customers would not be willing to pay a premium for a circular economy innovation when it is not yet in the mainstream, but they are willing to pay for a more beautiful foodware package, as typically, this budget is parked under marketing and branding. We think this early strategy of offering a value proposition that customers are willing to pay for, in the early stage of Restorative Innovation, is a critical factor to start the downward cost curve.

Prior to mainstream adoption, few customers would be willing to pay more for a 'sustainability' or 'do-good' value proposition. We dub this the 'Fancy Horse' strategy and offers a win-win arrangement for both the Restorative Innovation champions and their customers.

This approach is proving to work and as a result, TRIA has been able to convince larger industry players to jump on the bandwagon and go green. Following the Restorative Innovation growth pattern, the authors firmly believe this product adoption strategy will gradually improve TRIA's operational efficiencies and consequently, eliminate unnecessary cost and reduce their overall cost of production

With time, the Bio24 ecosystem will progressively enter the acceptable price zone and this will encourage more food service providers to make the switch and adopt these innovative solutions that cater to their latent needs.

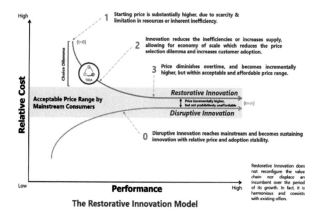

The Restorative Innovation Model

For now, the authors have determined that TRIA is close to the second point and s progressing well to be a successful Restorative Innovation.

1 Conclusion

o conclude this book chapter, let us remind ourselves that the world's population s soon reaching 10 billion, which will create unprecedented resource stress on our lanet. If we continue to singularly focus on innovation as a relentless pursuit of conomic growth, we may find ourselves depleting what nature intends for us, in erms of positive growth and change. We believe the time has come to encourage nore innovators and business leaders to reimagine value creation in terms of how ve can restore health, humanity, and the environment. The changes that are causing legative effects to our health, our societal values, and our environment have been dding up gradually, and if we are blinded by the gradualness of changes and evolution n our Darwinian world view, we may not be ready when a bifurcation event happens, uddenly and on a large-scale, as such changes almost always do in a complex system.

We explored the various models of innovations and provided an in-depth analsis of Apple's iPod, in how Apple is able to create an ecosystem to deliver the xponential benefit of music-on-the-go of own preference to mass consumers, at an ffordable cost, with speed, fidelity, and convenience. This required Apple to build nis ecosystem in stages—first by attracting an initial group of iPod consumers with ast file transfers with a superior UI/UX—then by adding in the ability to purchase ndividual music scores at an affordable unit-price. To accomplish this, required a econfiguration of the music supply chain to offer a lower cost distribution and the villingness of music publishers to forego a larger unit sales revenue in exchange f a promise of a larger total revenue from an exponentially larger audience. This romise is amply realized by Apple, to the delight of her shareholders.

We believe this model is instructive for how to drive larger adoption of Restora tive Innovations. Instead of relying on the do-good contributions from various stake holders under the "Corporate Social Responsibility" umbrella, Restorative Innova tion champions should take aim at intervening at any or all three critical points: (1 Increase supply availability, (2) Increase efficiency of supply chain, (3) Increase demand to arrive at a price premium that is not greater than 30% to existing mass-consumer equivalent.

We then looked at TRIA, a restorative champion in the initial stages of imple menting such interventions. One tactic to accomplish these interventions may involve use of a 'fancy horse' offer where the buyer is paying the premium for a non restorative feature. We look forward to contributors to add to this body of knowledge in offering other tactics to achieve the success that all restorative innovations deserve for the sake of our planet. We hope this book chapter, and especially Restorative Inno vation, will serve as an inspiration to ignite our collective imagination on what i possible and change the innovation narrative from relentless pursuit of economi growth to pursuit of value from a more balanced view on what is good for econom and what is good for us. Innovation should serve a purpose for society and shoul not be a solely monetary pursuit.

Questions

1. Why do most innovations and initiatives around sustainability ends up becomin, a Corporate Social Responsibility initiative or equivalent?
2. Why do organisations pursue incremental innovation more frequently than radica innovation?
3. What is Restorative Innovation?
4. Why is it important to learn and understand Restorative Innovation?
5. What are the key interventions impact-driven leaders need to introduce to driv their restorative innovations into a downward price cycle and increase adoption

Further Discussion Questions

1. What example(s) can you think of, a conscientious product that was introduce in recent years but have not reached mass adoption, and hypothesize on th reason(s)?
2. Can this product innovation benefit from a Restorative Innovation framework?
3. What intervention point(s) would you consider to be the most effective? an why?

Answers

1. Organizations tend not to invest into creating a long-term, economically robu: model for innovations and initiatives around sustainability. As a result, thes efforts often get side lined due to conflicting, higher priority, and/or short-terr profit-driven demands. To resolve this issue, the organization must conside embedding a mechanism that delivers a quantifiable economic benefit back t the organization.

2. As radical innovation often involves the translation of breakthrough research, it often comes with heavy resource commitments, including capital and labor, and the use of special resources which are not easily attainable. In addition, radical innovation also involves considerable regulatory hurdles to cross, substantial risks in the ability to deliver the expected product with stable performance and the unpredictable market receptivity to the innovation.

3. Restorative Innovation is an innovation model that aims to accelerate the adoption of innovative solutions designed to restore our health, humanity, and environment, with interventions aimed to drive down the economic costs. It explains how innovations that contribute to the betterment of our society goes to market and scale.

4. It is important for us to learn and understand Restorative Innovation as it illustrates the possibility of creating and delivering innovations that are capable of achieving continued economic growth while simultaneously contributing to the betterment of society.

5. Impact-driven leaders needs to work on (1) increasing the production availability and capacity, (2) improving the efficiency of their supply chain, and (3) driving the adoption of the innovation with targeted go-to-market strategies.

Suggested Reading

1. Christensen, C. M. (1997). The innovator's dilemma: When new technologies cause great firms to fail. *Harvard Business School Press*, Boston, MA.
2. Casadesus-Masanell, R., Kim, H., Reinhardt, F. L. (August 2010). Patagonia. *Harvard Business School Case*, 711–020, (Revised October 2010).

[Readers are highly encouraged to read up on various business case studies—like Patagonia—that involves companies pursuing both economic growth while simultaneously contributing to the betterment of society.].

References

1. Roberts, E. (1988). What we've learned: Managing invention and innovation. *Research-Technology Management, 31*(1), 11–29.
2. Intergovernmental Panel on Climate Change (2018). Global warming of 1.5 °C. Summary for policymakers. *World Meteorological Organisation & UN Environment*. Retrieved Sep 17, 2019, from https://report.ipcc.ch/sr15/pdf/sr15_spm_final.pdf.
3. Adner, R. (2012). Innovation success: How the Apple iPod broke all Sony's walkman rules. *[Blog] INSEAD Knowledge*. Retrieved Sep 23, 2019, from https://knowledge.insead.edu/blog/insead-blog/innovation-success-how-the-apple-ipod-broke-all-sonys-walkman-rules-2791.
4. Johnson, M., Christensen, C., & Kagermann, H. (2008). Reinventing your business model. *Harvard Business Review*.
5. Kahn, K. (2018). Understanding innovation. *Business Horizons, 61*(3), 453–460.
6. Christensen, C. et al. (2016). Know your customers' "Jobs to Be Done". *Harvard Business Review*.

7. Apple Inc. (2001). *Apple presents iPod*. Retrieved Oct 23, 2019, from https://www.apple.com newsroom/2001/10/23Apple-Presents-iPod/.

8. Cha, V., Cai, Y., & Tan, J. (2019, In Publication). *Innovation risk cube*.

9. Anderson, P. (2018). How do you spot a restorative innovation opportunity. *INSEAD*.

10. Yao, Y. (2019). How to clean up bike-sharing firms' mess. *China Daily*. Retrieved Oct 18 2019, from http://www.chinadaily.com.cn/cndy/2018-08/27/content_36828744.htm.

11. United Nations, Department of Economic and Social Affairs, Population Division (2019) *World population prospects 2019: Ten key findings*.

12. Ghosh, A. (2019). The circular economy is a golden opportunity. Don't let it go to waste. *World Economic Forum*. Retrieved Oct 9, 2019, from https://www.weforum.org/agenda/2019/01/the circular-economy-turns-waste-into-gold-so-lets-get-on-with-it/.

13. BASF. (2019). ecovio®. Retrieved Oct 19, 2019, from https://products.basf.com/en/ecovio html.

14. Smart Cities World. (2016). *Philips' light-as-a-service offering*. Retrieved Oct 19, 2019, from https://www.smartcitiesworld.net/news/news/philips-light-as-a-service-offering.

15. Wee, R. (2020). Insectta sees a big future from small insects. *Garage powered by the busi ness times*. Retrieved March 29, 2020, from https://www.businesstimes.com.sg/garage/insectta sees-a-big-future-from-small-insects.

16. Christensen, C. (1997). Innovator's dilemma: When new technologies cause great firms to fail (management of innovation and change series). *Harvard Business Review*, 174.

17. Nielsen Insights. (2018). *Tops of 2018: Organic*. Retrieved Oct 20, 2019, from https://www. nielsen.com/us/en/insights/article/2018/tops-of-2018-organic/.

18. Consumer Reports. (2015). *Cost of Organic Food—Consumer Reports*. Retrieved Oct 20, 2019 from https://www.consumerreports.org/cro/news/2015/03/cost-of-organic-food/index.htm.

19. Strzok, J., & Huffman, W. (2016). Willingness to pay for organic food products and organic purity: Experimental evidence. *Journal of Agrobiotechnology Management & Economics* *18*(3), 13. Retrieved Oct 20, 2019, from http://www.agbioforum.org/v18n3/v18n3a13-hu fman.htm.

20. Food and Agriculture Organization of the United Nations. (2003). *On-farm composting methods*. Land and Water Discussion Paper.

Jovan Tan has vast experiences in launching & consulting for both for-profit and non-profit enterprises. He is the co-creator of Restorative Innovation and the Founder of RIGHT Founda tion, a non-profit think tank dedicated to advance and proliferate the body of knowledge of Restorative Innovation. In addition he is also the Co-Founder of REAL IMPACT, an impact consul tancy firm, and the Chief Evangelist of TRIA where he fervently champions the company's cause, support its growth efforts, and advise the senior leadership team on key strategic issues.

Previously, he has led special projects and built business units from scratch to profitability. More recently, he was also involved in venture capital, teaching, and research. Today, Jovan is an active researcher pursuing his degree of Doctor of Philos ophy (Ph.D.) with the National University of Singapore. His research interest is in the areas of applied innovation and sustainable development, where Restorative Innovation is his most notable work. He has also been featured on Channel News Asia's Money Mind and CNA938 for his expertise in sustain ability and the circular economy.

Jovan actively gives back to the community by serving as the Chairman of the SAFRA Entrepreneurs' Club and a also an advisor to King Mongkut's University of Technology Thonburi's (KMUTT) STEAM Platform of Transformation. He is also represented in the Technical Committee on Food Services, under the Food Standards Committee of Singapore Standards Council. Jovan earned his first degree in Innovation and Entrepreneurship from the University of Adelaide, and his Master of Science in Management from the National University of Singapore.

Virginia Cha is a leading educator of Innovation & Entrepreneurship in Singapore with multiple appointments at Singapore's leading tertiary education institutions: Adjunct Professor at NUS Business School and at INSEAD; Adjunct appointments at SMART (Singapore MIT Alliance for Research and Technology) and Lean LaunchPad @Singapore. She has conducted numerous executive programs, with special original content on Entrepreneurial Mindset with emphasis on action-planning for Corporate Innovation Programs. Her research work in entrepreneurial logic has been published as a chapter in "The Entrepreneurial Behaviour. Unveiling the cognitive and emotional aspect of entrepreneurship", published by Emerald.

In her multi-faceted 40 year-long industry and academic career which spanned multiple countries, Virginia co-founded or was the sole-founder and CEO of multiple venture-funded, hi-tech companies in Singapore and China, with listings on NASDAQ and HKSE. She has co-authored a book "Asia's Entrepreneurs: Dilemmas, Risks and Opportunities" which captured Singapore's technology entrepreneurial history from 1995–2005. Additionally, Virginia is a member of the Future Council of World Economic Forum. In addition to teaching entrepreneurship, Virginia is also an active researcher, mentor, and angel investor in Singapore's entrepreneur ecosystem. Virginia has 16 companies in her angel investment portfolio, supporting start-up entrepreneurs with operations in Singapore, Vietnam, Indonesia, USA, Finland, and the UK.

Virginia's latest research into innovation is on the subject of restorative innovation (www.restorativeinnovation.com). This emergent framework uses an economic model on how sustainable goods can reach mass consumption through innovation along three dimensions: increased supply, supply chain efficiency, and increased adoption. This framework is now taught in leading Singapore tertiary institutions.

Virginia earned her Bachelor of Science in Information Computer Science, University of Hawaii, in 1980, and her PhD from National University of Singapore. She has lived in Hong Kong, Thailand, multiple cities in the U. S. and P. R. C., and continues to be based in Singapore.

The Business Opportunity of a Circular Economy

Ellen MacArthur Foundation

Abstract Business leaders and governments around the world are increasingly looking beyond the linear 'take, make, waste' model of growth, with a view to making a strategic move towards an approach fit for the long-term. Research by the Ellen MacArthur Foundation and others has demonstrated the potential of the circular economy—a model that decreases resource dependence and increases prosperity. In addition to creating direct economic benefits for businesses and households, a circular economy represents a significant opportunity to help tackle global challenges such as the climate crisis, biodiversity loss and land degradation. This chapter examines the business opportunities of the circular economy in three key sectors: the food system in India; the built environment in China's cities; and mobility in Europe. It further quantifies the economic, environmental and social benefits of these opportunities and explores what are the levers to bring them to scale.

Keywords Circular economy · Design out waste and pollution · Keep products and materials in use · Regenerate natural systems · Material cost savings · Decoupling economic growth from material consumption · Business models · Opportunities for the food system · Built environment · And mobility

Learning Objectives

Understand the limits of the current linear system
Learn what is the circular economy and its role as a solution framework to tackle global challenges
Understand the value creation opportunity of the circular economy
Learn how circular economy strategies can achieve a regenerative food system, the case of India
Learn what the circular economy strategies are for the built environment,

Ellen MacArthur Foundation (✉)
Ellen MacArthur Foundation, Cowes, UK
e-mail: info@ellenmacarthurfoundation.org

© Springer Nature Singapore Pte Ltd. 2021
L. Liu and S. Ramakrishna (eds.), *An Introduction to Circular Economy*,
https://doi.org/10.1007/978-981-15-8510-4_20

the example of China
Learn what the circular economy strategies are for the mobility sector,
the EU example
Understand the enabling factors to accelerate the transition towards a circular
economy.

1 Introduction

Our Current Linear System is Ripe for Disruption

Our linear 'take, make, waste' economy relies on fossil fuels and does not manage
resources such as land, water, and minerals for the long-term. Its extractive
and wasteful nature leads to greenhouse gases (GHG) emissions, pollution, and
contributes to biodiversity loss. These negative effects are set to worsen in future if
rapid industrialisation of emerging economies and mass consumption in developed
economies continues in a linear system. By 2050, the global population is projected
to reach 10 billion, with an emerging middle class expected to double its share of
global consumption. This welcome broad-based rise in prosperity will cause emissions to exhaust the available carbon budget by a large margin. The related impact
put further pressure on the other planetary boundaries, for example, biodiversity
loss. Recent studies have demonstrated that around a million species of animals and
plants are already at risk of extinction. Overall, resource extraction and processing
are responsible for more than 90% of land- and water-related environmental impact
(water stress and biodiversity loss), with agriculture being the main driver. A fundamental change in the way goods are made and used is required to address these global
challenges.

What is the Circular Economy?

Moving away from today's linear model towards an economy that is regenerative
by design is increasingly seen as required to meet the Sustainable Development
Goals (SDGs) and tackle global environmental challenges. In such an economy,
natural systems are regenerated, energy is from renewable sources, materials are
safe and increasingly from renewable sources and waste is avoided through superior
design of materials, products and business models. A circular economy offers a
positive way forward by redefining value creation to focus on society-wide benefit.
It addresses the shortcomings of the current system, while creating new opportunities
for businesses and society.

   In short, a circular economy aims to decouple growth from the consumption of
finite resources and build economic, natural and social capital. It is built on three
principles: design out waste and pollution; keep products and materials in use; and
regenerate natural systems (Fig. 1).

   The circular economy framework distinguishes between technical and biological
material cycles. In biological cycles, food and biologically-based materials—for
example, cotton, wood—feedback into the system through composting or anaerobic

DESIGN OUT WASTE AND POLLUTION **KEEP PRODUCTS AND MATERIALS IN USE** **REGENERATE NATURAL SYSTEMS**

Fig. 1 The 3 principles of the circular economy

digestion. These cycles regenerate living systems—for example, soil—which provide renewable resources for the economy. Technical cycles recover and restore products, components and materials through strategies including reuse, repair, remanufacture, refurbishment and in the last resort recycling. Digital technology has the power to support the transition to a circular economy by radically increasing virtualisation, dematerialisation, transparency and feedback-driven intelligence (Fig. 2).

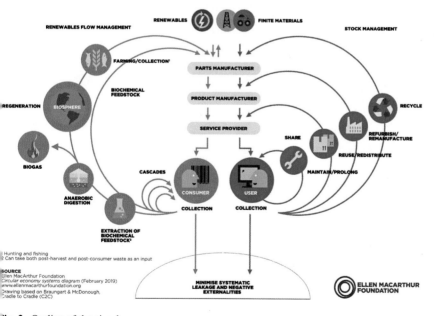

Fig. 2 Outline of the circular economy

2 The Circular Economy Is Increasingly Recognised as a Value Creation Opportunity and Solutions Framework to Address Global Challenges

Materials Cost Saving and Innovation Opportunities

In recent years, businesses from key industry sectors have made bold circular economy commitments. Since its launch in 2018, more than 450 organisations have signed the New Plastics Economy Global Commitment to work towards eliminating unnecessary plastic packaging, innovating what is needed so it is reused, recycled or composted rather than becoming pollution. The business signatories represent 20% of global plastic packaging use, with some having made bold new targets in the past year. Action is not restricted to plastics—it is happening across sectors. In agriculture, for instance, companies including Unilever, Danone and Nestlé joined the One Planet Business for Biodiversity coalition in September 2019, which recognises its members' strategic reliance on biodiversity and the urgent need to take action to restore it, notably through regenerative agriculture, a core circular economy opportunity. In textiles, the circular economy is also increasingly driving innovation both in terms of business models and materials: H&M plans to sell a garment made from an innovative material that uses recycled cotton from March 2020; and Adidas & Stella McCartney are piloting sweaters made with an innovative process to turn used cotton into new fibres.

A Solutions Framework for Addressing Global Challenges

A consensus is rapidly building that the circular economy offers a route to prosperity that responds to global imperatives. As a megatrend, this positive, innovation-centred vision for the future of the economy is increasingly on top of mind for business leaders being a strategic priority for some of the largest companies in the world. It is also increasingly high on the agendas of major international organisations, including the One Planet Summit, the UN Climate Change Conference (COP), the G7 and the World Economic Forum. A growing number of governments and international institutions view the circular economy as a delivery mechanism for wider goals. The European Green Deal Circular Economy Action Plan with an investment plan set up to mobilise at least EUR 1 trillion over the next decade, aims to promote 'just and inclusive' growth but also 'to protect the health and well-being of citizens'. The circular economy is now clearly seen as a solution framework to tackle climate change, alongside the energy transition. As of COP25, circular economy measures will be integrated into meeting countries' Nationally Determined Contributions to reducing emissions.

The circular economy can provide multiple value creation mechanisms decoupled from the consumption of finite resources. Shifting towards this model could deliver better outcomes for the economy, people and the environment. By embarking on circular economy transformations, countries could leverage high levels of growth and development. According to analyses by the Ellen MacArthur Foundation, the transformation could yield annual benefits for Europe of up to EUR 1.8 trillion in 2030 [1]. For China, activating broader circular economy solutions in cities could significantly lower the cost of access to goods and services and could save businesses and households approximately USD 11.2 trillion in 2040 [2]. For India, the annual benefits could amount to USD 624 billion in 2050, compared with the current development path [3].

A Circular Economy for Food in India: Scaling up Regenerative Practices

The Current Food System Has Many Negative Consequences

The food system meets vital societal needs as nutrition and provides ecosystem services, such as pollination. The food industry is a key pillar of many economies. It is the world's largest sector, accounting for around 10% of global GDP, and employing over 1 billion people. While the food system has made significant productivity gains over the past half-century, many negative consequences are already being felt. Growing food demand and environmental challenges associated with climate change, land degradation and biodiversity loss are putting increasing pressure on the system.

.1 Six Circular Economy Opportunities to Reshape the Indian Food System

The agricultural system is crucial to the Indian population and the economy. With some 854 million people living in rural areas, India houses the world's largest rural population and agriculture is the principal means of livelihood for 58% of rural households. In 2013, India became the world's sixth largest net exporter of agricultural products. A broad range of circular economy opportunities exists for India to consider when shaping the future of its food system and agricultural activities. Applying circular economy principles to the development of the Indian food system could create annual benefits of USD 61 billion in 2050, reduce greenhouse gas emissions, water usage, environmental degradation and play a fundamental role in securing long-term food supply [3].

Regenerative agriculture production
A regenerative agricultural system preserves the integrity of the natural system, phases out toxic materials and minimises nutrient leakage. The system uses practices like crop rotation and cover cropping and minimises tillage to retain natural capital. It often combines livestock with crop production to create additional nutrient loops. These practices can build ecological resilience to changing climate

conditions such as extreme heat or severe droughts and other environmental shock like pests and diseases outbreaks, while increasing yields for farmers. Organisa tions like INORA, Kalpavruksha Farm, Organic India and Govardhan Ecovillag are applying regenerative practices at different scales and report increasing yields health and income. Native, Brazil's largest sugarcane producer, cultivating ove 20,000 ha and realising profits of USD 10 million, demonstrates that regenerativ practices can succeed at large-scale. These practices come with additional benefits They can help mitigate climate change by increasing the amount of carbon store in soils and reducing the need to apply fossil fuel derived synthetic fertilisers.

– More efficient agriculture practice enabled by new technology
  Sophisticated agricultural approaches that leverage IT, big data, remote sensin and real-time environmental data can optimise returns, while reducing envi ronmental externalities. Non-digital technological solutions, such as systems t improve irrigation, can also create value. Digital technology is changing farmin by enabling increasingly sophisticated precision agriculture approaches, as we as whole farm management through the emergence of integrated data platforms Precision farming can increase the efficiency of conventional agricultural system but has proven, especially effective combined with regenerative practices. I India, small farm size limits the feasibility of such solutions, particularly i the short-term. But Indian farmers could profit from adopting low-marginal cost, cloud-based digital solutions that would scale easily to support many sma farmers.

– Asset and knowledge sharing via digital platforms
  Asset sharing can increase innovation, productivity and yield. High fragmentatio of land has limited the adoption of innovation in India, as farms are often too sma to justify the capital investment required to implement more efficient technologie and systems. As a result, average yield rates in the Indian agricultural secto are low by international standards. Sharing platforms can give farmers acces to machinery that they otherwise would not be able to afford. Digitally enable knowledge-sharing solutions are expected to encourage adoption of best practice increase yield and advance regenerative agriculture. These solutions would l farmers share local and traditional knowledge on a peer-to-peer basis and receiv information on innovative practices customised to their region and crops. Digit Green is a non-profit organisation that trains farmers in more than 2,000 village to make and show short videos that record their problems, share solutions an highlight success stories.

– Digitised food supply chains
  Indian farmers sell their produce pretty much as they have done for the la 50 years. The system relies on the government-regulated Mandi marketing channe that, according to the Task Force on Agricultural Development, is 'characterise by inefficient physical operations, excessive crowding of intermediaries, long an fragmented market chains and low scale, depriving farmers of a fair share of th price paid by the final consumer'. Digital supply chain solutions increase tran parency, decrease the high transaction costs attributable to multiple and variou

intermediaries, and better connect producers with customers. These solutions also enable better inventory management and self-organised production optimisation across small-hold farmers. Knowing the size and timing of demand, farmers can use digital technology to coordinate, adapt and optimise the supply of food to their region, reducing price volatility and increasing their income.

- Peri-urban and urban farming
Bringing food production and consumption closer together by increasing agricultural activity in and around cities reduces food transport and associated costs (such as food waste, fuel and environmental externalities) and tightens biological nutrient cycles, while increasing access to fresh, healthy food and creating new income streams. Specialised urban farming techniques (like vertical, hydroponic and aquaponic farms) can be more resource-productive than traditional cultivation techniques, saving on energy, water and fertiliser. Rapid urbanisation in India moves food demand closer to urban centres that are also experiencing problems with overheating. Peri-urban and urban farming could help overcome some of these challenges, while providing a new source of employment and income. While not all crops can grow in an urban environment, and scaling presents challenges, urban farming works well for highly perishable vegetables and herbs, delivering them to consumers fresh with little investment in resources and transportation.

- Nutrient looping
Not all biological nutrients that reach their place of consumption are actually used. Some end up as household or industrial food waste, others are consumed but not absorbed by the human body and discharged in human excreta. Processes like composting and anaerobic digestion can recover these nutrients for return to the agricultural system, and in the case of anaerobic digestion, produce energy. While avoiding food waste should be a priority along the supply chain, and several opportunities mentioned above can contribute to this reduction, residual food waste in industrial and household kitchens is unavoidable. Indian non-profits and businesses, in cooperation with municipalities and the informal sector, are already implementing composting and anaerobic digestion solutions to capture these nutrients.

.2 Benefits of a Circular Economy for Food in India

Application of circular economy principles would make agricultural production more regenerative, creating a more diverse and resilient food system that could supply fresh, healthy produce to India's growing population. This development path would preserve the integrity of the natural system, phase out toxic materials, and minimise nutrient leakages, reducing negative environmental and health externalities, while supporting rural livelihoods and incomes. Overall, following a circular economy development path could generate annual benefits of USD 61 billion in 2050, compared with the current development path. Following the circular path would also

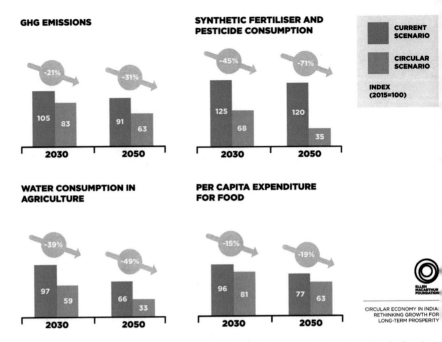

Fig. 3 Indian food system: comparison of potential development paths (food and agriculture)

reduce negative environmental impact with 31% less GHG emissions from agricul ture, 71% less use of synthetic fertilisers and pesticides and water consumption fo irrigation almost halved (Fig. 3).

4 Circular Economy in the Built Environment in China's Cities: Designing Out Waste and Increasing Asset Utilisation

Cities Play a Critical Role in Economic Growth But Face Many Challenges

By 2050, two thirds of us will live in cities. However, our urban centres are grapplin, with the effects of our current take-make-waste economy. Under this 'linear system' cities consume over 75% of natural resources, produce over 50% of global waste and emit 60–80% of GHGs. Although recent decades have seen tremendous progres in improving the energy efficiency of buildings and the liveability of cities, the buil environment sector remains incredibly wasteful for a sector considered as matur and optimised.

4.1 Six Circular Opportunities for the Built Environment in China's Cities

Many millions of houses will be built in the next two decades and how they are built will determine China's mid- to long-term development. Currently, the country is facing severe challenges from air pollution and construction waste. For instance, in 2013, China generated a billion tonnes of Construction and Demolition Waste (CDW). Applying circular economy principles at the design, construction, use and deconstruction stages would help address these issues and provide economic, environmental and societal benefits. Implementing circular levers in a coordinated way could unleash the remarkable potential in the built environment, creating benefits of USD 2 trillion in 2030, compared to the current development path [4].

- Design for flexibility
  Facing the wave of new construction needed in China over the next two to three decades, designing buildings for flexibility has the potential to create economic, environmental and societal benefits. New and advanced technologies, including innovative materials, products and services are being designed for modularity, repair, flexible upgrade and disassembly which could help to reduce maintenance costs and extend the economic viability of buildings and structures, ensuring such assets are used optimally over time. By designing flexible building cores, for instance, developers can enable a building to switch user and purpose later in its use cycle. Buildings designed to be modular and flexible are also easy to disassemble, allowing the materials used in them to retain their value by being reused, reducing waste and the consumption of virgin resources.

- Industrialise construction process
  The industrial manufacturing of standardised and modular building components, combined with on-site assembly, reduces construction costs, time to completion and waste. Modularisation could also contribute to lower operational and maintenance (O&M) costs at the end stage of a building, as all its components can be easily disassembled and/or replaced. The Broad Group, a Chinese constructor specialising in modular construction, has managed to increase efficiency in production, installation and logistics six to ten times with almost zero materials wasted and 40% lower total cost of construction through such design approaches. Another technology that could accelerate the shift to more efficient and industrialised construction is 3D printing. In 2014, the Chinese construction company WinSun 3D-printed and assembled ten 195 m^2 houses in 24 h. Using such technology decreases material use by 30–60%, thereby reducing costs and waste compared to conventional methods. Alongside new technologies, renewable and local materials could be widely employed in urban construction in the circular economy scenario. For example, bamboo is a fast-growing and sustainable material for houses. The cost of a bamboo house could be 60% lower than that of a concrete one and be built in a modular and adaptable fashion.

– Share space to increase asset utilisation
Space sharing could provide a good way of improving utilisation without requiring
new construction. The space-sharing market in China is growing fast. In 2016, the
total turnover in hospitality and office sharing was USD 3.9 billion—an increase of
131% from 2015. The public is no stranger to platforms, such as Airbnb, that adopt
a customer-to-customer (C2C) model to rent out underutilised residential space
and the adaptation of such models to the working environment is gaining ground
In the office-sharing market, co-working schemes provide options for freelancers
entrepreneurs and start-ups to have a small office in a central business district with
cheaper rents, flexibility and access to office facilities. A number of international
and domestic companies have entered this market and opened offices in large
cities, including SOHO 3Q, UR Work, Naked Hub, People Squared, WeWork
WE+ and Cowork. With the start-up boom in China (on average, there were
15,000 new company registrations every day in 2016), the office-sharing model
has rapidly expanded in major cities such as Beijing, Shanghai and Guangdong
province, and has now started to expand into smaller cities such as Suzhou and
Urumqi.

– Design buildings to improve energy efficiency
Building innovation offers great potential not only for buildings that have minimal
running costs, but also for enhancing the health and well-being of users. The bene
fits of 'green buildings' are reflected in lower energy and water usage and reduced
O&M costs. Indeed, energy efficient buildings have the potential to bring down
energy consumption by 30–50%. For example, Building Integrated Photovoltaic
(BIPV) and grid-connected generating technologies could be a solution for solar
energy utilisation in China's densely populated cities. The world's largest stand
alone building with an integrated PV system is in Shanghai Hongqiao Railway
Station. More than 20,000 solar panels on the 61,000 m^2 roof produce 6.3 million
kWh of electricity annually. Rolling this technology out further would form an
important part of the energy plan for a circular city. One concrete example of
green building in China is the passive house. Compared to traditional houses
passive houses save 80% of heating energy and 50% of energy for cooling and
dehumidification. Currently, only a minority of new buildings are passive houses
so there is a clear case for development here that could be economically and
environmentally beneficial.

– Improved digitisation
The concept of a 'smart city' is gaining traction in China thanks to the devel
opment of digital technology, in particular, the Internet of Things (IoT). Smart
building technology—such as sensors, data storage and computing services—is
a prerequisite of the smart city concept and is increasingly part of people's day
to-day lives. Smart meters for electricity and gas allow residents and owners to
make informed decisions about their energy usage and can result in substantial
cost savings. The use of such smart-equipment technologies could result in a 40%
reduction in electricity use and an 87% reduction in gas consumption. Scaling

up such technology could see measurable improvements in energy efficiency and productivity throughout China's built environment.

- Scale up reuse and recycling of construction and demolition waste
While the construction of a 10,000 m² building will create 500–600 tonnes of waste, its demolition will create 7,000–12,000 tonnes of waste, so there is a clear need to address the after-use phase of buildings. Discarded materials could be recovered and reused in a variety of ways, such as for gravel, road building materials or flooring. Maximising the use of repurposed materials and components in this way would reduce the need for virgin materials and would cut the pollution caused by CDW in the urban environment. In China, tiles, bricks and bamboo are traditional materials that could be used more, even in new buildings, and at a later point, the materials can be potentially recycled again.

4.2 Benefits of a Circular Economy Approach to the Built Environment of China's Cities

In a circular economy scenario, the identified opportunities impact different parts of the value chain, interacting and amplifying each other's effects on the overall built environment. The buildings are assumed to be designed for longevity, be shared widely, built by industrialised construction and make use of smart home/office technology. Flexibility, modularity and shareability are the key attributes of future buildings. In this scenario by 2030, 70% of urban new builds would be green buildings, 80% of CDW in cities would be recycled and reused and 25% of urban buildings would be built using industrialised construction processes such as prefabrication or 3D printing. In 2040, the cost of building construction in a circular scenario could be 41% lower than that in the current development path. This cost reduction is mainly due to improved resource productivity in the construction phase, driven by the adoption of industrialised construction technology. CO_2 emissions could decrease by 9% in 2030, and by 24% in 2040, compared with the current development scenario. The reduction is driven by promoting energy efficient buildings, smart buildings and space sharing. In a circular scenario, virgin material consumption drops 18% in 2030 and 71% in 2040, when compared with the current development path. In 2030, this reduction is driven by space-sharing opportunities, which increased building utilisation. In 2040, scaling up reuse and recycling of CDW would further decrease the level of raw material consumption (Fig. 4).

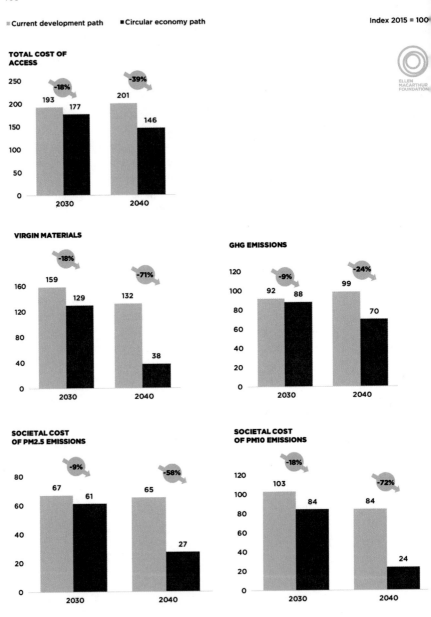

Fig. 4 A circular built environment: the benefits for China's cities

5 Circular Opportunity for the Mobility Sector in the EU: Keeping Products and Materials in Use

This sector has been vital to Europe's development, captivating customer imagination, spearheading economic growth and creating some of the most respected companies on the continent. Mobility is vital for the economy and for the quality of life of citizens as they enjoy their freedom to travel. But the current transport system is costly and tied to a linear depletive model. Cars expose 90% of city residents in Europe to air pollution at levels deemed harmful by the World Health Organisation (WHO). The core challenge is the waste embedded in the transport system. European cars are parked on average 92% of the time. When the car is used, only 1.5 of its 5 seats are occupied. As much as 50% of inner-city land is devoted to mobility but, even at rush hour, cars occupy only 10% of the average European road. Congestion cost approaches 2% of GDP in cities like Stuttgart and Paris [5]. These problems require a systemic approach to rethinking mobility in Europe.

5.1 Five Circular Economy Opportunities in the EU Mobility Sector

The convergence of disruptive technologies, social trends and new business models promises to disrupt mobility in Europe, and around the world. In the coming decade, at least four major levers—sharing, electrification, automation and materials evolution—look likely to transform the personal car, which accounts for more than 80% of the average European's motorised transportation on land today. A fifth lever—the system-level integration of transport modes—has yet to achieve scale but could allow users to shift between personal and public transportation.

Sharing mobility service
Mobility services and vehicle sharing businesses are thriving, thanks to smartphones, big data and the growing popularity of a sharing economy. In Europe car sharing grew 40% a year between 2010 and 2013. E-hailing or shared e-hailing: on-demand hiring of a private or shared-occupancy car via a service that matches passengers and driver is a booming market. Car sharing through a fleet operator offers on-demand, short-term rentals of cars owned and managed by the fleet operator. OEMs are entering this space rapidly. Peer-to-peer car sharing is a variation on the fleet model. Users share individually owned vehicles on an online platform. Drivy is the largest platform in Europe. App-enabled car-pooling such as BlaBlaCar links a non-professional driver with passengers to fill empty seats. Sharing models are also popping up in transit transportation.

Electrification
Electric Vehicles (EV) cost less to operate since their fuel (electricity) is much cheaper than petrol (about 30% lower); their powertrains are at least three times

more efficient, but suffer from conversion losses at the power plant, they also have fewer moving parts and other maintenance requirements. Maintenance cost can drop at least 50–70% as EVs need no transmission fluid, engine tune-ups, or oil changes and experience dramatically less brake wear due to regenerative braking. Their lower operating cost, therefore, lower total lifetime costs make EVs likely to dominate the high-utilisation world of shared mobility, which would also create significant environmental benefits. To get the full environmental benefit, Europe would need to supply the grid with more renewable energy. Hydrogen fuel cell might be another technology adopted widely.

– Autonomous driving
With sufficient penetration, autonomous vehicles could improve the mobility system. They have optimal acceleration and deceleration and can convoy with other autonomous vehicles, which could reduce congestion more than 50% by closing space between cars (1.5 m versus 3–4 car lengths today) and improve energy efficiency significantly. Autonomous and self-driven vehicles can reduce weight by removing unnecessary human interface equipment like brake pedal and can cut accidents 90%—saving lives, and nearly eliminating damage repair costs. People could use their time productively in transit. In spite of these large benefits, the adoption of autonomous and driverless vehicles may be slowed by regulatory barriers and consumer customs.

– Material evolution
Material evolution is already happening today as potential disrupters like River simple and incumbent OEMs like BMW are using carbon fibre to create lightweight vehicles with better aerodynamics and much longer life. Renault is planning to upgrade lifetime-dependent components to more durable and easily recyclable materials to get a longer life for the vehicle. Renault is investigating many types of materials for greater durability such as high-quality and thinner steel, aluminium chassis and powertrain parts, magnesium body panels, in addition to serial production solutions like plastic fenders, that could reduce vehicle weight and the mechanical stresses at the same time. Expensive and capital-intensive materials like aluminium, high-quality steel and carbon fibre create strong incentives for remanufacturing vehicles. Renault's disassembly and remanufacturing plant at Choisy-le-Roi is the company's most profitable industrial site. It reuses 43% of carcasses, recycles 48% in foundries to produce new parts and valorise the remaining 9%. Making remanufacturing and thereby upgradeability work at scale is likely to become a key driver of OEM performance, once more durable products are in the market.

– Integrated transport mode
The technology and digital revolution could anchor the integration of transportation modes that would let people shift between personal, shared and public transportation in an optimised mobility system. While the technology for developing efficient public transportation exists, city governments often have difficulty balancing stakeholder interests and implementing a modern public transport

system. But some European cities are taking steps towards an optimised system solution. Helsinki has launched a programme to make personal cars irrelevant by 2025, by implementing a comprehensive mobility on-demand system. Vienna is developing a prototype for an integrated mobility smartphone platform that integrates diverse mobility offerings into one option based on user needs. Lyon has seen the number of cars entering the city drop 20% over the last decade, encouraged by bike-sharing schemes and car clubs. Green spaces and parks have taken the place of car parks.

5.2 Benefits of a Circular Economy Approach to the Mobility System in the EU

The circular scenario lays out an optimistic vision, in which European towns, cities and businesses would recognise the huge potential of circular mobility, invest to overcome today's barriers to its development, and see the quality of the mobility system improve dramatically. The circular scenario would take advantage of the five levers that stand to transform mobility in Europe in an integrated way. This path would build an automated, multi-modal, on-demand system. The system would have multiple transportation options (like biking, public transit, ride-sharing and car sharing) at its core and would incorporate automated individual transport as a flexible, but a predominantly last-mile solution. A circular path could decouple the rebound effect, increasing consumer utility, GDP and jobs by adding passenger-kilometres, while reducing car-kilometres, congestion, GHG emissions and resource consumption. In the circular scenario, user benefits could increase at least by a factor of three by 2050. This increase in passenger-kilometres would not increase congestion or time cost for households. In fact, this opportunity cost through transport standards would decrease by a factor of five, from EUR 2,600 per household to EUR 475, as total car-kilometres (not passenger-kilometres) in Europe would drop 25%, thanks to greater use of public transportation and non-motorised transport and more passengers per car as shared services became more popular. Besides congestion costs, other indirect cash-out costs and opportunity costs include infrastructure and governance costs, the societal cost of CO_2, and costs related to pollution, noise and accidents. These costs could drop from EUR 3,350 to 1,330 per household in the circular scenario thanks to the overall decrease in car-kilometres, the shift to silent, non-polluting, renewable-powered EVs and almost accident-free autonomous cars. The circular scenario would create better environmental results. By 2030, with roughly half of passenger-kilometres delivered in a system-optimised way, emissions could be expected to fall 55%. By 2050, the sector could be almost entirely decarbonised (95%) as the vehicle and public transport fleet would be electrified and powered by renewable energy. Some minor emissions would likely remain in production, but would be reduced by extending the average car's lifetime and looping the materials, decreasing the extraction of virgin materials (95%) to achieve an almost fully circular system (Fig. 5).

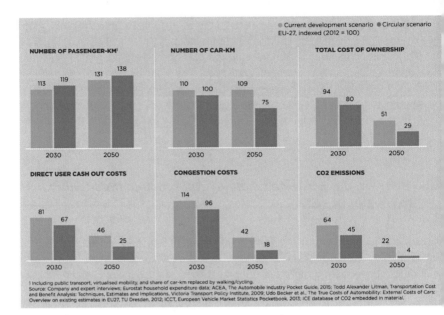

Fig. 5 EU mobility system: potential economic and environmental impact of current development scenario versus circular scenario for the mobility system

6   Accelerating the Transition

The circular economy's potential to generate value while responding to global imperatives can be included in company strategy at the highest level. When it comes to complex materials streams like plastics, textiles, or food, the whole value chain needs to cooperate and align around a common vision. High levels of commitment and incentives and actions at the pre-competitive level are needed from those with a stake in the way materials cycle in the economy. Additionally, through pilots, incubators and demonstration projects, circular business solutions can be tested, and a better understanding can be gained of the benefits they generate for business, society and the environment.

6.1   Enabling Conditions

Achieving the transition requires concerted action from multiple stakeholders including policymakers, the finance sector and academia to provide the necessary enabling conditions for business to scale circular economy opportunities.

- Policymakers

Circular economy ambitions can be integrated into supranational, national and city strategies, roadmaps and long-term targets, as it offers a viable contribution to the mitigation of negative externalities such as CO_2 emissions and biodiversity loss. Some EU member states are for example integrating circular economy measures in their National Energy and Climate Plans. At the national level, circular economy criteria can be included in public procurement tenders, which could incentivise market innovation, as well as support research. The French government, through its law on circular economy adopted in February 2020, introduced a ban on various single-use plastics and on the destruction of unsold products, including clothing, shoes and beauty products—a measure the German government has subsequently adopted. In the UK, the export of plastic waste to non-OECD countries is now restricted or banned to ensure greater management of materials at home. This latter policy echoes China's 2017 National Sword policy to ban the import of certain types of waste, which, as part of a decades-long effort to increase the circularity of the country's economy has had a massive impact on trade flows of used plastics globally. From a local perspective, fiscal measures can include tax benefits for circular economy products or businesses, tax increases on undesirable waste streams, tax reductions on the use of secondary materials and tax reductions for businesses that share, repair and recycle. Cities have a particularly important role to play in ensuring the effective recirculation of materials, products and nutrients in urban areas. Enabling this will require infrastructures such as asset-sharing infrastructure, waste collection systems, treatment facilities, material banks and disassembly and recycling centres.

Finance

Investors can play an essential role in directing more assets and capital to businesses that are capturing higher values in circular supply chains, through product innovation, upscaling efforts and developing markets for secondary materials and refurbished goods, thereby offering the opportunity to significantly reduce GHG emissions and generating greater resilience to climate change. Strategies that could increase the financeability of circular business models include: taking end-of-life value of products into account for a financial business case; determining the residual value of used products in second-hand markets; offering multiple forms of capital such as bank finance, venture capital, capital market financing and impact investing; cash-flow optimisation and shortening the pay-back period to manage the risk of circular business model contracts, for example, by charging higher fees in the first years of pay-per-use models, and offering contract opportunities in place of hold over assets for service-based business models. For instance, the Intesa Sanpaolo Bank and the European Investment Bank (EIB) are cooperating together to provide a EUR 1 billion credit facility to support circular economy projects carried out by mid-cap companies and Italian SMEs.

– Education

Embedding circular economy principles into teaching across all ages of learning supports a mindset shift that will enable future leaders and young professionals to gain circular economy insights, skills and capabilities which they can take forward within their careers. For example, University of Exeter offers a number of learning opportunities through its Centre for Circular Economy. As an engine for innovation applied research can provide the critical insights and knowledge required to initiate industry and policy shifts. Stimulating academic research on circular economy where many crucial topics remain unexplored or at an early stage will be vital to develop understanding and knowledge to support industry to act differently.

6.2 Measuring Progress

Non-financial accounting and the ability to measure company success based on it impact on the society and environment have been on the rise in the past decade Environmental, Social and Governance (ESG) measurement-based investments are worth USD 30 trillion, which is significant given that these methods have evolved in the past decade or so. More lately, studies have increasingly shown a tendency for companies which score well on environmental and social indicators also creating short and long-term value. Furthermore, early 2020, has given a boost to many of these non-financial indicator sets. In his letter to CEOs, Larry Fink, CEO of BlackRock, featured visibly Sustainability Accounting Standards Board (SASB) and Task Force on Climate-related Financial Disclosures (TCFD). The World Economic Forum International Business Council (IBC) issued the 'Compact for Responsive and Responsible Leadership', which curated a suggestion for common non-financial metrics. Circular economy is increasingly considered a key element of these indicator sets as it has been recognised as a concrete means to meeting many environmental aspirations such as emissions reduction and reaching a number of SDGs.

While circular economy is starting to gain a foothold in broader non-financial accounting discussions, the development of circular economy specific measurement is at an early stage and in full swing by multiple organisations. Examples of notable company level efforts in this space are Circulytics [6], by the Ellen MacArthur Foundation, the Circular Transition Indicators by World Business Council for Sustainable Development, Global Reporting Initiative's upcoming circular economy reporting guidelines in the context of waste and others.

The increased need from companies to measure performance in a circular economy and the growing number of solutions to address that need stem from the rapid maturity of the concept of circular economy. Circular economy is reaching a point where the social and environmental benefits have been established and its business case proven. The concept has been embraced in concrete actions (e.g. the New Plastic Economy Global Commitment). Organisations are now increasingly moving into implementing real-life changes to business processes to transition from a linear way of doing business to a more circular one. Implementing these changes can only be done effectively when we have a vision of a future state in terms of circular economy, understand the current state of affairs and are able to track how effectively we are moving between these two states. Measuring progress with indicators that describe these two states and having the ability to track changes is a must-have enabling factor for the transition to happen.

More specifically, measuring progress on a company level is necessary for strategic decision-making and required by third parties (e.g. financiers and policymakers) to provide enabling conditions (e.g. preferential financing terms for circular economy related projects, favourable regulations for circular economy) for the transition to a circular economy. However, we need tools to measure progress on other levels as well. The maturity of tools on these different levels varies and includes—a non-exhaustive list:

The impact circular economy has on broader environmental and social goals, such as GHG emissions reduction and the SDGs. Most publicly traded companies have such goals, and circular economy is recognised as an effective means to work towards both, but a concrete way of measuring this impact is currently lacking.

- Circular economy performance of products. Recirculation of materials is at the heart of circular economy, and companies' ability to make the right design and materials choices is key. Tools to help companies on this front exist (for example, the Material Circularity Indicator by the Ellen MacArthur Foundation and Granta Design; Circular Design Guide) and many are developed by companies themselves to suit their specific situations.

Circular economy performance of services. Services of different types enable the recirculation of materials. This can take many forms such as: services to increase the intensity of product use (e.g. car sharing), services to keep materials in use (e.g. refurbishing), enabling downstream players in a value chain to produce goods for a circular economy (e.g. chemistry for electric car batteries). Currently, there are no broadly adopted tools to measure services comprehensively, but the Ellen MacArthur Foundation's Circulytics is increasingly developing these elements for company level metrics.

Furthermore, when increasing the scope from company to the economy, it is important to be able to measure circular economy performance for policymakers as well. Here, indicators on a regional or national level enable the measurement of the systemic change that the transition to a circular economy requires. To ensure conformity of performance indicators on different levels, the core of these indicators needs to be the same on all levels of granularity from products to the economy. In this way, the aggregate impact from company level progress can be reflected on a national or regional economy level. It also allows straighter forward optimisation of policies to support circular economy. At the time of writing, economy (or country) level metrics are nascent, and we expect to see rapid progress within the EU and a select countries in other regions (e.g. Chile) during 2020, and coming years as national governments set circular economy targets, for instance, the Netherlands has announced to have a fully circular economy by 2050).

7 Conclusion

By embarking on this transformation, launching new circular economy initiatives and reinforcing existing ones, India, China and the EU could achieve resource-efficient and competitive economies that enhance natural capital and protect citizens' well being. But achieving the circular economy transformation will require concerted efforts—no organisation can do it alone.

Questions

1. How can a circular economy help tackle climate change?
2. List a few of the circular economy levers for the food system in India.
3. Who are the key actors that can enable a system-level approach?
4. Why is it important to be able to measure progress towards a circular economy

Answers

1. The circular economy can contribute to reducing carbon emissions by trans forming the way we make and use products. This can be done by increasing the use rates of assets such as buildings and vehicles, and recycling the materials used to make them. In the food system, using regenerative agriculture practices and designing out waste along the value chain can help mitigate climate change by increasing the amount of carbon stored in soils and reducing the need to apply fossil fuel derived synthetic fertilisers.
2. Digitally enabled knowledge-sharing solutions are expected to encourage adop tion of best practices, increase yield, and advance regenerative agriculture. Peri urban and urban farming can help reduce food transport and associated cost (such as food waste, fuel and environmental externalities) and tighten biolog ical nutrient cycles. Composting and anaerobic digestion solutions can recover nutrients for return to the agricultural system.
3. Investors can play an essential role in directing more assets and capital to busi nesses that are capturing higher values in circular supply chains. Policymakers can integrate circular economy strategies into supranational, national and city roadmaps and long-term targets. Academia can embed circular economy princi ples into teaching across all ages of learning to enable future leaders and young professionals gain circular economy insights, skills and capabilities.
4. Circular economy is moving into the implementation phase and it is critical to know the current state, the vision of a future state and measure progress between these two states in order to track implementation success. Measuring progress is also important to create an enabling environment for the transition, for example to unlock financing to high performing companies and inform policy decision making. On a more micro level, it is also important for making the right design and procurement choices for products and services.

References

. The Ellen MacArthur Foundation. (2015). Growth within: A circular economy vision for a competitive Europe.
. The Ellen MacArthur Foundation. (2018). The circular economy opportunity for urban and industrial innovation in China.
. The Ellen MacArthur Foundation. (2016). Circular economy for India: Rethinking growth for long-term prosperity.
. The Ellen MacArthur Foundation. (2018). The circular economy opportunity for urban and industrial innovation in China (and all section 4).
. The Ellen MacArthur Foundation. (2015). Growth within: A circular economy vision for a competitive Europe (and all section 5).
. https://www.ellenmacarthurfoundation.org/resources/apply/circulytics-measuring-circularity.

Ellen MacArthur Foundation The Ellen MacArthur Foundation is a UK-based charity, committed to the creation of a circular economy that tackles some of the biggest challenges of our time, such as waste, pollution, and climate change. A circular economy designs out waste and pollution, keeps products and materials in use, and regenerates natural systems, creating benefits for society, the environment, and the economy. The Foundation collaborates with its Strategic Partners and its wider network of businesses; governments, institutions, and cities; designers; universities; and emerging innovators to drive collaboration, explore opportunities, and develop circular business initiatives.

Circular Supply Chain Management

Charoenchai Khompatraporn

Abstract Modern supply chain management has been shifting away from the traditional linear supply chain model of "take-make-use-dispose" as it is not environmentally sustainable. The circular economy principle brings forth circular supply chain to cope with the ecological threats caused by the linear supply chain by addressing material circularity. The fundamental concept is to prolong material utilization and reduce material exploitation, while capturing and recreating new values of the products and services along the supply chain. This chapter describes the basic principles of circular supply chain, as well as different circular supply chain models. Various circular value creation guidelines are presented to encourage new circular businesses. Bio-base materials which follow a different cycle from the traditional industrial ones are examined. Different performance measures of circular supply chain to assess how well different parties in the circular supply chain have accomplished are introduced. Implementation strategies and various technologies for the future circular supply chain are also discussed.

Keywords Circular value creation · Circular supply chain models · Circular economy building blocks, Measures, and strategies · Digital technologies · Bio-based materials and energy circle · Product-as-a-service model · R principles

Learning Objectives

- Distinguish between traditional linear supply chain and circular supply chain.
- Be able to apply R principles in the context of circular supply chain.
- Understand different circular supply chain models, circular value creation, and energy cycle of bio-base materials.
- Identify building blocks, measures, and implementation strategies in circular supply chain.
- Design business model to achieve circularity.

C. Khompatraporn (✉)
Department of Production Engineering, Faculty of Engineering, King Mongkut's University of Technology Thonburi (KMUTT), Bangkok 10140, Thailand
e-mail: charoenchai.kho@kmutt.ac.th

© Springer Nature Singapore Pte Ltd. 2021
L. Liu and S. Ramakrishna (eds.), *An Introduction to Circular Economy*,
https://doi.org/10.1007/978-981-15-8510-4_21

1 Introduction

Supply chain management has traditionally been concerning how to efficiently an
effectively deliver goods from the sources to the consumers. Economic returns ar
the main focus of this *linear* economic model where the resources are extracte
or acquired to make products and then disposed of after their potential lives. Thi
model often ignores societal implications and conservation of scarce resources. It i
only recently that managing of the supply chain has been gradually moving awa
from the "take-make-use-dispose" practice and starts to include "reverse logistics" t
collect end-of-life products, recreate values of the returned products in "closed-loop
supply chain, and now recouple economic values of the returned materials as inpu
in "circular economy".

Figure 1 illustrates traditional linear supply chain where virgin material is acquire
and transported to the raw material producer. This raw material is then sent to th
product manufacturer to produce the finished goods, which is sold to the consume
After the end of the product useful life, the product is discarded as waste. The circula
supply chain begins similarly as in the linear supply chain. However, in the circula
supply chain, the used products can be reused by the consumer or go directly to
resell market (such as garage sell or an online platform like Craigslist). Some of ther
are collected and sorted by a collector. Certain used products could be refurbishe
or remanufactured, and so they can be used by the product manufacturer. Som
returned products need to be recycled or recovered by being physically transforme
into raw-material-like substances prior to being reused later on in the circular suppl
chain.

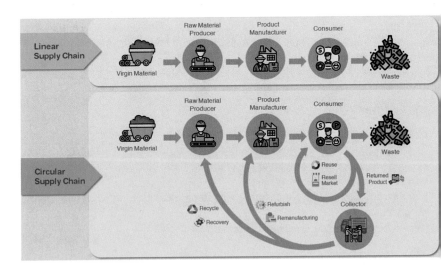

Fig. 1 Linear supply chain and circular supply chain (Credit: Illustration by Sawaros Thongkae
from STEAM Platform)

Circularity of material is the key feature in circular economy and the need for a new supply chain model is prominent. Materials such as gold, silver, zinc, indium, iridium, hafnium, and terbium which are vital to many industries could be depleted in the next fifty years if we remain business as usual. Commodity prices have been drastically increasing since the turn of this century, erasing all the price decline due to production and logistics efficiency of the entire previous century. An estimate of three billion additional middle-class consumers are entering the market by 2030, while the total world population is going to exceed nine billion by 2050. This growth of the global population accelerates increasing demands; at the same time, catalyzes the needs for radical changes in material efficiency, minimization of energy, as well as reduction of waste along the supply chain. Without any change, there may not be enough futile lands, materials, and other resources for future generations.

Despite various arguments on how the circular economy (CE) is rooted, it offers a new economic concept that is both restorative and regenerative by intension and design. CE encompasses environmental economics, industrial ecology, ecological economics, and "cradle-to-cradle". To realize the potentials of CE, changes in economic and social activities in all micro-level (individuals and within a firm), meso-level (network of firms), and macro-level (policies and regulations) are required. The following principles are examples of how changes may be formulated.

R Principles

In every supply chain, there are flows of materials, information, and fund. The three flows are intertwined through space and time. For example, an order triggers production and causes a request to transport the product to the customer, while there are financial transactions throughout the supply chain between the parties involved in the chain. Operation activities are usually monitored and data are collected and analyzed to improve the efficiency, transparency, and connectivity along the supply chain. By-products and wastes are treated, recycled, or disposed of.

In addition, CE includes simple yet effective actions for us as individual consumers and firms to lessen the burden for material input requirements. Those actions are Reduce, Reuse, Remanufacture, Recycle, and Recover.

Reduce: Reduction of material or resource requirements lessens the need for virgin material acquisition. To reduce the need for resources, it often entails rethinking or redesigning of the product or process. Examples of new designs can now be seen in everyday life such as juice and water containers, as well as food packages that use less materials but maintain or sometimes exceed their primary functions. New production processes are equipped with modern technologies in conjunction with information collection function to adjust parameters suitable for the inputs or production plan to not only reduce the amount of material inputs, but also save energy and cost.

- **Reuse**: Certain products can be used several times without losing their mai functions. Coffee shops that discount the drinks to customers who bring their owr reusable tumblers are commonly seen nowadays. To encourage shoppers to reuse shopping bags, supermarkets in several countries charge bag fees if customer request new plastic or paper bags from the stores.
- **Remanufacture**: When a product malfunctions, repairing, refurbishing or reman ufacturing can prolong the useful life of the product. Replacing certain parts whil retaining other well-functioning parts reduces the need for new materials and save cost. It is projected that new generations of mobile phones will be redesigned s that new models can be remanufactured using parts from the discarded ones Customers who turn-in their used mobile phones may be benefited by obtainin a discount to purchase a new phone.
- **Recycle**: If the products cannot be reused or remanufactured, another option i to try to recycle their materials. Recycling requires transforming physical phase of materials through energy with proper handling and processes. Metals, plastics and papers are common materials that are recycled. Certain recycled materials tha if they are used to produce the original products can exhibit inferior properties i comparison to virgin materials. These materials could be used for different bu yet valuable products. An example is a fabric made from recycled plastic bottles The fabric is now used to make raincoats and reusable shopping bags.
- **Recover**: Recovery is employed when waste is converted into resources such a heat, electricity, or even fuel. Once considered as wastes, biomasses from farming particularly in agriculture-base countries, are converted into sources of energy t be used in postharvest processing. Plastic waste is known to be able to be turne into fuel although its economic viability remains questionable.

Reduce, Reuse, Remanufacture, Recycle, and Recover (5R) principles shoul be applied by all the parties throughout the supply chain including the customer themselves. From the perspective of reducing the need for virgin materials, reduce i the 5R principles tends to be more effective as it directly decreases the demands an material needs. To reuse the product, the customer may need to wash or clean it whic may induce environmental impact from the cleaning process with a cleaning agen For remanufacture, recycle, and recover, the impacts to the environment increase respectively, when the principles are practiced. Thus, it is advised to start from reduc and then extend to other principles as shown in Fig. 2. Disposal is used only afte all the above options are exploited. Common disposal methods are landfilling an incineration.

3 Circular Supply Chain Models

For a long-lasting business, its operations must create economic value or the equi alence of financial return. Circular economy is relatively a new economic concep For the concept to work, it requires an innovative value creation with radic

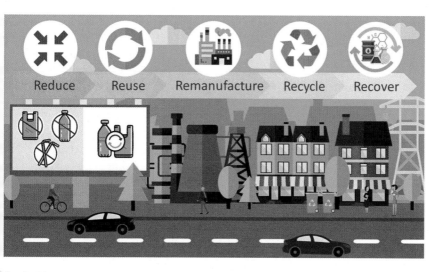

Fig. 2 Hierarchy of 5R principles (Credit: Illustration by Sawaros Thongkaew from STEAM Platform)

or incremental changes to deliver that value—through goods or services—to the customer.

For example, Google data centers have adopted the circular economy concept and are able to claim hundreds of millions of dollars per year in cost saving and new revenue earning. Circular economy practices at Google can be grouped into three dispositions: inventory management, re-sale/remarketing, and material recycling (see Google Data Centers example for details). Another example is a French car maker, Renault, who is also an early adopter of the circular economy model. Renault makes changes in several of its operations, including remanufacturing of used components, managing material flows, manufacturing service improvement, and new business model of electrical-vehicle battery ownership. These changes generate nearly $270 million dollars annually and saving of 20–88% in various operations along the company's supply chain (more details are in Renault example).

Google Data Centers [6]

Once servers at any Google's data centers are decommissioned, they are dismantled. Usable components such as CPUs, motherboards, hard disks, and memory modules that pass the quality inspection are kept as refurbished inventory. These refurbished parts are then used to build remanufactured servers and are deployed back into the centers. The centers thus utilize both the latest technology platforms, as well as older ones. Both new components and refurbished ones are equivalent once they are in the inventory.

Maintenance and repair are also practiced regularly to prolong the life expectancy of the servers. Google internally assesses component inventory every quarter. Any excess is sold to secondary markets through selected remarketing partners. In 2015 alone, Google resold almost 2 million units. Components such as hard drives and storage tapes that cannot be resold are crushed and shredded. They are securely processed as electronic waste and sent to recycling partners. Recycled materials can later be turned into reusable materials. The saving from these dispositions of circular economy throughout the supply chain of the data centers ranges from 19 to 75% o. Google's normal operations.

Renault [2]

Since the adaptation of circular economy, Renault has been realizing economic bene fits in several of its operations. The followings are examples of circular economy practices at the company.

Part Remanufacturing: The company invested in redesigning its automotive components to increase the reuse ratio and standardize components for easy sorting Although the labor cost has increased, the saving is realized in less machining of parts, higher material yield as there is less waste, as well as energy and water usage reduction.

Raw Material Stream Management: End-of-life vehicles are dismantled and quality materials are recycled. Design adjustments of several parts enable material from used vehicles to be turned into high-grade materials for new cars. Renault also works closely with the steel recycler to provide steady streams of raw materials.

Internal Service Improvement: Across Renault's supply chain, the company actively seeks collaborations with its suppliers to benefit from circular activities The cutting fluid used in the company's machining centers is an excellent example Renault asked the supplier to not only supply the cutting fluid, but also maintenanc and waste disposal services. The supplier then reengineered the cutting fluid and usage process, yielding 20% of ownership cost reduction and 90% less discharge volume which become savings in waste treatment.

New Battery Ownership Model: Renault was the first car manufacturer to leas batteries for its electrical vehicles. Renault's batteries are now fully traceable to enhance closed-loop collection rates for recycling.

3.1   Circular Value Creation

Value creation or economic realization in circular economy may be categorized into at least four models.

- **Reuse, Refurbishing, and Remanufacturing**: These simple activities have proven to yield fast and substantial saving in practice. Reuse, refurbishing, an remanufacturing often lead to less materials to process; thus they require les energy and water, and discharge less waste. These footprints are shared throughou the product's useful life. Associated externalities such as greenhouse gasse and environmental toxicity are also reduced. With less utility usage and wast treatment, cost saving could be realized.

- **Longer Circularity**: The idea of longer circularity is to maximize the life of the product in every cycle in order to avoid new product creation. A new product ownership model can encourage repair, reuse, or remanufacture actions of the new owner to extend the life expectancy of the product. Second-hand apparel is an example of prolonging product life cycle through change of ownership.

  Cascaded Usage: Once a product reaches its maximum usage in one function, it may be transformed or reused in another function. Fabric made from recycled plastic that is earlier mentioned is one example of a cascaded use. Another example is cotton clothing which is now reused as a substitute for virgin materials in fiber-filled upholstery.

- **Pure Inputs**: Uncontaminated materials increase collection efficiency and reuse ratio. Quality used materials are also easier to processed and recycled which helps increasing material productivity and extending material longevity. Collection of uncontaminated materials is still a challenging task and is driven by the market price mechanism where uncontaminated materials are priced higher than those contaminated ones. However, in the future, the used material collection could be executed more effectively through digital technologies and tracking information.

Noted that there are still the three flows (materials, information, and fund) in the circular supply chain but in certain areas, they are utilized for a different purpose, so-called Repurpose. For example, the data from product traceability which was difficult to trace from the customers is now used by the product manufacturer for repair and maintenance due to change of ownership in the business model. Financial benefits could be created and captured through applications of each model or their combinations. With the current technologies, most products are incinerated or put into landfill at the end of their useful lives (as shown in Fig. 3). With better technologies and changes in customer behavior, these burdens to our environment could be alleviated or completely eliminated.

3.2 Energy Cycle and Bio-Base Materials

An important driver of circular economy is renewable sources of energy such as solar, wind, wave, and biomass which in return significantly save energy cost for goods production. The cost of solar PV based electricity is now less expensive than that produced by fossil fuel since 2012, and the price continues to drop. In agricultural-based regions, organic wastes are processed through anaerobic digestion to produce biogas, usable to make heat, steam, and electricity which are utilized in postharvest production of several agro-industries.

Cassava starch production, for instance, has used its bio-base wastes to generate biogas, produce fertilizer, and mix with other ingredients in animal feed. The biogas obtained is processed to produce heat for the drying process, as well as to generate electricity to power the production plant. A similar practice is seen in the canned pineapple industry. The skin of fresh pineapples and pineapple cores are used to

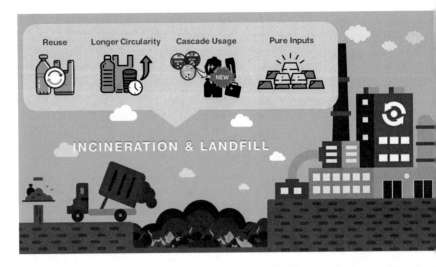

Fig. 3 Value creation models in circular supply chain (Credit: Illustration by Sawaros Thongkae"
from STEAM Platform)

produce methane (biogas). The wastewater collected from the pineapple cannin
processes is mixed with microorganisms to produce liquid fertilizer that is sometime
given back for no cost to neighboring pineapple farmers and communities. From
these examples, the energy needed to power the economic cycle can at least partiall
be migrated to more bio-nutrient sources to decrease fossil fuel dependency an
increase system resilience. Circularity of biological materials to the biosphere a
usually preferred over technical materials which are more difficult to decompose.

Through digital technologies and regulation changes, it is now possible
exchange and/or trade energy with a local network of energy producers, consumer
and prosumers (an entity that behaves as a producer at one instance and a consume
at another instance) in many parts of the world. One such technology is virtual pow
plant (VPP). Through VPP, an energy producer in a sunny area may sell excess elec
tricity to those in shady ones; or an agro-business may sell its biogas surplus durin
the harvest season but buy other forms of energy during other periods of the yea
VPP allows community energy trade without having to establish large power plant
Large power generation plants are not so welcome nowadays as they often lead
local pollutions to nearby communities.

3.3 Post-consumption Collection Schemes

After products are used or reach the end of the useful life, they are discarded. T
further reuse the products or certain parts of them, they need to be collected. The

are several schemes to systemically gather these post-consumption products. The followings are some examples of the collection schemes.

- **Advanced deposit fee**: When a product is purchased, there could be a fee charged at the price of the product. The fee is refunded to the customer only when the used product is deposited to a collection center such as a supermarket in the right condition. This collection scheme is often used with beverage cans and plastic bottles.
- **Take-back program**: In some countries, the manufacturers of certain products are required by law to take back the used products. For examples, waste electrical and electronic equipment (WEEE) and batteries in the European Union need to be collected by the producers because they may contain substances that are harmful to human health or have negative impacts to the environment. This legislation initiative is a part of extended producer responsibility (EPR) aiming to increase product recovery by passing the responsibility to the producers or polluters.
  Trade-in or rebate: Products that reach the end of their useful lives could sometimes be exchanged, or so-called traded in, for some discount to purchase new products. Although the discount may be devised primarily to attract sales, the used products get collected and treated. The returned products are processed to lessen effects on the environment and are sometimes recycled to make new products. Examples of products that have used this collection scheme are used clothes and mobile phones.
- **Pick-up system**: Household wastes are often seen using this collection scheme where the wastes are collected from customer locations in exchange for some fee. The collected wastes are then transported to local municipals to be sorted, treated, or sent to landfill locations.
  Public recycling facility: In some areas, the consumers are responsible to deposit a certain used products at a designated public recycling facility. The recycling service can be provided at no cost or for some small fee. The returned products get recycled at the center or sent out to other special recycling facilities.

The next section provides basic building blocks of the circular economy concept and suggests certain assessment guidelines.

Circular Economy Building Blocks and Measures

It is important to understand factors that influence the success of circular economy in existing organizations so that these enabling factors can be imitated and amplified in other firms. From the literature, key building blocks of circular economy can be grouped as follows.

Business and product design: Identify value creation, value capture, and value distribution in a given business context. Broader business models should be explored such as a service or performance-based one, value-added service, product return with intended incentive, and cascaded use of products.

Circular design: Identify means to extend the useful life of the products. Certain techniques can be deployed such as design for easy end-of-life disassembly and sorting, eco-design, standardized components, innovative ways to reutilize by products and wastes. Reengineering of materials in certain applications can prolong material recycling ability.

Forward and reverse supply chain: Identifies operations that contribute to efficiency and effectiveness in forward and reverse chain such as delivery logistics, customer services, returned product sorting, green purchasing, material leakage points, and risk assessment. These activities are potential cost savings, enhancing product collection and treatment system, as well as effective towards end-of-life product services.

Environmental and societal consideration: Consider environmental and social issues in business strategies and operations. These issues include health of both employees and communities, population growth, job creation, climate change, quality agriculture, conservation of renewable sources, and customer awareness. Environmental and social issue inclusion make circular economy more sustainable beyond material circularity and profit-making in traditional business.

System and infrastructure: Enable the system through digital transformation, new forms of partnerships and collaborations. Use new regulations as incentives to internally change the organization and rethink organizational characteristics and financial models.

Several instruments and models could be deployed to promote circular economy. The next issue is to assess how well the organization performs under the context of circular economy. There exist several ideas to measure the CE performances such as the following metrics (Table 1).

These metrics are not exhaustive. They are merely examples of measures that can be used to benchmark with other organizations. The user should be selective of the measures or develop new ones to fit the context of the organization. For instance, from the Google Data Center example, the company optimizes end-of-life or server based on the total cost of ownership instead of industry accounting standard. This keeps Google stay technologically competitive and yet economical.

Although there are several metrics for assessing circularity, economic value is one of the most common metrics. Economic values may not account for externality costs in some products, but these values could at least signal relative scarcity. The economic value could be estimated in several ways, but a simple one is from the market prices.

For a used product, it is possible that only certain parts or a fraction could be recirculated. Let c be product-level circularity, then the product circularity can be expressed from its economic value as

$$c = \frac{economic\ value\ of\ recirculated\ parts}{economic\ value\ of\ all\ parts} \tag{1}$$

We may think of a product as consisting many parts. Some parts are recirculatable (r), while the others are non-recirculatable (n). For any product or part i, then

Table 1 Circular supply chain metrics

Category	Metric detail
Resource efficiency	Measures relating to part/product over its lifetime **Metrics**: recycling efficiency, product longevity, supply risk and scarcity, conservation of value, value change and retention, resource productivity, energy efficiency, life cycle assessment, net material saving
Material stocks and flows	Measures relating to resources and usage **Metrics**: stock availability and concentration, down-cycling quality loss, recycling/remanufacturing potentials, resource productivity, cascading use, product traceability
Product-centric and environmental impacts	Measures relating to product, waste, and environment concerns **Metrics**: amount of virgin stock and secondary materials, process inputs, product toxicity, cradle-to-cradle certification, eco-labeling, waste disposal, greenhouse gas emission, reuse/recycling complexity, pollution reduction, shift to renewable energy, land productivity and soil health
Economic and other performances	Measures relating to product, waste, and environment concerns **Metrics**: financial growth, investment recovery, job creation, health impacts to animal and public, take-back rate, resource sharing, information exchange among partners (trust level), technology assisted circulation, price volatility reduction

$$c_i = \frac{r_i}{r_i + n_i} \qquad (2)$$

The term $r_i + n_i$ can be thought of the value of a new product (or part). Services involved in producing the product may be considered through the non-recirculatable term in Eq. (2). Using this assessment metric, the product circularity is always between 0 and 1.

Example A laser printer toner cartridge consists of a plastic casing and ink powder. The plastic casing can be reused after the cartridge is depleted, but the ink is not. Suppose that to refill the reused cartridge it requires some labor work of $5 per cartridge. The plastic casing is worth $40 each and the ink powder $30 per cartridge. Then the circularity of this cartridge is

$$c = \frac{plastic\ casing\ cost}{plastic\ casting\ cost + (ink\ powder\ cost + labor\ cost)}$$
$$= \frac{40}{40 + (30 + 5)} = 0.533.$$

In addition to economic and environmental impacts of circular economy, socia implications are perhaps the most difficult aspect to measure. Corporate report usually account for immediate financial consequences in their financial statements There is yet a diverse debate in the academic literature of appropriate methodologie to include social externalities in financial values in corporate financial reports. Fo example, food waste in inner-city areas that are once unwanted could be turned int energy for urban families through the circular economy concept. It does not only reduce negative environmental externalities, but potentially improve the resiliency o the communities in being more energy independent. Managing all the food waste ca be a challenge, but at the same time, it creates business opportunities for individual or possibly the entire communities.

Methods to explicitly measure, collect, and analyze data on social impacts ar still evolving. The consequences remain in including these social externalities i the fund flow of the circular supply chain. It is possible that social externalities ar contextual depending on activities, communities, industries, and the risks involved Without social impacts measured, however, assessment of circular supply chain i not completed, and the full potentials of circular supply chain may not be obtained The benefits of circular supply chain could be underestimated or completely ignored particularly the social implications.

5 Circular Supply Chain Implementation Strategies

To execute circular economy, strategic activities must be established. Clear *plan c action* is a fundamental requirement needed in the organization. A starting action ma be to identify certain materials, understand the flows of materials, information, an fund, and determine material leakage points together with their quantities—whicl from another angle could be considered as potential benefits. Then map out relatin, operations, barriers, and technological landscape to explore the most promising areas Product and material usage may need to be redefined from the perspective of intende use, alternative use, and use period. This can help create new businesses or facilitat replacement and financial return periods. Improvement of existing operations shoul be continually executed, while identifying stakeholders to support short, mediun and long-term goals.

Once there is a clear plan of action, enabling mechanisms can start. The *revers loop* may need to be set up if it does not already exist. Select the material fror a preferred short list that is suitable for the circularity. Identify the right partnen and agree on a business model that the benefits are shared. Follow mutually agree roadmap among the partners. Quantification of economic impacts may begin wit material flow or material circulation in comparison to no circularity. A focus coul be on inputs such as material, energy, water, labor, and/or outputs such as wastes an carbon emissions. Assessment may include innovation development and leakag point reduction.

Several successful companies who implement circular economy philosophy have expanded their business from suppliers to business solution providers by which idle assets are better utilized (see Philips Lighting as a Service for an example). Ownership and take-back models are being rethought over or taken control in the form of rental and leasing. This allows high-cost products to compete with low-cost ones over traditional sale model. At the same time, the products or assets are better utilized and the cost is shared among the parties including the customers to gain better services for the customers, better quality of returned products to the producer, as well as a more stable stream of revenue to the involving partners.

Collaboration and sharing can accelerate circular economy. *Collaborative platforms* may allow people to exchange intangible assets such as skills, experience, knowledge, and space. These platforms could be a marketplace for recruitment and outsourcing small jobs. Redistributive markets such as eBay and Craigslist are also possible through internet technologies. These virtual markets reduce shopper's travel trip and potentially save carbon emission. Information technology is a key enabler to mobilize circular economy in many industries. Reverse material data and Big Data analysis can be utilized to improve material circularity. Innovative materials that can endure circularity longer may need to be engineered with perhaps external experts or new partners.

A connection between digitization and circular economy is becoming prominent in recent years. We have seen that *digital technologies* have been adopted in several functions of the supply chain such as in inventory and warehouse management, material purchasing, enterprise resource planning (ERP), freight management, and financial transaction record keeping. Modern digital technologies may include machine learning and predictive analysis to project the future outcomes or detect certain behaviors in that function. For circular supply chain, in many cases, it is the digital technologies that initiate changes in the infrastructure that in turn facilitates circular supply chain transformation of the company. Digitization in circular supply chain is often used to connect data among suppliers, service providers, customers, and their devices. Data storage and data access could be executed through cloud. This can greatly benefit companies that operate in several locations to be able to share, synchronize, and analyze the same set of data in real time or almost real time. Consequently, once the data access and transparency is improved the decision-making of the entire organization can be made faster and with the most updated information. This would improve operational efficiency and optimize business impact.

Several technologies and digitization of supply chain operations are gaining popularity, including blockchain, big data, data visualization, e-commerce platform, smart mobility, cloud service, virtual reality, augmented reality, artificial intelligence, business intelligence, and optimization. Whichever of these technologies are selected, they should improve at least some of the circular supply chain activities such as product design, target customer attraction, product tracking, technical and maintenance support, and end-of-life activities. An area of concern for the digitized circular supply chain is data security. While sharing of data enables collaboration in the supply chain, some data may need to be securely isolated and shared only when necessary for

the purpose of competitive advantages. Data security measures and data privacy poli
cies should be put in place when digitization is implemented. Ethical issues shoulc
be clearly discussed early on when data digitization is employed to a personal o:
community level.

Philips Lighting as a Service [2]

Philips became a pioneer who shifted from selling light bulbs to lighting as a servic(
provider. Philips' business models redefine the notion of access versus ownership o■
lighting systems, but the business model was in fact generated from circular econom▸
principles. Through the service, Philips install, maintain, manage end-of-life ligh▸
bulbs. The customers no longer need to invest in the lighting system upfront, no
stocking light bulbs for emergency replacement. Since Philips retains the ownershi▸
of the products, this allows easier and better management of discarded light bulb
for recycling and reclaiming valuable materials. The lighting services include renta■
leasing, pay-per-use, and pay-per-service models.

The role of the consumers in circular supply chain is another area that is largel▸
unexplored and requires further investigation. Circular products need to be designe(
in such a way that they are appealing to general consumers. Educating the genera■
public of the quality of the circular products, as well as marketing and pricing, are a■
essential to move circular economy forward and make the whole supply chain mor▸
sustainable.

6 Conclusion

Climate change and environmental conditions are posting unprecedented threats t▸
the world's growing population. It is our duty to resolve those threats. Circula
economy is offering a solution to alleviate those threats while being economicall▸
viable. Its fundamental concept is to maximize material utilization to reduce the nee▸
for virgin material and maintain environmental balance, and yet continue economi
growth.

Several companies have adopted circular economy throughout their supply chain
and found success. Examples of different activities and areas where other companie
should concentrate on circular economy are given in the chapter. These could b
a starting point for novice circular economy adopters. Nevertheless, the context c
the individual business should also be considered and adaptation may be needec
Several measures to assess how well the parties in the circular supply chain ar
performing are suggested, although they too are context-dependent. Sectors tha
have already realized benefits from circular economy through their supply chain
include packaging, food, electronic and electrical equipment, transport, furnitur(
buildings, and construction. While new problems require innovative solutions an▸
determination to overcome, businesses and customers together need to be resolute ▸

mobilizing circular economy and circular supply chain to realize their true potentials and maintain the balance of our environment for future generations.

Questions

1. Compare the traditional and circular supply chain models in terms of involving parties, advantages and disadvantages, ease of implementation, etc.
2. Select an existing supply chain. Then apply the circular supply chain concept to it.

a) Discuss different business models, material leakage points, material flows, etc.
b) If the selected supply chain is already a circular supply chain, try to improve it based on different metrics.
c) Propose technologies that can improve the circular supply chain.

3. Select a product. Propose how to improve its collection and recyclability. Assess economic, environmental, and/or social impacts of your proposed solution.
4. Recommend relevant digital technologies that may improve the circularity of a product. Different technologies may be utilized for different parts of the supply chain of the product.
5. It is possible that a disruptive event such as a natural disaster or global pandemic (COVID-19 for instance) can affect a circular supply chain. Suggest preventive or corrective measures to alleviate impacts from such disruptive event.
6. It is still a debatable topic in how social externalities be measured in financial terms.

a) Discuss advantages and disadvantages of different assessment methods of social externalities. Select one method (or create one) and suggest how it should be integrated into circular supply chain, particularly in the financial flow.
b) Are there any business opportunities from the lack or incomplete assessment of social externalities in circular supply chain?

Suggested Reading

Bocken NMP, de Pauw I, Bakker C, van der Grinten B. Product design and business model strategies for a circular economy. Journal of Industrial and Production Engineering 2016; 33(5): 308–320.

Coase RH. The problem of social cost. The Journal of Law & Economics 1960; 3: 1–44.

Di Maio F, Rem PC, Balde K, Polder M. Measuring resource efficiency and circular economy: a market value approach. Resource, Conservation and Recycling 2017; 122: 163–171.

Ellen MacArthur Foundation. Artificial intelligence and the circular economy - AI as a tool to accelerate the transition. 2019.

Farooque M, Zhang A, Thurer M, Qu T, Huisingh D. Circular supply chain management: a definition and structured literature review. Journal of Cleaner Production 2019; 228: 882–900.

Wang H, Gu Y, Li L, Liu T, Wu Y, Zuo T. Operating models and development trends in the extended producer responsibility system for waste electrical and electronic equipment. Resources, Conservation & Recycling 2017; 127: 159–167.

Maatta M. Circular goes digital. Deliotte.

Remy N, Speelman E, Swartz S. The circular economy: Moving from theory to practice. McKinsey & Co. 2016.

STEAM Platform video on Circular Economy- the Cassava Story. Available online at https://www.youtube.com/watch?v=BSNlJBWmdb4.

References

1. Antikainena, M., Uusitaloa, T., & Kivikytö-Reponen, P. (2018). Digitalisation as an enabler of circular economy. *Procedia CIRP, 73,* 45–49.
2. Govindan, K., & Hasanagic, M. (2018). A systematic review on drivers, barriers, and practice towards circular economy: A supply chain perspective. *International Journal of Production Research, 56*(1–2), 278–311.
3. Masi, D., Kumar, V., Garza-Reyes, J. A., & Godsell, J. (2018). Towards a more circular economy: Exploring the awareness, practices, and barriers from a focal firm perspective. *Production Planning & Control, 29*(6), 539–550.
4. Ferguson, M. E., & Souza, G. C. (2010). *Closed-loop supply chains: New development to improve the sustainability of business practices.* New York: CRC Press.
5. Rana, S, Brandt, K. (2016). Circular economy at work in google data centers. Google & the Ellen MacArthur Foundation.
6. Jabbour, A.B.L.S., Jabbour, C.J.C., Filho, M.G., Roubaud, D. (2018). Industry 4.0 and the circular economy: A proposed research agenda and original roadmap for sustainable operations. *Annals of Operations Research, 270*(1), 273–286.
7. Geissdoerfer, M., Morioka, S. N., de Carvelho, M. M., & Evans, S. (2018). Business model and supply chains for the circular economy. *Journal of Cleaner Production, 190,* 712–721.
8. Linder, M., Sarasini, S., & van Loo, P. (2017). A metric for quantifying product-level circularity. *Journal of Industrial Ecology., 21*(3), 545–558.
9. Mishra, J. L., Hopkinson, P. G., & Tidridge, G. (2018). Value creation from circular economy led closed loop supply chains: A case study of fast-moving consumer goods. *Production Planning & Control, 29*(6), 509–521.
10. Parchomenko, A., Nelen, D., Gillabel, J., & Rechberger, H. (2019). Measuring the circular economy—A multiple correspondence analysis of 63 metrics. *Journal of Cleaner Production, 210,* 200–216.
11. Ripanti, E. F., & Tjahjono, B. (2019). Unveiling the potentials of circular economy values in logistics and supply chain management. *International Journal of Logistics Management, 30*(3), 723–742.
12. Unerman, J., Bebbington, J., O'dwyer, B. (2018). Corporate reporting and accounting for externalities. *Accounting and Business Research, 48*(5), 497–522.
13. World Economic Forum. (2014). Towards the circular economy: Accelerating the scale-up across global supply chains.

Charoenchai (Charlie) Khompatraporn is an Associate Professor at the Department of Production Engineering, Faculty of Engineering, King Mongkut's University of Technology Thonburi (KMUTT), Bangkok, Thailand. His research interests include supply chain and logistics management, operations under uncertainty and social consideration, data analytics as well as cross-disciplinary optimization and metaheuristics. He has been working closely with both public and private sectors on various projects in logistics and operations planning, disaster management, and competitive sustainable issues. He holds a Ph.D. from University of Washington, a master's degree from Georgia Institute of Technology, and a bachelor's degree from Rensselaer Polytechnic Institute, New York.

Circular Economy Business Models and Practices

Anna Itkin

Abstract Business model is at the heart of every business. It is the way a company creates, captures, and delivers value and it is a prerequisite for a strategy to be implemented throughout an organization. Traditionally, value was considered to be purely financial. However, with increasing global environmental, social, and political challenges, we need to create a more sustainable society where economic growth is decoupled from resource consumption. In order to enable the transition to circular economy, businesses are expected to play a significant role in the shift to a circular economy by implementing circular business models.

Keywords Business model · Circular business model · Value · Innovation · Design · Sustainability · Resource loops

Learning Objectives

What is a business model?
What is value?
Business models as a value creation mechanism
Circular Economy Business Models as an archetype of Sustainable Business Models

The original version of this chapter was revised: This chapter revised with correct figure captions and numbers and including author bio with photo. The correction to this chapter is available at https://doi.org/10.1007/978-981-15-8510-4_31

A. Itkin (✉)
The Inceptery Pte Ltd, A Sustainability-Led Innovation Practice Based in Singapore, Singapore, Singapore
e-mail: anna@theinceptery.com

1 Introduction

1.1 Introduction to a Business Model

Business, since it first began in pre-historic times, was about selling at profit. The business model (BM) concept became prevalent with the dawn of the Internet in the mid-1990s (Fig. 1). And in 2010 Alexander Osterwalder and Yves Pigneur had offered a comprehensive definition of a BM as "…the rationale of how an organization creates, delivers, and captures value" (Fig. 2). This definition is built on a broad underlying theory that represents company's interactions with its value chain and broader network of stakeholders, fulfillment of its existing business strategy, and can be used as a tool for innovation and communication.

Osterwalder and Pigneur have designed Business Model Canvas (BMC) (Fig. 3) that lays out and connects assumptions about key resources and activities of the value chain, value proposition, creation, delivery and capture, customer relationships, key channels, customer segments, cost structures, and revenue streams [1]. The proposed nine building blocks will have a specific configuration for each company that describe and define how it creates, captures, and delivers value. BM is not a static structure, rather it's a continuous work in progress, given a company wants to stay relevant.

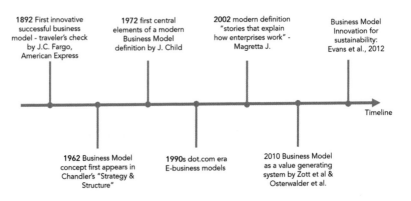

Fig. 1 Major Milestones in Business Model Development

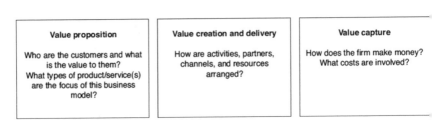

Fig. 2 Core pillars of the business model. Whalen, K [12], Ph.D. thesis

The Business Model Canvas

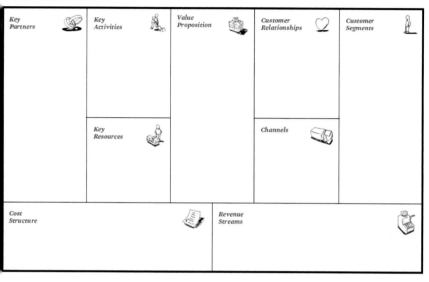

Fig. 3 Business Model Canvas (BMC) tool developed by Osterwalder and Pigneur [1] (strategyzer.com) build on the BMC, Ellen McArthur foundation in collaboration with IDEO have added prompts and questions to facilitate design of a circular business model [24]

and profitable in the long-term. Similar to the scientific method–it starts with a hypothesis, which is then tested and revised with changing business environment. In increasingly *volatile, uncertain, complex,* and ambiguous (VUCA) world, companies can use this approach to become resilient and to remain relevant. Consideration of a wide range of stakeholder interests – including environment and society among them – in the BM, serves as an important driver of business innovation that can help to embed sustainability into a business purpose, processes and strategy, and serve as a key to creation of competitive advantage. How can we encourage companies to significantly change the way they operate to ensure more responsible production and consumption? Business Model Innovation (BMI) is an approach that helps to deliver sustainability [2] and circularity through BM reconfiguration and broader value creation.

2 Value Creation Through a Business Model

From a value creation perspective, BMs traditionally focused on satisfying the customer needs, economic return to shareholders, compliance, regulation or legislation requirements, long-term economic viability or continuity of the firm. Sustainable business models (SBM), however, widen the notion of value creation to include environmental and social value. Environmental value creation covers sustainable use of natural resources, biodiversity conservation, avoidance of waste and pollution or thei safe recycling and elimination, regeneration of ecological services such as natura climate regulating systems, pollination, and soil fertility. Social value creation i concerned with issues such as all stakeholder inclusion and participation in decision making, responsibility, labor standards, human rights, community relations, welfare culture, poverty alleviation, and equal opportunity to access resources [2]. In order t shift from traditional BM to SBM, companies undertake BMI activities (Fig. 4) tha enable them to generate competitive advantage whilst also creating environmenta and social value and contributing to sustainable development.

A definition of SBM by Schaltegger et al. states that "A business model for sustain ability helps describing, analyzing, managing, and communicating (i) a company' sustainable value proposition to its customers, and all other stakeholders, (ii) how i creates and delivers this value, (iii) and how it captures economic value while main taining or regenerating natural, social, and economic capital beyond its organizationa boundaries" [3].

One subset of SBM are Circular Business Models (CBM) – these are BMs an industrial processes which embed extended use of products, their parts, and material by means of designing out waste and pollution. Beyond immediate environmenta benefits at its core, the circular economy is about economics and competitivenes says Professor Walter Stahel. In 1970s, he pioneered work on extending the lif cycle of products and later coined the term **Performance Economy** (or Function Service Economy), which focuses on selling performance/services instead of good in a circular economy, internalizing all costs (using closed loops, cradle to cradl approach) [4]. In 1981, Stahel articulated these ideas in his paper "The Produc Life Factor" and identified selling service instead of products as the ultimate SBM of a circular economy: selling service enables to create sustainable profits withou an externalization of the costs of risk and costs of waste [5]. From the busines perspective, longer product-life strategies and the impact of circular industrial desig deliver competitive advantages to companies when compared to simply recyclin Taking plastic as an example, McKinsey & Company calculated that plastics reus and recycling could generate profit-pool growth of $60 billion for the petrochemica and plastics sector [6]. The "The New Plastics Economy: "Rethinking the futu of plastics" report by World Economic Forum, Ellen MacArthur Foundation an McKinsey & Company calculated that plastic packaging material value loss translate into an annual value loss of $80–120 billion [7].

1.3   Expanded Schools of Thought

A small number of thought leaders and scholars devised additional schools of though around what circular economy is about (Fig. 5). Professor John T. Lyle proposed th idea of **regenerative design** where all systems, from agriculture onwards, could b conceived and executed in a regenerative manner-that processes themselves rene or regenerate the sources of energy and materials that they consume [25].

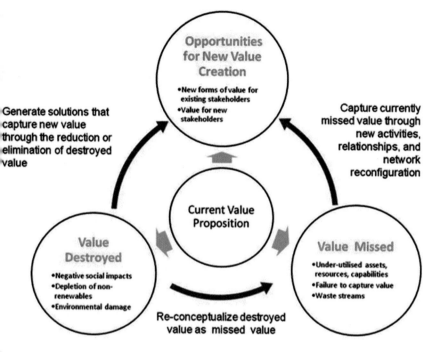

Generate solutions that capture new value through the reduction or elimination of destroyed value

Capture currently missed value through new activities, relationships, and network reconfiguration

Re-conceptualize destroyed value as missed value

Fig. 4 A value mapping tool for sustainable business modelling [26]

Michael Braungart and Bill McDonough have developed the **Cradle to Cradle**™ concept [27] and certification process [28]. The Cradle to Cradle framework focuses on design for effectiveness in terms of products with positive impact, which fundamentally differentiates it from the traditional design focused on reducing negative impacts.

Industrial ecology is a man-made ecosystem that operates in a similar way to natural ecosystems, where the waste or by product of one process is used as an input into another process, thus closing the loop on the notion of waste with an emphasis on natural capital restoration. Industrial ecology adopts a systemic point of view, designing production processes in accordance with local ecological constraints while looking at their global impact such that they perform as close to living systems as possible. Principles of industrial ecology can also be applied in the services sector with a focus on social wellbeing [8].

Janine Benyus defines her approach—**Biomimicry**—as "a new discipline that studies nature's best ideas and then imitates these designs and processes to solve human problems". Biomimicry relies on three key principles where nature serves as model to emulate forms, processes, systems, and strategies to solve challenges; as an ecological standard to judge sustainability of all innovations; and as a mentor to learn from rather than a reservoir of resources to exploit [9].

Natural capitalism was described by Paul Hawken, Amory Lovins, and L. Hunter Lovins in their book "Natural Capitalism: Creating the Next Industrial Revolution" as

Fig. 5 Extended Schools of
Thought (the image by
Sawaros)

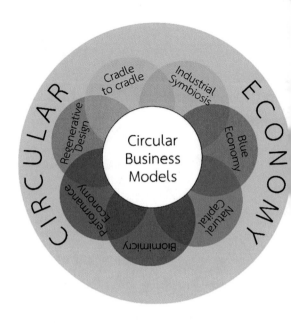

a global economy in which business and environmental interests overlap, recognizin
the interdependencies that exist between the production and use of human-mad
capital and flows of natural capital [10]. The idea of natural capitalism relies on fou
principles:

1. Radical change in design, production, and technology that prolongs the use o
   natural resources resulting in savings in cost and time, and therefore, increase
   productivity.
2. Eliminate the concept of waste, inspired by nature where all systems ar
   circular—an output of one process becomes and input of another.
3. Product as service business model where value is created through continuou
   flow of services rather than product sales.
4. Restore and regenerate natural capital—the world's stocks of natural asse
   including soil, air, water, and all living things.

Blue Economy initiated by former Ecover CEO and Belgian businessman Gunte
Pauli. The idea of Blue Economy is based on 21 founding principles using locall
available resources and sufficiency. It is driven by innovative business mode
that provide competitive advantage through innate virtues and values connectin
untapped local potential [11].

1.4   Overview of Circular Business Models

Existing CBMs are difficult to assess because of the diverse value constellations an
various extent of contributions towards achieving Circular Economy. It was foun

hat firms contribute to a circular economy through two main overarching circular business model strategies: "Extending Product Value" and "Extending Resource Value" [12] (Table 1).

The sole theoretical classification of CBMs currently available in academic literature can be seen in Fig. 6 [13]. Building on this concept, Whalen [12], classified the archetypes according to their value proposition, creation & delivery, and capture (Table 2). The classical definition states that CBMs capture, create, and deliver value by *slowing, closing,* and *narrowing resource loops* [12]. *Slowing resource*

Table 1 .

	Extending Product Value	Extending Resource Value
Focus of value propositions	offer products or components (durable or previously obsolete)	offer recovered resources or production by-products
		offer products made from recovered resources
	offer services related to products and their recovery	offer services related to resource recovery
		offer energy from recovered resources
Approaches to value creation & delivery	provide long life products or offer warranty guarantees	use recovered resources in new production
	perform product maintenance	
	enable product recontextualization	facilitate resource recovery
	undertake repair/ refurbishment/ remanufacturing	undertake resource recovery
Sources of value capture	revenue from resale of previously obsolete products and resources (e.g., new revenue streams)	
	cost savings	
	service contracts	
	partnership agreements	
	premium pricing	
	environmental or social benefits	

Adopted from Whalen (2020) Proposed a framework to illustrate circular business models elements. Developed from an empirical review of circular business models. *Note* value capture spans both columns, as similar approaches apply to both categories [12]

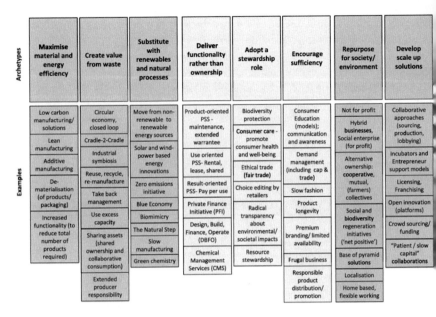

Fig. 6 Circular Business Models Archetypes adapted from Bocken et al. (2016)

loops includes product-life extension strategies such as maintenance, repair, refur
bishment, and remanufacturing, while *closing resource loops* is linked to capturin
the residual value from by-products or "waste" through business model innovatio
[14]. *Narrowing resource loops* focuses on reducing the demand for energy an
resources, thereby reducing demand for primary extraction of resources, reducin
waste to landfill, CO_2 emissions, and other pollutants. This latter approach contribute
towards system-wide reduction of resource consumption, efficiency in material, an
energy. However, optimization and efficiency implemented in silo, without regardin
the system in which companies operate, generates rebound effects [15], and wide
negative impacts. In addition, since productivity and efficiency in manufacturin
often mean automatization and novel technology adoption, these trends have le
to increased unemployment and resulting social sustainability issues [16]. Hence
CBM need to incorporate wider environmental and social considerations and aim a
positive social impacts alongside environmental sustainability.

Case Study 1

Lehigh Technologies

Lehigh Technologies is a specialty chemicals company that produces highly engi
neered, versatile raw materials called micronized rubber powders (MRP) that ca
replace up to 20% of traditional oil-derived and rubber-based feedstocks in a wid
range of applications. High-quality, micron-scale MRPs are suitable for a numbe
of high-performance consumer and industrial applications, including tires, plastic

Table 2 Theoretical archetype framework for CBMs

Archetype		Value proposition	Value creation & delivery	Value capture
CBM1	Access/Performance Model	Delivery of functionality or use instead of ownership	Lower cost of ownership to consumer; offers firms economic incentives to slow resource loops (perform maintenance, undertake repair)	Pay per unit of service; pay for functionality; pay per use
CBM2	Extending Product Value	Use and/or recovery of used products or components	Collection or redistribution or used products; perform repair; refurbishment; remanufacturing	New forms of value; reduced material or overall costs
CBM3	Classic Long Life	Offer long-lasting products	Durable, high-quality product and long-term customer service	Usually premium price
CBM4	Encourage Sufficiency	Encourage reduced consumption	Quality products, high levels of customer service; firm takes stance against obsolescence	Premium pricing due to slower sales
CBM5	Extending Resource Value	Use and/or recovery of wasted resources	Take-back; collection, recycling; recovery; and/or use of waste; often takes place at product level	Reduced material of overall costs
CBM6	Industrial Symbiosis	Waste outputs used as inputs	Outputs used as feedstock for different use; often takes place at manufacturing and process level to benefit business	Reduced operating cost; new business lines

Adapted from Bocken et al. (2016) and created by Whalen [12]

asphalt, and construction materials among others. Importantly, MRPs are a lower cost, sustainable alternative to virgin rubber and oil-based materials, providing a boost to the bottom line.

Lehigh Technologies developed an approach to value recovery from a waste stream using proprietary cryogenic (freeze-dry) turbo mill technology that turn end-of-life tires into competitive feedstock for new products. As of April 2018, the company has manufactured over 500 million new tires using its circular model reducing the amount of tires and post-industrial rubber discarded to landfills.

Every pound of Lehigh's MRP helps save: 10kwH of energy and 40% of the CO produced with traditional alternatives [17].

The company is not only a manufacturer of specialty chemicals, but also technology company and a provider of consulting services to their customers.

The company used the approach of closing resource loop for value proposition creation and capture, and a narrowing resource loop during its manufacturing process

Focus of value proposition here is Extending Resource Value—the company offer products made from recovered resources (i.e. old tires). The value is created and delivered via manufacturing of new products from recovered resource and the source of value captured is many fold—revenue from previously obsolete product (old tire destined to landfill), cost savings, service contracts, partnership agreements, and environmental and social benefits.

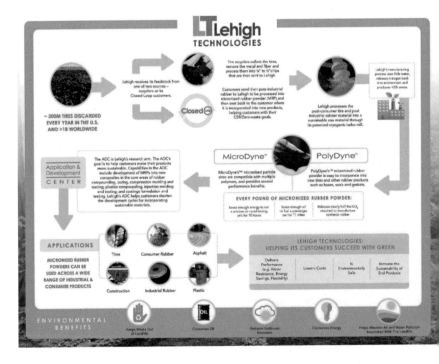

Credit: Lehigh Technologies business model (http://www.lehightechnologies.com)

Case Study 2

"Since 1959, Vitsœ has made long-living furniture, always striving to be better rather than newer"[18]. **Vitsœ** is a shelving and storage manufacturer and supplier. It was founded as a radical design-driven company, introducing a modular and timeless design philosophy to product design. Its vision is to manufacture furniture to last as long as possible, be adaptable and infinitely reusable, and not subject to fashion trends. The company specifically avoids built-in obsolescence and avoids fashionable furniture trends. Customers receive bespoke service through allocated personal planner to ensure that every detail is taken care of and… it's free of charge! The key ingredient is trust: customer trusts that the company has their best interest in mind, the product purchased now will be around for years to come, and they can purchase additional units as they need (when redecorating or moving to a different house). Company's products reflect longevity, durability, modularity, and long-term close relationships with customers. The furniture is made directly at their factory and delivered directly to the customer worldwide. Vitsœ takes a stand against inflated prices and following "discounts". In the absence of a middlemen, their markup is lower than the industry norm allowing customers to receive higher-quality furniture for this price via fair, honest price lists—"…the only difference between the price lists is the inclusion of the cost of our administration and packaging depending on the customer's location in the world".

This case illustrates Cradle to Cradle framework approach where network/system perspective is taken in design of the business model. At the basis of it is not only long-lived and durable furniture design manufactured in sustainable manner (reduced resource consumption and minimal environmental impact), but importantly, consumer education and awareness that encourages sufficiency—business that seeks to moderate overall resource consumption by curbing demand through education and consumer engagement. The company has 5 locations and ships its products to 80 countries around the world.

This company offers an example of a circular business model where sustainability is embedded throughout the business, with vision, value, and organizational culture (norms, values, and governance) driving the initiatives on sustainable consumption and production [2]. "Our reason for being: allowing more people to live better, with less that lasts longer. When you visit Vitsœ, you will find that we are walking the talk—everywhere. As I type, I can see our chef making use of the leftover food for another meal. How do you measure that? There are hundreds of similar examples—many too small to bother with. But they all add up."—says Mark Adams the Managing Director of Vitsœ.

Credit: Vitsœ

Case Study 3

Time Magazine wrote about them: *"The best part: no sizes."* The revolutionar
process, offered in San Francisco and Hong Kong, eliminates inventory, fabric wast
and ill-fitting standardizations to provide the *only actually* sustainable model tha
exists.

Unspun®'s mission is to permanently alter design, manufacturing, and consump
tion of fashion utilizing technology. While doing so, they aim to reduce global carbo
emissions by 1% (note: in 2019, global CO_2 emissions reached 36.8 Gt) [19]. O
their website you can find the following statement: "We strive for global change an
massive impact so ensure the planet continues to self-regulate and support all life"

What are they doing differently? First, there is no inventory—by focusin
on designing timeless garments that won't go out of style, and are made onl
when someone asked for them (i.e. on-demand manufacturing). Second, they ar
implementing localized production, and third, use low impact fabrics. With thes
steps they are creating a future of closed loops, zero-waste supply chains whe
jeans can be disassembled and assembled again, meaning garments won't go t
landfill/incineration after they've been worn down or no longer wanted.

Being only in the beginning of their journey, Unspun®'s jeans manufacturin
process already emits **24% less CO_2**, compared to traditionally produced pair c
jeans, based on Cradle-to-Grave Life Cycle Assessment (LCA) [20]. The dissolvabl
thread collaboration launched with Resortec® and paired with a 3D weaving proje
(not yet released) eliminates cut waste entirely and allows to dismantle and reus
the denim material. In the future, the company estimates that these innovations wi
reduce the CO_2 impact by 53%. That's over half the CO_2 per one pair of jeans.

The focus of Unspun®'s value proposition is offering durable products, as well as offering services related to these products and their recovery. Value creation and delivery is centered on "wear more, keep longer, and buy less" principles. Jeans are custom-made using the highest quality materials so that they last. If each person doubled the amount of times they wear a garment, Green House Gas (GHG) emissions would be 44% lower, which saves 36% of carbon per pair of jeans [21]. The company also offers repairs and alterations to ensure the longest wear.

The company captures value from premium pricing, environmental and social benefits, service offerings, and partnership agreements.

Credit: Unspun®

Case study 4

PramShare and **PramWash** were founded in Singapore, by newly minted parent: that wanted to address challenge faced by many new parents—stranded prams, wasted materials, expensive baby gear, frequent equipment replacement, and maintenance— these pain points drove the founders of PramShare and PramWash to use a different approach to prams and their maintenance in Singapore, and provide parents with an enhanced user experience.

PramShare offers long-term and one-time rentals of high-quality prams, strollers and car seats. Parents select the kind of gear they want to rent via the website, make a payment, and collect their rental at a designated self-pickup point or opt for delivery While at PramWash, specialists do detailed cleaning of prams, strollers, baby carriers car seats, play pens, and even soft toys.

A product-as-a-service business model that provides parents with high-quality prams, free of hassle of cleaning and maintaining, buying new gear, and stocking the old one, all while reducing waste and keeping materials in the value chain for as long as possible. Usually, at the end of the use of a pram, parents will dispose of it where it will either end up in a landfill or will be incinerated. However, increasingly parents choose to sell their prams or even give them away for free to other parent or companies like PramShare for continuous use. At the end of life, about 80% of a pram's materials can still be reclaimed and recycled into a new product.

The business model of PramShare draws inspiration from the sharing economy where customers rent products for the period of time they need them instead of owning the product. Through this approach, PramShare is able to maximize the use of their fleet of prams instead of the less efficient conventional business model where prams are sold to consumers, used for several months, and then are either disposed or left idling for months or even years.

PramShare applies the concept of circular economy to create value from underused products where they reuse and up-cycle used prams to fully working, as good-as-new condition. Recovering, recycling, and repairing used prams reduces environmental impacts by avoiding the need to extract and process raw materials in order to manufacture brand new prams. This value is captured through creation of new revenue stream from underused resources, results in cost savings for the company and customers generates service contracts, and provides social and environmental benefits (original case study was published here [22].

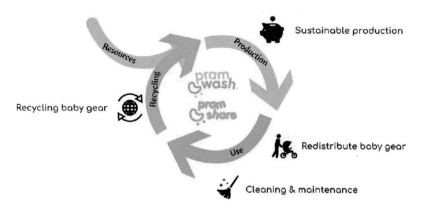

Credit: PramShare

Questions

1. How to create a Circular Business Model?
2. How can circularity at the level of an organization can be assessed?

Answers

1. Shifting from linear to circular business models requires a fundamental system change. This process is called business model innovation. There are numerous tools available to assist with such endeavor. They range from generic a BMC tool (Fig. 3), tools for eco-design, for sustainable business modeling to game-based tools.
2. Earlier this year, Ellen McArthur Foundation released Circulytics [23] tool, calling it "the most comprehensive circularity measurement tool." It uses wide range of indicators across its entire operations and assigns a score card with comprehensive results that helps companies to track their results and level up their aspirations.

Suggested Reading

1. Dennis Meadows, D. M., & Jorgen, R. (2004). *Beyond the Limits. The 30-year update.* Chelsea Green Publishing.
2. Stahel, W. R. (2019). *The circular economy. A user's guide.* (1st ed.). Routledge.
3. Braungart, W. M., a. M. (2002). *Cradle to cradle—remaking the way we make things.* (1st ed.), (Vol. 193).North Point Press.
4. Anderson, R. (2011). *Business lessons from a radical industrialist.* (Vol. 336). St. Martin's Griffin.
5. Benyus, J. M. (2002). *Biomimicry: Innovation inspired by nature.* Harper Perennial.

6. Hawken, P. (2010). *The ecology of commerce revised edition: A declaration of sustainability (collins business essentials).* Revised Edition (Vol. 224). Harper Business.
7. Lovins, L. H., Lovins, A., & Hawken, P. (2017). *Natural capitalism: The next industrial revolution.* (2nd ed.). Routledge.

References

1. Osterwalder, A., & Pigneur, Y. (2010). *Business model generation: A handbook for visionaries, game changers, and challengers.* Hoboken, NJ: Wiley.
2. Rana, P., Short, S. W., Bocken, N. M. P., & Evans, S. (2013) Towards a sustainable business form: A business modelling process and tools.
3. Schaltegger, S., Hansen, E. G., & Lüdeke-Freund, F. (2016). Business models for sustainability, Origins, present research, and future avenues. *Organization and Environment, 29,* 3–10.
4. Stahel, W. (1982). *Cradle to Cradle.* http://www.product-life.org/en/cradle-to-cradle.
5. Walter, S. (1982) The product-life factor. *NA.*
6. Hundertmark, T., Mayer M., McNally, C., Simons, T. J., & Witte, C. (2018). How plastics waste recycling could transform the chemical industry *McKinsey Insights.*
7. MacArthur, D. E., Waughray, D., & Stuchtey, M. R. (2016). The new plastics Economy–rethinking the future of plastics. *World economic forum.*
8. McArthur, E. (2013). Towards the circular economy. *Economic and business rationale for a accelerated transition.*
9. Benyus, J. *Biomimicry.* https://biomimicry.org/.
10. Hawken, P., Lovins, A. B., & Lovins, L. H. *Natural Capitalism.* https://natcapsolutions.org/.
11. Pauli, G. (2010–2013) *Blue Economy.* https://www.theblueeconomy.org/.
12. Whalen, K. (2020). *Circular business models that extend product value: Going beyond recycling to create new circular business opportunities.* Ph.D. thesis, Lund University.
13. Bocken, N. M. P., Short, S. W., Rana, P., Evans, S. (2014). A literature and practice review to develop sustainable business model archetypes. *Journal of Cleaner Production, 65,* 42–56.
14. Nancy Bocken, K. M., Evans, S. (2016). In *Conference "New Business Models"—Exploring Changing View on Organizing Value Creation.* Toulouse, France.
15. Binswanger, M. (2001). Technological progress and sustainable development: What about the rebound effect? *Ecological Economics, 36,* 119–132. https://doi.org/10.1016/S0921-80 9(00)00214-7.
16. Nicholas, A., Ashford, R. P. H., Robert, H. A. (2012) The crisis in employment and consumer demand: Reconciliation with environmental sustainability. *Environmental Innovation and Societal Transitions, 2,* 1–22.
17. *Lehigh Technologies.* (2007). https://lehightechnologies.com/commitment_to_sustainability.
18. Vitsoe. (2020). https://www.vitsoe.com/rw.
19. *Global Carbon Project.* (2001–2020). https://www.globalcarbonproject.org/.
20. Ammar, R. (2019). *Resortecs.* LCA-Rebirth.
21. Unspun. (2020). https://unspun.io/pages/climate-positive.
22. Piya Kerdlap, A. I. (2019). *Bringing circularity to baby Gear: Prams as a service for modern parents.* https://www.theinceptery.com/post/prams-as-a-service.
23. Foundation, E. M. (2017). *Circulytics—measuring circularity. circulytics*^{TM} *is the most comprehensive circularity measurement tool for companies.* https://www.ellenmacarthurfoundation. org/resources/apply/circulytics-measuring-circularity.
24. IDEO, E. M. *A Circular Business Models.* https://www.circulardesignguide.com/post/circular business-model-canvas.

25. Lyle, J.T. (1996). Regenerative design for sustainable development, John Wiley & Sons Inc, ISBN 978-0-471-17843-9
26. Bocken, N., Short, S., Rana, P. and Evans, S. (2013). A value mapping tool for sustainable business modelling, Corporate Governance, Vol. 13 No. 5, pp. 482-497
27. Braungart, M., McDonough, W., (2002), Cradle to Cradle: Remaking the Way We Make Things, North Point Press, ISBN 0-86547-587-3
28. Cradle to Cradle Products Innovation Institute, https://www.c2ccertified.org/

Anna Itkin , PhD has spent nearly two decades at the vanguard of scientific research and innovation in both academia and industry with extensive experience in innovation, systems thinking and cross-disciplinary work. In 2016 Anna co-founded The Inceptery Pte Ltd—a sustainability-led innovation consulting firm focusing on business model innovation and looking to address the ever more pressing need for businesses to simultaneously become economically, socially, and environmentally responsible, resilient and regenerative, thereby creating a long-term positive impact. Anna lives in Singapore anna@the inceptery.com https://www.linkedin.com/in/annaitkin/

Economic Instruments and Financial Mechanisms for the Adoption of a Circular Economy

Santiago Enriquez, Ernesto Sánchez-Triana, and Mayra Gabriela Guerra López

Abstract The transition to a circular economy requires the mobilization of resources and investments that support the adoption and upscale of ecological design and new technologies and business models. Policy instruments, including economic instruments, play a key role in ensuring that the prices of goods and services reflect the economic damages or benefits associated with their production and use, thereby creating incentives for circularity. Such incentives can help create a more playing level field for innovative technologies and business models and play a key role in freeing and reallocating resources that are currently used in the linear model. Finance instruments will be instrumental to scale up funding in new business models to support the transition toward a circular economy.

Keywords Circular economy · Economic instruments · Finance · Incentives · Policy instruments · Externalities · Subsidies

Learning Objectives

- What are economic instruments and financial mechanisms and why are they important for a circular economy?

  What are the main policy instruments that can be used to create incentives that facilitate a transition toward a circular economy?

  How should policy makers choose among the different available policy instruments?

S. Enriquez (✉)
International Consultant, Mexico City, Mexico
e-mail: senriquez@gmail.com

E. Sánchez-Triana
The World Bank, Washington, D.C., USA
e-mail: esancheztriana@worldbank.org

M. G. Guerra López
International Finance Corporation, Sao Paolo, Brazil
e-mail: mguerralopez@worldbank.org

© Springer Nature Singapore Pte Ltd. 2021
L. Liu and S. Ramakrishna (eds.), *An Introduction to Circular Economy*,
https://doi.org/10.1007/978-981-15-8510-4_23

- What are the key trends and available tools that can help to mobilize private sector investments into circular economy business models and technologies?

1 Introduction

Economic growth and industrialization have increased demand for resources and waste generation, amplifying pollution and emission of greenhouse gasses, environmental degradation and biodiversity loss. Economic instruments are incentives that aim to incorporate environmental costs into the budgets of households and enterprises and encourage environmentally sound and efficient production and consumption through full-cost pricing [1].

This chapter examines how environmental policy instruments and financial mechanisms can support the transition toward a circular economy model. Section 2 describes the broad range of environmental policy instruments that are available to support the transition to a circular economy, including phasing out of harmful subsidies that consume significant government resources and represent a major disincentive to promote a circular economy approach, as well as the adoption of other instruments that can provide clear price signals and economic incentives favoring circularity. Section 3 focuses on opportunities and challenges to maximize finance for a circular economy and Sect. 4 presents our conclusions.

2 Environmental Policy Instruments for Circular Economy

Circular Economy shifts away from the current "take-make-dispose" model and aims at decoupling economic growth from resource consumption by retaining as much value as possible from products, parts, and materials and organizing the economic activities into a closed-loop process of "resource-production-consumption-regenerated resource" (Fig. 1).

Circular economy encourages a circular flow of materials and energy, based on waste reduction, repair, reuse, remanufacturing, and recycling practices. The transition to a circular economy implies the adoption of measures that reduce the extraction and use of natural resources; and includes ecological design, new business models, an adequate political context, economic instruments, and financing (Fig. 2). Environmental policy instruments and financing enable investments in eco-design and the adoption and scaling up of new technologies and business models. They include incentives to free up and reallocate resources that are currently used in the linear model, as well as mobilizing new funding to support the transition toward a circular economy.

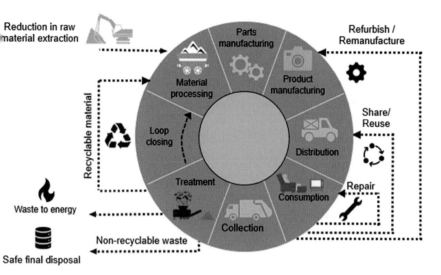

Fig. 1 Circular economy cycle. *Source* Authors

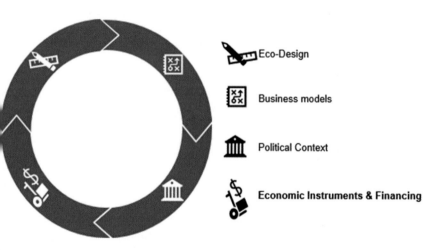

Fig. 2 Circular economy needs. *Source* Authors

Institutional, market, and policy failures such as the absence of secure property rights, environmental externalities,[1] and subsidies, distort prices of natural resources and environmental assets. As a result, producers and consumers do not receive the correct price signals which in turn leads to overproduction and overconsumption of products/services that are resource depleting or environmentally harmful. Economic

---

[1] According to the OECD [2], externalities refer to situations when the effect of production or consumption of goods and services imposes costs or benefits on others which are not reflected in the prices charged for the goods and services being provided.

instruments such as the removal of environmentally harmful subsidies, property rights, pollution taxes, tradeable emission permits, and refundable deposits aim to correct these failures, by reflecting environmental impacts and resource scarcity in prices so that producers and consumers can respond appropriately [3, 4].

This section will explore how economic instruments, including market-based instruments (MBIs), can support the transition toward a circular economy.

2.1 Phasing Out Subsidies

Subsidies are deliberate policy actions that maintain consumer prices artificially low leading to higher consumption of subsidized goods, which in turn results in higher natural resources extraction, consumption, pollution, and GHG emissions [5].

Subsidies for goods and services can result in severe environmental degradation. Examples of harmful subsidies include fossil fuels that increase greenhouse gas emissions, air pollution and congestion, and discourage energy efficiency; agricultural subsidies that can lead to the overuse of pesticides and fertilizers; and water subsides that can promote its overexploitation. By not pricing the waste management system and bearing the costs it generates, governments implicitly subsidize waste generation. Subsidies may also lead companies to under-invest in more efficient and environmentally friendly technologies, discouraging innovation and moving away from adopting a circular economy approach [5, 6].

Restructuring environmentally harmful subsidies represents a "win–win" opportunity for the economy and the environment [7]. For instance, removing fossil fuel subsidies would reduce greenhouse gas emissions and airborne pollutants and would increase government revenues. According to the International Monetary Fund (IMF), fossil fuel subsidies were estimated at $4.7 trillion or 6.3% of global GDP in 2015 and $5.2 trillion or 6.5% of GDP in 2017. Efficient prices would have decreased carbon emissions by 28% and would have avoided 46% of the deaths caused by fossil fuel-related air pollution in 2015 (Fig. 3) [8]. Subsidies also result in lower prices for other fossil fuel-based products, such as diesel, gasoline, or plastics.

Mexico successfully phased out subsidies to gasoline and diesel between 2007 and 2011. During this period, subsidies were reduced a few cents every month. The gradual reduction of fuel subsidies resulted in savings of an estimated 112 million tons of carbon dioxide [9]. An economy-wide assessment conducted in 2013, found that elimination of all energy subsidies would be associated with a 1.5% higher GDP growth over the long term because resources that were being used to pay for subsidies could be used instead to increased government expenditure, potentially including the expansion of public healthcare [10].

Even though energy subsidies are often well-intentioned, they usually do not benefit the poor or even the lower middle class. Wealthier people have bigger homes to heat, own more and bigger cars, and use more energy. Therefore, resources that were meant to protect the poor are ultimately transferred to the wealthier part of society. For example, in Mexico, the bottom 20% of the income distribution purchased only

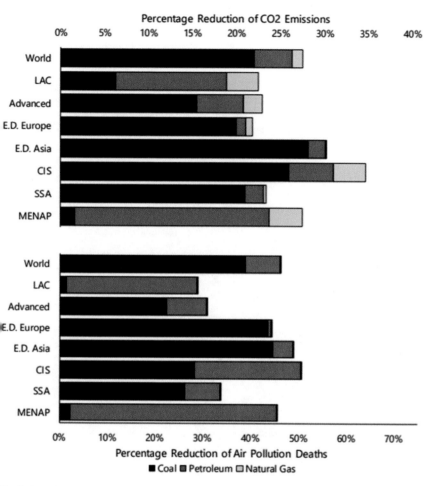

Fig. 3 Environmental gains from removing energy subsidies, 2015. *Source* Coady et al. [8]

% of the gasoline and diesel in 2010 (Fig. 4). Given that the subsidy was applied per unit of fuel, 97% of the assistance went to the top 80% of income earners in Mexico [11]. In Indonesia, half of all petroleum subsidies were transferred to the wealthiest 10% of the population (Fig. 5) [12].

In addition, when governments provide subsidies, they spend scarce resources that could have been invested in projects, such as health, social protection, environment, education or infrastructure, which could genuinely benefit those who are most in need [13]. A comparison of 109 Low-and-Middle-Income-Countries found that an increase in the amount spent of energy subsidies equivalent to 1% of GDP was associated with public expenditures in education and health that were on average lower by 0.6% of GDP [14].

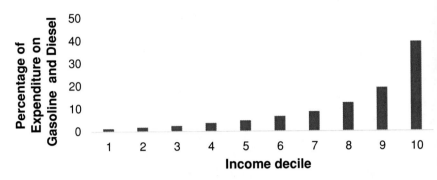

Fig. 4 Mexico's distribution of spending on gasoline and diesel by income decile in 2010. *Source* Authors based on data from Mexican the Secretariat of Finance and Public Credit

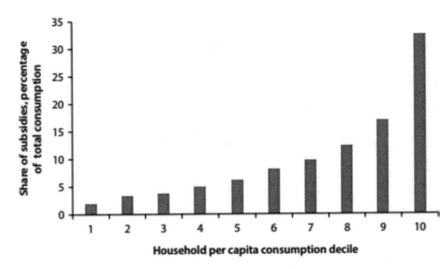

Fig. 5 Distributive impact of fuel subsidies in Indonesia, 2014. *Source* National Socioeconomi Survey of the Central Statistics Agency, Indonesia

Subsidies have clear and well-documented negative economic, environmental and social impacts. Yet, they are difficult to reform because they are extremel popular politically and attract groups willing to mobilize politically to resist thei restructuring [12]. Political risks that are possible to occur as a result of subsidie reforms can be reduced through a combination of communication and compensatio strategies. Communication with stakeholders on the costs and opportunities of reform can determine whether it turns out to be successful or not. Communication ca build support and acceptance, trust, and understanding of the political decisions tha underpin the reform. Disseminating information on the negative effects of subsidie and how those resources will be spent in the future is key to gaining social suppo for subsidy reform [15]. A compensation strategy should ensure that the poorest d not end up suffering the most from the subsidy removal/reduction by, for example, implementing a better-targeted social protection system [16].

Subsidy reforms are not a single event but a process that takes times and effort to implement. Creating social assistance programs to compensate the poor may take years and often includes administrative and technological innovations. Ideally, social assistance should be implemented in the form of direct transfers of cash that vulnerable groups can then decide how best to allocate [12]. Lessons from energy reforms undertaken by countries around the world suggest that reforms are more likely to be successful if they are part of a comprehensive long-term strategy for the energy sector, developed in consultation with stakeholders. In addition, the reduction of subsidies should be based on automatic and transparent pricing mechanisms with the aim of depoliticizing the issue [17].

2.2 Circular Economy and Environmental Policy Instruments[2]

As mentioned in the previous section, subsidies can be harmful because they artificially lower the prices of products and activities. However, even in the absence of subsidies, prices generally fail to reflect the environmental damage and natural resource depletion caused by production and consumption of goods and services because of externalities, which arise when the agent making the production or consumption decision does not bear all of the costs or benefits of this decision [2].

Addressing externalities is central to promote a circular economy. Disposal of industrial effluent into a waterway may be a low-cost solution to waste disposal for the polluter, but firms and individuals downstream may suffer consequences through higher costs from lost fishery production, higher water treatment costs, lower amenity values (for recreation), or loss of critical drinking water supplies. A similar situation could exist with environmentally beneficial decisions. For example, a firm that cleans polluted intake water and then discharges clean water after using it in its internal process would be creating a positive externality. Whereas the example of the negative externality is illustrative of the predominant take-make-dispose model, that of the positive externality would be expected in a circular economy.

A number of environmental policy instruments are available to correct negative externalities and promote positive ones, with the aim of facilitating the transition toward a circular economy. At one extreme, such instruments include fines or sanctions that are linked to traditional command-and-control (CAC) regulations.[3] At the other extreme, they include laissez-faire approaches that require consumer advocacy

---

This section draws largely from Sánchez-Triana et al. [18].

According to Giner [19], command and control regulations focus on preventing environmental problems by mandating standards and technologies to control pollution. This approach generally relies on emissions standards, ambient standards, and technology-based and performance-based standards in conjunction with enforcement programs.

or private litigation to act as incentives for improving environmental management In between are the more familiar tax-and-subsidy approaches, as well as the les familiar mechanisms relying on traded property rights [18].

There is no single standardized definition of an incentive-based or market-base instrument (MBI), but the commonly held understanding and the definition employe here is that an MBI must, foremost, attempt to align private costs with social costs t reduce externalities [3]. Within this definition, the particular strength of an MBI the depends on the degree of *flexibility* that a polluter or resource user has in achievin a given circularity target. A very weak environmental policy instrument essentiall dictates through regulation the type of technologies that firms must use, or the targe they must meet. This is the inflexible CAC approach—which also entails an economi incentive to the extent that failure to comply can result in monetary sanctions. A ver strong MBI allows market signals rather than explicit directives to determine the bes way to meet a given standard or goal.

Flexibility is operationalized by equating it to the level of decentralization tha occurs in transferring social (or state) decisions to the private (individual) level. / strong MBI decentralizes decision-making to a degree that the polluter or resourc user has a maximum amount of flexibility to select the production or consumptio option that minimizes the social cost of achieving a particular level of environ mental quality. When the polluter or resource user is driven by profit- or utility maximizing behavior, a strong MBI also generates the lowest social cost outcom for the achievement of a given policy objective.

Table 1 illustrates the broad spectrum of instruments that might be available all of which implicitly or explicitly have some incentive effect. They fall across continuum ranging from very strict CAC approaches to decentralized approache that rely more on the market or legal mechanisms.

As illustrated by Table 1, there is a broad range of environmental policy instru ments available to address externalities and promote a circular economy. A ke consideration in selecting the most appropriate instrument is *cost-effectiveness*. Fc example, the asymmetry of information often implies that individual agents, privat firms, or community associations are more likely than governments to identify th most cost-effective means for achieving a given environmental goal, such as les water withdrawal, less water pollution, or more forest coverage. This forms the basi for the common theoretical result that—if one focuses entirely on private costs– strong forms of MBIs are more cost-effective than their weaker counterparts or tha CAC approaches ([20]).

Another advantage of stronger MBIs is that, to the extent that they provid economic agents flexibility to choose how they achieve a given environmental objec tive, they provide incentives for *innovation*, such as the development and adoptio of better abatement technologies [21]. In a seminal article, Porter and van der Lind [22], argued that more stringent but well-designed environmental regulations trigg innovation to such an extent that it can increase the competitiveness of the regulate firms. This argument, known as the Porter Hypothesis, suggests that pollution is ofte a waste of resources and that innovation aiming to increase efficiency in their us may lead to an improvement in the productivity with which resources are used [23

Table 1 Classification of circular economy and environmental policy instruments based on flexibility in individual decision-making

Minimum flexibility				Maximum flexibility
Greater government involvement				Greater private initiative
Control oriented	Market-oriented			Litigation-oriented
Regulations and sanctions	Charges, taxes, and fees	Market creation	Final demand intervention	Liability legislation
General examples				
Standards government restricts nature and amount of pollution or resource use for individual polluters or resource users. Compliance is monitored and sanctions imposed (fines, closure, and jail terms) for noncompliance	Effluent or user charges: government charges fees to individual polluters or resource users based on amount of pollution or resource use and nature of receiving medium. Fee is high enough to create incentive to reduce impacts Subsidies: Government provides subsidized inputs to encourage their adoption	Tradable permits: government establishes a system of tradable permits for pollution or resource use, auctions or distributes permits, and monitors compliance. Polluters or resource users trade permits at unregulated market prices	Performance rating: government supports labeling/performance rating program that requires disclosure of environmental information on the final end-use product. Performance based on adoption of ISO 14,000 voluntary guidelines: zero pollution discharge, mitigation plans submitted; pollution prevention technology adopted, reuse policies and waste recycling	Strict liability legislation: the polluter or resource user is required by law to pay any damages to those affected. Damaged parties collect settlements through litigation and the court system
Specific examples				

(continued)

Table 1 (continued)

Minimum flexibility				Maximum flexibility
Greater government involvement				Greater private initiative
Control oriented	Market-oriented			Litigation-oriented
Regulations and sanctions	Charges, taxes, and fees	Market creation	Final demand intervention	Liability legislation
• Pollution standards • Licensing of economic activities • Land use restrictions • Zoning and setback requirements • Water use quotas • Construction impact regulations for roads, pipelines, ports, or communications grids • Fines for spills from port or land-based storage facilities • Bans applied to materials deemed unacceptable for solid waste collection services	• Noncompliance pollution charges • Source-based effluent charges to reduce downstream water treatment requirements • Royalties and financial compensation for natural resources exploitation • Performance bonds to ensure construction standards • Subsidies to construct Common Effluent Treatment Plants • Tipping fees on solid wastes • User charges for water	• Payment of ecosystem services to forest owners to ensure water protection ecosystem services • Designation of property rights to farmers to improve irrigation water and drainage management • Deposit-refund systems for solid and hazardous wastes • Tradable permits for water abstraction rights, and water and air pollution emissions	• Consumer product labeling (eco-labels) relating to production practices, energy efficiency, and so forth • Supply chain intervention where intermediate buyers insist on installation of Effluent Treatment Plants for upstream product production processes • Education regarding recycling and reuse • Disclosure legislation requiring manufacturers to publish solid, liquid, and toxic waste generation • Blacklist of polluters	• Damages compensation to plaintive • Liability placed on guilty firm's managers and environmental authorities • Long-term performance bonds posted for potential or uncertain hazards from infrastructure construction • "Zero net impact" requirements for infrastructure projects

Source Sánchez-Triana et al. [18]

Multiple empirical studies have been developed to test the Porter Hypothesis. ome of these studies find that environmental regulation is associated with increased esearch and Development by firms, particularly when regulation is stable and flexble [24]. They also find that environmental regulation can increase productivity, as n the case of refineries located in Los Angeles, California, which achieved signifiantly higher productivity than other U.S. refineries despite their need to comply with more stringent air pollution regulation than those located in a different geographic rea [25]. In a different study, stricter regulations led to modest long-term gains n productivity in a sample of 17 manufacturing sectors in Quebec. Productivity lighthly fell in the first year of the regulation's adoption, but steadily increased in ubsequent years, resulting in a net increase in overall productivity. This finding is onsistent with the Porter Hypothesis assumption that innovation may take time and nderscores the need to ensure the stability of regulations [26].

In addition to addressing externalities and ensuring cost-effectiveness, policy nakers often have a third goal when designing an appropriate economic incenve system: *revenue generation*. Stronger MBIs have the advantage over convenonal CAC regulations of delivering a double dividend, meaning that, in addition) advancing environmental goals, they also generate government revenues. This is articularly the case of charges, taxes, and fees, and potentially of tradeable permits n which the initial allocation of permits is auctioned. The government may use evenues collected through these instruments to increase its expenditure and investent in socially desirable areas, such as further environmental protection [20]. The evenues collected by taxing "bads" such as pollution can be used to lower existing istortionary taxes on desirable activities such as labor, thereby establishing clear nd consistent price signals that incentivize greater circularity.

There are, however, practical tradeoffs to consider between revenue generation nd incentive effects. For example, it would be possible to levy a very high charge that ffectively discourages all polluting activity. Abatement levels would be very high in uch a case, but no revenue would be generated. Similarly, very low charges would enerate little revenue and generate little abatement because there is no incentive or firms to reduce pollution. Typically, revenue is maximized at some intermediate evel of abatement. A policy decision must be made relating to how much additional evenue (beyond the maximum) a government is willing to give up to generate higher evels of abatement. The answer to this policy question should be related to the arginal benefits of pollution abatement. However, it is typically more a function f government budgetary realities that regard such taxes as a convenient means for nderwriting environmental management efforts.

The following subsections provide more details and examples of the available rcular economy and environmental policy instruments.

2.1 Command and Control Regulations, Fines, and Penalties

ommand and control approaches relying extensively on regulatory guidelines, ermits, or licenses have traditionally been the preferred mechanisms for controlling environmental impacts in urban areas. For instance, many countries have recently adopted regulations banning single use plastics. As of July 2018, 83 countries had adopted bans on free retail distribution of plastic bags and 61 countries had banned their imports [27].

Command and control regulations can be separated into two types: technology based and performance-based [21]. Technology-based regulations specify the methods and equipment that firms must use to meet the preestablished target Conversely, performance standards set an overall target for each firm, or plant, and let firms decide how to meet the standard but hold them to a uniform level across the industry.

Although it is technically simple to impose regulations with specific fines for noncompliance, the problems associated with implementing them and achieving compliance are insurmountable for many developing countries. First, regulated entities might need to dedicate significant time and resources to comply with existing regulations, instead of using them to grow their businesses. As a result, regulations can drag economic development. Second, the capacity to implement regulations is often limited because of inadequate human resources, or inadequate supportive infrastructure such as environmental information or monitoring networks. Third, local financing constraints arise because the authority for environmental regulation is often delegated to lower (local) levels of government without adequate sources of financing for implementing and monitoring the regulations. Fourth, conflicting standards often prevail where individual ministries or departments have been responsible for setting environmental regulations within their own departments; lack of coordination often leads to conflicting or overlapping regulations. This is often the case for water-related issues, because of the numerous stakeholders involved in water use. Finally, conflict of interest within government programs exists where government agencies are themselves the implementing or investing authority; self-regulation becomes problematic under such circumstances and seldom are their built-in incentives to ensure compliance. This is especially a problem with common infrastructure facilities that typically are a government mandate.

In addition, command and control regulations give the manufacturer little incentive to improve efficiency and environmental performance. While these regulations were successful in securing the first tranche of emissions reductions from previously unregulated industries, they are now viewed as increasingly burdensome [28]. Esty [29] argues that the twenty-first century sustainability challenges call for a new policy framework that goes beyond government mandates, prioritizes innovation, and includes regulatory strategies that engage the public and businesses in environmental problem-solving.

2.2.2 Pollution Charges and Taxes

As other economic instruments, pollution charges seek to influence a producer or consumer behavior through a monetary incentive and can be applied in different ways: (i) charges to emissions or effluents; (ii) charges to users; and (iii) charges to products. These instruments are based on the "polluters pay principle" where the polluting party pays for the damage done to the natural environment [30].

Emission or effluent charges are applied to emissions that are released into the air, water or soil and are most suited for large stationary sources. They can be levied on emissions that are directly metered, on a proxy source (for instance, water consumption can be used as a proxy for wastewater emissions), a presumptive pollution level or in the form of a flat rate [31]. When a presumptive pollution level is considered, a firm is compelled to pay the charge with no specific monitoring conducted. If the firm wishes to reduce its tax burden, it must conduct monitoring at its own expense (but still subject to regulatory audit) to demonstrate that its actual pollution loads are less than the presumed loads. Emissions charges can be used to address local challenges: for example, taxes for airborne pollutants such as Nitrogen Oxides (NOx) and Sulphur emissions and water pollution provide a continuous incentive to implement pollution-abatement options and encourage innovation. For instance, Sweden has accomplished reductions in NOx emissions through emissions charges. In 1990, the Swedish Parliament passed a legislation introducing NOx charges emitted from energy generation at combustion plants. The charge is applied to measured emissions, or to presumptive emissions levels and plant operators may choose to pay the charge on the basis of presumptive emissions levels or by installing measuring equipment. By 2004, Sweden achieved a 65% reduction in NOx emissions compared to 1990 [32].

User charges are payments for specific environmental services, such as collective or public treatment of effluents or waste disposal. These payments are intended to reflect the costs of providing the service. Taxes on waste, based on weight or quantity, can promote reuse and recycling practices while diverting waste from landfills [33]. The Netherlands uses a combination of charges to finance its water-related services. Three different water levies are currently applied in the Netherlands: (i) a water systems levy to cover the costs of flood protection measures and to provide surface water; (ii) a wastewater treatment levy related to the cost of treatment; and (iii) a pollution levy for direct discharges to surface waters. The three are earmarked taxes that provide the Regional Water Authorities (RWAs) with sufficient funds for necessary investments and maintenance. These charges have proven to be effective to reduce water pollution. For instance, in 1981, untreated sewage had a nitrogen removal rate of 53% and by 2014, the removal rate was 86% [17].

Product charges are applied to products that pollute in the manufacturing, consumption, or disposal phase and can be based either on the characteristics of the product or on the product itself. These charges can create more favorable market conditions for "cleaner" products and less favorable conditions for polluting products. In 1986, Denmark introduced a tax on domestic manufacturers and importers of pesticides sold for agriculture. The tax amount started at 3% of the wholesale price for all pesticides and increased over the years. By 1997, the country reached a 47% reduction in pesticides consumption [34].

Before fixing the tax rate, governments need to carefully analyze the country's economic, social, and political context. If not correctly designed and implemented, pollution taxes may sometimes cause undesired effects or disproportionately affect low-income groups. To prevent these cases from happening, models can be useful in predicting potential leakage or distributional impacts and mitigation measures [35]

2.2.3 Market Creation: Tradable Permits

At a more complex level, market-oriented approaches can include some form of market creation. The most complex system involves tradable permits where regulators establish an allowable level of pollution which is distributed among firms in the form of permits. Polluters that need important amounts of money to reduce their emissions can buy emissions allowances from polluters that have cheaper abatement costs [36], and companies that manage to keep their emissions below their allocated level can sell their surplus allotment to other firms or use them to cover excess emissions in other parts of their facilities [37]. Emission trading systems enable emission reductions where it is cheapest to achieve them [38].

One of the most well-known cases of tradable permits is the European Union Emission Trading System (EU ETS) that was set up in 2005, as the world's first international ETS and is the major carbon market worldwide. The EU ETS works on the "cap and trade" principle where a cap is set on the total amount of greenhouse gasses that can be emitted by installations covered by the system. In this context companies receive or buy emission allowances that can be traded among the different actors. Each year, a firm must prove that it has enough allowances to cover all its emissions in order to not get fined. If a company reduces its emissions, it can keep the spare allowances to cover its future needs or else sell them to another company that is short of allowances. The cap reduces with time to achieve a reduction of emissions [39]. Emissions from stationary installations have declined by around 29% between 2005 and 2018 [40].

Instruments to set a price on carbon have gained prominence in the fight against climate change. As of 2019, 46 national and 31 sub-national governments had adopted or were analyzing the adoption of taxes, emissions trading systems, or a combination of both to promote the transition to a decarbonized economy [32].[4]

Another example of the application of tradable permits is the Regional Clean Air Markets (RECLAIM) program in California. RECLAIM was established in 1994 to reduce nitrogen oxide and sulfur dioxide emissions in the Los Angeles area and is the world's first comprehensive market program for reducing air pollution. The program sets a factory-wide pollution limit for each business and lets them decide the most cost-effective way to meet their emission limits. As in the case of the EU ETS allowable emission limits decline a specific amount each year and companies that can reduce emissions more than required can then sell excess emission reductions to other firms. Each firm participating in the program receives trading credits equal to its annual emissions limit that are based on past peak production and the requirements of existing rules and control measures. By 2018, RECLAIM achieved a 73% reduction

---

[4]Data from the Carbon Pricing Dashboard: https://carbonpricingdashboard.worldbank.org/map data.

or NOx emissions and a 70% reduction for SOX emissions compared to 1994 levels
[41].

One potential advantage of tradable permit systems is that they may reduce bureaucracy and government participation in the process. Such decentralization of decision making is particularly important in high growth economies where regulatory drag might otherwise be a problem.

2.4 Market Creation: Deposit-Refund Systems

In a deposit-refund system, regulators impose a monetary sum that is paid when a product is purchased. This deposit is refunded when the item or its packaging are returned for recycling or appropriate disposal. This model has advantages over taxes that are applied directly to waste and has been successfully implemented in many countries. A deposit-refund system can be effective in coping with some of the challenges that arise from waste management taxes that are applied directly to households. For instance, households might turn to waste burning or illegal dumping when waste disposal is directly taxed. Conversely, rebates for products that are returned constitutes an incentive for appropriate waste management.

In Sweden, the private company Returnpack is responsible for the deposit-refund system of metal cans and recyclable PET bottles. The system is regulated by a Deposit Ordinance decided by the Swedish Parliament that states that "Anyone who professionally fills plastic bottles or metal cans with ready-to-drink beverages or who professionally imports ready-to-drink beverages to Sweden in plastic bottles or metal cans shall ensure that the bottles and cans are included in an approved recycling system." In 2018, the recycling rate reached 85% for both aluminum cans and PET bottles. More than 2 billion cans and bottles were recycled or the equivalent to 201 packages per person. Containers are collected in return machines that can be found in supermarkets, sports clubs, and cafés, and are returned to a plant that reprocess the materials into new products.

In the United States, deposit-refund systems are applied to different products as lead-acid batteries, pesticide containers and tires. The Battery Council International (BCI), recommends retailers to charge a $10 deposit on all batteries sold, the fee is reimbursed when the customer returns the used battery for recycling within 30–45 days of purchase. According to the BCI, lead batteries have a recycling rate of 99.3% and are the most recycled consumer product in the US. The rate of recycling is attributed to industry investment in a state-of-the-art closed-loop collection and recycling system which keeps more than 129 million lead batteries away from landfills annually [42].

Deposit-refund systems have been used extensively to promote recycling. Around 63 countries have adopted requirements for taking back of single-use plastics through deposit-refund schemes, most of them focusing on beverage bottles [27]. Such schemes are also appropriate for difficult problems such as toxic and hazardous waste management.

2.2.5   Extended Producer Responsibility

Extended Producer Responsibility (EPR) refers to a policy that aims to make producers responsible for the environmental impacts of their products throughout the value chain, from the design to the post-consumer phases [38]. EPR systems have been adopted with the aim of alleviating municipalities' burden for managing waste, reducing the amount of waste destined for final disposal, and increase rates of recycling. Also, by shifting the responsibility upstream toward the producer, EPR systems aim to provide incentives to producers to take into account environmenta considerations when designing their products. Hilton et al. [43] identified about 400 EPR systems in operation worldwide, most of them in OECD member countries Electronics, tires, and packaging are the products for which a larger number of EPR systems have been adopted worldwide.

EPR systems may include a physical responsibility for adequately collecting and treating the waste and/or a financial obligation to pay for such collection and treat ment. The deposit-refund systems described above are generally used by producer to comply with their financial responsibility.

2.2.6   Market Creation: Property Rights

Another potentially important type of market creation involves conferring some form of property right (either individual or collective) to environmental assets.

This instrument is generally regarded as the solution to a situation, generally known as the "Tragedy of the Commons" [44]. This refers to a situation in which resources are shared among a very large group and there is no way of excluding anyone from consuming such resources. Each individual obtains all the benefits from consuming the resources, but the loss resulting from overconsumption is shared by all group members. As a result, every individual has an incentive to overconsume the resource. In contrast, in a situation with property rights, right holders have the incentive to manage resources sustainably and resource depletion is internal to the owners/users. The consequence of this internalization is that the owners will not engage in resource extraction unless the price of the resource commodity covers not only the extraction cost but also the depletion or user cost, which is the foregone future benefit as a result of present use. With secure property rights, the price of resource commodities such as minerals, oil, and timber would reflect the resource depletion cost and provide the right signals for efficient use and conservation. Property rights are particularly applicable to land and soils (land rights), water resources (water rights), minerals (mining rights), and other natural resources that can be parceled out or easily demarcated.

Sri Lanka's coastal fisheries have applied property rights in the form of rights of access for a long time. In earlier times, beach seine owners controlled the access to coastal waters and had associated fishing rights that were inherited by descent or marriage. Although at the start, each beach seine owner had his own beach for which he had exclusive rights to operate, each of his children had only a fraction lot of his beach, but of his right to fish off the beach along with his brothers and brothers-in-law. Outsiders are not allowed to anchor or beach the fishing boats along the shoreline of the community, and labor is not recruited from outside the village 3].

2.2.7 Final Demand Intervention

Final demand interventions provide information to investors, consumers, and the general public and decentralize decision-making to the final consumer. These tools can encourage behavioral change of consumers toward long-term sustainability, encouraging producers to adopt more environmentally friendly approaches.

2.2.8 Final Demand Intervention: Eco-labeling

Eco-labeling has been used as an effective instrument to make production and consumption patterns more environmentally friendly. Labeling gives information about the environmental impacts associated with the production or use of a product and might shift consumption and production toward socially responsible and sustainable alternatives [45].

An eco-labeling certification is often voluntary but may be mandatory in certain instances, such as for identifying toxic ingredients. The International Organization for Standardization (ISO) recognizes three types of voluntary labels: (i) type I labels are based on a pass-fail multi-criteria approach that indicates the overall environmental performance of a product and are verified by a third-party; (ii) type II labels are defined as "self-declared" environmental claims made by manufacturers and businesses without set criteria or quality checks; (iii) type III labels are third-party verified under established programs based on a product's life cycle assessment [46].

ENERGY STAR is a voluntary labeling program established by the United States Department of Energy (US DOE) and the United States Environmental Protection Agency (US EPA). The program aims to reduce energy consumption and greenhouse gas emissions by power plants and helps purchasers identify and purchase energy-efficient products. Products can earn the ENERGY STAR label by meeting the energy efficiency requirements set forth in ENERGY STAR product specification. Since 1992, ENERGY STAR products have achieved over 3 billion metric tons of greenhouse gas reductions. In 2017, $30 billion in energy costs have been avoided in the United States [47].

2.2.9 Final Demand Intervention: Environmental Certification

Unlike eco-labeling, which gives information on the impacts associated with a particular product, environmental certification programs assess the overall environmental policy and management of a company and provide information on its environmental impacts, processing, and production methods, including resource use, production techniques, and emissions. There are two internationally accepted environmental certification schemes: the ISO Environmental Management System (EMS) standard and the Eco-Management and Auditing Scheme (EMAS). These certifications provide overarching comparisons between industries and outline fundamental environmental codes for the industry. Additionally, there are sector-specific environmental certification schemes that can provide more specialized and detailed guidance to companies within the same industry.

2.2.10 Final Demand Intervention: Disclosure Requirements

A more aggressive form of final demand intervention involves promulgating disclosure requirements. While labels convey a signal of how environmentally friendly a product is, disclosure requirements provide information on the environmental performance of a firm, including emissions to air, wastewater discharges, waste, and compliance with standards. Principle 10 of the Rio Declaration, established during the United Nations Conference on Environment and Development in June 1992 reinforces the principle of public access to environmental information in achieving sustainable development. Public access to environmental information can empower people to make decisions relating to environmental issues [45].

There are no sanctions attached to such disclosure, but it gives consumers the choice of how to deal with the products of particular firms. Public disclosure schemes have been applied in many countries around the world. Philippines introduced the program Eco-watch in 1996, to provide incentives to industries to comply with environmental regulations. Through it, the government was able to set up an environmental grading system to categorize companies' environmental performance using color labeling system. A black label was used for firms with no pollution control or causing serious damages to the environment, blue for firms that met all environmental standards and required procedures, and gold for firms that met environmental standards for three years in a row and conducted at least two environmental programs such as waste reduction and recycling projects. In 1997, before the program came into effect, over 92% of plants were found to be non-compliant; by 1998, the number of compliant plants increased to 58% [48].

In Colombia, the Cauca Valley Corporation adopted a water pollution charge program. The effluent fees paid by firms were publicly disclosed and the publicity influenced the reputations and decisions of company leaders reducing the discharge of water pollutants [49].

2.2.11 Final Demand Intervention: Supply-Chain Management

Supply chain management (SCM) refers to the process through which a company manages all the processes related to the planning, sourcing, processing, manufacturing, and delivery of goods and services in order to meet consumers' demand with more efficient use of resources. As consumers become more concerned with the environmental and social sustainability of the products they consume, more producers have integrated environmental and social criteria into their supply chain. Through supply chain management, private sector companies can extend their influence in pollution control and abatement beyond their own operations by collaborating with partners throughout the value chain [45]. Supply chain management is also key to improve a company's sustainability performance because it accounts for more than 80% of a typical consumer company's total GHG emissions and more than 90% of its impacts on air, land, water, biodiversity, and geological resources [50].

Firms are increasingly sensitive about the environmental and social context in which their suppliers operate and have introduced environmental and social criteria into their supply chain. In such cases, firms downstream in the supply chain intervene in the upstream production processes of their intermediate products by insisting that certain environmental protection activities are undertaken during production. This type of intervention has resulted in upstream firms installing pollution control equipment to satisfy their buyers' sourcing criteria.

A number of frameworks are available to assess the sustainability of key supply chains. For instance, WWF offers more than 50 performance indicators that help assess the security and governance, environmental, social, and economic and financial risks in the value chain. WWF has worked with several companies to apply its framework and achieve specific goals, including helping McDonald's to increase the amount of agricultural raw materials for its food and packaging products that come from sustainable sources; supporting Johnson & Johnson to assess the environmental and social risks of sourcing palm oil and other naturally derived materials; and helping EDEKA to improve sustainability performance along its value chain through measures spanning ecosystem and biodiversity protection, soil and water management, waste management, responsible and reduced use of agrochemicals, and ensuring occupational safety and health.[5]

7.3 Liability Legislation

Liability for environmental harm is designed to compensate affected individuals or groups, with a particular focus on restoring or replacing damaged resources and/or compensating lost value [51]. Litigation-oriented approaches to environmental management require only that legislation be in a place that confers relatively straightforward rights and obligations to resource users. These approaches form a legal umbrella for court cases, which then consider the nature and extent of environmental damages on a case-by-case basis. Advocates of liability legislation argue that it is a highly efficient instrument to address externalities because it only requires monitoring of specific incidents, rather than a need to monitor behavior, as is required by regulation [52]. Most of these approaches are relatively new and have seen very

---

https://supplyrisk.org/.

limited application in developing countries (quite often because legal systems are themselves weak in such countries). India offers several examples of cases in which the courts have ordered polluters to compensate for the environmental damages they caused, including the cost of reforestation and cleanup of polluted rivers. In an emblematic case, one of India's largest smelters was found to be operating without a valid environmental permit. The Supreme Court ruled that the company had to pay 10% of its profits before depreciation, interest, and taxes for the 15 years it operated without a permit, or $15.5 million [51].

Common challenges faced in developing countries include difficulties in esti mating damages and the required compensation and limited accountability for ensuring that recoveries intended for restoration are actually spent to that end [51] Even in industrial countries, the use of liability legislation is hampered by the analyt ical difficulties of establishing cause and effect, or of ascribing blame or negligence One significant objection to using litigation-oriented mechanisms is neither environ mental nor economic: it is social. Because such systems assume that all have equal access to the courts, the mechanisms often discriminate against the poor and others with limited access to legal recourse.

3   Maximizing Finance for Circular Economy

The transition to a circular economy requires the development of new business models, and therefore, financing opportunities. A McKinsey analysis for the Ellen MacArthur Foundation found material cost savings worth up to $630 billion per year by 2025, in EU manufacturing sectors [53]. Accenture has identified a $4.5 trillion global opportunity before 2030, through avoiding waste, making businesses more efficient and creating new employment opportunities [54]. The Ellen MacArthur Foundation also found that opportunities in India amount to $218 billion per year by 2030 [53].

Through its different products and services, the World Bank Group has already financed several projects supporting the transition to a circular economy. Between 2004 and 2018, the World Bank committed more than $49 billion targeting pollution management and circular economy interventions, covering sectors from solid waste management to water pollution control, air quality management, management of chemicals and toxins and policy reforms, among others.

The World Bank assistance to client countries includes: (i) Investment Project Financing, providing loans, grants, and guarantee financing to governments for activ ities that create infrastructure; (ii) Development Policy Financing that supports policy and institutional actions designed by client countries; (iii) Program-for-Results that links disbursement of funds directly to the delivery of defined results; (iv) trust fund and grants that allow scaling up of activities; and (v) Private Sector Financing.

In addition to the loans and technical assistance that explicitly promote environ mental goals in line with a circular economy, the World Bank requires that client countries manage environmental and social risks of all projects through its Environmental and Social Framework, that proactively integrates standards to improve the environmental and social performance of projects. These include standards on resource efficiency and pollution prevention and management that are fully aligned with a circular economy approach. A project implemented in Morocco, illustrates the Bank support to the modernization of waste management, including at sites like Oum Azza, near Rabat, where traditional trash-pickers now operate a recycling collective in improved conditions. The International Finance Corporation (an organization of the World Bank Group) is supporting the private sector on recycling actions. For example, it invested $33 million in a recycling facility in Mexico, that will convert post-consumer plastic bottles into food-grade, recycled resin that will be sold to the Mexican soft-drinks bottling industry. The success of the project relied on a stable domestic supply on the one hand, and the commitment from a large soft-drinks company as the main off-taker of the production on the other hand.

While the contributions from international development organizations are valuable, they will be insufficient to support a global transition toward a circular economy. A key challenge is working with the financial sector to adapt and reassess existing options and perspectives.

Circular economy presents several financing opportunities that range from consumer lending and leasing to large project financing, green bonds[6] and equity capital [55]. For instance, many of the key opportunities identified by the Green Bond Principles cover elements of circular economy, including renewable energy and energy efficiency; sustainable waste management and land use; clean transportation and clean water [56]. The World Bank issued the first green bond in 2008, paving the road for today's green bond market. Since then, the World Bank has raised more than US$13 billion through almost 150 green bonds in 20 currencies for institutional and retail investors all over the globe.

By 2019, the green bond global market had issued USD 257.7 billion through 788 green bonds from 496 issuers in 51 jurisdictions [57]. Other bonds, as blue bonds or the Breathe Better Bond Initiative can also help finance circular economy-related projects. The Breathe Better Bond will be issued by local governments in developing countries and its proceeds will be used to invest in projects that reduce both air pollution and greenhouse gas emissions [58]. In 2018, the World Bank helped issue the first blue bond for USD 15 million in the Seychelles to provide financing for marine and ocean-related activities. While interest in sustainable bonds has increased, they represent only a small portion of the global bond market. In 2018, the Global Bond Market reached USD 102 trillion while the Global Equity Market reached USD 74.7 trillion [59].

Investors are also increasingly interested in sustainable investing, an investment approach that considers environmental, social, and governance (ESG) factors in portfolio selection and management, which has potential linkages with circular economy. By early 2018, sustainable investing assets in the five major markets worldwide stood at $30.7 trillion [60]. Reasons for ESG investing might include investors' interest

---

Green bonds are standard bonds created to finance environmentally beneficial projects.

in aligning their portfolio with their norms or beliefs or wanting to use her capital to trigger change for social or environmental purposes, such as decarbonization of the economy. Interestingly, ESG investment has also become a strategy to improve the risk-return characteristics of a portfolio. Companies with a strong ESG profile may attract more investors because they are more competitive as a result of more efficient use of resources, innovation management, or human capital development. These companies also typically have comparatively higher risk control and compliance standards across the company and within their supply chain management. As a result, they suffer less frequently from incidents that can impact the company's value. In addition, these companies may be less vulnerable to systematic market shocks; for example, an energy-efficient company is less vulnerable to changes in energy prices than their less efficient competitors [61]. While rapid growth in ESG investment is encouraging, there is still a significant challenge in terms of finding metrics that will accurately reflect whether such investment is in fact contributing to advance environmental and social goals.[7]

The transition toward a circular economy requires deep transformations of supply chains and consumption patterns. New technologies and business models are becoming available to promote this transition by encouraging the use of renewable, recyclable or biodegradable resources, extending the life of a product or offering it as a service and recovering resources at the end of a product life cycle [62]. Flows of money are also changing along with business models. For example, the pay for-use model has a different cash flow structure to the traditional pay-for-ownership approach, generating a direct impact in the cost structure of a company, and therefore its financing requirements [63].

Even though banks are becoming more interested in circular economy and are already providing services to meet this emerging demand, finance is still considered a critical barrier in the transition toward a circular economy ([63, 55, 15, 56, 64, 65]). The financial sector can be an important enabler for the transition toward a circular economy and although there are several financial offerings that can provide companies with financial opportunities, the sector also needs to mobilize different forms of capital, adopt a different view of existing offerings and develop new competencies and ways of thinking. For instance, a circular business model has a different risk to that of a linear model. Financers may need to reassess the risks of a circular model but also the risks posed by the existing linear model to, for example, include externalities [63]. Therefore, understanding how circular business models differ from traditional business models and what their barriers to financing are is highly relevant for financiers [66].

---

[7]For a critical view on the existing scoring systems to assess firms' performance based on ESG factors, see the remarks by US Security Exchange Commissioner Hester M. Peirce before the American Enterprise Institute, made on June 18, 2019, in Washington, D.C. https://www.sec.gov/news/speech/speech-peirce-061819.

4 Conclusions

The transition to a circular economy implies the adoption of measures that reduce the extraction and use of natural resources and includes ecological design, new business models, an adequate political context, economic instruments, and financing. Economic instruments and financing enable investments in eco-design and the adoption and scaling up of technologies and new business models. They include incentives to free up and reallocate resources that are currently used in the linear model, as well as mobilizing new funding to support the transition toward a circular economy.

Phasing out subsidies on fossil fuels could make significant contributions to move from a linear model into a circular economy model. By artificially reducing consumer prices, subsidies encourage overconsumption of products such as fuels and plastic, and result in increased waste and pollution. They are also regressive and cause severe strains of governments' finances.

A complement to phasing out subsidies is the adoption of policies that address externalities. Several instruments can help to reflect in the prices of goods and services the cost of environmental damage and natural resource depletion caused by the production and consumption of such goods and services. These range from very strict command and control regulations to economic instruments such as pollution charges, emissions trading systems, and deposit-refund systems, to final demand interventions and liability legislation. Adopting these instruments can incentivize individuals and firms to change their consumption and production patterns, leading the way in the transition to a circular economy.

The financial sector can be an important enabler for the transition toward a circular economy. Although there are several financial offerings that can provide companies with financial opportunities, the sector also needs to mobilize different forms of capital, adopt a different view of existing offerings, and develop new competencies and ways of thinking.

Questions

. What are economic instruments?
.. What are subsidies and why are they harmful to the environment?
.. What are tradable permits? Provide one example.
-. How can disclosure requirements improve a firm's environmental performance?

Answers

. Economic instruments are incentives that aim to correct institutional, policy and market failures by incorporating environmental costs into the budgets of households and enterprises so that producers and consumers can respond appropriately.
. Subsidies are deliberate policy actions that maintain consumer prices artificially low. Low prices increase production and consumption which result in higher natural resources extraction, increased pollution and GHG emissions.

3. Tradable permits are market-based instruments where regulators establish an allowable level of pollution which is distributed among firms in the form of permits. These permits can be traded among companies. Companies that manage to keep their emissions below their allocated level can sell their surplus allotment to other firms or use them to cover excess emissions in other parts of their facilities. One of the most well-known cases of tradable permits is the European Union Emission Trading System that is the major carbon market worldwide.

4. Disclosing information on a firm's environmental performance to the public encourages companies to adopt more environmentally friendly production approaches.

References

1. United Nations. (1997). *Glossary of environment statistics.* New York: United Nations.
2. OECD. (2001). *Environmentally related taxes in OECD countries: Issues and strategies.* Paris: OECD.
3. Panayotou, T. (1994). *Economic instruments for environmental management and sustainable development.* Cambridge, MA: Harvard Institute for International Development.
4. Sterner, T. (2003). *Policy instruments for environmental and natural resources management.* Resources for the Future.
5. Enriquez, S., Larsen, B., & Sanchez-Triana, E. (2018). Good practice note 8: Local environmental externalities due to energy price subsidies: A focus on air pollution and health. *Energy subsidy reform assessment framework.* Washington, D.C.: The World Bank.
6. Coady, D., Parry, I., Sears, L., & Shang, B. (2016). How large are global fossil fuel subsidies? *World Development,* 11–27.
7. OECD. (2005). *Environmentally harmful subsidies.* OECD.
8. Coady, D., Parry, I., Le, N.-P., & Shang, B. (2019). *Global fossil fuel subsidies remain large: An update based on country-level estimates.* International Monetary Fund.
9. Muñoz-Piña, C., Rivera Planter, M., & Montes de Oca, M. (2011). *Subsidios a las Gasolinas el Diesel en México: Efectos Ambientales y Políticas Públicas.* Mexico City: Instituto Nacional de Ecología.
10. The World Bank. (2013). *United Mexican States. Reducing fuel subsidies: Public policy options.* Washington, D.C.: The World Bank. Retrieved from https://documents.worldbank.org/curated/en/447861468281079252/pdf/ACS37840P1129500use0only0900A9R29E6.pdf.
11. Plante, M. D., & Jordan, A. (2013). *Getting prices right: Addressing Mexico's history of fuel subsidies.* Dallas: Southwest Economy.
12. Inchauste, G., & Victor, D. G. (2017). *The political economy of energy subsidy reform.* Washington DC: World Bank.
13. Rzeczpospolita, X. D. (2014). Subsidies encourage waste. Washington, D.C.: The World Bank. https://www.worldbank.org/en/news/opinion/2014/04/17/subsidies-encourage-waste.
14. Ebeke, C., & Lonkeng Ngouana, C. (2015). *Energy subsidies and public social spending: Theory and evidence.* Washington, D.C.: International Monetary Fund.
15. Worley, H., Pasquier, S. B., & Canpolat, E. (2018). Good practice note 10: Designing communication campaigns for energy subsidy reform. *Energy subsidy reform assessment framework.* Washington, D.C.: The World Bank.
16. Yemtov, R., Moubarak, A. (2018). Good practice note 5: Assessing the readiness of social safety nets to mitigate the impact of reform. *Energy subsidy reform assessment framework.* Washington, D.C.: The World Bank.

17. Vollebergh, H., & Dijk, J. (2016). *Taxes and fees of regional water authorities in the Netherlands.* Institute for European Environmental Policy.

18. Sánchez-Triana, E., Enriquez, S., & Siegmann, K. (2020). Environmental impact assessment for integrated coastal zone management. In E. Sánchez-Triana, J. Ruitenbeek, S. Enríquez, & K. Siegmann (Eds.), *Opportunities for environmentally healthy, inclusive, and resilient growth in Mexico's Yucatán Peninsula* (pp. 83–98). Washington, D.C.: The World Bank.

19. Giner, F. (2012). Environmental regulation and standards, monitoring, inspection, compliance, and enforcement. In The World Bank (2012). *Getting to green. A sourcebook of pollution management policy tools for growth and competitiveness.* Washington D.C.: The World Bank.

20. Hahn, R., & Stavins, R. (1992). Economic incentives fro environmental protection: Integrating theory and practice. *American Economic Review, 82.*

21. Stavins, R., & Whitehead, B. (1992). Pollution charges for environmental protection: A policy link between energy and environment. *Annual Review of Energy and the Environment, 17.*

22. Porter, M., & Van der Linde, C. (1995). Toward a New Conception of the Environment-Competitiveness Relationship. The Journal of Economic Perspectives, 9(4), 97–118. Retrieved November 18, 2020, http://www.jstor.org/stable/2138392.

23. Want, P., Poe, G., & Wolf, S. (2017). Payments for ecosystem services and wealth distribution. *Ecological Economics, 132,* 63–68.

24. Ambec, S., Cohen, M., Elgie, S., & Lanoie, P. (2011). *The porter hypothesis at 20: Can environmental regulatin enhance innovation and competitiveness?* Washington, D.C.: Resources for the Future. Retrieved from https://media.rff.org/documents/RFF-DP-11-01.pdf.

25. Berman, E., & Bui, L. (2001). Environmental regulation and productivity: Evidence from oil refineries. *Review of Economics and Statistics, 83*(3), 498–510.

26. Lanoie, P., Patry, M., & Lajeunesse, R. (2008). Environmental regulation and productivity: Testing the porter hypothesis. *Journal of Productivity Analysis, 30*(2), 121–128.

27. UNEP and WRI. (2018). *Legal limits on single-use plastics and microplastics: A global review of national laws and regulations.* Nairobi: UNEP.

28. Austin, D. (1999). *Economic instruments for pollution control and prevention.* Washington DC: World Resources Institute.

29. Esty, D. (2017). Red lights to green lights: From 20th century environmental regulation to 21st century sustainability. *Environmental Law, 47*(1), 1–80.

30. Sánchez-Triana, E., Enriquez, S., Afzal, J., Nakagawa, A., & Khan, A. S. (2014). *Cleaning Pakistan's air. Policy options to address the cost of outdoor air pollution.* Directions in Development. Washington, D.C.: The World Bank.

31. The World Bank Group; United Nations Environment Programme; United Nations Industrial Development Organization. (1999). *Pollution prevention and abatement handbook.* Washington, D.C.: The World Bank Group.

32. Agency, S. E. P. (2006). *The Swedish charge on nitrogen oxides.* Stockholm: Naturvårdsverket.

33. European Environment Agency. (2000). *Environmental taxes: Recent developments in tools for integration.* European Environment Agency.

34. Sjöberg, P. (2005). *Taxation of pesticides and fertilizers.* Luleå: Luleå University of Technology.

35. Partnership for Market Readiness. (2017). *Carbon tax guide: A handbook for policy makers.* Washington DC: The World Bank.

36. The World Bank. (2014). What is carbon pricing? Washington DC.

37. Stavins, R. N. (2003). *Handbook of environmental economics.* Washington DC: Elsevier Science.

38. OECD (2016). *Extended producer responsibility: Updated guidance for efficient waste management.* Paris: OECD. https://www.oecd.org/environment/waste/Extended-producer-responsibility-Policy-Highlights-2016-web.pdf.

39. European Commission. (2019). *EU emissions trading system.* Retrieved from European Commission: https://ec.europa.eu/clima/policies/ets_en.

40. Healy, S., Graichen, V., Graichen, J., Nissen, C., Gores, S., & Siemons, A. (2019). *Trends and projections in the EU ETS in 2019.* Boeretang: European Topic Centre on Climate Change Mitigation and Energy.

41. Illes, G., Sanford, B., Hynes, C., Burleigh Sanchez, B., & Maxwell, R. (2020). *Annual reclaim audit report for 2018 compliance year*. Diamond Bar: South Coast Air Quality Management District.

42. SmithBucklin Statistics Group. (2019). *National recycling rate study*. Chicago: Battery Council International.

43. Hilton, M., Sherrington, C., McCarthy A., & Börkey, P. (2019). *Extended producer responsibiltiy (EPR) and the impact of online sales*. OECD Environment Working Papers No. 142. Paris: OECD. https://doi.org/10.1787/cde28569-en.

44. Hardin, G. (1968). The tragedy of the commons. *Science, 162*(3859), 1243–1248.

45. The World Bank. (2012). *Getting to green: A sourcebook of pollution management, policy tools for growth and competitiveness*. Washington DC: The World Bank.

46. Suttie, E., Hill, C., Sandin, G., Kutnar, A., Ganne-Chédeville, C., Lowres, F., & Dias, A. (2017). *Environmental assessment of bio-based building materials*. Woodhead Publishing Limited.

47. United States Environmental Protection Agency. (2019). *About Energy Star 2018*. United States Environmental Protection Agency. Retrieved from Energy Star: https://www.energystar.gov about.

48. Kathuria, V. (2008). *Public disclosures: Using information to reduce pollution in developing countries*. Chennai: Environment Development and Sustainability.

49. Sánchez-Triana, E., & Ortolano, L. (2005). Influence of organizational learning on water pollution control in Colombia's Cauca Valley. *Water Resources Development, 21*, 493–508. https:// doi.org/10.1080/07900620500139168.

50. Titia-Bové, A., & Swartz, S. (2016). Starting at the source: Sustainability in supply chains. *Sustainability & resource productivity*. Mckinsey & Company. https://www.mckinsey.com business-functions/sustainability/our-insights/starting-at-the-source-sustainability-in-supply-chains.

51. Jones, C. A., Pendergrass, J., Broderick, J., & Phelps, J. (2015). Tropical conservation and liability for environmental harm. *Environmental Law Reporter, 45*, 11032–11050.

52. Shavell, S. (2013). A fundamental enforcement cost advantage of the negligence rule over regulation. *The Journal of Legal Studies, 42*, 275–302.

53. Ellen Macarthur Foundation. (2013). *Towards the circular economy: Economic and business rationale for an accelerated transition*. https://www.ellenmacarthurfoundation.org/assets/dow nloads/publications/Ellen-MacArthur-Foundation-Towards-the-Circular-Economy-vol.1.pdf.

54. Lacy, P., & Rutqvist, J. (2015). *Waste to wealth*. Palgrave Macmillan.

55. Fredell, L.-L. (2019). Financing the circular economy is a new opportunity for banks. *Global Banking and Finance Review*.

56. World Business Council for Sustainable Development. (2018). *Circular economy practitioner guide*. Retrieved from https://www.ceguide.org/Strategies-and-examples/Finance/Green-bonds.

57. Fatin, L. (2020). *Green bond highlights 2019: Behind the headline numbers Climate bonds market analysis of a record year*. Retrieved from Initiative Climate Bonds: https://www.climatebonds.net/2020/02/green-bond-highlights-2019-behind-headline numbers-climate-bonds-market-analysis-record-year.

58. Global Innovation Lab for Climate Finance. (2019). *Breathe better bond initiative*. Retrieved from The Lab: https://www.climatefinancelab.org/wp-content/uploads/2019/03/Breathe-Better-Bond_Instrument-overview.pdf.

59. SIFMA. (2019). *Capital markets fact book: 2019*. SIFMA.

60. Global Sustainable Investment Alliance. (2019). *2018 global sustainable investment review*. Global Sustainable Investment Alliance.

61. Giese, G., Lee, L.-E., Melas, D., Nagy, Z., & Nishikawa, L. (2019). Foundations of ESG investing: How ESG affects equity valuation, risk, and performance. *The Journal of Portfolio Management, 45*(5), 1–15.

62. ING Economics Department. (2015). *Rethinking finance in a circular economy*. ING.

63. FinanCE Working Group. (2016). *Money makes the world go round*.

4. Fischer, A., & Achterberg, E. (2016). Create a Financeable Circular Business in 10 Steps. *Circle Economy and Nederland Circulair!*. https://circulareconomy.europa.eu/platform/en/knowledge/create-financeable-circular-business-10-steps.

5. van Eijk, F. (2015). Barriers and Drivers towards a Circular Economy: Literature Review. A-140315-R-Final. The Netherlands: Acceleratio B.V. https://circulareconomy.europa.eu/platform/sites/default/files/e00e8643951aef8adde612123e824493.pdf.

6. Achterberg, E., & van Tilburg, R. (2016). *6 guidelines to empower financial decision-making in the circular economy*. Nederland Circulair.

7. The World Bank. (2019). *State and trends of carbon pricing 2019*. Washington DC.

8. International Monetary Fund. (2013). *Energy Subsidy Reform: Lessons and Implications*. Washington, D.C.: IMF.

Santiago Enriquez is an international consultant with 20 years of experience in the design, implementation, and evaluation of policies relating to the environment, climate change, and clean energy. He has developed analytical work for the World Bank, United States Agency for International Development, and the Inter-American Development Bank on topics that include mainstreaming of environmental and climate change considerations in key economic sectors, institutional and organizational analyses to strengthen environmental management, and policy-based strategic environmental assessments. From 1998 to 2002, he worked at the International Affairs Unit of Mexico's Ministry of Environment and Natural Resources. Enriquez holds a master's degree in public policy from the Harvard Kennedy School.

Ernesto Sánchez-Triana is the Global Lead for Environmental Health and Pollution Management for the World Bank. He has worked on projects in numerous countries, including Afghanistan, Argentina, Bangladesh, Bhutan, Bolivia, Brazil, Ecuador, India, the Lao People's Democratic Republic, Mexico, Pakistan, Panama, Paraguay, and Peru. Before joining the World Bank, he served as Director of Environmental Policy at Colombia's National Department of Planning and as President of the Board of Directors of the Cundinamarca Environmental Protection Agency. Sanchez-Triana has led the preparation of numerous policy-based programs, investment projects, technical assistance operations, and analytical works. He holds an engineering degree from Universidad de Los Andes (Colombia), and two master's of science degrees and a Ph.D. from Stanford University.

Mayra Gabriela Guerra López is an engineer with experi
ence in international development and the private sector. She
currently works at the International Finance Corporation as an
Associate Environmental and Social Development Specialist for
the Latin America and the Caribbean region. Before joining
the IFC, she served at the World Bank as an Environmental
Specialist, focusing on projects related to land-based pollution
water pollution control, air quality management and circular
economy. Previously, Mayra worked as an environmental engi
neer for the multinational Oil & Gas company Total. She holds a
master's degree in engineering and project development from the
French Petroleum Institute in France and has worked on projects
in several countries, including Bolivia, France, Lao PDR and
Mexico.

Life Cycle Greenhouse Gas Emissions for Circular Economy

Thumrongrut Mungcharoen, Viganda Varabuntoonvit, and Nongnuch Poolsawad

Abstract Greenhouse gas emissions (GHG) during the life cycle of a product, a process, or a service are the major cause of global warming and climate change. The circular economy principle has been proven to help increase resource efficiency and reduce GHG emissions. From the life cycle concept, it is known that closing loops or circularity doesn't always generate positive environmental consequences. To ensure that circular activities are actually environmentally beneficial, the calculation of life cycle GHG emissions reduction is essential. The calculation methodology based on Intergovernmental Panel on Climate Change (IPCC) guidelines and life cycle assessment standards is described in the chapter. Case studies on opt out of plastic cutlery and upcycled fashion footwear, both from the Asia- Pacific region, are provided to illustrate the detailed calculation procedure and the GHG emissions reduction from the actual circular economy practices.

Keywords Circular economy · Life cycle greenhouse gas (GHG) emissions · GHG emissions reduction · GHG emission factor · Upcycling

Learning Objectives

Understand the circular economy concept and its relation to the GHG emissions
Be able to identify the sources of GHG emissions from human activities based on IPCC guidelines
• Understand life cycle thinking and activities related to GHG emissions

. Mungcharoen (✉) · N. Poolsawad
National Science and Technology Development Agency, 111 Thailand Science Park, Phahonyothin Road, Khlong Nueng, Pathumthani 12120, Thailand
e-mail: thumrongrut@nstda.or.th

N. Poolsawad
e-mail: nongnucp@mtec.or.th

. Mungcharoen · V. Varabuntoonvit
Department of Chemical Engineering, Kasetsart University, 50 Ngamwongwan Road, Ladyao, Chatuchak, Bangkok 10900, Thailand
e-mail: fengvgv@ku.th

© Springer Nature Singapore Pte Ltd. 2021
. Liu and S. Ramakrishna (eds.), *An Introduction to Circular Economy*,
https://doi.org/10.1007/978-981-15-8510-4_24

- Understand the emission factor and the GHG emissions due to the circular economy activities
- Be able to assess the life cycle GHG emissions reduction and circularity in economy activities.

1 Introduction: Circular Economy and GHG Emissions

The circular economy aims to balance economic development with environmental and resource protection through closing the loop by maximizing the use of renewable resources, recirculating of resources and products, and also designing waste out of the economic system. The transition to a circular economy, in principle, can reduce GHG emissions. For example, the design for disassembly, modularity, repairability, flexibility, and biodegradability supports the activities of reuse, remanufacturing, refurbishment or regeneration, and resulting in waste minimization. The use of recirculating materials in the system can reduce GHG emissions from avoiding the virgin material production and the end-of-life treatment, such as landfill or incineration. The use of renewable resources, such as agriculture products, can not only reduce GHG emissions, but also store carbon into soils and plants. So, the circular economy activities can contribute significantly to the GHG emissions reduction, support the global action to combat climate change, and impact the sustainable development goals (especially SDG 13).

2 GHG Emissions from Human Activities—Based on IPCC Guidelines

According to the Intergovernmental Panel on Climate Change (IPCC), the GHG emissions from human activities are categorized into 4 sectors which are Energy; Industrial Processes and Product Use (IPPU); Agriculture, Forestry and Other Land Use (AFOLU); and Waste. The calculation methodology for all direct GHG emissions based on the 2006, IPCC Guidelines for National Greenhouse Gas Inventories (see also 2013 revised supplementary methods and 2019 refinement to the current 2006 IPCC guidelines) with the tier 1 emission factor (EF) for each sector can be described as follows.

2.1 Energy

The Energy sector is by far the biggest source of the world GHG emissions which mainly comes from fossil fuel combustion. From the combustion stoichiometry, the hydrocarbon converts to carbon dioxide (CO_2) and water (H_2O) for complete combustion. In the real situation, it is always incomplete combustion, resulting in the additional emission of other GHGs such as methane (CH_4) and nitrous oxide N_2O). GHG emissions are mostly from the stationary combustion, mainly from energy industries including power plant and refineries. The mobile combustion from road and other modes of transportation is accounted as the second share of GHG emissions from energy sector.

The GHG emissions from energy sector can be calculated as shown in Eq. (1). The tier 1 IPCC default emission factor can be used, if there is no specific emission factor based on country fuel characteristic or combustion technology). The emissions of all GHGs shall be quantified together in term of carbon dioxide equivalent (CO_2eq) using the Global warming potential (GWP) value of each GHG.

$$Emissions_{GHG,fuel} = FuelConsumption_{fuel} \times EF_{GHG,fuel} \qquad (1)$$

where: $Emissions_{GHG,fuel}$ = emissions of a given GHG by type of fuel (kgGHG).
$FuelConsumption_{fuel}$ = heat of fuel combusted (TJ)
$EF_{GHG,fuel}$ = emission factor of a given GHG by type of fuel (kgGHG/TJ).

2.2 Industrial Processes and Product Use (IPPU)

The IPPU sector covers the GHG emissions resulting from various industrial activities that produce emissions, not directly the result of energy consumed during the process, and the use of GHGs in products [1], such as refrigerators, foams or aerosol cans.

The process emissions link directly to the production process reactions in chemical process to polymerize or change form of plastic, the calcination reaction in cement production, and the process emissions from glass, ammonia, iron, steel, and aluminum production.

The general methodology to estimate GHG emissions associated with each industrial process involves the product of activity data such as the amount of material produced or consumed, and an emission factor (mostly tier 1 IPCC default emission factor) per unit of production according to the Eq. (2). The emission factor is related to the product and technology of production process. The emissions for each product may come from several GHGs. Therefore, the specific GWP value of each GHG should be used to calculate the amount of GHG emissions in term of CO2eq.

$$E_{GHG,non-energy} = Amount of Production \times EF_{production-based} \qquad (2)$$

where: $E_{GHG,non-energy}$ = emissions of a given GHG from IPPU (kgGHG)
$EF_{production-based}$ = emission factor of a given GHG for product (kgGHG/unit of product).

2.3 Agriculture, Forestry, and Other Land Use (AFOLU)

The activities in agriculture, forestry, and other land use sector are mostly from the sources and removals by sinks from managed lands. The GHG emissions are calculated based on the accounting of biomass, dead organic matter, and soil carbon stock change in land use and the biomass burning. The non-CO_2 emissions from soil management and biomass burning, and livestock population and manure management are also included. The carbon stock change is estimated from land area (e.g., climate zone, soil type, management regime, etc.) including stock change of above-ground biomass, below-ground biomass, deadwood, litter, soils, and harvested wood products. The default emission factors for each type of area; i.e., forest land, cropland, grassland, wetland, settlement, and other lands; can be found in the IPCC guideline

Other non-CO_2 emissions from variety of sources are generally determined by the appropriate specific emission factor related to the activity sources such as area (e.g., for soil or area burnt), population (e.g., for livestock) or mass (e.g., for biomass or manure).

2.4 Waste

The waste sector covers the GHG emissions from solid waste disposal, solid waste treatment (e.g., biological treatment, incineration, open burning, and landfill), and the wastewater treatment and discharge. The GHG emissions from waste can be determined from the emission factor depending on the carbon content of waste and waste disposal method. The wastewater GHG emissions depend on BOD/COD concentration and wastewater treatment method.

3 Life Cycle GHG Emissions of the Circular Economy Activities

The life cycle thinking is a systematic framework that considers all the stages of a product life cycle from "Cradle to Grave". Life cycle assessment is a tool which supports the integration of sustainability to avoid shifting the environmental burden from one stage to another (e.g., from production to consumption). Activities related to circular economy are the effort to re-circulating resources in the life cycle stage. The circular activities aim to minimize waste and also avoid extraction and processing of the virgin resources, along the life cycle stages as shown in Fig. 1. Even though the circular economy activities can reduce GHG emissions from waste disposal and virgin resources extraction (called embedded GHG emissions), these activities also emit GHG due to energy and others activities involved. Hence, quantification of the

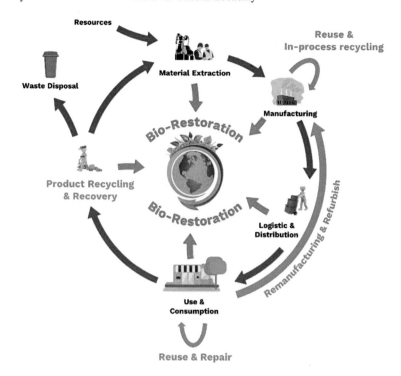

Fig. 1 Life cycle of the circular economy activities

life cycle GHG emissions is very important to justify the benefits of the circular economy activities.

Calculation of GHG Emissions Reduction from Circular Economy Activities

The GHG emissions reduction from circular economy activities are evaluated comparing to the baseline or typical linear economy activities using the life cycle consideration (Fig. 2). Direct GHG emissions from each stage of the life cycle should be included. To cover all stages of the life cycle, the background data can be used according to the life cycle assessment methodology. The GHG emission factors (embedded GHG emission factors of resources, processes, and products) from the life cycle inventory database are needed for calculation.

To compare the GHG emissions from two or more product or process systems, the functional unit has to be the same for all the compared systems. The functional unit should describe and quantify at least 3 product properties, i.e., quantity, quality, and durability. Therefore, the product with extended service life should have less direct

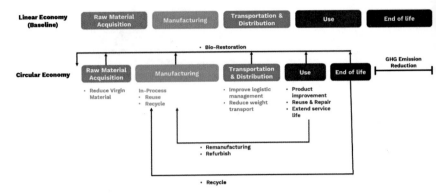

Fig. 2 Linear economy and circular economy life cycle GHG emissions

environmental impact than the original short-life product (with the same quantity and quality) considering the functional unit of durability such as "one year of use".

The GHG emissions reduction can be evaluated using the methodology of life cycle assessment with 4 main steps; goal and scope definition, life cycle inventory analysis (LCI), life cycle impact assessment (LCIA), and interpretations (*see chapter* "Life Cycle Thinking in a Circular Economy" *topic 5 for more details*).

Goal and scope definition: The step is to set the goal of the evaluation; e.g. evaluation GHG emissions reduction throughout the life cycle, to compare GHG emissions of the original product/process with the new product/process using circular economy concept. The scope has to be defined in this step, such as system boundary, functional unit, reference flow, data requirement, allocation procedure, assumption, limitation, impact method, etc.

Life cycle inventory analysis: This step is to collect related inventory data (input and output) from each process in the defined life cycle stages. The primary data are usually collected from the main process, while the secondary data can be used for the other life cycle stages. The direct GHG emissions can be calculated using IPCC methodology described in Sect. 2. Some parts of the circular activities, e.g., recycling, refurbishing, remanufacturing may emit GHGs due to some additional processes that consumed resources. These emissions have to be included in the inventory.

The use of biological resources are encouraged according to circular economy concept. The carbon dioxide (CO_2) emissions from plantation, harvest, digestion, fermentation, processing, and combustion of biological resources are considered as biogenic carbon which would eventually return to the environment from its sequestration. According to the biogenic carbon concept, for bio-resources, the emission of CO_2 are neglect, but the emissions of other GHGs, such as methane (CH4) and nitrous oxide (N2O), should be calculated and accounted the same as others direct GHG emissions.

Life cycle impact assessment: This step is focused on the Climate Change impact based on IPCC methodology. The GWP values are reported in the IPCC Assessment Report (AR). Table 1 shows the GWP values of selected GHGs from different

Table 1 The GWP values from IPCC Assessment Report	Greenhouse gas	GWP (100 years model), CO_2eq for each gas		
		AR2	AR3	AR4
	CO_2	1	1	1
	CH_4	21	23	25
	N_2O	310	296	298
	SF6	23,900	22,200	22,800
	HFCs	124–14,800		
	PFCs	7,390–12,200		

Assessment Reports. So, it is very important to specify which models and which ARs are used in GHG emissions calculation.

The GHG emissions reduction (or avoided GHG) can be evaluated in terms of CO_2eq by subtracting the total GHG emissions of circular economy project from the total GHG emissions of baseline or original scenario, as shown in Eq. (3).

$$AvoidedGHG = GHGemissions_{baseline} - GHGemissions_{CE} \qquad (3)$$

where: $GHGemissions_{baseline}$ = life cycle GHG emissions of the baseline or original scenario or linear economy (kgCO_2eq)

$GHGemissions_{CE}$ = life cycle GHG emissions of circular economy project or after implementation of circular activities (kgCO_2eq)

Interpretation: This step is to communicate the results on GHG emissions reduction together with the economic benefit, and also to identify the hotspots and suggestions for further improvements in the context of circular economy activities and GHG mitigation actions.

5 Case Study I: Opt Out of Plastic Cutlery—Company A

Company A is a global brand food delivery service in 10 countries throughout the Asia-Pacific. In Thailand, company's customers are between 25 and 50 years old, representing the working-age population with hustling lifestyle. There are more than ,000 partner restaurants spreading across 8 major cities in Thailand. With over 00,000 orders per month, almost all of the cutlery sent out is plastic.

It is realized that most people order food delivery when at home or at the office where they have access to durable cutlery. So, for these customers, there are no need for plastic cutlery sets which will then turn to plastic wastes and result in unnecessary (and avoidable) environmental costs. Company's initiative to reduce plastic consumption and also plastic waste is to build an opt out system which customers can decide to reject plastic cutlery (Fig. 3).

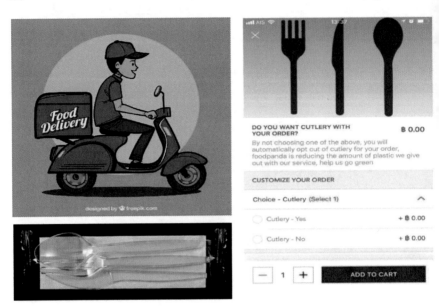

Fig. 3 Company A: Opt out of plastic cutlery

This circular economy strategy would help businesses procure fewer plastic cutlery sets to meet the reduced demand and also encourage consumers to carefully consider their plastic needs to reduce resource consumption and waste.

Calculation of GHG Emissions Reduction

The project is aimed to reduce plastic consumption (conserve resource) and also plastic waste by Opt out of plastic cutlery platform. Company A would like to evaluate the GHG emissions (or carbon footprint: CF) reduction from this initiative comparing to the conventional platform of giving plastic cutlery by default to all delivery orders. Figures 4 and 5 show the system boundary of each platform. The functional unit is defined as one month period of food delivery.

It is assumed that the single use plastic cutlery and the sachet are made from polypropylene (PP) which will end up in landfill. The customer who opts out of plastic cutlery would use durable cutlery and then clean with water and washing agent after use. Diesel is used for transportation. Assuming that the amount of diesel used for transportation of PP cutlery to customers are the same for both platforms.

The primary input and output data were collected during the first trial one month of the opt out platform (in 2018), i.e.;

- Amount of PP cutlery set delivered to customer and went to landfill; 230,000 sets for conventional platform and 226,320 sets for the opt out platform
- Average weight of PP cutlery set; 7 grams pet set
- Washing of durable cutlery set (including wastewater treatment); use 5 grams of detergent and 0.002 m³ of water per set.

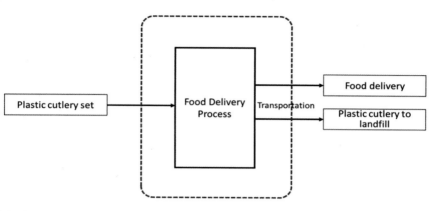

Fig. 4 System boundary of the conventional platform

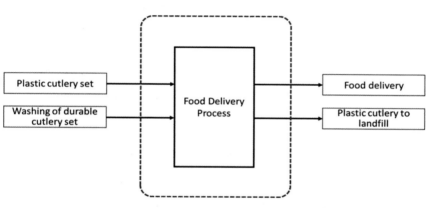

Fig. 5 System boundary of the opt out plastic cutlery platform

The secondary data and the simplified GHG (CO_2eq) emission factors for materials and processes relevant to both platforms are taken from the Thai national LCI database and other references, i.e.;

- PP cutlery set (cradle-to-gate data, calculated from raw material extraction to production)
- Transportation of PP cutlery set as the waste to landfill site (EF, in $kgCO_2$eq/tkm, based on a mean distance of 40 km using 16-ton truck and 100% effective load-carrying capacity) [6]. [Note: tkm or ton-kilometer is a unit represents the transport of 1 ton of goods over a distance of 1 km]
  Landfill (waste management) of PP waste

The results of "cradle-to-grave" GHG emissions from both platforms can be calculated by multiplying the material or activity data and their associated GHG emission factors (Tables 2 and 3).

Table 2 GHG emission from using the plastic cutlery of food delivery in one month

Materials or processes	Amount (per month)	Unit	EF (kgCO$_2$eq/unit)	GHG emissions (kgCO$_2$eq/month)
PP plastic cutlery set (pack of spoon, fork and knife)	230,000	set	0.0226[a]	5,198.0000
Diesel for transportation of PP cutlery waste	64.40[b]	tkm	0.0472[a]	3.0397
Landfill of PP waste (230,000*0.007)	1,610	kg	0.1590[c]	255.9900
Total GHG per month				**5,457.0297**

[a]*Thai national LCI database* [5]; [b]*TGO* [6]; [c]*Shonfield* [3]

Table 3 GHG emission from opt out of plastic cutlery of food delivery in one month

Materials or processes	Amount (per month)	Unit	EF (kgCO$_2$eq/unit)	GHG emissions (kgCO$_2$eq/month)
PP plastic cutlery set (pack of spoon, fork and knife)	226,320	set	0.0226	5,114.8320
Washing of durable cutlery set	3,680	set	0.0002[a]	0.7360
Diesel for transportation of PP cutlery waste	63.37	tkm	0.0472	2.9911
Landfill of PP waste (226,320*0.007)	1,584	kg	0.1590	251.8560
Total GHG per month				**5,370.4151**

[a]*EF of washing process of durable cutlery set (calculated from Thai national LCI database)*

For the opt out of plastic cutlery strategy, refer to Eq. 3, the GHG emissions (pe functional unit) can be reduced by 86.6 kgCO$_2$eq per month or approximately 1,03 kgCO$_2$eq per year, which translates into a 1.6% reduction in the demand for plasti cutlery compared to the baseline. This reduction may be quite small due to a ver new approach of the food delivery sector and not acquaintance to the customers, a well as the lack of through public relations during this trial period.

However, company A has a vision to change the default of giving plastic cutlery t without cutlery unless customers choose to request cutlery during the food ordering With this new plan, company A could offer a special discount or promotion t customers who choose to opt out of plastic cutlery. With the aggressive plan an action, it is expected that company A would receive opt out plastic cutlery order more than 80% of the total orders and could reduce the GHG emissions more tha 50,000 kgCO$_2$eq per year. Several partner restaurants of company A currently go b eco-friendly packaging (e.g., biodegradable boxes).

6   Case Study II: Upcycled Fashion Footwear

Company B is a fashion-based social enterprise with circular economy concept in the Philippines. Realizing that fashion industry is the second largest polluter in the world, one of the company missions is to reduce textile wastes. Statistics show that clothing production accounts for 10% of global GHG emissions. Instead of throwing textile wastes into landfills, the company collects and turns them into higher value goods such as footwear, fashion accessories, and lifestyle pieces (Fig. 6). The company also tries to raise awareness among customers on choosing sustainable fashion products to help reduce the GHG emissions. On the social aspect, the company has the policy to cooperate with disabilities, unemployed footwear craftsmen, and local designers. Rather than calling its business "recycling", which generally means a down-cycling of value, the company prefers "upcycling"-more valuable products than the original waste from the textiles.

Calculation of GHG Emissions Reduction

Presently, virgin materials extraction, resources consumption, and excessive wastes generation are critical issues in the linear economy. Company B aims to tackle these issues by replacing virgin textile and substituting with recycled textile and fabric waste to produce new fashion footwear (called "upcycled footwear"). It is required to evaluate the GHG emissions reduction from this circular activity, i.e., producing upcycled footwear comparing to conventional footwear. Figures 7 and 8 show the system boundary of each footwear production. The functional unit is defined as 100 pairs of footwear produced in one month.

To simplify the calculation, it is assumed that (1) the weight of virgin textile and fabric waste to produce a footwear are the same; (2) the GHG emissions from the transportation to the production site of fabric waste and virgin textile are the same; (3) the energy use during the production process of both types of footwear are the same (4) the GHG emission factors from the Thai national database will be used if necessary; (5) there are no significant difference in properties between the conventional and the upcycled footwear.

The primary data provided by company B is the amount of virgin textile to produce footwear, i.e., 10 m^2 of lightweight cotton per 100 pairs of footwear (in 2018).

Fig. 6   Company B: upcycled products

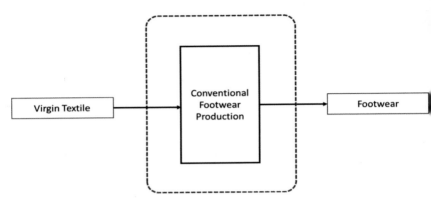

Fig. 7 System boundary of the conventional footwear production

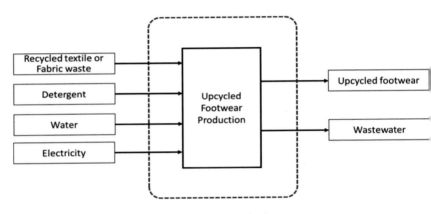

Fig. 8 System boundary of the upcycled footwear production

The secondary data and the simplified GHG (CO_2eq) emission factors for materials and processes relevant to both cases are taken from the Thai national LC database and other references, i.e.;

- Area density of lightweight cotton; 135.62 g/m^2 [4]
- Washing 1 kg of fabric waste; use 0.009 kg detergent, 0.01176 m^3 water, 0.14 kWh electricity, and emit 11.765 L (L) wastewater (*ref.* 2)
- Ironing 1 kg of washed fabric waste; use 0.27 kWh electricity (*ref. EGAT Label No. 5*).

The results of "cradle-to-grave" GHG emissions of both cases can be calculated by multiplying the material or activity data and their associated GHG emission factor (Table 4).

For this upcycling activity of company B, referring to Eq. 3, the GHG emission (per functional unit) can be reduced by 17.5 kgCO_2eq per month or approximately 210 kgCO_2eq per year compared to the baseline. It should be emphasized that with

Table 4 GHG emissions from producing of conventional and upcycled footwear (100 pairs each)

Materials or Processes	Amount (per 100 pairs)	Unit	EF (kgCO$_2$eq/unit)	GHG emissions (kgCO$_2$eq/100 pairs)
Conventional footwear				
Virgin textile (Lightweight cotton 10 m^2 per 100 pairs of footwear)	1.3560[a]	kg	13.1705[b]	17.8592
Total GHG per 100 pairs of footwear				**17.8592**
Company B (upcycled) footwear				
Fabric waste	1.3560	kg	0	0
Fabric cleaning (including washing and ironing)	1.3560	kg		
Detergent	0.0122	kg	0.1454[b]	0.0018
Water	0.0159	m^3	0.8006[b]	0.0127
Electricity (for both washing and ironing)	0.5574	kWh	0.6933[b]	0.3864
Wastewater	15.9557	L	0.0001[b]	0.0016
Total GHG per 100 pairs of footwear				**0.4025**

Light weight cotton 135.62 g/m^2 [4]; [b]Thai national LCI database [5]

this circular activity of company B, the virgin textile used is reduced by 100% and he GHG emissions reduction is approximately 0.18 kgCO$_2$eq per a pair of footwear produced.

Besides upcycled fashion footwear, Company B produces other higher valued fashion accessories and lifestyle pieces from textile wastes using creative design circular economy concept. With the global movement toward sustainable lifestyle, sustainable fashion has gone mainstream with big and small brands producing more upcycled products. The pollutants, especially the GHG emissions from the fashion industry would decrease significantly in the future.

Questions

. What are the main sources of GHG emissions from human activities according to the IPCC guidelines?

. Is circular economy a synonym of "circularity"? Does a circular activity always generate positive environmental impacts? List two examples of circular activities which generate negative impact on the environment.

. How does a circular economy respond to greenhouse gas emissions?

. What is a GHG emission factor? How can you get the GHG emission factors for materials or processes? How are the GHG emissions calculated?

. To calculate the GHG emissions reduction from the circular economy activities comparing to the baseline, what do you need to do to ensure the fair comparison?

6. What is the difference between recycled product and upcycled product?

Further Reading

1. Deloitte Sustainability. (2016). Circular economy potential for climate change mitigation. *Electricity generating authority of Thailand (EGAT)*, Label No. 5 Retrieved April 30, 2020, from http://labelno5.egat.co.th/new58/?page_id=1506.
2. Ellen MacArthur Foundation. (2019). *Completing the picture: how the circular economy tackles climate change.*
3. Inaba, A. et.al. (2016). Carbon footprint of products. Chapter 2 "Special types of life cycle assessment". In M. Finkbeiner (Ed.), *LCA compendium*. Springer.
4. IPCC. (2014). *2013 Revised supplementary methods and good practice guidance arising from the Kyoto Protocol.*
5. IPCC. (2019). *2019 Refinement to the 2006 IPCC guidelines for national greenhouse gas inventories.*
6. Keidanren (Japan Business Federation). (2018). *Contributing to avoided emissions through the global value chain—A new approach to climate change measures by private actors.*
7. Mungcharoen, T., & Olarnrithinun, S. (2014). Life cycle inventory database and its applications to support public policy. In: *Proceedings EcoBalance 2014*, Tsukuba, Japan.
8. World Business Council of Sustainable Development (WBCSD). (2019). *CEO guide to the circular bioeconomy.*
9. World Resources Institute. (2019). *Working paper—Estimating and reporting the comparative emissions impacts of products.*

Acknowledgements The authors would like to thank company A, company B, and the UN Environment Programme (Asia-Pacific Low-Carbon Lifestyles Challenge project) for the information related to the two case studies and also thank Ms. Ruethai Onbhuddha for the support on data collection.

References

1. IPCC. (2006). *2006 IPCC guidelines for national greenhouse gas inventories.*
2. LG Electronics (Thailand) Co., Ltd. LG WF-T9076TD *User manual*. Retrieved April 30, 2020 from https://manualsbrain.com/en/manuals/1522905/.
3. Shonfield, P. (2008). LCA of management options for mixed waste plastics. *WRAP material change for a better environment*. ISBN: 1-84405-397-0.
4. Tash. (2015). *Understanding fabric weight in order to choose the right fabric*, Nov 2015 Retrieved Paril 30, 2020, from https://blog.fabricuk.com/understanding-fabric-weight/.
5. Technology and Informatics Institute for Sustainability, National Science and Technology Development Agency. (2019). *Thai national LCI database*, Oct 2019. Retrieved April 30, 2020, from http://thaicarbonlabel.tgo.or.th/admin/uploadfiles/emission/ts_f2e7bb377d.pdf.

6. Thailand Greenhouse Gas Management Organization (TGO). (2018). *Guidelines for the quantification of the carbon footprint of products*, August 2018. Retrieved May 2, 2020, from http://thaicarbonlabel.tgo.or.th/admin/uploadfiles/download/ts_cececc6f1e.pdf.

Thumrongrut Mungcharoen is a Sustainable Development Advisor at the National Science and Technology Development Agency and an associate professor at Kasetsart University. He is also a president of the foundation for Asia Pacific Roundtable on Sustainable Consumption and Production (APRSCP). For more than 25 years, he has involved as an expert in several projects on cleaner production, 3Rs, life cycle assessment, eco-design, and sustainable consumption & production (SCP) for several local and international organizations. He is among the key persons who have started the Thai National Life Cycle Inventory Database project since 2005, Thai Carbon Footprint Label project since 2009, Science Technology & Innovation for SDGs initiatives since 2016, and Thai SCP Network since 2018. He has authored more than 220 scientific publications in books, papers, and conferences. Recently, he has been appointed chairman of the subcommittee on Circular Economy at Program Management Unit for Competitiveness (PMU-C) under the Ministry of Higher Education, Science, Research, and Innovation.

Dr. Thumrongrut obtained his Ph.D. in Chemical Engineering from the University of Texas at Austin, USA, and Bachelor's degree with honors from Chulalongkorn University, Thailand.

Viganda Varabuntoonvit is an assistant professor at Chemical Engineering Department, Faculty of Engineering, Kasetsart University, Thailand. Her areas of specialization and researches include life cycle assessment, energy efficiency, energy and carbon intensity, energy management system, carbon footprint, water footprint, and SDGs. She obtained her doctoral, master's, and bachelor's degrees from Kasetsart University. During the study for master degree, she received a student exchange scholarship from the Association of International Education, Japan (AIEJ), for studying at Mie University.

In addition to the teaching career at the university, Dr. Viganda is also a consultant and auditor for several leading organizations, mostly in the energy and petrochemical sectors. Her interests and researches have led her to be a pioneering contributor to the development of Thai National Life Cycle Inventory Database of electricity, petroleum, and petrochemical sectors.

Dr. Viganda used to be a Climate Change Working Committee member of the Ministry of Energy. She is currently a Carbon Footprint Technical Committee of Thailand Greenhouse Gas Management Organization (Public Organization) and Eco-Factory and Water Footprint Technical Committee of the Federation of Thai Industries. Recently, she has been appointed to sit in the Eco-Efficiency Technical Subcommittee of the State Enterprise Policy Office, Ministry of Finance.

Nongnuch Poolsawad has joined the Technology and Infor matics Institute for Sustainability (TIIS), National Science an Technology Development Agency (NSTDA) since 2019, as researcher. From 2014–2019, she was a researcher at the Life Cycle Assessment (LCA) Laboratory, Environment Research Group under the National Metal and Materials Technolog Center (MTEC), NSTDA. As a member of the Intelligen Systems (IS) research group focusing on data mining, she i particularly interested in the predictive modeling and comple data handling. Her research interests include the multidisci plinary approach of environment and sustainability, mainly o recording, processing, classifying, and prediction using compu tational and statistical techniques with related software. Dat analytics is the major area of her study and research whic focuses on the development of effective model and selectio method to improve the prediction accuracy based on understand ings of environmental data including resource use, emission profile, consumption pattern, and lifestyle behavior.

Dr. Nongnuch obtained her Ph.D. in Computer Science a Faculty of Science and Engineering from the University of Hul UK.

Life Cycle Costing: Methodology and Applications in a Circular Economy

Piya Kerdlap and Simone Cornago

Abstract Life cycle thinking is important for holistically measuring the environmental and economic costs and benefits of activities in a circular economy. This helps to avoid shifting problems among different stakeholders and different stages of the life cycle. Through methods such as life cycle assessment (LCA) and life cycle costing (LCC), the environmental and monetary flows of a product or service that passes through different life cycle stages and stakeholders can be mapped. This chapter focuses on LCC, a method for quantifying the economic performance of products, services, and other activities in a circular economy. We provide a brief introduction on the three types of LCC which are conventional LCC, environmental LCC, and societal LCC and their respective relevant stakeholders. This chapter mainly focuses on conventional LCC as its methodology can be applied directly in economic decision-making for consumers and businesses. The principles of LCC are explained and a step-by-step procedure is provided on how to conduct a conventional LCC of a product or service. We provide two case studies of a conventional LCC to help understand how the LCC methodology explained in this chapter is applied. Finally, this chapter discusses the relationship between LCC and LCA.

Keywords Financial analysis · Cost–benefit analysis · Economic evaluation · Business planning

Learning Objectives

Understand the principles of life cycle costing,
Understand the differences between conventional LCC, environmental LCC, and societal LCC,
Identify and calculate the total revenue, costs, and value added of a product or service,
Calculate the present and future value of revenue and cost,

P. Kerdlap (✉) · S. Cornago
National University of Singapore, Singapore, Singapore
email: piyakerdlap@u.nus.edu

Singapore Institute of Manufacturing Technology, Singapore, Singapore

© Springer Nature Singapore Pte Ltd. 2021
L. Liu and S. Ramakrishna (eds.), *An Introduction to Circular Economy*,
https://doi.org/10.1007/978-981-15-8510-4_25

- Calculate the net present value,
- Understand the different stakeholder perspectives of LCC,
- Understand the differences and similarities between LCC and LCA,
- Be able to apply LCC in an economic analysis of a product or service or an activity in a circular economy.

1 Introduction

Life cycle thinking is crucial for quantifying the sustainability of circular economy activities. Some examples of activities that take place in a circular economy are product reuse, repair, refurbishment, remanufacturing, recycling, and product-as-a service. Tools such as life cycle assessment (LCA) and life cycle costing (LCC) allow us to measure the environmental and economic performance of such activities in a circular economy. LCA measures the impacts on the environment of an entire product or service system across its life cycle in a circular economy. LCC measures the flows of money associated with a product or service across its life cycle. A life cycle is composed of the interlinked processes involved in the product or service system under study. It is typically composed of the upstream processes such as raw material extraction and processing, the production (core processes), the distribution, the use, and the end-of-life. In a circular economy, wastes can be reused or recycled across different product systems that become linked together. This means that LCA and LCC should carefully consider a wider system to avoid unaccounted environmental or economic burden shifting across connected life cycles. In open-loop recycling, a waste produced by one product system is used as a resource in a different product system. In closed-loop recycling, a waste produced in one product system is used as a resource in a previous step within the same product system. Mapping the environmental and monetary flows across different life cycle stages in different or the same product system is therefore important in understanding the costs and benefits of activities in a circular economy.

The concept of the circular economy has been widely promoted due to the potential environmental benefits it offers. Although the environmental dimension of circular economy activities is important, measuring the economic feasibility of such activities is necessary. Businesses and consumers will ultimately want to know if changing the product they buy or modifying their individual lifestyle to become more circular is financially beneficial to them. LCC enables measurement of the economic costs and benefits of circular economy activities across many different actors across one or more product system. This chapter introduces the principles of LCC and provides the steps on how to conduct the analysis.

2   Life Cycle Costing Methodology

LCC is a method for evaluating the economic dimension of sustainability of a product or service. In LCC, the monetary costs of a product or service across its entire life cycle are calculated. LCC is versatile in that it can be used for a wide range of purposes and projects at different stages and scales to help in decision-making. There are three types of LCC.

1. Conventional LCC: Also called financial LCC and is synonymous with total cost of ownership (TCO).
2. Environmental LCC: Aligned with the LCA in terms of goal and scope, functional unit, system boundaries, and methodological steps. It can include the monetary value of environmental impacts (externalities).
3. Societal LCC: Includes the monetary value of externalities which are environmental impacts and social impacts.

Each type of LCC is useful for specific stakeholders. Conventional LCC is useful for stakeholders such as consumers, manufacturers, or project managers who are only interested in analyzing the cash flows they directly incur. Environmental LCC includes all stakeholders in the value chain or life cycle and is used for analyzing both the environmental impacts and economic costs. Societal LCC is useful for stakeholders working in the government and other public authorities. In a societal LCC, the monetary value of impacts to the environment and society are included in the analysis. This chapter will focus on explaining how to conduct a conventional LCC for the purpose of supporting consumer and business decision-making in the context of a circular economy.

2.1   Definitions

Table 1 provides an overview of the different terms that are often used in LCC.

2.2   Types of Cash Flows

In LCC, revenue and costs are the two types of cash flows that are used to represent a transaction. A cost is the amount of money spent in a transaction. Costs can include money spent on buying resources, products, equipment, or paying fees. Revenue is the amount of money gained from a transaction. Revenue can include money gained from selling resources, products, equipment, or receiving fee payments. Value added, which is often referred to as profit or margin, is the difference between the total revenue and total cost. There are two types of costs and revenue: fixed and variable.

Fixed costs are cash flows that are not affected by changes in activity level over a feasible range of operations. Examples of fixed costs are equipment such as machines,

Table 1 Terms and definitions used in LCC

Term	Definition
Price	The amount of money that is used to buy a specific product or service
Transaction	The act of buying or selling something
Revenue	The amount of money gained from a transaction such as selling a product or service
	Synonyms: Income, cash-in, cash inflow
Cost	The amount of money lost from a transaction
	Synonyms: Expense, cash-out, cash outflow
Internal cost	Costs that are borne by actors who are directly involved in the life cycle of the product or service system being studied
External cost	Costs that are borne by actors outside the system (also called externalities) and occur as a side-effect of an economic activity, such as buying a product or service
Variable cost	A cost that occurs periodically or increases according to the volume of product or service provided
	Synonyms: Operational cost, operational expenditure (OPEX)
Fixed cost	A cost that does not occur periodically and does not increase according to the volume of product or service provided
	Synonyms: Capital cost, capital expenditure (CAPEX), equipment cost
Value added	The difference between the costs incurred and the revenue generated from the sale of a product(s) or service(s)
	Synonyms: Profit, margin
Life cycle costs	The sum of value added over the life cycle of a product system
Total cost of ownership	The sum of all the costs a stakeholder incurs from using or producing a product
Present value	The value of money in the current period of time
Future value	The value of money in a future period of time
Net present value (NPV)	The difference between the present value of cash inflows and the present value of cash outflows over a period of time
Payback period	The length of time required for an investment to recover its initial outlay in terms of profits or savings
Inflation rate	The change in prices of a good or service over a period of time
Exchange rate	The currency conversion between different currencies

trucks, or physical structures at a factory. These fixed costs typically occur only one time and the item purchased lasts for many years until it needs to be replaced. The cost of disposing a product can be considered as a fixed cost if it occurs only once at the end of its life cycle. Revenue can also be fixed as well. For example, a company can receive an amount of money from the government or a private investor in a specific year, but it does not receive that amount of money in other years.

Variable costs occur on a periodic basis and increase or decrease depending on the quantity of output. Material, energy, and water are examples of variable costs. For example, a company that produces plastic chairs will require plastic materials, electricity, and water in the manufacturing process. As more plastic chairs are produced, the amount of plastic material, electricity, and water required will increase as well. Taxes can also be viewed as a variable cost if a company is taxed for every unit of output produced. Revenue can also be variable. When more products are produced, the revenue will increase with each additional product sold.

2.3 Costs and Revenue in a Circular Economy

In a circular economy, there are some unique activities that take place in contrast to a traditional linear economy. Activities in a circular economy can involve reuse, repair and refurbish, recycle (upcycling and down cycling), and product-as-a-service. Although these activities take place in a circular economy, the way they are accounted for as costs and revenue in an LCC is not so different. Table 2 provides examples of different activities in a circular economy and how stakeholders would treat the activity as either a cost or revenue.

2.4 Time Dimension

In LCC, since costs and revenues occur over a lifespan, the analysis needs to consider how monetary flows occur at different times. The monetary value of a product or service can increase or decrease depending on changes that occur in the market. In order to consider the time dimension in an LCC, the analysis needs to compare costs in a chosen reference year and all future costs must be adjusted to the reference year when doing the comparison. The cost in a reference year is often called the present value. To adjust future costs to a reference year, inflation rates need to be used. An inflation rate represents the change in price of a good or service over a period of time. Equation 1 shows how to calculate the price P of a product in time t (in years) at an assumed inflation rate of r. $P(0)$ is the price of the product at the reference year $t = 0$).

$$P(t) = (1 + r)^t P(0) \qquad (1)$$

For example, a computer is priced at $1,000 ($P(0)$) in 2020 ($t = 0$) and the inflation rate is 2%. In 2021, the computer would cost

$$P(1) = (1 + 0.02)^1 (\$1,000) = \$1,020$$

Table 2 Examples of transactions in a circular economy and the stakeholders involved

Activity in a circular economy	Transaction	How a stakeholder treats activity as a cost	How a stakeholder treats activity as revenue
Reuse	Passing a used product to another user	The buyer spends money on buying a used product	The seller gains money from selling the used product
Repair and refurbish	Repairing and refurbishing a used product back to good-as-new condition	A repair and refurbish company pays for the cost of labor and materials to take a used product and bring it back to a good-as-new condition	A repair and refurbish company receives revenue for fixing the used product and selling it back to another user
Recycle (downcycling)	Converting waste into a resource	A waste conversion company spends money to convert the waste into a resource that can be used in another product system	The waste producer generates revenue from selling the converted waste to a recycling company
Recycle (upcycling)	Converting waste into a new product	An upcycling company spends money on labor and resources to convert waste into a brand new product	The waste producer generates revenue from selling the waste to an upcycling company. The upcycling company generates revenue by selling the brand new product.
Product-as-a-service	Paying for only the use of a product	The consumer spends money to rent a product to be used for a certain amount of time	The product-as-a-service company generates revenue from renting out the product

For a cost that takes place in the future, it needs to be normalized to the reference year. Suppose it is 2020 and we know that in 2022, a computer will cost $1,040.40. The inflation rate was 2% between 2020 ($t = 0$) and 2022 ($t = 2$). We want to find out what the cost of that computer was in 2020. To determine the cost, Eq. 1 would be used in the same way.

$$P(2) = \$1,040.40 = (1 + 0.02)^2 P(0)$$

To find the price of the computer now in 2020 ($P(0)$), Eq. 1 would be rearranged as follows

$$P(0) = (1 + 0.02)^{-2} P(2) = (1 + 0.02)^{-2} \$1,040.40 = \$1,000$$

Thus, the computer would have cost $1,000 in 2020 for it to cost $1,040.40 in 2022, assuming an inflation rate of 2% between 2020 and 2022.

2.5 Internal and External Costs

Internal costs are the costs that are directly borne by the actors involved in the life cycle of a product. In addition, there may be external costs (also referred to as externalities) that are borne by actors who are indirectly affected by the activities of a product's life cycle. An example of an external cost is the cost of health services needed to support people who have been affected by water or air pollution due to a project such as constructing a highway or building a factory. These external costs are not included in the price of a product even though they occur as a result of an economic activity. Externalities are not accounted for in conventional LCC. However, they are accounted for in environmental and societal LCC. Businesses may or may not include these externalities depending on the goal and scope of their LCC and their intended audience. To include external costs in an environmental LCC or societal LCC, the externalities need to be monetized. There are several different methods that have been proposed for monetizing externalities in an LCC. Many methods often try to determine the willingness-to-pay for a particular benefit, or in contrast, the willingness to accept a payment due to a loss or disbenefit that occurred due to an economic activity. Some other methods take a more direct approach and quantify the money that would have to be paid to avoid or counterbalance the externality. Readers can refer to J.-M. Rödger et al. (2018) for more information about the different methods for monetizing external costs.

2.6 Calculation

To carry out the LCC, the total costs and revenue need to be calculated. Profit is the difference between total revenue and total cost. The way to express total cost, total revenue, and profit are shown in Eqs. 2, 3, and 4.

$$\text{Total cost} = \text{Total fixed cost} + (\text{Demand} \times \text{Variable cost}) \qquad (2)$$

$$\text{Total revenue} = \text{Total fixed revenue} + (\text{Demand} \times \text{Price}) \qquad (3)$$

$$\text{Profit} = \text{Total revenue} - \text{Total cost} \qquad (4)$$

In order for the life cycle costs to be calculated, the total cost, total revenue, and profit must be calculated for each year of the life cycle of the product or service being analyzed. The total costs and total revenue for each year should be adjusted based on the fixed and variable costs and revenue that occur during the specific period of time.

For example, suppose a small company decides to buy a 3D printer to produce plastic toys for a period of one month (30 days). The 3D printer has a fixed cost of $500. The variable costs for producing a plastic toy are the plastic filament and electricity. Production of one plastic toy requires five pieces of plastic filament and 1 kWh of electricity. Each piece of plastic filament costs $0.50 and each kWh of electricity costs $0.16. Therefore, to produce one plastic toy, the company spends $2.50 for plastic filament and $0.16 for electricity. Each toy is sold at $7 per piece to generate revenue. The 3D printer is able to produce 10 toys each day and sells all the toys produced each day. The total cost, revenue, and profit over a period of 30 days would be calculated as follows.

$$
\text{Total cost} = \$500 \text{ for 3D printer} + \left(30 \text{ days} \times \frac{10 \text{ toys}}{\text{day}} \right.
$$

$$
\left. \times \left(\frac{\$2.5[\text{filament}]}{\text{toy}} + \frac{\$0.16[\text{electricity}]}{\text{toy}} \right) \right.
$$

$$
= \$500 + \$750 + \$48 = \$1,298
$$

$$
\text{Total revenue} = \frac{\$7}{\text{toy}} \times \frac{10 \text{ toys}}{\text{day}} \times \frac{30 \text{ days}}{\text{month}} = \$2,100
$$

$$
\text{Profit} = \text{Total cost} - \text{Total revenue} = \$2,100 - \$1,298 = \$802
$$

3 LCC Case Studies

This section provides two conventional LCC case studies to help understand how the LCC method discussed above can be applied in decision-making. The first case study compares the LCC of using an electric motorcycle versus a conventional gasoline motorcycle for five years. The second case study compares the LCC of choosing to own a pram or rent a pram to move a child around during the first three years of life.

3.1 Electric Versus Gasoline Motorcycles

Electric motorcycles (eMCs) offer an alternative mode of transport that is higher in transport energy efficiency, has zero tailpipe emissions, and minimal sound pollution. However, instead of gasoline, eMCs consume electricity which can come from a wide range of clean to high-polluting energy sources. Furthermore, instead of a traditional internal combustion engine, eMCs require the use of a battery that needs to get replaced every five years if it is a lithium-ion battery or every single year if it is a lead-acid battery. In this example, we examine the case of a person who wants to run his/her own food delivery service for five years. The person needs to decide whether he/she should invest in purchasing an electric or gasoline motorcycle to provide the service. To help in making a decision between the two vehicle options, a conventional LCC is conducted to answer the following questions:

1. When will the investment pay itself back?
2. Which motorcycle option has a higher NPV?

3.1.1 Fixed Costs and Revenue

In this example, there are two motorcycle options. For both options, the only fixed cost is the purchase of the vehicle. The electric motorcycle costs $4,091 and the gasoline motorcycle costs $1,418. There is no fixed revenue for either option over the period of five years.

3.1.2 Variable Costs and Revenue

Both the electric and gasoline motorcycles have variable costs and revenue during the lifetime of five years. It is assumed that both motorcycles travel a total of 50 km each day to deliver food to various customers. For both options, the variable revenue is the same. Although the number of deliveries can vary each day, it is assumed that on average the person can make six deliveries per day, six days a week. It is assumed that a person makes $8 per delivery.

For the eMC, the vehicle consumes electricity to travel. The fuel economy of the motor is 31.25 km/kWh and the cost of electricity is $0.16/kWh. This translates to a cost of $0.01/km traveled. Every five years, the lithium-ion battery needs to be replaced. The rating of the battery pack is 48 V and 40 Ah and so the capacity of the battery pack is 1.92 kWh. The market price of lithium-ion batteries is assumed to be $300/kWh. Thus, one lithium-ion battery pack will cost $576.

For the gasoline motorcycle, the efficiency of the engine is 37.5 km/L gasoline and the cost of gasoline is $1.46/l. This translates to a cost of $0.04/km traveled. Also, once every year, the gasoline motorcycle needs to be sent to the shop for inspection and repairs which costs $25 per appointment.

Table 3 summarizes all the fixed and variable costs and revenue of both motorcycle options being considered.

3.1.3 LCC Calculation Results and Discussion

Tables 4 and 5 list the costs, revenue, profit, present value, and net present value of both motorcycle options over a period of five years.

Table 3 Cost and revenue values for electric and gasoline motorcycles

Item	Unit	Electric motorcycle	Gasoline motorcycle
Vehicle	USD	4,091	1,418
Lithium-battery pack	USD/battery pack	576	N/A
Fuel economy	km/kWh, km/L	31.25	37.5
Cost of electricity	USD/kWh	0.16	N/A
Cost of gasoline	USD/kWh	N/A	1.46
Fuel cost per kilometer traveled	USD/km	0.01	0.04
Annual repair cost	USD/year	N/A	25
Revenue per food delivery	USD/delivery	8	8

In both options, the interest rate is assumed to be 2.5%.

Table 4 LCC of electric motorcycle

Year	Fixed cost	Variable cost electricity	Revenue	Profit	Present value of profit	Net present value
0	$(4,090.91)	0	0	$(4,091)	$(4,091)	$(4,091)
1	0	$(79.87)	$14,976	$14,896	$14,533	$10,442
2	0	$(79.87)	$14,976	$14,896	$14,178	$24,620
3	0	$(79.87)	$14,976	$14,896	$13,833	$38,453
4	0	$(79.87)	$14,976	$14,896	$13,495	$51,948
5	$(576.00)	$(79.87)	$14,976	$14,320	$12,657	**$64,605**

Table 5 LCC of gasoline motorcycle

Year	Fixed cost	Variable cost		Revenue	Profit	Present value of profit	Net present value
		Gasoline	Repair				
0	$(1,418)	0	0	0	$(1,418)	$(1,418)	$(1,418.18)
1	0	$(607.36)	$(25.00)	$14,976	$14,344	$13,994	$12,576
2	0	$(607.36)	$(25.00)	$14,976	$14,344	$13,652	$26,228
3	0	$(607.36)	$(25.00)	$14,976	$14,344	$13,320	$39,548
4	0	$(607.36)	$(25.00)	$14,976	$14,344	$12,995	$52,542
5	0	$(607.36)	$(25.00)	$14,976	$14,344	$12,678	**$65,220**

The LCC shows that the payback period of both motorcycle options is less than one year. In year 0, the NPV is negative, but then becomes positive in year 1. Thus, the person does not have to worry about a long period of time for recovering the investment costs of either vehicle option. However, the NPV of the gasoline motorcycle was higher than the electric motorcycle by a difference of $615 at the end of five years. One of the reasons the electric motorcycle had a lower NPV is because at the end of the 5th year, a new lithium-ion battery pack had to be purchased. Thus based on just the NPV, the person should invest in the gasoline motorcycle to deliver food for a period of five years. However, if the global cost of lithium-ion battery packs for electric motorcycles becomes cheaper in the future, this would reduce the initial investment cost of the vehicle as well as the cost of the replacement battery pack and therefore increase the NPV of the electric motorcycle option.

8.2   Owning Versus Renting a Pram

Prams, also known as baby carriages or strollers, are used by parents to push their children around outdoors. A parent can either buy and own a pram or rent one from a company. Renting prams is an example of the product-as-a-service business model. In this type of business model, instead of owning and using a product, consumers simply rent a product to be used for a specific period of time, similar to renting a movie or book from a library. This business model is technically feasible for certain products such as prams that are durable for a long period of time. Parents in general use prams for 3–4 years until their child no longer needs it. In this LCC, a set of parents just gave birth to their first child. They plan on having their child use a pram for three years. They seek to decide whether they should purchase and own a pram for three years or use a rental service for the same amount of time. The parents conduct an LCC of both pram options to answer the following questions:

. Which option has lower costs?
. When is it advantageous to rent a pram?

8.2.1   Fixed Costs and Revenue

In the pram rental option, there are no fixed costs. In the pram ownership option, the fixed cost is the initial purchase of the pram which is $200.

8.2.2   Variable Costs and Revenue

In the pram rental option, the only variable cost is the rental fee of the pram from the company which is $30/month. For the pram ownership option, the only variable cost is water and paper towels for light cleaning of the pram every three months, which is four times per year. It is assumed that the cost of each instance of cleaning is $2. Table 6 summarizes the cost values for both pram options.

8.2.3   LCC Calculation Results and Discussion

Tables 7 and 8 list the costs, present value, and net present value of both pram options over a period of 3 years.

Table 6 Cost values for pram ownership and rental options

Item	Unit	Pram ownership	Pram rental
Vehicle	USD	$200	N/A
Rental rate	USD/month	N/A	$30
Cleaning costs	USD/year	$8	N/A

In both options, the interest rate is assumed to be 2.5%.

Table 7 LCC of pram ownership option

Year	Pram	Cleaning	Total cost	Present value	Net present value
0	$(200.00)	–	$(200.00)	$(200.00)	$(200.00)
0.5	–	–	–	–	$(200.00)
1	–	$(8.00)	$(8.00)	$(7.80)	$(207.80)
1.5	–	–	–	–	$(207.80)
2	–	$(8.00)	$(8.00)	$(7.61)	$(215.42)
2.5	–	–	–	–	$(215.42)
3	–	$(8.00)	$(208.00)	$(7.43)	$(228.85)

Table 8 LCC of pram rental option

Year	Rental	Total cost	Present value	Net present value
0	–	–	–	–
0.5	$(180)	$(180)	$(178)	$(178)
1	$(180)	$(180)	$(176)	$(353)
1.5	$(180)	$(180)	$(173)	$(527)
2	$(180)	$(180)	$(171)	$(698)
2.5	$(180)	$(180)	$(169)	$(867)
3	$(180)	$(180)	$(167)	$(1,035)

The results of the LCC show that overall, it is far less costly to purchase a pram and own it for three years compared to renting a pram for three years. However renting a pram can be economically advantageous if parents only plan on using pram for a total of six months. At year 0.5 (6 months), the pram rental option had lower cost compared to purchasing a pram. Thus, by calculating the NPV at smaller time intervals, such as every half-year in the case study, the analysis can show at what point in time one product option can be more economically advantageous over another.

4 Relationship Between LCC and LCA

Life cycle assessment (LCA) is a methodology to quantitatively assess the environmental impacts of products, processes, or organizations. It is regulated by the standards ISO 14040 and ISO 14044. Figure 1 represents the four phases of LCA:

It should be noted how the four phases cannot be listed linearly, as the process has an iterative nature which particularly involves the interpretation phase.

The main difference between LCC and LCA is that LCA usually includes multiple environmental impact categories that have to be evaluated. This is because different environmental issues are not necessarily correlated, hence the need to assess more than one indicator. For example, the Environmental Footprint certification of the European Commission considers a set of sixteen impact indicators. It is important to account for a comprehensive set of impact indicators to be aware and avoid burden shifting among different impact categories.

Particularly in the circular economy context, it is important to mention the concepts of multi-functionality and of allocation. If a product is recycled, its product system will serve more than one function. The first is the main function of the product, while the second is the production of recycled material. This is a classic example of multi-functionality. However, this raises the question: "How to define the portion of burden that the studied product is responsible for in the life cycles that uses the recycled material?" More generally, the issue is to limit the system boundaries in a circular context which tends to expand them by linking together different life cycles. This is the problem of allocation. To sum up, systems with multi-functional processes require the application of allocation rules. To choose among the available allocation rules, the ISO 14044 promotes the following hierarchy: sub-division, system expansion, physical allocation, and economic allocation. The definitions of these terms and more are provided in Table 9 according to the ISO standards and other literature.

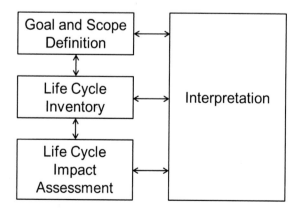

Fig. 1 The iterative process of LCA from ISO 14040

Table 9 LCA terms and definitions (ISO 14040, 2006; ISO 14044, 2006; Hauschild et al. 2018)

Term	Definition
Life cycle assessment	A systematic tool that allows for analysis of environmental loads of a product in its entire life cycle and assessment of their potential impacts on the environment (ISO 14040)
Goal definition	States the purpose of conducting the LCA
Scope definition	Describes the product system, the function of the product system, the product system boundaries, and data category
Functional unit	A measure that allows quantification of the function that is defined. It should represent the performance of the functional outputs of the product system. It provides a reference to which inputs and outputs are related
Reference flow	The amount of product(s) that is necessary to fulfill the function of the system
System boundaries	Defines the processes of the assessed product system that are included in the LCA
Cradle-to-grave LCA	An assessment of the potential environmental impacts of a product from raw materials extraction (cradle) to disposal/recycling (grave)
Cradle-to-gate LCA	An assessment of the potential environmental impacts of a portion of the product life cycle which includes the stages of raw materials extraction (cradle) to final production at the factory gate (i.e., before it is transported to the consumer)
Life cycle stage	A stage that takes place during the life cycle of a product or service. Examples are raw material extraction and processing, manufacturing, use, transportation, and disposal
Life cycle inventory	Data that quantifies the inputs and outputs (e.g., materials, energy, water) associated with a product system in the LCA study
Characterization factor	A number that expresses how much a single unit of an emission to the environment or a single unit of a resource consumed from the environment contributes to an impact category
Environmental impact category	Represents the environmental issue of concern to which the results of the life cycle impact assessment are assigned to
End-of-life	The stage that involves processes for treating a product that needs to be disposed. Examples include demolition, landfill, incineration, recycling
Allocation	Rule that determines how environmental impacts are divided among different life cycle stages or among different but connected product systems. ISO 14044 recommends adopting the following 4-step hierarchy of solutions
Sub-division	Allocation is avoided by increasing the level of detail of the modeling and of the data collection
System expansion	The system boundaries are expanded to include the additional functions of the co-products

(continued)

Table 9 (continued)

Term	Definition
Physical allocation	Inputs and outputs of the multi-functional process are partitioned with a ratio obtained from a physical relationship among the co-products
Economic allocation	Like physical allocation, but the partitioning ratio is obtained from the economic costs of the co-products
Cut-off	Case of allocation, in which the partitioning ratio is 100% for one co-product and null for the others. Often used to exclude all environmental burdens from recycled materials, apart from those generated in the waste collection and recycling processes

5 LCA Case Studies

Due to the possibility to quantify the environmental impacts along the life cycle of products, the application of LCA is particularly useful in the field of circular economy. Through LCA, the environmental dimension of sustainability can be examined holistically and the results can be used to complement the economic analysis for consumer and business decision-making. In this section, an LCA is conducted for the previous two LCC case studies about electric versus gasoline motorcycles and owning versus renting a pram. The case studies have been simplified for the purpose of focusing on how the results can be used to complement the LCC analysis.

5.1 Electric Versus Gasoline Motorcycles

In the LCC analysis, it was determined that the gasoline motorcycle was economically advantageous compared to the electric motorcycle for the food delivery business. This was because the gasoline motorcycle had a higher NPV at the end of five years. An LCA is conducted to compare the life cycle environmental impacts of both motorcycle options to provide the food delivery service an environmental perspective in the motorcycle decision-making process.

5.1.1 Goal and Scope Definition

The goal of this comparative LCA is to quantify the life cycle environmental impacts of an electric motorcycle versus a gasoline motorcycle. The intended audience of this LCA are food delivery service providers and motorcycle manufacturers who seek to understand the environmental impacts of the type of motorcycle food delivery service providers choose to drive. The functional unit of this study is a single motorcycle that is driven 50 km each day, six days a week, to deliver food to customers for five years. Averaged throughout an entire year, the motorcycle is able to make six food deliveries

Table 10 Simplified LCI of electric and gasoline motorcycles

Component	Electric motorcycle	Gasoline motorcycle
Mass of vehicle	118 kg (inclusive of battery)	112 kg
Mass of 1 battery pack	16.24 kg	N/A
Fuel economy	31.25 km/kWh	37.5 km/l gasoline
Distance traveled during life cycle (5 years)	78,000 km	78,000 km
Total electricity consumed	2,496 kWh	N/A
Total gasoline consumed	N/A	2,080 L
Total waste landfilled	134.24	112 kg

each day. The system boundaries of this LCA is cradle-to-grave. Therefore, this LCA includes all life cycle stages which are raw material extraction, manufacturing, product use, and disposal. At the end of each motorcycle's lifetime of five years, the vehicle is disposed at a sanitary landfill. The environmental impact categories considered within the scope of this LCA are climate change, metal depletion, and ecotoxicity.

5.1.2 Life Cycle Inventory

To conduct the LCA, a life cycle inventory (LCI) is developed. The LCI lists out the physical flows in terms of input of resources and materials and outputs of products, emissions, and wastes. As stated previously in the LCC case study, both motorcycles drive 50 km each day, six days a week for five years. Each day, both motorcycles are able to deliver six meals. The electric motorcycle uses an electric motor and a lithium-ion battery and has a fuel economy of 31.25 km/kWh. The lithium-ion battery pack has to be replaced at the start of the fifth year. The gasoline motorcycle uses gasoline as its energy source and has a fuel economy of 37.5 km/l gasoline. Both motorcycles are sent to a sanitary landfill at their end-of-life after five years. Table 10 provides a simplified LCI of both motorcycle options.

5.1.3 Life Cycle Impact Assessment

As stated in the goal and scope, the impact categories considered in this LCA are climate change, metals depletion, and ecotoxicity. Climate change is an impact category used to represent the potential environmental impacts of emitting greenhouse gases (GHGs) into the atmosphere and is measured in units of kg CO_2-equivalent. Metal depletion is an impact category that represents the potential loss of metallic materials from the environment and is measured in units of kg Fe(iron)-equivalent. Ecotoxicity is an impact category used to represent the potential environmental impacts of emitting toxic materials into the environment which is measured in units of kg 1,4 dichlorobenzene (1,4-DB) equivalent. To quantify the environmental impacts impact assessment is conducted to translate the physical flows of both motorcycles into environmental impacts. In the category of climate change, the emissions of GHGs are calculated. In the category of metals depletion, metal depletion potential (MDP) is calculated. In the category of ecotoxicity, the ecotoxicity potential (ETP) is calculated. Tables 11 and 12 show the life cycle environmental impacts of the different life cycle processes for each motorcycle using data from a study conducted by Kerdlap and Gheewala [2]. Cradle-to-gate production includes all the processes that take place starting from raw material extraction all the way to final production of the product that exits the factory gate.

By using the information in Tables 10, 11, and 12, the final life cycle environmental impacts in the three impact categories can be calculated. The impacts of each process listed in either Table 11 or 12 are to be multiplied by the physical flows of the respective motorcycle, which can be referred to in Table 10. For example, the steps of the method for calculating the life cycle GHGs of the electric motorcycle would be as follows.

Table 11 Potential environmental impacts of electric motorcycle processes

Process	Reference flow	GHGs (kg CO_2-eq)	MDP (kg Fe-eq)	ETP (kg 1,4-DB-eq)
Cradle-to-gate production of motorcycle frame (does not include battery pack)	1 motorcycle frame	346.50	137.62	0.00557
Electricity use	1 kWh	0.42	0	0
Cradle-to-gate production of battery pack	1 battery pack	291.62	221.89	0.00897
Disposal in sanitary landfill	1 kg waste disposed	0.51	0.00077	1.03E−05

Table 12 Potential environmental impacts of gasoline motorcycle processes

Item	Reference flow	GHGs (kg CO2-eq)	MDP (kg Fe-eq)	ETP (kg 1,4-DB-eq)
Cradle-to-gate production of motorcycle frame	1 motorcycle frame	310.66	75.19	0.00697
Gasoline production and consumption	1 L of gasoline	0.23	0	0
Road emissions	1 km driven	0.05	0	0
Disposal in sanitary landfill	1 kg waste disposed	0.51	0.00077	1.03E−05

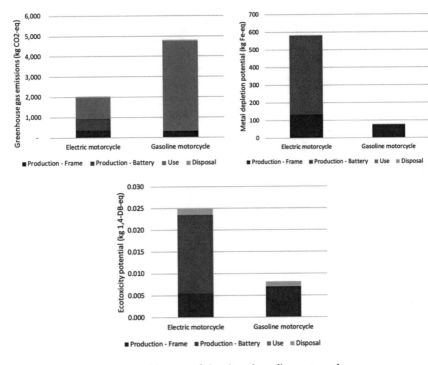

Fig. 2 Life cycle environmental impacts of electric and gasoline motorcycles

Step 1: Calculate the GHG emissions of cradle-to-gate production of one electric motorcycle frame without the battery pack (346.50 kg CO_2-eq).

Step 2: Calculate the electricity used to drive the motorcycle six days a week for five years (2,496 kWh) and its associated GHG emissions (1,046 kg CO_2-eq).

Step 3: Calculate the GHG emissions of cradle-to-gate production of two lithium ion battery packs (583 kg CO_2-eq).

Step 4: Calculate the GHG emissions of disposing the mass of the electric motorcycle frame and the two lithium-ion batteries (68 kg CO_2-eq).

By following the same steps above for the impact categories of metal depletion and ecotoxicity, the LCA results for both the electric motorcycle and the gasoline motorcycle can be calculated which are shown in Fig. 2.

5.1.4 Interpretation

The final LCA results show that the electric motorcycle has higher life cycle environmental impacts in two out of the three impact categories considered within the scope of the study. The electric motorcycle had lower impacts on climate change compared to the gasoline motorcycle, but had higher impacts on metal depletion and ecotoxicity. The reason the electric motorcycle had higher impacts in metal depletion is because the two lithium-ion battery packs consume precious metals such as nickel and cobalt. The metals used in the batteries of the electric motorcycle are highly toxic to the environment. Therefore, the electric motorcycle has a high ecotoxicity potential compared to the gasoline motorcycle.

The LCA results in this case study demonstrate how trade-offs can take place among different environmental impact categories. Thus, if the food delivery person chose to use an electric motorcycle, the person would avoid potential impacts to climate change, but in exchange would increase the impacts to metal depletion and ecotoxicity. From an environmental perspective, the food delivery person would need to decide which environmental impact categories should be prioritized when choosing which motorcycle to use. If climate change is the only impact category of concern, then the electric motorcycle should be chosen. If metals depletion or ecotoxicity are the impact categories of concern, then the user should select the gasoline motorcycle. Conducting an LCA helps identify opportunities for reducing environmental impacts. This LCA case study showed that the lithium-ion battery pack had the highest contribution to impacts to metal depletion and ecotoxicity. Thus, the company that produces electric motorcycle would know that attention should be focused on extending the useful life of the lithium-ion battery packs to reduce the impacts on metals depletion and ecotoxicity.

5.2   Owning Versus Renting a Pram

In the LCC analysis, it was determined that owning a pram for three years is economically advantageous compared to renting a pram for the same period of time. However, if a parent seeks to use a pram for six months or less, renting a pram has lower costs compared to buying and owning one. An LCA is conducted to compare the life cycle environmental impacts of both pram service options to provide parents with an environmental perspective in the decision-making process.

5.2.1   Goal and Scope

The goal of this comparative LCA is to quantify the life cycle environmental impacts of owning a pram versus renting a pram. The intended audience of this LCA are parents with children between the ages of 0 and 4 years who seek to understand the life cycle environmental impacts of their choice to either own a pram or use one from a rental service. The functional unit of this LCA is the use of the service of a pram to move a child around as needed during the first three years of life. The service of a pram is used every day for three years. The system boundaries of this LCA is cradle-to-grave. Therefore, this LCA includes all life cycle stages of the pram which are raw material extraction, manufacturing, product use, and disposal. At the end of each pram's lifetime, the vehicle is disposed of at an incinerator with energy

Table 13 Simplified LCI of pram ownership versus rental

Component	Ownership	Rental
Mass of pram	9.6 kg	9.6 kg
Number of prams	1	1
Years of use	3 years	3 years
Lifetime of pram	3 years	6 years
Number of times cleaned	12 times	24 times
Total waste incinerated	9.6 kg	9.6 kg

recovery. The environmental impact categories considered within the scope of this LCA are climate change, metal depletion, and water depletion. Water depletion is an impact category that is used to represent the amount of water consumed from the environment which is measured in units of cubic meters of water.

5.2.2 Life Cycle Inventory

Table 13 shows a simplified life cycle inventory of owning or renting a pram.

As stated in the LCC case study, the activities in the life cycle of both prams are different depending on whether the parents own the pram or rent it from a pram rental company. In the case of owning a pram, the pram is manufactured, the parents purchase it, use it for three years and then dispose of it through an incinerator with energy recovery. During the ownership of the pram, light cleaning is done every three months through the use of some paper towels, water, and a small amount of cleaning detergent. By the end of three years, the pram would have gone through light cleaning 12 times.

In the case of renting a pram, the pram rental company purchases the pram and owns it for a total of six years. The pram is rented out to the parents for a total of three years. During those three years, the parents must send back the pram to the pram rental company for heavy cleaning every three months. Heavy cleaning must be done so that the pram is still maintained in good condition and can last for a total of six years. Once the parents no longer need the pram after three years, they return it back to the pram rental company. The pram rental company continues to rent the same pram for another three years for a different child and still does heavy cleaning every three months. At the end of six years, the pram is no longer functional due to high usage by different customers. It is therefore disposed of at an incinerator with energy recovery. By the end of six years, the pram would gone through heavy cleaning 24 times.

5.2.3 Life Cycle Impact Assessment

As stated in the goal and scope, the impact categories considered in this LCA are climate change, metals depletion, and water depletion. In the impact assessment

Table 14 Potential environmental impacts of pram life cycle processes

Process	Amount	GHGs (kg CO$_2$-eq)	MDP (kg Fe-eq)	Water depletion (m^3 water)
Cradle-to-gate pram production	1 pram	78.3	7.7	3.3
Light cleaning	1 cleaning cycle	0.008	0.001	0.00028
Heavy cleaning	1 cleaning cycle	1.42	0.085	0.01
Disposal through incineration	1 pram disposed	2.6	−0.15	0.0046

GHGs is used to measure impacts to climate change, MDP is used to measure impacts to metal depletion, and the amount of water consumed is used to measure impacts to water depletion. Table 14 shows the life cycle environmental impacts of different processes during the life cycle of the pram using data from a study conducted by Kerdlap et al. [1]. Cradle-to-gate production of the pram refers to all the processes that take place starting from raw material extraction all the way to final production of the pram that exits the factory gate.

The steps of the method for calculating the life cycle GHG emissions of the pram options are as follows.

Step 1: Calculate the GHG emissions of cradle-to-gate production of one pram (78.3 kg CO$_2$-eq).

Step 2: Calculate the GHG emissions from the number of cleaning cycles done. The GHG emissions of cleaning is dependent on the number of times the pram is cleaned in both cases (0.09 kg CO$_2$-eq for 12 cycles of light cleaning) and the rental case (17 kg CO$_2$-eq for 24 cycles of heavy cleaning).

Step 3: Calculate the GHG emissions of disposing the entire mass of one pram (2.6 kg CO$_2$-eq).

The total life cycle environmental impacts of the pram ownership option over three years are the sum of the impacts of (1) producing one pram, (2) 12 light cleaning cycles, and (3) disposal of one pram through incineration. To calculate the impacts of the pram rental option over six years, the sum of the impacts of (1) producing one pram, (2) 24 cleaning cycles, and (3) disposal of one pram through incineration is calculated. Then these total impacts over six years must be divided by two. This is because the functional unit of the study states that the parents use the service of the pram for only three years. In the pram rental option, the pram has a lifetime of six years because the pram rental company takes care of the pram when the parents no longer need it and rent it out to other customers. Thus, the parents are only responsible for the life cycle environmental impacts of the first three years, which is half of the pram's lifetime in the rental option.

By following the steps above for the impact categories of metal depletion and water depletion, the life cycle environmental impacts of the pram ownership and rental options can be calculated which are illustrated in Fig. 3.

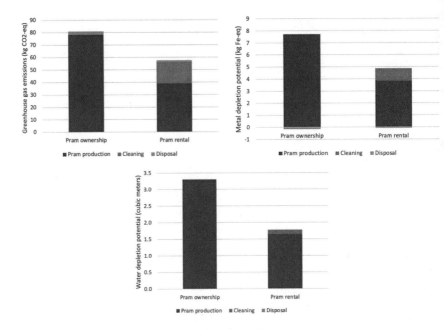

Fig. 3 Life cycle environmental impacts of owning and renting a pram

5.2.4   Interpretation

The results in Fig. 3 show that renting a pram for three years has lower life cycle envi ronmental impacts to climate change, metal depletion, and water depletion compared to buying and owning a pram for the same amount of time of three years. The impacts in all three categories were lower for the rental option because the pram rental company extended the use of the rental pram to another set of parents after the first three years of use. In contrast, in the pram ownership option, the parents only used the pram for three years and then disposed of it. Thus, this LCA reveals that from an environmental perspective, renting a pram that is shared between multiple parents has lower environmental impacts compared to owning a pram and disposing it within 1–3 years. The pram ownership option could reduce its life cycle envi ronmental impacts if the parents pass down the used pram to another set of parent thereby extending the pram's useful life.

6   Conclusion

LCC is a versatile method for conducting economic analysis of products and service from a life cycle perspective. The method is able to account for fixed and variable costs and revenue and the time value of money. The results can be used to answer different questions for a variety of stakeholders regarding the economic performance of products and services. NPV and payback period are economic indicators that can be determined by conducting an LCC. There are three different types of LCC. These are conventional LCC, environmental LCC, and societal LCC. Convention LCC only accounts for direct flows of cash from one stakeholder perspective. In environmental LCC, the study is done in alignment with the goal and scope of an LCA. Furthermore, externalities can also be included in the environmental LCC. Externalities are accounted for in the LCC by putting a monetary value on the environmental impacts of a product or service. In a societal LCC, the intended audience is usually stakeholders working in the government and other public authorities. The monetary value of impacts to the environment and society are included in a societal LCC. This chapter focused on providing step-by-step instructions for conducting a conventional LCC for the purpose of supporting consumers and businesses in making economic decisions related to the circular economy.

To demonstrate how LCC can be used to support consumer and business economic decision-making in the circular economy, two case studies were examined. The first case study looked at the choice of purchasing either an electric motorcycle or a gasoline motorcycle to be used for running a food delivery service for five years. The results of the LCC showed that the gasoline motorcycle had a higher NPV compared to the gasoline motorcycle at the end of five years. The main reasons the electric motorcycle had a lower NPV is because of the higher initial cost of the vehicle and the replacement of the lithium-ion battery pack that had to be purchased in the fifth year of the life cycle. The NPV of the electric motorcycle could be reduced if the global cost of lithium-ion battery packs decreases in the future. To complement the economic analysis, an LCA was conducted to quantify the life cycle environmental impact of both motorcycle options. The LCA results showed that the electric motorcycle had lower impacts to climate change, but had higher impacts to metal depletion and ecotoxicity compared to the gasoline motorcycle. This was because the lithium-ion battery pack uses precious metals such as nickel and cobalt that had a high contribution to metal depletion. Furthermore, the other materials and processes involved in producing the battery had a high contribution to ecotoxicity. This case study demonstrated how LCA can identify trade-offs that take place between different impact categories. When considering both the environmental and economic perspectives, the choice to use either an electric or a gasoline motorcycle becomes challenging. The food delivery person would need to decide whether to prioritize the economic or environmental performance of both motorcycle options. However, through LCC and LCA, the food delivery person has full visibility of the factors that contribute to the economic and environmental performance of both options from a holistic perspective.

In the second case study, the LCC examined the choice of owning a pram for three years or renting a pram from a company for the same amount of time. In the pram ownership option, the parents dispose the pram after three years. In the pram rental option, the pram has a lifetime of six years with multiple users since it is taken care of by the pram rental company. The results of the LCC showed that the pram ownership option had a lower total cost compared to the rental option over a period of three years. However, if a set of parents were to only use a pram for six months or less, the rental option would have a lower total cost. To complement the LCC, an LCA was conducted to quantify the life cycle environmental impacts of both pram options. The LCA results showed that the pram rental option had lower environmental impacts compared to the pram ownership option in all three impact categories. This was because the pram in the rental option had a longer lifetime of six years with a second user after the first three years. In the ownership option, the parents use the pram for three years and then dispose of it. Thus, by extending the useful life of the pram, the environmental impacts for the parents for using the pram are reduced. The life cycle environmental impacts of the pram ownership option can be reduced as well if the used pram is passed down to another set of parents after the first three years of use. When examining the LCC and LCA results of the pram case study, the parents would have to decide whether they prioritize the environmental or economic performance when it comes to selecting to own or rent a pram. However passing down a used pram to a second set of parents after the first three years of use would avoid this trade-off between environmental and economic performance. If a used pram is passed down to another set of parents for free, the second set of parents would have lower life cycle environmental impacts of using the pram. Also the second set of parents would have lower life cycle costs because the cost of the used pram was free.

The two case studies show how the LCC methodology can be applied to support in economic decision-making for products and services in a circular economy. As the environmental perspective is one of the main motivations for transitioning to a circular economy, conducting an LCA can support in quantifying the life cycle environmental impacts of those products and services. This helps to identify potential burden shift and opportunities for improving the environmental performance of different product and services and support individual and business decision-making.

7   Take-Home Messages

Listed below are several important take-home messages regarding LCC, its relation ship with LCA, and applications in a circular economy.

1. Conventional LCC is useful for individuals and companies who seek to examine only the direct monetary flows of a product or service across its life cycle.
2. In an environmental LCC, the LCC is aligned with LCA in terms of consistency with the functional unit, goal and scope, and system boundaries. In addition to the direct monetary flows of the product or service, the monetary value of the externalities may also be included in an environmental LCC.
3. Societal LCC is used to support governments and public authorities in decision making. This includes quantifying the monetary value of the environmental impacts of the product or service being examined.

4. In a circular economy, there are many different actors at different life cycle stages. Defining from which stakeholder perspective to analyze in the LCC is therefore important. This is because in activities such as waste-to-resource conversions and remanufacturing, certain transactions that are a cost for one stakeholder can be a form of revenue for another stakeholder.
5. To include externalities in an LCC, the externalities need to be monetized using a methodology that is suitable for the goal and scope of the study.
6. Conventional LCC can be applied to determine the NPV and payback period. When calculating the NPV, the time value of cash flows at specific interest rates need to be factored in.
7. Conducting an LCA is important for holistically evaluating the environmental performance of a product or service in a circular economy.
8. Through conducting an LCA, one can have full visibility of which processes have the highest contribution to the total life cycle environmental impacts. This visibility helps in determining which processes should be focused on to improve the environmental performance.
9. An LCA should assess a comprehensive set of impact indicators that pertain to the goal and scope of the study and its intended audience. This helps to avoid burden shifting to non-accounted categories.
10. Extending the useful life of a product through multiple users can help to reduce the life cycle environmental impacts a user or business incurs as opposed to single use and disposal of a product.

Depending on the level of detail of the LCC, the process of finding monetary data can be time consuming. Table 15 provides some databases with free access to LCC data.

Table 15 Publicly accessible sources of LCC data (J.-M. Rödger et al. 2018)

Data type	Organization	Link
Crude oil	International Energy Agency	www.iea.org/statistics/topics/priceandtaxes
Plastics	The Plastic Exchange	www.theplasticsexchange.com
Marine fuel oils	Ship and Bunker	www.shipandbunker.com/prices
Chemicals	Independent Commodity Intelligence Service	www.icis.com/chemicals
Metals	London Metal Exchanges	www.lme.com
Commodities	United Nations	www.comtrade.un.org/data
Inflation	World Bank	www.data.worldbank.org
Wages	International Labour Organization	www.ilo.org
Currency exchange rates	World Bank	www.data.worldbank.org
Power, gas, coal, oil	European Stock Exchange	www.eex.com/en

Table 16 Publicly accessible sources of LCA data

LCA data type	Organization	Link
Various	openLCA Nexus	https://nexus.openlca.org/databases
Various	Life Cycle Initiative	https://www.lifecycleinitiative.org/resources-2/global-lca-data-network-glad/
Plastic materials	Plastics Europe	https://www.plasticseurope.org/en

Some LCA data is also available for free. Several of these free resources are listed in Table 16.

To learn more about the LCC and LCA methodology and its applications, readers are encouraged to read through the resources listed in Further reading.

Further Reading

British Standard, BS 8001:2017 Framework for implementing the principles of the circular economy in organizations—Guide. The British Standard Institution, 2017.

ISO. 2006. ISO 14040 Environmental management—life cycle assessment—principles and framework. International Organisation for Standardisation.

ISO. 2006. ISO 14044 Environmental management—life cycle as—sessment—requirements and guidelines. International Organisation for Standardisation.

Hauschild MZ, Rosenbaum RK, Olsen SI (2018) Life cycle assessment, Life Cycle Assessment: Theory and Practice. Springer. https://doi.org/10.4324/978131 5778730.

Niero M, Schmidt Rivera XC (2018) The role of life cycle sustainability assessment in the implementation of circular economy principles in organizations. Procedia CIRP 69, pp. 793–798. https://doi.org/10.1016/j.procir.2017.11.022.

Reddy VR, Kurian M, Ardakanian R (2015) Life-cycle cost approach for management of environmental resources: a primer. Springer. https://doi.org/10.1007/978-3 319-06287-7.

Rödger J-M, Kjær LL, Pagoropoulos A (2018) Life cycle costing: an introduction. In: Hauschild MZ, Rosenbaum RK, Olsen SI (Eds) Life cycle assessment: theory and practice. Springer, pp. 373–400. https://doi.org/10.1007/978-3-319-56475-3.

Schaubroeck T, Petucco C, Benetto E (2019) Evaluate impact also per stakeholder in sustainability assessment, especially for financial analysis of circular economy initiatives. Resour Conserv Recycl 150:104411. https://doi.org/10.1016/j.resconrec 2019.104411.

Swarr TE, Hunkeler D, Klopffer W, Pesonen H-L, Ciroth A, Brent AC, Pagan I (2011) Environmental life cycle costing: a code of practice. Society of Environmental Toxicology and Chemistry.

References

. Kerdlap P., Gheewala, S. H., & Ramakrishna, S. (2020). To rent or not to rent: A question of circular prams from a life cycle perspective. *Sustainable Production and Consumption, 26,* 331–342. https://doi.org/10.1016/j.spc.2020.10.008.
. Kerdlap, P., & Gheewala, S. H. (2016). Electric motorcycles in Thailand: A life cycle perspective. *Journal of Industrial Ecology, 20,* 1399–1411. https://doi.org/10.1111/jiec.12406.

Piya Kerdlap is a PhD Candidate at the National University of Singapore. His research is focused on multi-level life cycle environmental and economic performance evaluation of industrial symbiosis. Piya's research is being conducted in collaboration with the Singapore Institute of Manufacturing Technology and is supported by A*STAR—Agency for Science, Technology, and Research.

Simone Cornago holds a BSc and a MSc in environmental and land planning engineering, both from the Polythecnic University of Milan. After a research fellowship at the National Research Council of Italy, he is now a PhD Candidate at the National University of Singapore. He focuses on life cycle assessment, absolute sustainability and emission reduction target setting, mainly applied to the manufacturing sector. His research is being conducted in collaboration with the Singapore Institute of Manufacturing Technology and is supported by A*STAR—Agency for Science, Technology and Research.

Towards Sustainable Business Strategies for a Circular Economy: Environmental, Social and Governance (ESG) Performance and Evaluation

Rashmi Anoop Patil, Patrizia Ghisellini, and Seeram Ramakrishna

Abstract This chapter seeks to give a foundational overview of Environmental, Social, and Governance (ESG) metrics. The definition of individual ESG factors is first introduced to highlight sustainability considerations in businesses and how these can be considered and support circularity in business operations. The chapter further develops the concept that ESG reporting serves as an enabling tool with which the business operations can drive circularity and remedy the existing limitations of the linear economy in practice. The rise in ESG reporting from companies, and the ESG considerations of companies based on their disclosures are discussed. Incorporation of ESG factors into business operations is also evidenced through real-world case studies. The impact of ESG performance and the growing awareness of sustainability among businesses, consumers, and investors on the current investing trends and its contribution to embracing circularity is presented as the conclusion of the chapter.

Keywords ESG · Sustainability · Business · Circularity · Circular economy

R. A. Patil (✉) · S. Ramakrishna
Circular Economy Task Force, National University of Singapore, Singapore, Singapore
e-mail: rashmi.anoop33@gmail.com

S. Ramakrishna
e-mail: seeram@nus.edu.sg

P. Ghisellini
Department of Sciences and Technologies, Parthenope University of Naples, Naples 80143, Italy
e-mail: patrizia.ghisellini@uniparthenope.it

S. Ramakrishna
Department of Mechanical Engineering, National University of Singapore, Singapore 117576, Singapore

Learning Objectives

- To understand ESG metrics, meant for business sustainability.
- To understand why businesses seek to enhance their ESG performance.
- To appreciate the link between ESG performance and circularity envisioned by the circular economy.
- To understand the concept of how an ESG score is calculated.
- To understand how companies enhance their ESG performance.

1 Circularity Assessment and the ESG Context

Companies across all industrial sectors represent a significant part of our economy. Majorly, the economic patterns have been linear—take-make-use-dispose—and have dominated the business operations for long [1, 2]. This linear approach has been agreeably unsustainable [1–4]. The rising concerns about the impacts of the linear economy on the environment and society have spurred the companies to firmly rethink the sustainability implications of their operations [1, 5]. During the previous decade this sustainability consciousness has established a direct correlation with the circular economy (CE) principles. Many companies across the world, therefore, are moving towards circular business approaches[1] and it is evidenced that the CE know-how is concentrated in large companies and less diffused across small medium businesses [6]. This transition to CE, however, needs a comprehensive assessment of the extent of circularity achieved and the potential economic benefits.

Circularity assessment is relatively new and is currently being explored by companies intending to transit towards circularity [7, 8]. Having an assessment tool for circularity that could be used as a yardstick with which companies can evaluate their circularity and understand the gaps for improvement is crucial. The CE scholarship has developed several methods and tools for assessing circularity. For example, Circulytics[2] is a comprehensive tool developed by the Ellen MacArthur Foundation for circularity measurement of companies that also highlights the areas for improvement in circularity performance. Various organizations such as the Cradle to Cradle Products Innovation Institute,[3] the alchemia-nova,[4] and the ecopreneur provide assistance in assessing the materials circularity of the products, services and/or entire supply chains with their proprietary approaches and tools. Such tools enable companies to tap into new opportunities and stay relevant in the competition by adopting circular thinking into their core business strategy.

The concept of environmental, social, and governance (ESG) metrics for business assessment was in place much before the advent of circular economy principles.

---

[1]https://www.ellenmacarthurfoundation.org/our-work/activities/ce100.

[2]https://www.ellenmacarthurfoundation.org/resources/apply/circulytics-measuring-circularity.

[3]https://www.c2ccertified.org/get-certified/product-certification.

[4]https://www.alchemia-nova.net/services/circular-business/.

[5]https://ecopreneur.eu/circularity-check-landing-page/the-circularity-check-explanation/.

It can be considered as a measure of the overall sustainability of the business that depends on the environmental, social, and governance factors. However, achieving circularity is a winning approach to enhance ESG performance while becoming more sustainable [9].

From this point onwards, for ease of communication, the ESG factors are referred to as a single entity in some instances and the terminology *ESG* is used to represent this in the text. In the next sections, the individual ESG factors (**E**, **S**, and **G**) are briefly explained followed by a discussion on how CE principles influence the ESG pursuit of businesses evidenced by company case studies.

2 Understanding ESG

In general, a business needs material resources and energy to function and produces waste in some form or the other. In practice, such a business is dependant on labor, within a broader societal setup involving stakeholders such as investors and consumers. For the smooth functioning of the business, a set of practices, rules, and regulations, and operational procedures are followed, that decide how decisions are made in the company, mostly regarding governance and general operational management. Thus, the businesses we see around are fundamentally interweaved with various ESG factors in some capacity. Its imperative to first briefly understand the individual factors of ESG (Fig. 1) and their interdependencies.

- The **E** in ESG, environmental aspect, encompasses the energy consumption of the company, the resources/raw materials consumed, the waste discharged, and the impact of these on the ecosystems. Carbon emissions and the contribution to climate change are the two most important and common criteria which represent the ecological footprint of the company.

- The **S** in ESG, social criteria, includes the reputation and relationships that the company has earned with the employees, consumers, and institutions in the community where the business has been established and is being run. It mainly represents the inclusive culture and diversity in human capital[6] of a company to suit the societal requirements.

- The **G** in ESG, governance, is the in-house system of controls and protocols a company follows to govern itself, in order to make effective decisions, abide by the laws of the land, and meet the needs of the stakeholders. This is essential for every company to function smoothly in the long run.

These individual factors are also interconnected with each other and usually, work in combinations for business operations. For instance, when a company is trying to comply with the environmental law of the state in lowering the carbon emissions, such compliance requires the company's governance factors to abide by the law, overlapping with the social factors that address the broader concerns about

---

https://www.investopedia.com/terms/h/humancapital.asp.

Fig. 1 Schematic illustration of the ESG factors for business sustainability. The important param eters that are used to determine the environmental, social responsibility and governance perfor mance of a business are highlighted. The ESG factors can be either individual determinants of business performance or serve as a collective metric as they are intertwined with each other. Desig *adapted* from a template; Copyright PresentationGO.com

sustainability. From a CE perspective, our focus is mostly on the environmenta and social factors, however, governance can never be hermetically separated from these two. Indeed, excelling in environmental and social criteria needs expertise in governance as exemplified by being aware of and taking measures to address the legal issues, and maintaining a transparent rapport with the concerned government bodies [10].

2.1 ESG Over the Years

Even though ESG is a millennial concept, it has evolved over the decades with history that dates back to the post World War II period [11]. During the post-war period (1945—early 1970s), besides the economic expansion, a shortage of worker triggered the fight for employee rights by labor unions and as a result, such right were included in the scope of business governance. Also, in the 1960s and 1970s several public movements led to the social issues such as consumer and civil right gaining traction, acquiring the attention of the businesses (Fig. 2).

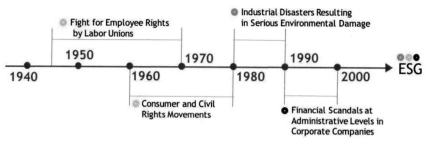

Fig. 2 A timeline of major events leading to the development of ESG. It began with the struggle for labor rights just after World War II and continued with the social movements for consumer and civil rights. This was followed by a series of industrial disasters such as Bhopal gas tragedy, the Chernobyl nuclear disaster, and the Exxon Valdez oil spill that led to serious environmental concerns. Then came the financial scandals at the higher management levels of companies such as Enron and Tyco which made investors skeptical about transparency in business governance. All these collectively gave rise to the concept of ESG

Later in the 1980s, a series of industrial disasters such as the Bhopal gas tragedy, the Chernobyl nuclear disaster, and the Exxon Valdez oil spill led to serious environmental concerns and an increasing consciousness regarding the need for a common evaluatory metric on the environmental integrity of business operations (Fig. 2). During the same time, the EIRIS (Ethical Investment Research Services) Foundation was established in England that systematically rated companies based on their social and environmental responsibilities which turned out to be an important criterion for investors [11].

Corporate governance has always been important to the executive boards and employees of a company, and investors. However, in the 1990s, the financial scandals at the higher management levels of companies such as Enron and Tyco made investors and employees more skeptical about the financial transparency of businesses in general (Fig. 2). Many a time, this has led to conflicts of interest between the investor community and company management. Even today, institutional investors engage in a detailed dialogue with the company boards regarding governance issues before making investment decisions.

Circular Economy Enhances ESG Performance

In recent years, there has been a rise in sustainability and circularity consciousness to address the scarcity of raw materials, resources, rising commodity prices and environmental pollution threatening our ecosystems and economies. Currently, many countries are transiting from a traditional linear economy to a CE, one that is regenerative and waste-free by design. To facilitate such a transition and steer towards circularity, governments are enframing relevant legislation and policies for circular manufacturing and business operations [12, 13]. The objective of the transition is to absorb as much value as possible from resources, products, and services to create a system that promotes materials circularity and renewable energy. There is also a temporal dimension to this objective. In that, the goal of product design and manufacturing is keeping products, components, and materials at their highest utility for the longest possible times [1]. The introduction of CE-related legislation drive businesses to put the principles of CE in practice.

The CE can be perceived as a model to achieve a green economy,[7,8,9] where the economic growth pattern is modified to take into account the limits of the natural environment. It emphasizes on the environmental responsibility of businesses and proposes that (i) recovery of materials be factored in from the initial stages of the manufacturing processes (such as ideation and design) and the outputs fed back as inputs into the manufacturing cycle [2, 14]; (ii) resource consumption be reduced [2, 14], and (iii) renewable energy be utilized in the manufacturing processes [2, 14]. Although adopting such practices in businesses is a meritorious path to achieving sustainability, it is not easy to implement the same in an established industrial ecosystem. It certainly inspires business owners to opt for deconstructing end-of-life products and reuse them as inputs for production, possible if each stage of the manufacturing process could be adapted to achieve such a result.

Besides the financial risks associated with the current linear economy [4], companies are also facing reputational challenges such as brand association with environmental degradation due to raw material extraction, resource depletion, and postconsumer pollution which eventually give rise to social challenges such as adverse effects on human and animal health. Companies should also anticipate regulatory challenges such as landfill closures, material bans, and extended producer responsibility (EPR) policies, springing up globally. By applying the principles of the CE, companies can maximize resource productivity, minimize pollution and waste generation, decouple economic growth from virgin natural resource consumption at a macro level in the global economy, and complement it with the use of renewable energy consumption as shown in Fig. 3. This will reduce the companies' environmental impact and avoid social and governance issues.

Demand for products and services is increasing in tandem with the increasing global population and consumer affordability. Commodity and raw material prices continue to increase and remain volatile. At the same time, the environmental and social capital costs related to the extraction of non-renewable resources and waste disposal are directly impacting the overall cost of business operations. As the CE is much more than just materials recycling and using renewable energy, companies stand to gain economically by following the CE principles in addition to reducing

---

[7]UNEP Green Economy https://www.unenvironment.org/regions/asia-and-pacific/regional-initiatives/supporting-resource-efficiency/green-economy.

[8]The 5 Principles of Green Economy https://www.greeneconomycoalition.org/news-analysis/the-5-principles-of-green-economy.

[9]Green Economy and Circular Economy: targets and prospects http://www.wiretechworld.com/green-economy-and-circular-economy-targets-and-prospects/.

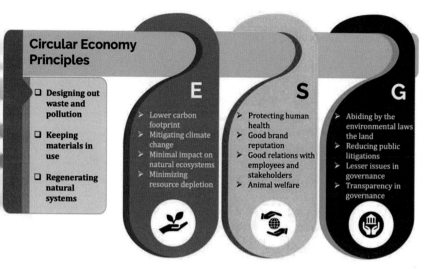

Fig. 3 Schematic illustration of the link between CE principles and ESG. The companies seek to adopt the principles of the CE mainly to enhance their environmental performance, that leads to better social credibility and governance. Design *adapted* from a template; Copyright PresentationGO.com

ESG-related risks. Embracing circularity by companies not only optimizes operational costs through judicious utilization of resources but also avoids expenses on litigations and fine payments for violating CE related legislation. Therefore, adopting circular mindset, practising circular business model, and complying with the laws of the land for a sustainable economic model and reducing business costs has gradually become a necessity for companies.

The ESG evaluation of companies assists in identifying and understanding the different aspects of the business with scope for adopting circular principles. This will directly aid in improving the environmental performance of the business, eventually leading to enhanced social credibility and better governance (Fig. 3). This progression towards sustainability not only improves the total ESG performance of businesses but also attracts investors, especially, institutional investors interested in sustainable businesses. Such financing is necessary for evolving businesses to become more circular through research and innovation. The gradual embracing of CE principles, therefore, will not only enable sustainability in the long run but may also improve the circularity of businesses.

For example, consider the fashion industry that uses virgin feedstock (nearly 97%) such as cotton, plastic, and other fibers along with resources such as water and energy. As a result of fast trends in this industry, nearly three-quarters of the used textiles have been landfilled or incinerated in the past decade. There is also considerable loss of virgin materials during both the production and recycling of fashion products. The study conducted by the Ellen MacArthur Foundation estimates that $500 billion worth of raw material is lost every year due to clothing being barely used and rarely recycled [15]. In addition to environmental pollution caused by these processes, the loss of materials and the resources used in the production add up to the total carbon footprint of the industry, thereby contributing to climate change. If this trend continues, a study by the Ellen MacArthur Foundation estimates [15] that by 2050, the fashion industry will be responsible for using a quarter of the world's carbon budget! This has become a major concern for popular fashion brands as they are being associated with a growing negative environmental impact leading to social and governance concerns. Therefore, some brands are revisiting their designs to make them circular. One such approach involves using recycled feedstock and fibers from renewable resources such as natural fibers and recycled plastic. A few start-up fashion brands such as Rapanui have even come up with business models to collect back their products that have reached their end-of-life and upcycle them into new products. These initiatives make the business sustainable by reducing the negative impact on the environment locally and on a global scale [16]. Moreover, such circular business approaches also address governance issues such as adhering to a cap on carbon emission, landfill closures, and mandatory use of recycled feedstock. Though this transformation seems difficult initially, it is profitable in the long run.

4   ESG Performance and Evaluation

ESG was a niche concept until recent years. Incorporating the environmental, social, and governance factors into investing and lending is moving into the mainstream, transforming the operations and management dynamics of companies. According to the KPMG Survey of Corporate Responsibility Reporting (2017) [17], around 93% of the world's largest 250 corporations (G250) report on their sustainability performance.[10] Many companies are integrating non-financial aspects of business performance represented by ESG in their annual reports. In the context of global sustainability, such reporting by the companies is crucial as it demonstrates their strategic contributions and communicates their values and governance ethics to the stakeholders. This reduces the asymmetric information between the company and its stakeholders and provides the opportunity for stakeholders in choosing companies based on their ESG performance. In doing so, the stakeholders (such as investors, consumers, and suppliers) contribute indirectly to improve global sustainability. From a company's perspective, non-financial reporting provides them an opportunity to reflect on their ESG performance, and recognize and seize new opportunities and manage changes towards sustainability [18, 19]. When viewed through a CE lens, the ESG reporting, provided it's transparent, assists in selecting the better companies that are implementing the CE framework and the 6R framework[11]: reduce, reuse, recycle, recover, redesign, and remanufacturing. However, adoption of ESG practices in operations and management of a company is still voluntary (and there is general consensus that it should be made mandatory).

---

[10]https://www.globalreporting.org/information/about-gri/Pages/default.aspx.

[11]https://www.ellenmacarthurfoundation.org/circular-economy/concept/infographic.

To report their ESG performance, currently, companies are following existing international standards and frameworks set by organizations such as the Global Reporting Initiative (GRI),[12] the Sustainability Accounting Standards Board (SASB),[13] and the Climate Disclosure Standards Board (CDSB).[14] The most popular and widely adopted reporting standard is the GRI Sustainability Reporting Guidelines [20], a comprehensive framework that includes references to other widely recognized standards, such as the OECD Guidelines for Multinational Enterprises, the United Nations Global Compact Principles, and the International Organization for Standardization (ISO). Many companies are also following the ESG-related ISO standards listed in Table 1) and are seeking the necessary certifications. Incorporating such standards makes the information disclosures by the companies more credible and makes it easier to compare with their peers following the same standards. Although ESG reporting is a remarkable development towards global sustainability, it has to be noted that the diversity in standards has resulted in nonuniform disclosures.

To support the allocation of capital to sustainable finance, market data providers such as Refinitiv[15] (formerly the Thomson Reuters, Finance and Risk Business), Sustainalytics Inc.,[16] and MSCI[17] are providing ESG information on companies. For example, Refinitiv offers one of the most comprehensive ESG data offerings in the industry, covering over 70% of global market cap, across more than 450 different ESG metrics, with a history going back to 2002. ESG metrics comprise of ESG scores, raw metrics, standardized and analytic data. The ESG space consists of 9,000 companies, covering 23 global and regional indices. The database and ratings are updated weekly, with a fully transparent and objective methodology. ESG news and controversies are updated continuously as and when such events occur and get picked up by global media. In addition to scores, analytics, and raw metrics for companies, Refinitiv houses one of the largest databases of green bonds. Refinitiv also provides hundreds of other data sets that can be integrated with ESG content, such as but not limited to Mergers and Acquisitions, Deals, Company Financials, Estimates, and Ownership. Such an extensive database serves as a good starting point for analysts to perform equity research, screening, and quantitative ESG analysis and identify risks and opportunities that are not detectable by conventional analytical methods. From the perspective of companies being evaluated, such raw datasets provide insights into the areas of operation and management where there is scope for improvement.

Different methodologies and weightings are often used by data providers to provide an ESG rating. These range 0–100 in the case of scores or AAA-CCC for ratings [21]. Access to only a combined ESG risk score limits the exploration of the critical components of ESG performance impacting business sustainability. The ESG-scoring is rules-based and, mostly quantitative and less qualitative. Depending

---

12 https://www.globalreporting.org.
13 https://www.sasb.org/.
14 https://www.cdsb.net/.
15 https://www.refinitiv.com/en.
16 https://www.sustainalytics.com/.
17 https://www.msci.com/.

Table 1 ESG related ISO standards

ISO standard	Title/Focus	Description
ISO 14000 Series[a]	Environmental management systems	Provides a framework for setting up an effective environmental management system in a company or organization
ISO 50001[b]	Energy management systems	Provides a practical framework for improving energy use within an organization
ISO 45001[c]	Management system for occupational health and safety	Provides occupational management standards for minimizing work-related accidents and diseases in an organization
ISO 9000 Series[d]	Quality management systems	Provides a universal reference for organizations seeking to improve the quality of the products and services and be in sync with customers' expectations
ISO/IEC 27000 Series[e]	Information security management system	Provides an enabling platform for organizations seeking to maintain/manage the security of their assets and critical information related to finance, intellectual property, employees, customers and/or a related third party
ISO 26000[f]	Social responsibility Guidelines	Provides clarity on what social responsibility is and helps businesses to understand and put into action, translational practices towards social responsibility
ISO 37001[g]	Anti-bribery management system	Provides a regulatory guidance system for anti-bribery practices in an organization
ISO/TC 322[h]	Sustainable finance	Provides an enabling framework for standardization in the financing of business activities in general, based on sustainability considerations inclusive of **E, S,** and **G** practices

[a]https://www.iso.org/iso-14001-environmental-management.html
[b]https://www.iso.org/iso-50001-energy-management.html
[c]https://www.iso.org/iso-45001-occupational-health-and-safety.html
[d]https://www.iso.org/iso-9001-quality-management.html
[e]https://www.iso.org/isoiec-27001-information-security.html
[f]https://www.iso.org/iso-26000-social-responsibility.html
[g]https://www.iso.org/iso-37001-anti-bribery-management.html
[h]https://www.iso.org/committee/7203746.html

on the industry sector in consideration, a set of relevant performance indicators for ESG parameters is chosen for assessment. The chosen indicators can apply to every ESG and CE other industry or be specific to a particular industry. For example, analysts do not evaluate the water usage and waste profile of financial institutions such as banks.

In this case, parameters such as board governance, community lending practices, work-life balance, and employee compensations are considered as key indicators. In contrast, for production industries such as electronics and personal care products, parameters such as water usage, waste generation and management, carbon and energy profiles, and impact on natural ecosystems and communities become important. These indicators are individually scored using publicly available data such as annual reports and corporate social responsibility (CSR) reports. The individual indicators are normalized relative to the industry peers and then combined to obtain a final ESG score.

4.1 ESG Scoring Procedure

A brief explanation of ESG scoring methodology used by Refinitiv [22] is given below as an example, for a basic understanding of how ESG scoring is performed for an industry group. The procedure used in practice by other data providers do vary based on the firm's rules and methodologies in place. The ESG score guides published by Refinitiv and Sustainalytics Inc. can serve as representative examples for a detailed study of ESG evaluatory procedures [22, 23].

An Example of ESG Scoring Procedure: Refinitiv's Methodology[18]
Refinitiv provides a set of two overall scores on a scale of 0–100 comprising of:

. **ESG score** which is a measure of the company's ESG performance based on publicly reported verifiable data.
. **ESG Combined (ESGC) score** which is the ESG score overlaid with the ESG controversies impacting the company materially.

To arrive at these scores, Refinitiv follows a 5-step process as shown in Fig. 4, which begins with considering 186 ESG-related data points (a subset of the 450 + data points calculated) based on comparability, impact, and data availability. Depending on the relevance to the industry group considered, data points used for calculation varies from 70 to 170. These data points are grouped into 10 categories (or parameters) that belong to the three pillars (or factors) E, S, and G as shown in Table 2. The categories are further divided into themes and based on the various indicators' values, the themes are evaluated to calculate the category scores. Then, the E, S, and G pillar scores which are a relative sum of category scores are calculated. Finally, the overall ESG score is calculated as a sum of the pillar values. Furthermore, the ESGC score is

---

[18] The explanation provided here reflects Refinitiv's ESG scoring methodology [22].

evaluated considering 23 categories of ESG controversies and combining it with the ESG score. For a detailed explanation of the calculation process and representative illustrations, please refer to the Refinitiv publication [22].

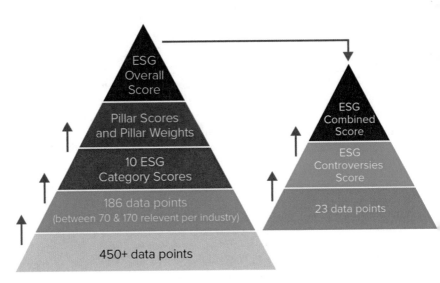

Fig. 4 Schematic illustration of the Refinitiv ESG scoring methodology. The scoring procedure can be summarized and illustrated by means of a five-step process flow (marked by grey arrows). Reproduced from Refinitiv publication [22] Copyright 2020, Refinitiv

Table 2 List of various categories and themes for ESG scoring considered by Refinitiv [22]

Pillar	Categories	Themes
Environmental	Emission	Emissions, waste, biodiversity, and environmental management systems
	Innovation	Product innovation, and Green revenues/R&D/capex
	Resource use	Water, energy, sustainable packaging, and environmental supply chain
Social	Community	Community engagement and responsibility
	Human rights	Human rights protection
	Product responsibility	Responsible marketing, Product quality, and Data privacy
	Workforce	Diversity and inclusion, Career development and training, Working conditions, and Health and safety
Governance	CSR strategy	CSR strategy, and ESG reporting and transparency
	Management	Structure (independence, diversity, committees), and Compensation
	Shareholders	Shareholder rights and Takeover defenses

5　Case Studies

Many companies have positively modified their environmental and social impact and altered their internal governance to enhance their ESG scores. Such efforts by the Coca-Cola Company, the Intel Corporation and Apple Inc. are discussed as case studies in the following subsections. The case studies are discussed here to give an idea of how the leading companies from various industry sectors have modified and revamped their operations to enhance their ESG performance.

5.1　Case Study 1: The Coca-Cola Company's ESG Performance

The Coca-Cola Company is a leading multinational corporation producing and marketing a variety of non-alcoholic beverages since 1886. It has a product portfolio of more than 3500 beverages that include soft drinks, dairy and plant-based drinks, energy and hydration drinks, tea, coffee, and bottled water marketed in more than 500 brands. With nearly 225 bottling partners (~900 bottling plants) in over 200 countries, it is critically dependent on resources such as water, raw materials such as agricultural produce, energy, and packaging materials.

On the other end, somewhere during its operations, the company's processes are contributing to global warming, climate change, and packaging waste (that pollute the local ecosystems) and affecting the health of its consumers (causing obesity and/or diabetes). Therefore, to lower the company's negative impact on the ecosystems and the society, and to have smooth and uniform governance across all its facilities, in 2015, the Coca-Cola Company board and administrative team prioritized several ESG issues to achieve a sustainable business [24]. These priority issues and strategies categorized into individual ESG factors appear in Fig. 5. This is followed by a brief explanation of the company's strategies to mitigate its negative impact on the environment and consequently, enhance its ESG performance.

Many of the issues listed in Fig. 5 are interconnected and need immediate attention. To transform the total beverage company into a sustainable business globally, the company has set a long-term vision. The major actions undertaken by the Coca-Cola Company in enhancing its ESG performance mainly include the efforts to reduce the negative impact of the company, related to ecological and social factors [24]. And this pursuit focused on three critical **E** (environmental) issues: water, carbon, and waste [24].

- *Water Stewardship*: The production of beverages is critically dependent on the availability of high-quality water. Also, communities living in the vicinity of the company's business locations depend on the nearby freshwater resources. The growing threats to the quality and availability of freshwater across the globe due to various reasons such as overuse of water resources, industrial pollution, and unsustainable agricultural practices will further strain the beverage

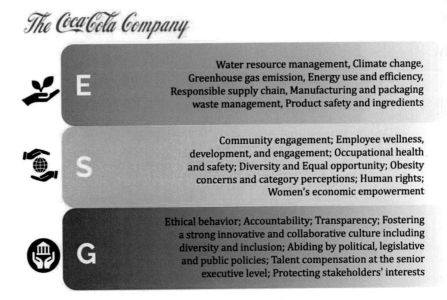

Fig. 5 Illustration of the key ESG parameters as prioritized by the Coca-Cola company board and administrative team for an enhanced ESG performance. These parameters are categorized into environmental, social responsibility and governance factors, and are considered to lower the company's negative impact on the environment and society and ensure a uniform governance across all its facilities

business. To tackle this, the Coca-Cola Company is trying to ensure that the company's operations contribute positively to maintaining freshwater resources and the natural water cycle, indirectly benefiting the communities around. The company's strategy is to use water judiciously and improve water security in the local environment through Water Stewardship Programs. The company has set targets to replenish more than 100% of the water they consume annually and increase water usage efficiency by 25% by 2020 compared to that in 2010 [24].

2. *Circular Approach for Lower Carbon Footprint*: The increase in carbon emissions has resulted in climate change effects such as disruption in the weather patterns and more frequent and severe natural calamities. This directly impacts the business operations of the company along with the surrounding communities. In the coming years, such climate aberrations will continue to affect agricultural activities and create food security issues. Besides the environmental and social concerns, the local governments are also introducing policies that incentivize the reduction in carbon emissions. Such changes motivate and drive the company to assess carbon emissions regularly in their value chain and reorient their supply chain to the one having a lower carbon footprint. Therefore, the company is taking responsibility to reduce its carbon footprint and achieve the climate change goal of the Paris Agreement.[19] This is being achieved by reducing emissions from

---

[19]https://unfccc.int/process-and-meetings/the-paris-agreement/the-paris-agreement.

manufacturing processes by improving energy efficiency and moving to renewable energy sources, and evaluating and making changes in operations throughout the Coca-Cola system value chain (manufacturing processes, packaging formats, delivery fleet, refrigeration equipment, and ingredient sourcing). The company's target is to reduce carbon emissions by 25% by 2030 considering 2015 as the baseline [24].

3. *Circular Approach for Zero-waste*: Post-consumer packaging waste is a major global issue. Besides the pollution issues related to the oceans and coastal regions, plastic packaging has also become a major constituent of the landfills. Such packaging waste not only pollutes the environment but also increases the carbon footprint of the product. To address the waste issue resulting from plastic packaging, many governments are introducing EPR policies. Consumers are becoming aware of the cost of material inefficiencies in linear production models. Therefore, the Coca-Cola Company has made zero packaging waste as one of its core strategies. The company plans to achieve this by setting three major goals: 100% recyclable packaging by 2025, 100% recycling rate by 2030, and using 50% of recycled materials for packaging by 2030 [24]. They are trying to eliminate low-value multi-layered packaging and other non-recyclable materials, increase the collection rates of clear PET bottles and introduce innovative methods to recycle colored bottles and dirtier waste streams.

To enhance the **S** performance, the company focused on product innovation for a healthier diet [24]. People's preferences and tastes keep evolving. As a consumer-centric business, the beverage company keeps altering and refashioning its product portfolio. Consuming less sugar as a part of the healthier diet has gained traction around the world. Hence, the company is making efforts to gradually reduce sugar (and calories) across its entire product portfolio. They have been aggressively reformulating the recipes to reduce sugar and promoting low/no sugar beverage options such as organic tea, coconut water, and natural fruit juices. The company has also introduced smaller packages to control sugar intake and new local favorites with nutritional and hydration benefits. Every product is provided with nutritional labels to assist consumers in making informed choices.

In addition to this, the company has set a goal of investing at least 1% of the annual operating income back into the local communities across the globe. It is also focused on increasing the number of women entrepreneurs across its global value chain as an effort to empower women. Since 2010, its target is to empower 5 million women by 2020 [24].

The internal governance or the **G** factor at the Coca-Cola company upholds principles and practices to promote an innovative and collaborative culture. The company's various administrative and executive committees are committed to ethical and transparent governance. Human rights principles are imbibed into their business culture and it is evident in their interactions with employees, bottling partners, suppliers, customers, consumers, and the communities. The principles start with employee safety and health at the workplace through minimizing the risk of accidents, injury, and exposure to health hazards across all of its business facilities. Stakeholder engagement is a priority for the company and hence, stakeholders are engaged in regular transparent communications and their different views and values are respected in decision making. As a global business, diversity and inclusion are critical for long-term sustainability. It encompasses diverse partners, gender and cultural diversity in the workforce and an inclusive environment to attract, recruit, develop, engage, and retain diverse talents.

It is also actively involved in its supply chain management. As the beverage industry is heavily dependent on agricultural products such as fruit juices, coffee, tea, herbs, sugar, and soy, the company has promoted procurement of ingredients from sustainable agriculture. During 2013–2018, there has been an increase from 8% to 44% in ingredients certified to a sustainability standard [24].

With all the aforementioned efforts from the Coca-Cola Company to enhance its ESG performance, its ESG risk score provided by Sustainalytics is 26 on a scale of 0–100 (at 40th percentile) as of January 2020. This means that the financial risk factor is medium for the shareholders. The individual scores of **E, S** and **G** are 9.2, 11.5 and 5.4 respectively. The company is also facing a significant number of controversies with a controversy score of 3 on a scale of 0–5.[20]

5.2   Case Study 2: Intel Corporation's ESG Performance

The Intel Corporation, popularly known as 'Intel' is a multinational company that manufactures semiconductor electronic components and devices since 1968. Today the company has evolved from being PC-centric to a data-centric business. This evolution was driven by the rapid growth of cloud data usage, artificial intelligence (AI) internet-of-things (IoT), data analytics and more recently the transition to 5G. With this, Intel is also reinventing itself on various crucial fronts such as environmental sustainability, responsible supply chain management, positive social impact, and key governance measures such as transparency, employee welfare, diversity and inclusion to achieve a sustainable business [25]. A summary of how Intel is prioritizing ESG in its business operations (Fig. 6) is presented below.

Intel's commitment to enhance its **E** performance is visible through its companywide set targets related to reducing greenhouse gas emissions, green energy portfolio, restoration of water resources, and waste management [25]. The company is working aggressively towards treating and restoring 100% of water consumed for its manufacturing processes by 2025. It is also expanding green power consumption in its global operations and its efforts in increasing energy efficiency have been duly recognized and awarded [25]. All the new buildings since 2015 are designed to

---

[20]The Coca-Cola Company (KO) ESG Risk Ratings by Sustainalytics Inc. (Last updated on 1/2020) https://sg.finance.yahoo.com/quote/KO/sustainability/.

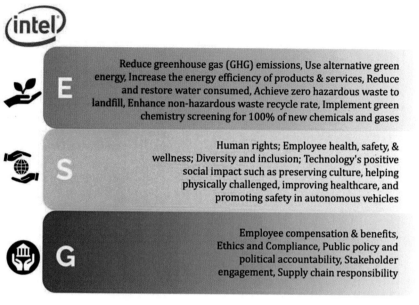

Fig. 6 Illustration of the key ESG parameters as prioritized by the Intel corporation's board and administrative team for an enhanced ESG performance. These parameters are categorized into environmental, social responsibility and governance factors, and are considered to lower the company's negative impact on the environment and society and ensure an uniform governance across all its facilities

be greener and energy-efficient (LEED[21]-certified).[22] The company has committed to CE principles in recycling manufacturing waste, especially hazardous waste, to achieve a 'zero hazardous waste to landfill' status quo by 2020. It has also implemented several programs to reduce, reuse and recycle office furniture and other non-hazardous waste to achieve a 90% non-hazardous waste recycling rate by 2020 [25]. Intel has also demonstrated the implementation of CE principles in its product design by eliminating harmful materials such as lead and halogenated flame retardants.

Besides such efforts, Intel has also made significant investments in reducing its ecological footprint, thereby playing its role in achieving the United Nations Sustainable Development Goals [25]. Intel technologies are supporting other leading companies in reducing their respective environmental footprints and driving sustainable consumption and production, as well as assisting several governments and non-profit organizations (working on environmental conservation projects) in protecting the natural ecosystems. For example, Intel's new AI technology developed with support from the Leonardo DiCaprio Foundation, RESOLVE (non-profit) and the National Geographic Society is helping fight illegal poaching of wildlife in Africa [25].

What is LEED? https://www.usgbc.org/help/what-leed.

Intel Building Certifications http://www.gbig.org/collections/14125/activities.

Driven largely by its vision of addressing global challenges and enhancing its credibility, Intel has been investing in technologies and innovations that empowe individuals and society in general [25]. Some of the examples of such technologica innovations of Intel are listed below.

1. Wheelchairs equipped with Intel's AI technology is maximizing mobility an independence of physically challenged people with spinal-cord injuries.
2. Intel is collaborating with start-ups to develop an advanced breast cance diagnosis system based on AI, using data analytics and machine learning.
3. With its drone technology and AI, Intel is helping to renovate the Great Wall o China and preserve its cultural legacy.
4. Intel is also working on improving the safety aspects of autonomous vehicle using its AI technology.

Intel also takes care of its employees' health, safety, and wellness by providin onsite health centers, fitness classes, and facilities. Intel is also committed to creatin social impact through three different programs [25] as listed below.

1. It encourages the employees to volunteer in engaging the local communitie (inclusive of schools, non-profit and non-government organizations) for solvin local societal and environmental issues.
2. The company is also empowering communities by providing exposure to tech nologies and learning opportunities, thereby inspiring young minds and th potential next-generation innovators.
3. The Intel Foundation invests in Science, Technology, Engineering, and Manage ment (STEM) education programs, and organizations providing humanitaria relief to survivors of natural calamities.

Intel has been striving to perform well in governance (**G** factor) and believes i the inclusion of diversity in its workforce as diverse perspectives can often lea to creative and innovative outcomes [25]. Such efforts include employee hirin and retention, encouraging women and underrepresented minorities to pursue thei careers in high technology, doing business with diverse suppliers, and diversifyin its venture portfolio.

The company follows a strict supply chain policy to eliminate forced and bonde labor, engage its suppliers to develop their corporate responsibility strategies, s high ethical standards, and be transparent about their performance. As a foundin member of the Responsible Business Alliance (RBA), Intel collaborates extensivel with supply chain-related organizations to help set electronics industry-wide bus ness standards and expects its suppliers to comply with the Intel Code of Condu and the RBA Code of Conduct (such as employee working hours, safety standard and minimal environmental impact). Through communication, assessments, an capability-building programs, Intel has developed a supply chain that is resilien responsible, and respectful of human rights [25].

Intel is trying to reduce it's environmental and social footprint with effectiv governance strategies. This has been reflected in the company's ESG risk score.

Sustainalytics rates Intel ESG performance as 17 (at 11th percentile) on a scale of 0–100 as of January 2020, which implies quite low financial risks. The individual scores of **E, S** and **G** are 4.9, 5.2 and 6.9 respectively. It also has a controversy score of 3 on a scale of 0–5. This shows that the company is encountering quite a significant number of controversies although it is lesser than the peer average.[23]

5.3 Case Study 3: Apple's ESG Performance

Apple Inc. being the largest IT company in the world in terms of the total assets and revenue is a designer, manufacturer, and retailer of mobile communication and media devices, personal computers, and portable digital music players. The company also has a business portfolio of its products related software, services, accessories, networking solutions, and third-party digital content and applications.

Apart from expanding its business and product portfolios, the company has also been making a conscious effort to reduce its carbon footprint and improve its **E** performance (Fig. 7), for over a decade now [26]. To meet this environmental objective, the company has set strict goals, focusing on three key priority areas listed below.

1. Climate change: Mitigating the impact of the company on climate change by using renewable energy resources and enhancing the energy efficiency of products, production facilities, and supply chain.
2. Resource Conservation: Minimizing the environmental impact by using recycled materials in manufacturing and reducing the consumption of virgin materials from nature.
3. Safer materials: Designing in such a way that safer materials are used in the product design and manufacturing processes.

To reduce its total carbon footprint (contributed by manufacturing, transportation, materials suppliers, and millions of customers), the company is using clean energy in its supply chain and at its facilities, increasing the energy efficiency of its business operations in the form of green buildings and consumption of recycled aluminium during manufacturing [26]. The company has already reached its target of 100% renewable (in 2018) and clean energy consumption in all its facilities worldwide inclusive of its offices, retail stores, and data centers [26]. As Apple consumes a significant quantity of aluminium in manufacturing, it has prioritized the use of aluminium smelted using hydroelectricity rather than fossil fuels and reengineered the manufacturing process to reincorporate scrap aluminium. It has also taken initiatives to reduce the packaging material by making it smaller and lighter, and collecting end-of-life products for recycling. Such efforts since 2011 have reduced Apple's carbon emissions by 54% even though its energy consumption tripled during this

[23]Intel Corporation (INTC) ESG Risk Ratings by Sustainalytics Inc. (Last updated on 1/2020) https://finance.yahoo.com/quote/INTC/sustainability/.

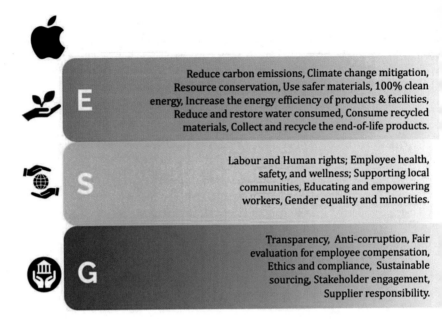

Fig. 7 Illustration of the key ESG parameters as prioritized by the Apple's board and administrative team for an enhanced ESG performance. These parameters are categorized into environmental social responsibility and governance factors, and are considered to lower the company's negative impact on the environment and society and ensure an uniform governance across all its facilities

period, thereby preventing nearly 2.1 million metric tons of CO_2e from entering the atmosphere [26]. In the process of transiting to renewable energy, Apple has also invested significantly in new renewable energy projects and approximately 66% of the renewable energy Apple procures comes from such projects [26]. Also, worthy of mention is the fact that the company encourages its employees to avoid using single occupancy vehicles and instead increase the use of electrically charged vehicles and bicycles to get to work.

Apple is also taking initiatives to save water, rare minerals and other natural resources that are consumed directly or indirectly. As a part of its water-saving program, besides reducing the usage of water, the company is supporting the use of alternatives to freshwater such as recycled water, reclaimed water, and harvested rainwater. To reduce the electronic waste generated and reuse the material resources, Apple offers recycling in almost all the countries it operates and has diverted more than 500 million pounds of electronic waste from landfills to the manufacturing line since 2008 [26].

Apple has been actively implementing CSR programs for enhancing its **S** performance since 2011 [26]. The "Global Volunteer Program" encourages the company's employees to volunteer in philanthropy projects of their choice and contribute to local communities. The company also helps the victims of natural disasters through fundraising and donations. In addition, Apple has educational and training programs for its employees and suppliers in order to improve their efficiency at work. The company has also expanded its employee benefits program by allowing its female employees to take a total of 18 weeks of leave during childbirth and post-pregnancy, and father/non-birth parents are also allowed 6 weeks of parental leave. Apple has strict labor and human rights policies and insists on its suppliers to comply with the same. It includes a maximum work week of 60 h (higher compared to the average maximum workweek of 48 h in the European Union), employee health and safety training, employee compensation, and ergonomic research to better design the workstations. The company's workforce is diverse and inclusive, with women involved at all levels of technical and administrative work. It also made a historic move in 2018 by hiring 50% of its new employees from underrepresented ethnic groups for technical jobs.[24]

Setting high corporate governance standards for a strong **G** performance is what makes Apple a reliable company for all its stakeholders. That is why the executive and non-executive committees at the top levels of management are committed to transparent and fair governance. They take into consideration the interests of all the stakeholders including investors, suppliers, consumers, and the community before making operational decisions. They follow rigorous audits, evaluations for employee compensation, anti-corruption and business conduct policies, guidelines to address conflicts of interest, stock ownership, and many bylaws for managing organizational operations.[25]

Apple is taking numerous measures to better its ESG risk rating, especially in the past decade. However, its ESG risk score provided by Sustainalytics Inc. is 24 (32nd percentile) on a scale of 0–100 as of January 2020.[26] The individual scores of **E**, **S** and **G** are 0.6, 13.0 and 10.3 respectively. This implies that the company is doing very well in environmental sustainability, however, it has to take steps to improve its social responsibility and better the governance aspect too. Apple also has a controversy level 3 (on 0–5 scale) similar to that of The Coca-Cola Company and Intel.

5 Current Outlook

Practicing CE principles in business operations and management aids companies to improve their overall ESG ranking. Adopting CE also helps companies in reducing operational costs and therefore, ESG practice has evolved gradually as a cost-anchored approach for the companies. The main value of the CE from an operational perspective for businesses is that it aids to reduce the costs of the supply chain or

---

[4]Apple Corporate Social Responsibility (CSR) by John Dudovskiy https://research-methodology. net/apple-corporate-social-responsibility-csr/.

[5]Leadership and Governance https://investor.apple.com/leadership-and-governance/.

[6]Apple Inc. (AAPL) ESG Risk Ratings by Sustainalytics Inc. (Last updated on 1/2020) https://sg. finance.yahoo.com/quote/AAPL/sustainability/.

generate value from byproducts, consequently making the business more competitive in addition to decreasing the negative environmental impact of companies. Implementing circularity could improve companies' environmental performance while creating value and increasing revenue. Most companies incorporate CE-based ESG practices to enhance their overall "green" image to attract sustainability investors. Therefore, ESG scoring is far from being an absolute truth in determining how sustainable companies are. It is when ESG concepts get embedded into the core business of a company, ESG practice will become more engaged with the idea of a CE. Besides, for systemic change towards a CE, ESG should not be limited to publicly traded companies. Government institutions and NGOs must also embrace measure, and perform well under ESG.

Reporting on environmental sustainable goals, social responsibility, and governance factors is an important practice for communicating the efforts of a company in achieving the ESG goals other than their conventional economic goals (e.g. profits) Consequently, embracing ESG as a part of the mainstream business has lately gained traction and ESG-oriented investing has experienced a meteoric rise. According to McKinsey 2014 estimates, globally, more than $13 trillion were invested in the businesses that incorporated ESG criteria in their operations [10]. In 2019 alone, the global sustainable investment topped $30 trillion—a 68% increase since 2014 and tenfold since 2004 [27]. This increase in the investments on sustainable businesses is a consequence of heightened social and governmental attention on the broader impact of companies on the environment and society, as well as the foresight of investors and financial advisors who are realizing that a strong ESG proposition can safeguard and promote a company's long-term success. There is empirical evidence of better returns resulting from investments based on ESG scores [11]. In particular, portfolios with high ratings for ESG criteria such as eco-efficiency, employee relations, and transparency in corporate governance have set new milestones in the stock markets and investing in general, with comparatively lesser downside risks. A good ESG performance by companies can also result in institutional investors such as banks investing in such companies through green bonds.[27] Also, companies with better ESG scores have benefited with higher credit ratings and lower costs of debts translating to better profits.

With the commercial relevance stated above, ESG related knowledge is still abstract and is not being imparted as a part of the curriculum in mainstream business management studies. This has resulted in an inadequate appreciation of ESG metric and their significance in the current corporate set up, and scant consideration of ESG datasets by average analysts and investors. Such a situation may put companies and investors in significant financial risks in this fast-paced market. It, therefore, become imperative to introduce ESG in mainstream engineering and business education, as a result of which business management (inclusive of operations and risks) and financial investments are focused on enabling sustainable businesses.

Questions

---

[27]https://www.investopedia.com/terms/g/green-bond.asp.

1. Bring out the interconnection of E, S, and G factors (as explained in the first section) through an example.
2. Explain with an example, how adopting CE principles can improve the ESG performance of a company.
3. List the existing international standards for sustainability reporting.
4. Explain why non-financial sustainability reporting became important for companies in recent years.
5. Write an ESG case study of a G250 company.
6. Listed below are a few top ESG-rated companies from various industry sectors. Find out the ESG ratings of these companies as given by MSCI.

Company	Industry sector
Edwards lifesciences	Medical products
Cadence design systems	Computer software-design
Microsoft	Computer software-design
Texas instruments	Electronics-semiconductor Mfg.
Applied materials	Electronics-semiconductor equipment
Procter and gamble	Cosmetics/personal care
ResMed	Medical products
Agilent technologies	Medical-research equipment/services
Alphabet	Internet content
Adobe	Computer software-desktop
PepsiCo	Food and beverage
Methode electronics	Electronics-parts

7. List the relevant indicators belonging to the resource use parameter for the fashion industry.
8. Explain the role of corporate governing bodies in enhancing E and S performance.
9. Calculate the percentile scores for the given values of an indicator for a set of companies.

Company	Indicator value
A	0.00016684373
B	0.00017996645
C	0.00027148692
D	0.00029749855

10. List the E, S, and G risk ratings of the Coca-Cola Co and Intel. Suggest three measures these companies can take to improve their E performance.
11. How can Apple improve its corporate social responsibility to enhance its S performance? Suggest up to three measures.

12. Explain the underlying reasons for Apple's superior E performance.
13. State the differences between a company's ESG score and ESG risk rating.
14. What is the green-washing phenomenon? Has the ESG scoring methodology contributed to this issue? Explain briefly.
15. Define green economy. Conceptually, how does a circular economy differ from a green economy?
16. What is a green bond? Briefly describe its connection to ESG performance.

Suggested Reading

1. Rajesh, R. (2020). Exploring the sustainability performances of firms using environmental, social, and governance scores. *Journal of Cleaner Production, 247* 119600.
2. Aureli, S., Gigli, S., Medei, R. et al. (2020). The value relevance of environmental, social, and governance disclosure: Evidence from Dow Jones sustainability World index listed companies. *Corporate Social Responsibility and Environmental Management, 27*(1), 43–52.
3. Orazalin, N. (2020). Do board sustainability committees contribute to corporate environmental and social performance? The mediating role of corporate social responsibility strategy. *Business Strategy and the Environment.*
4. Brogi, M., & Lagasio, V. (2019). Environmental, social, and governance and company profitability: Are financial intermediaries different?. *Corporate Social Responsibility and Environmental Management, 26*(3), 576–587.
5. Lagasio, V., Cucari, N. (2019). Corporate governance and environmental social governance disclosure: A meta analytical review. *Corporate Social Responsibility and Environmental Management, 26*(4), 701–711.
6. Braun, A. (2019). *ESG and circular economy (CE): Do they know each other Arctic values.* http://avkaksi.arctic-values.com/?page_id=601.

Acknowledgements The authors wish to acknowledge Dr. Lerwen Liu (STEAM Platform) and Ms. Julia Walker (Refinitiv) for their insights and guidance on this chapter. The second author wishes to thank the Italian Ministry of Foreign Affairs and International Cooperation (MAECI High Relevance Bilateral Projects), for the project "Analysis on the metabolic process of urban agglomeration and the cooperative strategy of circular economy" (2018–2020). The authors also thank Ms. Claudia Freed (EALgreen, IL, USA), and Dr. Anoop C. Patil (The N.1 Institute for Health, National University of Singapore) for their valuable comments.

References

1. Ellen MacArthur Foundation. (2013). Towards the circular economy. *Journal of Industrial Ecology, 2,* 23–44.
2. Ellen MacArthur Foundation. (2015). *Towards a circular economy: Business rationale for an accelerated transition* (pp. 1–20).

3. Sariatli, F. (2017). Linear economy versus circular economy: A comparative and analyzer study for optimization of economy for sustainability. *Visegrad Journal on Bioeconomy and Sustainable Development, 6*(1), 31–34.
4. Ramkumar, S., Kraanen, F., Plomp, R. et al. (2018). *Linear risks (joint project between circle economy, PGGM, KPMG, EBRD, and WBCSD)* (pp. 1–14).
5. Lewandowski, M. (2016). Designing the business models for circular economy towards the conceptual framework. *Sustainability, 8*(1), 43.
6. Stahel, W. R. (2016). The circular economy. *Nature, 531*(7595), 435–438.
7. Corona, B., Shen, L., Reike, D., et al. (2019). Towards sustainable development through the circular economy—A review and critical assessment on current circularity metrics. *Resources, Conservation and Recycling, 151*(104), 498.
8. Ellen MacArthur Foundation and Granta Design. (2015). *Circularity indicators—An approach to measuring circularity* (pp. 1–12).
9. Jawahir, I., & Bradley, R. (2016). Technological elements of circular economy and the principles of 6R-based closed-loop material flow in sustainable manufacturing. *Procedia CIRP, 40*(1), 103–108.
10. Henisz, W., Koller, T., & Nuttall, R. (2019). Five ways that ESG creates value. *McKinsey quarterly,* (Nov issue) (pp. 1–12).
11. Hoepner, A. G. F. (2013). *Environmental, social, and governance (ESG) data: Can it enhance returns and reduce risks?* (pp. 1–18).
12. Domenech, T., & Bahn-Walkowiak, B. (2019). Transition towards a resource efficient circular economy in Europe: Policy lessons from the EU and the member states. *Ecological Economics, 155*, 7–19.
13. Zhijun, F., & Nailing, Y. (2007). Putting a circular economy into practice in China. *Sustainability Science, 2*(1), 95–101.
14. Bocken, N. M., De Pauw, I., Bakker, C., et al. (2016). Product design and business model strategies for a circular economy. *Journal of Industrial and Production Engineering, 33*(5), 308–320.
15. Ellen MacArthur Foundation. (2017). *A new textiles economy: Redesigning fashion's future* (pp. 1–147).
16. Castellani, V., Sala, S., & Mirabella, N. (2015). Beyond the throwaway society: A life cycle-based assessment of the environmental benefit of reuse. *Integrated Environmental Assessment and Management, 11*(3), 373–382.
17. KPMG International. (2017). *The road ahead—The KPMG survey of corporate responsibility reporting 2017* (pp. 1–58).
18. Ballou, B., Heitger, D., & Landes, C. (2006). The rise of corporate sustainability reporting: A rapidly-growing assurance opportunity. *Journal of Accountancy, 202*(6), 65–74.
19. Brown, H. S., de Jong, M., & Levy, D. L. (2009). Building institutions based on information disclosure: Lessons from GRI's sustainability reporting. *Journal of Cleaner Production, 17*(6), 571–580.
20. del Mar, A.-A. M., Llach, J., & Marimon, F. (2014). A closer look at the global reporting initiative sustainability reporting as a tool to implement environmental and social policies: A worldwide sector analysis. *Corporate Social Responsibility and Environmental Management, 21*(6), 318–335.
21. Liern, V., & Pe´rez-Gladish, B. (2018). Ranking corporate sustainability: A flexible multidimensional approach based on linguistic variables. *International Transactions in Operational Research, 25*(3), 1081–1100.
22. Refinitiv. (2020). *Environmental, social and governance (ESG) scores from Refinitiv* (pp. 1–25).
23. Sustainalytics, Inc. (2019). *The ESG risk ratings methodology* (pp. 1–16).
24. The Coca-Cola Company (2018) *The Coca-Cola Company Business and Sustainability Report* (pp. 1–72).
25. The Intel Corporation. (2019). *Corporate responsibility at Intel* (pp. 1–68).
26. Apple Inc. (2018). *Environmental Responsibility Report*, 2018 Progress Report Covering Fiscal Year 2017 (pp. 1–76).

27. Lang, K., Electris, C., Voorhes, M. et al. (2018). Global sustainable investment review. *Globa Sustainable Investment Alliance*, 1–29.

Rashmi Anoop is a circular economy enthusiast and an enginee by profession with a Bachelors in Electronic Engineering from Visveswaraiah Technological University, India. As a membe of the Circular Economy Task Force at the National Univer sity of Singapore (NUS) (led by Prof. Seeram Ramakrishna Chair, Circular Economy Task Force, NUS), she is currentl researching on circular economy concepts. She is passionate about sustainability and ecofriendly businesses.

Patrizia Ghisellini is a postdoctoral researcher in the researc group headed by Prof. Sergio Ulgiati, affiliated to the Depart ment of Science and Technology at the University of Naples Parthenope. She graduated in Economics of Public Adminis trations and International Institutions and further specialized a academic level in the topics related to the sustainable develop ment and management of environmental systems as well as i life cycle assessment method at the National Agency for th New Technologies, Energy and Environment. She received he PhD from the Alma Mater Studiorum—University of Bologn in Agricultural and Food Economics and Policy. She is currentl involved at the University of Naples, Parthenope in a high rele vance project "Italy-China" funded by the Italian Ministry o Foreign Affairs and International Cooperation and other project funded by the EU and Local Italian Public Administrations. Th main goal of these projects is evaluating the implementatio of the circular economy at national and local level in priorit economic sectors such as C&DW, waste electric and electroni equipment, municipal solid waste, agriculture and food industr urban forestry. She has published several articles on circula economy. Besides the circular economy, her research interest include sustainable development, urban forestry, and life cycl sustainability methods.

Professor Seeram Ramakrishna *FREng, Everest Chair* (https://www.eng.nus.edu.sg/me/staff/ramakrishna-seeram/), is among the top three impactful authors at the National University of Singapore, NUS (https://academic.microsoft.com/institution/165932596). NUS is ranked among the top five best global universities for engineering in the world (https://www.usnews.com/education/best-global-universities/engineering). He is the Chair of Circular Economy Taskforce. He is a member of Enterprise Singapore's and ISO's Committees on ISO/TC323 Circular Economy and WG3 on Circularity. He also the Chair of Sustainable Manufacturing TC at the Institution of Engineers Singapore and a member of standards committee of Singapore Manufacturing Federation (http://www.smfederation.org.sg). He is an advisor to the Ministry of Sustainability & Environment—National Environmental Agency's CESS events, (https://www.cleanenvirosummit.sg/programme/speakers/professor-seeram-ramakrishna; https://bit.ly/catalyst2019video; https://youtube.com/watch?v=ptSh_1Bgl1g). European Commission Director-General for Environment, Excellency *Daniel Calleja Crespo, said, "Professor Seeram Ramakrishna should be praised for his personal engagement leading the reflections on how to develop a more sustainable future for all"*, in his foreword for the Springer Nature book on Circular Economy (ISBN: 978-981-15-8509-8). He is a member of UNESCO's Global Independent Expert Group on Universities and the 2030 Agenda (EGU2030). He is the Editor-in-Chief of the Springer NATURE Journal Materials Circular Economy—Sustainability (https://www.springer.com/journal/42824). He is an Associate Editor of eScience journal (http://www.keaipublishing.com/en/journals/escience/editorial-board/). He is an opinion contributor to the Springer Nature Sustainability Community (https://sustainabilitycommunity.springernature.com/users/98825-seeram-ramakrishna/posts/looking-through-covid-19-lens-for-a-sustainable-new-modern-society). He teaches ME6501 Materials and Sustainability course (https://www.europeanbusinessreview.com/circular-economy-sustainability-and-business-opportunities/). He also mentors Integrated Sustainable Design ISD5102 project students. Microsoft Academic ranked him among the top 25 authors out of three million materials researchers worldwide based on H-index (https://academic.microsoft.com/authors/192562407). He is named among the World's Most Influential Minds (Thomson Reuters) and World's Highly Cited Researchers (Clarivate Analytics). Listed among the top three scientists of the world as per the Stanford University researcher study on career-long impact of researchers or c-score (https://drive.google.com/file/d/1bUJrvurVVBbxSl9eFZRSHFif7tt30-5U/view). He is an Impact Speaker at the University of Toronto, Canada Low Carbon Renewable Materials Center (https://www.lcrmc.com/). He is a judge for the Mohammed Bin Rashid Initiative for the Global Prosperity (https://www.facebook.com/Make4Prosperity/videos/innovation-inclusive-trade/479503539339143/). He advises technology companies with sustainability vision such as TRIA (www.triabio24.com),

CeEntek (https://ceentek.com/), Green Li-Ion (www.Greenli-ion.com) and InfraPrime (https://www.infra-prime.com/vis ion-leadership). He is a Vice-President of Asian Polymer Association (https://www.asianpolymer.org/committee.html) He is a Founding Member of Plastics Recycling Association of Singapore (PRAS). His senior academic leadership roles include University Vice-President (Research Strategy), Dean of Faculty of Engineering; Director of NUS Enterprise and Founding Chairman of Solar Energy Institute of Singapore (http://www.seris.nus.edu.sg/). He is an elected Fellow of UK Royal Academy of Engineering (FREng), Singapore Academy of Engineering and Indian National Academy of Engineering He received PhD from the University of Cambridge, UK, and The TGMP from the Harvard University, USA.

Circular Economy Practices in India

Prasad Modak

Abstract Circular economy is considered as one of the important strategies to address the Sustainable Development Goals (SDGs). Its multi-stakeholder platform encouraging a partnership approach towards resource conservation, resource efficiency and resource recycling has been found to be promising. In many countries, policies and strategies on circular economy are formulated to build the recycling infrastructure, promote new business models and spur innovations, especially in responsible product design. India has been on a pace of economic growth. Transiting to circular economy is therefore very relevant to India to achieve its development goals without compromising on the resource security. The various missions launched by the Government such as Make in India, Zero Defect India and programmes like Smart Cities and *Swachh Bharat Abhiyan* (Clean India) resonate with the principles of circular economy. This chapter introduces the concept and evolution of circular economy. It explains the key 6Rs, i.e. reduce, repair, refurbish, remanufacture, reuse and recycle. Challenges faced in the upstream and downstream of the material flows are also described. Case studies that present successes in moving towards circular economy are included with a focus on India. Finally, the chapter ends summarizing the status and way ahead in India's circular economy.

Keywords India · Circular economy · Sustainability · Resource security · Recycling · Product redesign · Inclusive growth

Learning Objectives
Understand the evolution of circular economy
   Get an overview of the international scenario
   Get introduced to the business models that drive circular economy
   Case studies in circular economy focusing on India
   Understand the challenges faced
   Need to move towards regional circular economy
   Status and way ahead in India's circular economy

P. Modak (✉)
Environmental Management Centre LLP and Ekonnect Knowledge Foundation, Mumbai, India
e-mail: prasad.modak@emcentre.com

© Springer Nature Singapore Pte Ltd. 2021
L. Liu and S. Ramakrishna (eds.), *An Introduction to Circular Economy*,
https://doi.org/10.1007/978-981-15-8510-4_27

1 Introduction

India faces several environmental challenges today. Our resources are under threat due to intensive depletion and serious degradation. Further, we realize that risks related to our resource security are compounded due to looming threats of climate change. Policies and strategies for responding to these challenges need mainstreaming of sustainability across all developmental sectors.

Strangely, and oddly enough, the national governments, particularly the Ministries of Environment across the world focused more on the management of residues rather than management of the resources. Legislation was evolved to establish limits on the residues that were required to be met prior to disposal but not much attention was paid on the limits of extraction of resources and resource pricing. Government of India is not an exception to this bias.

Resource extraction across the world is getting increasingly intensive. Material flows of both virgin and used/secondary materials are getting skewed. Some of the important factors responsible for the shift are market globalization, presence of perverse subsidies (i.e. unrealistic resource pricing) and unevenness in environmental governance.

In early 2019, India's apex policy body National Institution for Transforming India (NITI) Aayog came up with an action plan to accelerate Resource Efficiency (RE) and circular economy. Ministry of Finance, Government of India set up a Task Force on Sustainable Public Procurement (SPP). Such policy decisions have pushed the agenda of building domestic recycling infrastructure, develop and promote sustainable product designs and support innovative business models toward circular economy.

2 About Circular Economy

Circular economy offers a platform for all stakeholders to get involved for sustainable and inclusive development. In addition to addressing environmental sustainability, circular economy improves the businesses' competitiveness, generates employment, increases green investment flows. Circular economy is built on partnerships and helps in establishing a transparent and inclusive governance. Essentially, circular economy aims to redesign the production and consumption systems by closing the loop. It is therefore an opportunity for the Indian government to leverage this concept and meet its economic, environmental and social objectives. Figure 1 depicts the evolution of the concept of the circular economy.

Figure 2 shows the key elements of such as transition between the linear model of growth and the growth models that encourage circularity.

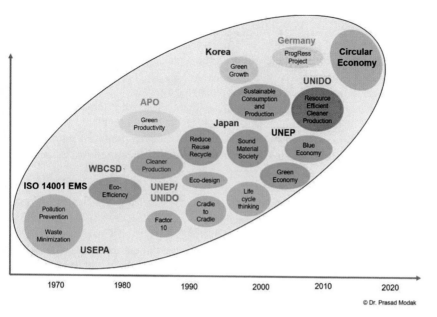

Fig. 1 Evolution of the concept of circular economy

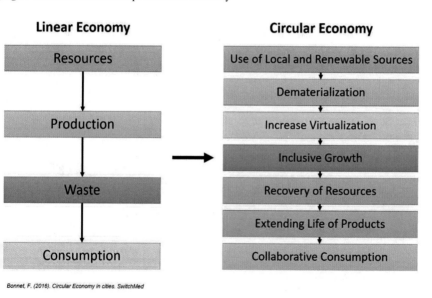

Fig. 2 Transition from linear to circular economy

3 International Scenario

One of the landmark legislation to push the agenda on circular economy was the Sound Material Society Act (2000) in Japan. This act followed the mantra of 3Rs viz. Reduce, Reuse and Recycle and demonstrated a decoupling between GDP growth with waste materials reaching the landfills. China legislated Circular Economy Law in 2007 focusing on 'circular' industrial estates and setting up large scale Material Recycling Facilities (MRFs). The government of Korea promoted the concept of circular economy through its Green Growth initiative stressing low carbon growth in 2009.

The European Union came up with country-specific targets, indicators and reporting requirements on the circular economy. Germany launched programmes Progress-I and Progress-II that focused on increasing Resource Efficiency (RE) and thus the Domestic Material Recycling Rates (DMR). The Government of South Australia developed a strategic action plan for the circular economy with impressive ground results. Gradually, several countries across the world as well as large corporations started transiting towards the circular economy.

4 Understanding the 6Rs of Circular Economy

The concept of circular economy added additional 3Rs namely—Repair, Refurbish and Remanufacture. These additional 3Rs strengthen three significant components viz. social (employment and engagement, especially of the informal sector), innovation (for the start-ups) and green investments. Box 1 describes the characteristics of these additional 3Rs.

> **Box 1: Repair, Refurbishing and Remanufacturing**
>
> Repair is the restoration of a broken, damaged, or failed device, equipment, part, or property to an acceptable operating or usable condition. Repair can involve replacement.
>
> Refurbishing is refinishing and sanitization (beyond repair) to serve the original function with better aesthetics. Repaired and refurbished products, although in good condition, may not be comparable with new or remanufactured products.
>
> In remanufacturing, the product is resold with performance and specifications comparable to new products.

Figure 3 explains the 6Rs in the form of outer and inner circles of circular economy. When we follow these circles, it helps to save money, conserve our resources, generate employment and come up with innovations.

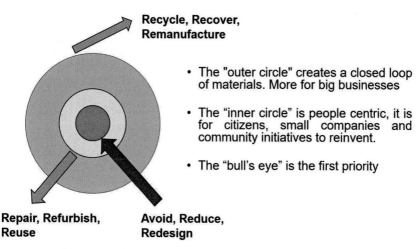

**Recycle, Recover,
Remanufacture**

- The "outer circle" creates a closed loop of materials. More for big businesses

- The "inner circle" is people centric, it is for citizens, small companies and community initiatives to reinvent.

- The "bull's eye" is the first priority

**Repair, Refurbish,
Reuse**   **Avoid, Reduce,
Redesign**

Fig. 3 Outer and inner circles of the circular economy

The 'outer circle' approach creates a closed loop of materials through recycling, recovery and remanufacturing. In the case of electronic goods, this means recovering of precious metals lodged in our gadgets, something only feasible with a sophisticated technology, requiring a scale and investments. Here, the medium to large companies make profit.

The 'inner circle' approach is essentially following the route of repair, refurbishing and reuse. It is the inner circle approach where we transform our living from the single-use and throw away culture. We extend product's life cycle through reuse. The inner circle is people-centric; it is for citizens and supports small companies. We need both the circles but the bull's eye to avoid, reduce and redesign should be the priority as it is intimately related to sustainable consumption.

5   Business Models in Circular Economy

To implement circular economy, new business models need to be identified and nurtured. Figure 4 describes a typology of the business models that map the life cycle.

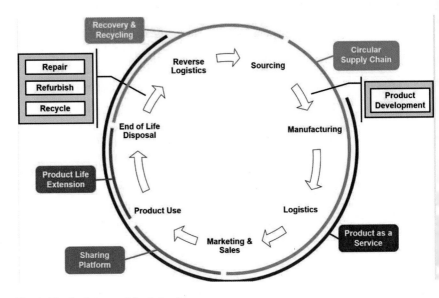

Fig. 4 Five business models of circular economy

Box 2 illustrates few examples of the five business models of circular economy

---

Box 2: Examples of Circular Business Models

Circular Supply Chain

Johnson Controls uses a circular supply chain and reverse logistics network to design, make, transport, recycle and recover vehicle batteries. It has reached 99% recycling rate for conventional batteries in North America, Europe and Brazil, whilst their sold batteries are now made up of 80% recycled materials [1].

Product/Packaging Development

Dell in India replaced foam, paper and plastic cushioning with bamboo cushions. Bamboo packaging helped reduce environmental impact, as bamboo is grown locally, it grows quickly and provides a strong and durable packaging [2].

Recycling

Interface, Inc. is a global commercial flooring company with an integrated collection of carpet tiles and resilient flooring, including Luxury Vinyl Tile (LVT) and nora® rubber flooring [3]. The carpets are designed in a modular form so if one part is bad it can be replaced with other. Further, the parts used are recyclable and made from renewable materials.

Product as Service

Michelin Solutions has adopted product as a service model by offering tire as a service. The company leases sensor-enabled tires to fleet customers, effectively selling a service which is monetized per kilometre driven. Customers do not own the tires and therefore don't have the responsibility of maintenance. Extending the lifecycle of tires saves consumption of more than 400 million tires and save 35 million tons of emissions globally.

Sharing Platform

BlaBlaCar connects drivers with vacant seats to co-travellers looking for a ride. People travelling to the same place can share their journey with many people sharing the cost of travel as well. More than 10 million people use BlaBlaCar every quarter. Approximately 1.6 million tonnes of CO_2 were saved by BlaBlaCar car-poolers in 2018. With only 1.6% more cars on the road, carpooling enables the transport of two times the number of passengers in cars (+210%), whilst reducing CO_2 emissions by 26% [4].

Product Life Extension

Repair

Antara Mukherji co-founded Repair Café Bengaluru on November 2015 along with Purna Sarkar. Since its inception, Repair Café Bengaluru has organized 19 workshops where adults pay a programme fee and learn how to repair household things ranging from iron to an induction top. The organization claims to have repaired more than 800 products and saved about 1,300 kg of waste from ending up in landfills [5].

Refurbish

Green dust is an online shopping site that offers customers and bulk buyers the option to purchase unused, branded factory seconds, surplus, overstock and refurbished products at the lowest prices. Green dust sells refurbished products the goods that are defective in the appearance of the product includes scratches and smudges, hence fail to meet the rigorous inspection criteria, but not affecting the functionality of the product [6].

Caterpillar India has been remanufacturing and repairing construction equipment under its Cat Reman business. The model is integrated into the entire value chain with incorporation of modular design principles and setting up of reverse logistics to collect used equipment in return for customer credit. The remanufacturing business employs 4000 people across 17 locations worldwide, refurbishing millions of components, thereby using 93% lesser water, 86% lesser energy and emitting 61% lesser [7].

Reconditioning

Tata Motors launched 'Tata Motors Prolife' service programme to recondition engine long blocks, aggregates and parts to ensure quality reconditioning which would result in superior performance of the vehicle. The service programme aimed to help extend the life of aggregates using a reduced quantity of materials compared to that required for a new part or aggregate [8].

Unfortunately, many of the case studies on the successful implementation of circular economy do not quantify, record and communicate the environmental and social benefits due to product/packaging redesign or by extending the product life due to repairing or refurnishing.

6 Role of Digital Technologies

Emergence of digital technology is expected to play a major role in circular economy especially in India, given its strength in the IT sector.

Attero was launched as India's first integrated end-end electronic waste recycling facility. Found in 2007, the Noida-based company has developed patented state-of-the art recycling technology to recycle and extract valuable materials viably, even with smaller e-waste volumes. The company offers refurbishment and reconditioning services to extend the useful life of electronics and a digital portal to enable take back from the end consumers. The company has developed a robust reverse logistics network backed by IT with collection centres in 22 states. There are several city-based initiatives too [9].

Mahindra's Trringo is India's first of its kind tractor and farm equipment rental and sharing platform, launched to improve asset utilization and address the equipment gap in Indian agriculture. It operates through a dual model, a digital platform-based B2B model where tractors are given out to franchisees to set up local hubs and a C2C model where large farmers can rent out underutilized equipment to other farmers. Trringo currently has over 100,000 registered users across five states in India [10].

Internet of Things (IoT) enabled waste collection and transportation can bring in significant advantages in the overall implementation of waste management solutions. The Urban Local Bodies (ULBs) can check contractor's effectiveness using IoT technologies. Deployment of smart bins, tracking of garbage pickup trucks as well as the sanitation workers, route optimization for trucks, cross-checking of garbage weight etc., can efficiently address the challenges of enforcement and transparency. Similarly, IoT-enabled sensors can also monitor the amount of alternate fuel generated from the processed waste.

BioEnable in India provides smart waste bin sensor which can be used in any container to monitor any types of waste real time. It can be used to monitor the fill level of the container. It provides complete solution with manpower that makes waste collection and dispatch operations both efficient and cost-effective. These tools eliminate hours spent on manual routing, maximize productivity, optimize equipment and staff allocations, and allow you to gain better control over your solid waste management operations. BioEnable technology has helped Hyderabad and Pune municipal corporations [11].

7 Strategies and Challenges to Adopt Circular Economy

Circular economy requires a change in the way we live. Given the rising rate of urbanization, the increasingly prosperous middle class (especially in Asia) and the promotion of consumerism through media, it is extremely difficult to expect this change will ever happen! If you say no to a product because you feel there is no need, someone will simply dump the product on you (as a free trial or as a friendly gift) to trap you or enslave you! Taming wasteful consumption is therefore the first step.

We therefore need to educate the citizens on the consumption itself and guide them to make 'green choices', i.e. avoiding the use of products to the extent possible that use harmful chemicals and non-biodegradable materials in the first instance. This will ensure circularity in the downstream. The production patterns should be influenced by responsible consumption. The manufacturers will need to extend their involvement beyond the factory gates and take responsibility across the life cycle. This is the principle of Extended Producers Responsibility (EPR). Like in many countries, especially in the European Union (EU), India has also legislated EPR focusing on electronic and plastic wastes.

The second important strategy is product/packaging redesign. Companies need to exhibit out-of-the-box thinking to find ways to reduce material and energy intensity and increase recycled content in their products. Products need to be redesigned to reduce/eliminate hazardous substances, increase recyclability (and improve safety during recycling) and make remanufacturing possible with most of the components getting reused. This requires innovation, risk appetite and top management commitment—and this cannot be achieved overnight. In India, today there are only a handful eco-design schools.

But working only on the 'upstream' alone is not sufficient, years of inefficiencies in manufacturing practices, rising waste volumes across the world need to focus on recycling. The 'downstream' industry addresses recycling and recovery—extracting metals, biosolids, refuse or solid derived fuels, biogas, syngas, heat, electricity, engineered materials, etc., from and reversing material flows and thereby reducing the consumption of virgin resources. The global waste recycling industry today supports significant employment—both in formal and informal sectors. Millions of poor people in the world's largest cities earn their livelihood because waste is around. India's recycling industry is supported by more than 4 million waste-pickers, waste collectors (*kabaddiwalas*) and small scale informal recyclers.

Recycling has many benefits. Firstly, it conserves natural resources as the extraction of virgin materials is reduced. Further, recycling diverts waste that is to be sent to incinerators and landfills. Landfills take up valuable space and emit methane, a potent greenhouse gas; and although incinerators are not as polluting as they once were, they still produce noxious emissions. Unless you segregate waste at source you cannot do effective recycling. So, segregation of waste at source and recycling must go hand in hand.

The waste recycling industry has seen numerous innovations. To address the challenge of plastic waste menace, an rPET initiative was launched globally at the Textile Exchange Recycled Polyester Round Table of 2017. Taking this cue, Reliance Industries in India launched a fashion brand R-Elan that uses used PET bottles. A pair of jeans of R-Elan is estimated to save 24% of GHG emissions, 3,218 litres of water, 33% of consumption of pesticides and divert 15 PET bottles to the landfill [12] (Fig. 5).

Sadly, the waste recycling industry wants more waste to be produced—so that the waste recycling business can grow and survive. Therefore, the strategy of 'Reduce' at 'upstream' can affect the 'downstream' opportunities for recycling and recovery.

There are examples where a CEO of a waste-to-energy plant who used to hate bans on plastics as they would reduce the calorific value of waste. A Common Effluent Treatment Plant (CETP) company discouraged members of the CETP to reduce the effluent volumes by specifying in the contract a guarantee for effluent supply. So

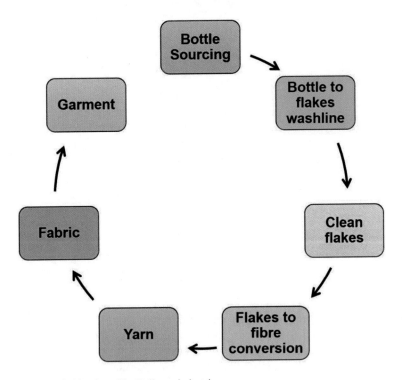

Fig. 5 Circular fashion brand by Reliance industries

has been the case in many Public-Private Partnership (PPP) contracts for managing Municipal Solid Waste wherein waste supply guarantee is an essential precondition. 'Don't you ever reduce waste that you have committed', the PPP partner warns.

There are many such examples of conflicting interests between the stakeholders involved 'upstream' and 'downstream'. In 'Reduce', top management, product designers and consumers play a dominant role, whereas in 'Recycle', waste pickers, community and waste processing specialists have a greater interest. The two groups rarely have a dialogue. This defeats sustainability and does not encourage circular economy in true sense. An integrated approach is the need of the hour.

According to economists, activists and many in the design community, the solution therefore is to get smarter about both the design and disposal of materials, and shift responsibility away from local governments and into the hands of manufacturers [13]. This is where legislation on EPR plays a role. Products as well as packaging need to be designed with recycling in mind. Waste generation should be considered as a design flaw. Remedying this problem may require a complete rethinking of industrial manufacturing. We also need to develop standards on recycled products and the recycle content. While India has initiated this process, there is a long way to go. The Task Force set up on Sustainable Public Procurement (SPP) and the reconstituted committee on India's Eco-Mark ecolabel may perhaps accelerate effort in this direction.

8 Case Studies from India

Case Study—1: Banyan Nation

Banayan Nation was founded by Mani Vajipey in 2013. Under the mentorship of Ron Gonen a pioneer in the waste management, he incubated Banyan Nation at Columbia. With arrival of Raj Madangopal, his batchmate from the University of Delaware, Mani could get assistance from key investors and advisors and set up operations in Hyderabad in India. Today, Banyan Nation is in the spotlight for being one of the finalists at the World Economic Forum's Circular Economy Awards.

Banyan Nation is a creative, plastic waste recycling company which is among the first's vertically integrated recycling companies. Banyan nation produced near-virgin-quality recycled plastic granules, Better Plastic™, from different post-consumption and industrial goods. Consumer goods, food and beverage, and automotive companies can use Better Plastic™ to manufacture more sustainable products and packaging.

Banyan's Nation's technology platform integrates many informal recyclers into its supply chain. It has pioneered India's first bumper to bumper closed-loop recycling initiative with Tata Motors. It helped in initiating a unique bottle to bottle recycling programme for a cosmetics brand. By 2017, it recycled over 500 tons of plastic and reduced over 750 tons of carbon dioxide and diverted over 1,000 tons of plastic from landfills [14].

Banayan Nation created data tracking platform where it could monitor waste flows in Hyderabad covering over 1,500 stationery recyclers. The IoT platform provided a bird's eye view of waste streams in Hyderabad.

Key points

- Use of frontier technologies and application of IT platform helps to design a powerful reverse logistic system to convert waste into resources.
- In countries like India, social engineering is equally important, and it is necessary that an ecosystem is established with the informal waste pickers for the collection the waste.
- Such business models lead to significant benefits such as providing liveli hoods to the waste pickers, reducing consumption of virgin resources without compromising quality and leading to reduction of GHG emissions.

Case Study—2: Partnership Projects on Waste Recycling in Indian Cities

Hindustan Coca-Cola Beverages (HCCB) Pvt. Ltd and Hindustan Unilever Ltd (HUL) are India's largest FMCG manufacturing and distribution companies.

The UNDP India and HCCB and HUL formed a partnership to create a model that ensured a circular economy for plastic waste by generating value out of used plastic UNDP played a role as a Program Manager by inducting NGOs and entrepreneur for organized sorting of the mixed municipal waste. UNDP also supported creation of awareness materials for citizens on waste segregation and management, educating the citizens on not to litter. The municipal corporation provided space, water and electricity to store, sort and stock the separated waste streams. HCCB and HUL funded the programme that included equipment such as weighing machines, shredders, bailing machines, etc., and supported training of the waste sorters. The waste after sorting and bailing is sold to the *kabadiwalas* and the industries to earn money that is used to pay the daily wages of the waste sorters and maintain the facility [15]

The goal is to operate such integrated recycling programmes in 50 major cities while improving the socio-economic conditions of waste-pickers, especially women The objective is to manage over 85,000 tonnes of plastic waste per year and improve lives of over 37,500 waste pickers or *safaii mitras*. The programme already operate in 17 cities. The idea is to create an environment where every player in the recycling value chain, starting from the waste collectors to recyclers are connected. These integrated recycling programmes will help establish multi-stakeholder partnership to create an end-to-end ecosystem for plastic recycling in the country.

Figure 6 Illustrates the stakeholder engagement in plastic waste recycling.

Similar to the UNDP-HCCB-HUL partnership, Indian Tobacco Company (ITC launched a programme Wellbeing out of Waste (WOW). The initiative is operational in over 10 States and 562 municipal wards. It has generated livelihood for around 14,500 waste collectors and ragpickers. It has impacted around 77 lakh citizens and collects around 50,196 metric tonnes of dry waste and more than 5000 tonnes of multi-layered laminates and thin films.

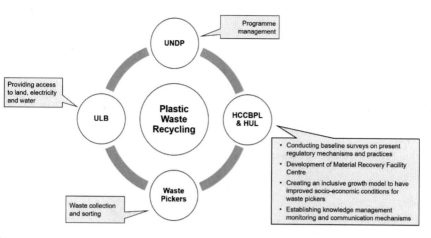

Fig. 6 Multi-stakeholder partnership in plastic waste recycling in Indian cities

WOW works in collaboration with different stakeholders, with each stakeholder having a role and responsibility. It optimises resources—using existing infrastructure where available and creating new where required. Designed to be a viable business model, it creates value for all players and generates employment boosts incomes and supports livelihoods for economically backward groups [16]. Figure 7 illustrates the stakeholder engagement in the ITC WOW initiative.

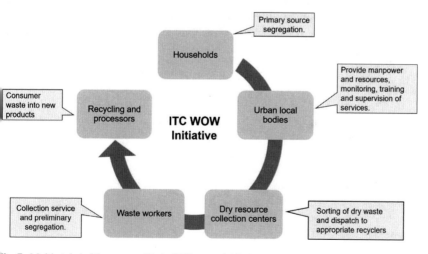

Fig. 7 Multi-stakeholder partnership in ITC's wow initiative

Key points

- Given the complexity of the canvas of circular economy and the interests of diverse stakeholders, a partnership approach is appropriate to implement circular economy projects.
- The examples described on waste recycling in India show successful partnership between Urban Local Bodies (ULBs), international agencies like UNDP, the private sector like HCCB, HUL, ITC and the social enterprises.
- The circular economy projects not only lead to economic and environmental benefits but also result in inclusive social development such as creation of livelihoods and addressing health and safety.

Case Study—3: Rethinking Product Design

India uses over 800,000 disposable plastic meal trays per day. About 99% of plastic meal trays end in landfills. The trays are made of mixture of a few polymers like polypropylene (PP) that is non-recyclable.

As the market is shifting towards biodegradable material especially due to plastic ban enforced in various states of India, an eco-friendly substitute for plastic meal tray was needed. A company named Bizongo created a corn-starch based, spill proof, 3 compartment tray [17].

Bizongo has 550 restaurants as clients across India which used plastic disposable meal trays on daily basis. The clientele in consideration was thus the urban market, the food delivery take-away joints, caterers and restaurants which primarily target the fitness-aware population residing in the metro cities of India.

The idea was not only to address a sustainable packaging design but ensure that is functional that would enable the storage, transport and support of Indian cuisine. Following was the consideration in the design:

- The material and the process: The major factor was to confirm that the product was biodegradable as well as cost effective so that it is easy to manufacture and efficiently produced. A composition of 71% corn-starch and 29% PP was used for the same product. The process of thermoforming was used to produce it.
- The Form and Communication: The product needed to possess a degree of resemblance, Grooves were made along the edges as well as between compartments so that the various elements of an Indian meal (often gravy-based) do not mix or spill. The colour of corn-starch is kept constant. The natural colour of corn-starch was kept intact. The form is made softer and more rounded. To inform customers of the decision of using eco-friendly product, a single line or text is added in the labelling of the product.

In 2018, Bizango won the prestigious Dieline Award for creating India's first biodegradable, spill-proof meal tray

Bakeys Edible Cutlery was founded by Narayana Peesapati with an idea of finding alternative solutions to the plastic used in restaurants. The cutlery is available in three flavours—savoury (salt and cumin), sweet (sugar) and plain. It can complement an

food. Since plastic is not biodegradable nor edible, Bakeys offers edible cutlery made from flours.

Bakeys has collaborated with Café Coffee Day as their prime supplier of spoons. As per the company, the spoons can decompose within 4–5 days and has a shelf life of three years. As of 2017 Bakeys sells, 1.5 million spoons per year to catering companies serving food at weddings and other events. The manufacturing unit employees 12 people for production and packaging work who produce around 30,000 spoons in 24 h.

The cutlery is priced more than the providers of mainstream product which adds up to 1.86 USD 100 pieces of plastic spoons as opposed to Bakey's price of Rs. 4.66 USD for the same quantity, fulfilling the first characteristic of the model.

Key points

- Banning is often not the solution to the problem of waste generation and subsequent environmental and social risks. Innovation in the product design and customer education are often the answers.
- Sometimes the costs of sustainable products may be high. But if all externalities are considered then the green products may turn out to be rather cost-effective. To establish this advantage, a comparative Life Cycle Assessment (LCA) is needed followed by communication on product sustainability to the consumers.
- Once a market demand for green products is created, then the sustainable products may become cheaper. Development of sustainable products that are lean on virgin resources and fossil-based energy, biodegradable and easy to disassemble for recycling help in reducing risks to the humans and environment and make the material flow circular.

9   Need to Move to Regional Circular Economy

Transition to circular economy is possible when policies and strategies are evolved at a regional scale. Implementing circular principles at project level alone does not often lead to the desired upscaling and replication to achieve the impact on a sustained basis.

Strategizing circular economy on a regional scale is done by formulating an enabling framework and by ensuring a multi-stakeholder involvement for delivery. When conceived at a regional level, a coordination between various departments and agencies also becomes necessary. Such a coordination helps to ensure harmony and synergy between various circular economy-related interventions. A system of performance or outcome indicators needs to be evolved that helps in adaption of policies and strategies.

Achieving circular economy for water on a regional scale for example may include the following strategies.

Set appropriate water pricing to discourage overconsumption of water

- Ask for the conduct of annual water audits at highly water-consuming industries
- Set benchmarks for water consumption and reward/punish if there are major deviations
- Support water-efficient technologies and show preference while granting environmental approvals and loans
- Sponsor research on technologies for wastewater recycling
- Make low flow plumbing fixtures mandatory in new buildings
- Insist on implementation of rainwater harvesting schemes in all public places
- Advise corporates to spend their CSR funds on watershed development
- Come with financing schemes to promote drip irrigation in agriculture
- Set up business models for water vending especially where piped delivery of water is not possible.

When we think of an enabling framework on a regional scale, we must address all the interconnections, e.g. in the case of water, we must widen the canvas by factoring the nexus between water, agriculture, food, waste and energy. Apart from recognizing the nexus, a consideration is also needed to account for impact of climate change. Climate change, for instance, can affect the water availability as well as water quality, requiring additional set of mitigation and adaptation measures. We expect to see the future of regional circular economy in these perspectives.

10 Summing up

India is estimated to become the fourth-largest economy in the world in about two decades. This economic growth is, however, going to come with challenges such as urbanization with increased vulnerability (especially due to climate change), poor resource quality and scarcity and high level of unevenness in the socio-economic matrix due to acute poverty. India, if it makes the right and systemic choices, has the potential to move towards positive, regenerative and value-creating development. Its young population, growing use of IT, increasing emphasis on social and financial inclusion can make this happen. For this, developing a national policy framework on the circular economy makes sense.

The recent report by the Ellen MacArthur Foundation on India [10] shows that a circular economy path to development could bring India annual benefits of ₹40 lakh crore (US$ 624 billion) in 2050 compared with the current development path—a benefit equivalent to 30% of India's current GDP. Following a circular economy path would also reduce negative externalities. For example, Greenhouse gas emissions (GHGs) would be 44% lower in 2050 compared to the current development path, and other externalities like congestion and pollution would fall significantly, providing health and economic benefits to Indian citizens. This conclusion was drawn based on a high-level economic analysis of three focus areas: cities and construction, food and agriculture, and mobility and vehicle manufacturing.

The Ministry of Environment and Forests and Climate Change (MoEF&CC) of Government of India set up the India Resource Panel (InRP) in 2016 to examine the material and energy flows across key sectors following a life cycle approach and to assess Resource Efficiency. Sectors such as Construction, Automobiles, Iron & Steel and Metals were considered, and key cross-cutting areas were examined. Recommendations of InRP were taken up by India's NITI Aayog leading to a paper on Strategy for Resource Efficiency. More recently, NITI Aayog released four sectoral publications on Steel, Aluminium, Construction and Demolition Waste and Waste from Electronic and Electrical Goods. In addition, an overarching report on the status on RE was produced with 32 recommendations addressing both 'inner' and 'outer' circles, emphasizing strengthening of the informal sector, forming a remanufacturing council and harmonizing waste management related regulations following a life cycle approach. Promotion of innovation and green investment flows following a PPP approach were also included as key interventions. Building a vibrant recycling industry in India was stressed given the recent green fencing of waste materials by China. Currently, MoEF&CC is finalizing a national policy on Resource Efficiency and Circular Economy on this basis. Box 3 shows some of the highlights of the national policy on resource efficiency.

Box 3: Highlights of Draft National Resource Efficiency Policy by MoEF&CC [18–22]

National Resource Efficiency Policy (NREP), 2019 seeks to create a facilitative and regulatory environment to mainstream resource efficiency across all biotic and abiotic resources, sectors and life cycle stages. This can be achieved by fostering cross-sectoral collaborations, development of policy instruments, action plans and efficient implementation and monitoring frameworks. Some of the highlights of the NREP are

- It describes the cost benefits of resource efficiency and the relevance of resource efficiency for sustainable development goals.
- For fostering resource efficiency, it emphasizes on the creation of a dedicated institution known as 'National Resource Efficiency Authority (NREA)' which would draw its power from Environment (Protection) 24 Act, 1986, to provide for the regulatory provisions of this policy. The NREA would have a collaborative institutional structure with members from line ministries, state governments, government agencies and stakeholders. An inter-ministerial National Resource Efficiency Board (NREAB) would provide necessary guidance on the aspects critical to the implementation of resource efficiency across all sectors.
- The policy encourages preferential procurement (for e.g. Green Public Procurement) of products with lower environmental footprints, by large public or private organizations that can be used to aggregate demand and create scale for products made from secondary raw materials. This would therefore bolster market demand.

- It highlights establishment of Material Recovery Facilities (MRF) equipped with best available technology systems for efficient end-of-life collection, followed by efficient sorting, and then the optimum suite of physical separation and metallurgical technologies for an economically viable recovery of metals.
- The NREP acknowledges Strengthening of Extended Producer Responsibility (EPR) systems to reduce the cost of end of life management of the products borne by taxpayers and municipalities by integrating sustainability measures into the design of products, including design for value recovery.
- The Policy advocates orienting Research and Development (R&D) support towards producing resource-efficient solutions and the development of resource-efficient products and services such as

  - Develop sound methodologies to carry out the inventorization and characterization of major waste streams
  - Industrial training networks to deliver the required courses and skill sets needed for implementing a resource efficiency and achieving a low-carbon economy
  - Improving knowledge and decision base for the secondary raw material sector with databases and dynamic forecasting models.

Whilst, policies are getting formulated through research and consultation, India has been following a regime of banning or announcing restrictions in manufacturing and us of certain materials and products. More than 24 States in India have banned certain categories of plastic, and national level target to phase single-use plastic by 2022 has been declared by India's Prime Minister. A year after China banned plastic waste imports, the Indian government also banned import of plastic waste in the country. These directives will also push the agenda on circular economy.

The Government of India has embarked on several iconic projects to improve and expand its infrastructure (transport, cities and energy) and undertake ecological modernization of important sectors such as water, agriculture and food. In these megaprojects, foreign direct investment is encouraged, and these investors are asking for good practices on Environmental and Social Governance (ESG) apart from conventional compliance. The 100 Smart Cities programme, Make in India initiative, *Swatch Bharat Abhiyan* (Clean India), *Namami Gange* (Ganga River Action Plan), Interlinking of Rivers, Climate Resilient Agriculture etc. are a few examples. In all these projects, an application of the principles of the circular economy is extremely relevant. It is, however, necessary that leadership on the circular economy is built in cities, industries, investors, project developers, and policymakers and regulators.

Circular economy is a concept that brings management and resources and residue together in the interest of economy, livelihoods and the environment. If implemented well, then it will spur innovation and stimulate green investments. For rapidly growing economies such as India, transiting to circular economy will be critical as it will help the country to progress towards the Sustainable Development Goals.

Questions

Q1 Will circular economy ensure sustainability?
Q2 Should circular economy be legislated?
Q3 What are the barriers to introduce changes in the upstream, e.g. reduce and redesign?
Q4 List few financial incentives and disincentives to promote circular economy
Q5 List three most important lessons learnt based on India's experience and efforts towards circular economy

Answers

A1

It is not necessarily true. Circular economy emphasizes resource efficiency but not always the resource sufficiency.

A2

China introduced Circular Economy Promotion law in 2007. There are mixed results of this legislation. Most believe that a circular economy is best realized through partnerships, technology innovations, social engineering and economic instruments

A3

Resistance to change in sourcing, production and product distribution systems.
Risks perceived on quality and functionality of the newly designed product
Material availability
Challenges on rebranding
Pricing

A4

Reducing taxes on recycled products or on products that have high recycle content
Providing grants or concessional loans to support innovative projects that demonstrate circularity
Levying high charges on waste sent to landfills

A5

A national policy on circular economy is needed to ensure coordination between the line ministries
For the purpose of leveraging, circular economy should be mainstreamed in the national missions
Partnership models play an important role in the effective implementation of circular economy projects. Role of informal sector is very important to ensure inclusive growth.

References

1. Foundation, J. (2019). *Johnson controls circular economy of automotive batteries and beyond* [online]. U.S. Chamber of Commerce Foundation. Retrieved September 30, 2019, from https://www.uschamberfoundation.org/johnson-controls-circular-economy-automotive-batteries-and-beyond.
2. ai, D. (2019). *Bamboo—Nature's eco-friendly packaging solution* [online]. Dell. Retrieved September 30, 2019, from https://www.dell.com/learn/ai/en/aicorp1/corp-comm/bamboo-packaging.
3. Finchandbeak.com. (2019). *How interface makes the circular economy come to life: Finch & beak consulting*. Retrieved September 30, 2019, from https://www.finchandbeak.com/1325.how-interface-makes-the-circular-economy.htm.
4. BlaBlaCar. (2019). *About us—BlaBlaCar* [online]. Retrieved September 30, 2019, from https://blog.blablacar.in/about-us.
5. Retrieved September 30, 2019, from https://www.crunchbase.com/organization/greendust#section-overview.
6. Accenture.com. (2019) [online]. Retrieved September 30, 2019, from https://www.accenture.com/t20150523t053139__w__/us-en/_acnmedia/accenture/conversion-assets/dotcom/documents/global/pdf/strategy_6/accenture-circular-advantage-innovative-business-models-technologies-value-growth.pdf.
7. Saha, P. (2019). *Year-end special: Repair economy 2.0* [online]. https://www.livemint.com Retrieved September 30, 2019, from https://www.livemint.com/Leisure/tPvH11Xns5lt9vXsb.9orJ/Repair-economy-20.html.
8. Tata Sustainability Group. (2018). *Closing the loop*. Retrieved September 30, 2019, from http://tatasustainability.com/images/NewsLetter/Files/40_circular%20economy%20bro_Final%2019%20June%202018.pdf.
9. Innovative Business Models and Technologies to Create Value in a World without Limits to Growth. (2019) [online]. Retrieved September 30, 2019, from https://www.accenture.com/t20150523t053139__w__/us-en/_acnmedia/accenture/conversion-assets/dotcom/documents/global/pdf/strategy_6/accenture-circular-advantage-innovative-business-models-technologies-value-growth.pdf.
10. Ellen Macarthur Foundation. (2016). *Circular economy in India: Rethinking growth for long term prosperity*. Retrieved September 30, 2019, from https://www.ellenmacarthurfoundation.org/assets/downloads/publications/Circular-economy-in-India_5-Dec_2016.pdf.
11. Smartbin Sensors—BioEnable. (n.d.). Retrieved September 30, 2019, from https://www.bioenabletech.com/smartbin-sensors.
12. R-Elan. (2019). *Raymond launches Ecovera in collaboration with Reliance industries*. Retrieved September 30, 2019, from https://www.ril.com/getattachment/0444c3ff-b7ce-488c-b7c4-de44a0caf480/Raymond-launches-Ecovera-in-collaboration-with-Rel.aspx.
13. Westervelt, A., *Is it time to rethink recycling?* Retrieved September 30, 2019, from https://www.vox.com/2016/2/13/10972986/recycling.
14. Mitter, S., *Banyan Nation is out to clean India and the world is taking note*. Retrieved September 30, 2019, from https://yourstory.com/2018/01/banyan-nation-clean-india-world-taking-note.
15. HUL. (2018). *Annual progress report plastic waste recycling management: A partnership*. Retrieved September 30, 2019, from https://info.undp.org/docs/pdc/Documents/IND/Annual%20Progress%20Report%202018_Plastic%20Waste%20Management.pdf.
16. ITC Limited. (2016). *Initiatives in solid waste management towards a circular economy*. Retrieved September 30, 2019, from https://www.itcportal.com/world-environmentday/pdf/WOW_Brochure_Text%20PDF_%20June%202018.pdf.
17. BIZONGO. (2017). *Biodegradable spill-proof meal tray*. Retrieved September 30, 2019, from https://bizongo.com/blog/wp-content/uploads/2018/06/Biodegradable-Spill-Proof-Meal-Tray.pdf.
18. MOEFCC. (2019). *National resource efficiency policy, 2019 (Draft)*. Retrieved September 30, 2019, from http://moef.gov.in/wp-content/uploads/2019/07/Draft-National-Resource.pdf.

19. Nike Sustainability. (2019). *Circular innovation challenge* [online]. Retrieved September 30, 2019, from https://purpose.nike.com/circular-innovation-challenge.
20. Patagonia Works. (2019). *Patagonia wins circular economy multinational award at world economic forum annual meeting in Davos—Patagonia works* [online]. Retrieved September 30, 2019, from http://www.patagoniaworks.com/press/2017/1/17/patagonia-wins-circular-economy-multinational-award-at-world-economic-forum-annual-meeting-in-davos.
21. Furlenco. (2019). *Furlenco* [online]. Retrieved September 30, 2019, from https://www.furlenco.com/about-us.
22. Button, F. (2019). *Sharing: A road to a circular economy* [online]. Floow2.com. Retrieved September 30, 2019, from https://www.floow2.com/news-detail/~/items/sharing-a-road-to-a-circular-economy.html.

Dr. Prasad Modak is currently Executive President, Environmental Management Centre LLP and Director, Ekonnect Knowledge Foundation. Earlier he was a Professor at Indian Institute of Technology, Bombay.

Dr. Modak has worked with almost all key UN, multi-lateral and bi-lateral developmental institutions and intergovernmental organizations in the world. Apart from Government of India and various State Governments, his advice is sought by Governments of Bangladesh, Egypt, Indonesia, Mauritius, Thailand and Vietnam.

He is a Member of Indian Resources Panel at MoEFCC, Member of Task Force on Sustainable Public Procurement at the Ministry of Finance. Dr. Modak drafted the Report on Resource Efficiency and Circular Economy—Current Status and Way Forward for NITI Aayog in India. He was a contributor to UNEP's Green Economy report, a Co-author of the Global Waste Management Outlook and Chief Editor of Asia Waste Management Outlook for the United Nations Environment.

Dr. Modak has published several books on environmental management. His book on EIA for Developing Countries was published by UN University. His latest book 'Environmental Management Towards Sustainability' was released in December 2017 with CRC Press of Taylor and Francis Group. Dr. Modak has been recipient of the Distinguished Alumni Award of AITAA in 2010 for Significant Contribution to International Affairs.

Circular Economy in Taiwan-Transition Roadmap and the Food, Textile, and Construction Industries

Hui-Ling Chen, Ya-Hsuan Tsai, Chiao-Ling Lyu, and Yu-Lan Duggan

Abstract Taiwan has played a pivotal role in the global supply chain. However, the pride of "Made in Taiwan" is built on the backbone of imported raw materials and environmental cost. Thus, a transition to a circular economy is necessary. A transition roadmap is introduced to help explain how businesses can transition from a linear economy to a circular economy. Case studies describing the development of a circular economy in various industries, particularly food, textile, and construction are also provided.

Keywords Taiwan · Circular economy · Policy · Transition roadmap for enterprises · Food · Textile · Construction · Business model · Supply chain · Innovation · Industrial symbiosis

Learning Objectives

- Understand the context and policy framework of circular economy in Taiwan
- Learn why, what and how businesses can make the transformation towards a circular economy
- Learn from cases in Taiwan from food, textile, and construction industries

1 Introduction

1.1 A Densely Populated Nation with Insufficient Resources

As one of the top twenty most populous nations, Taiwan is an island that hosts 23 million people within its 36,000 km². Whether it's assessed from the Human Development Index (HDI) [1] or Gross Domestic Product (GDP) perspective, Taiwan is highly rated in both social and economic development. Yet Taiwan is a nation that

H.-L. Chen (✉) · Y.-H. Tsai · C.-L. Lyu · Y.-L. Duggan
Circular Taiwan Network, Taipei, Taiwan
e-mail: info@circular-taiwan.org

© Springer Nature Singapore Pte Ltd. 2021
L. Liu and S. Ramakrishna (eds.), *An Introduction to Circular Economy*,
https://doi.org/10.1007/978-981-15-8510-4_28

is deficient in resources, importing roughly 70% of its resources (with metal, energy, fertilizers, and feed exceeding 90%) and 60% of its food [2]. As Taiwan is bound by the sea on all four sides and limited in landmass, waste management and resource supply challenges remain as Taiwan's most pressing issues.

1.2 From Trash Nation to Waste Management Genius

As Taiwan's economic growth rapidly took off in 1960 [3], so did the pileup of wastes. Garbage ended up on the streets due to a lack of landfill spaces, and parts of the country looked as if it was engaged in a trash battle. To solve this problem, the government began to install incinerators and promote recycling in 1987. Coupled with various other government leadership and educational initiatives, Taiwan's recycling rate rose to 60.6% in 2018 [4]. Such progress led the Wall Street Journal to refer to Taiwan as the "World's Geniuses of Garbage Disposal" in a recent article [5].

The "Four-in-One Recycling Program" is one of the most important policies that propelled the change. This unique recycling system has achieved great success as it is able to create a built-in economic incentive to encourage recycling. The program consists of four components: (1) community residents sorting their own trash; (2) local government collecting recyclables and waste separately; (3) recycling centers processing recyclables; and (4) a recycling fund that subsidizes recycling centers who then awards citizens, trash collectors, and businesses that participate in recycling. The recycling fund is created with the fee that manufacturers and/or importers of thirteen product categories tagged by the government must pay under the Extended Producer Responsibility (ERP) Scheme. This program has also made recycling a popular social activity. If you were to visit Taiwan today, you would see citizens at specific times throughout the day waiting with their sorted trash and recyclables chatting with each other as they wait for the garbage or recycling truck that is blasting Beethoven's music to arrive.

As the industrial waste accounts for 70% of the total waste volume, the Taiwanese government also began to seriously tackle industrial waste in 1997 by mandating GPS tracking on every truck that transports industrial waste. In addition, the government announced the "Waste Reuse Methods," [6] which is a directive that encourages companies toward recycling by imposing heavy fees for sending waste to the incinerator and/or landfill. This private-public partnership was able to help Taiwan achieve an industrial waste reuse rate of 75% as of 2018 [7].

2 Policy Development of the Circular Economy

Despite its success in driving recycling, in recent years, Taiwan's efforts have reached a plateau because many products are still constructed in a way that is hard to recycle. In addition, as recyclable items are often downcycled, they cannot meet the quality standards of manufacturers for new products. Therefore, Taiwan must redesign its economic system in a circular fashion more than any other nation to continue its productivity. It can also help Taiwan gain economic benefits by using raw materials more efficiently while reducing the externalities. Shifting toward a circular economy would allow Taiwan to reduce its dependency on raw materials, and to gain better control over its economic development and create exciting new prospects [8].

True to her inaugural speech where she said that "We cannot continue as we have done in the past, endlessly squandering natural resources and our citizen's health," President TSAI Ing-Wen led the Taiwan government to begin promoting the "Five Plus Two Innovative Industry Plan" in 2016. The five in the plan refers to the five major industries in Taiwan—Asia Silicon Valley, green energy, biomedicine, intelligent machinery, national defense and aerospace, and the two refers to two transformative strategies—circular economy and new agriculture. For Taiwan, a circular economy could bring synergy between industries, provide more local employment opportunities, minimize the differences between cities and rural areas and help Taiwan become a highly inclusive society. To help the country shift towards a circular economy, the Environmental Protection Agency has reformed the management framework while the Ministry of Economic Affairs has proposed an implementation program to ensure that both the environmental and commerce perspectives would be considered.

2.1 Environmental Protection Agency (EPA): From Waste Management to Resource Management

In 2018, the EPA has established a holistic framework that highlights four lifecycle stages, namely production, consumption, waste management, and market for secondary materials [9]. Established with twelve strategies and four performance indicators, the plan created a set of indexes, such as the[1] Resource Productivity and Cyclical Use Rate,[2] to measure how efficiently resources are used. This plan demonstrates the Taiwanese government's willingness to **move from end-of-life waste management to resource management**. Moreover, initiated by the EPA, 5 Green Deals autonomously established by four industries including the Waste Electrical and Electronic Equipment (WEEE), plastic, construction, and solar panels, in a hope to increase the resource circularity through public and private partnership.

Resource Productivity = GDP/DMC (Domestic Material Consumption).

Cyclical Use Rate = Recycling Material/(Recycling Material + Direct Material Input).

2.2 Ministry of Economic Affairs: Circular Economy Promotion Plan

The Ministry of Economic Affairs is the agency mainly responsible for carrying out the "Circular Economy Promotion Plan" [10]. The main thrust of the plan is to embed circular economy concepts and innovative sustainability thinking into the nation's economic activities. Specific strategies and measures include

- Establish a R&D center to promote circular technologies and materials innovation
- Establish a new circular economy demonstration park
- Encourage green consumption and green procurement
- Promote industrial symbiosis to integrate energy and resources in the industrial parks

From a resource-deprived trash nation to a garbage disposal genius, the government steered Taiwan toward a circular economy with relevant policies. In the next section, we will elaborate on how industries practice circular economy.

3 Circular Economy Transition Roadmap for Enterprises

To help industries shift their perspectives and implement changes more systematically, Circular Taiwan Network (CTN), a non-profit organization dedicated to advancing circular economy in Taiwan, created the "Circular Economy Transition Roadmap for Enterprises" that uses WHY, WHAT, and HOW to explain different steps of the transformation [11] (Fig. 1).

3.1 Taiwan Is Uniquely Positioned

Despite being a small island whose environment is not a great fit for linear manufacturing, Taiwan remains an important manufacturing base for the world due to it solid foundation and complete supply chain. Excelling in the areas of semiconductors, solar energy, telecommunications, plastics, textiles, and machinery, Taiwan i listed among the top 10 countries in the 2016 Global Manufacturing Competitiveness Index published by Deloitte [12]. In addition, Taiwan is ranked 12th in the world for competitiveness by the World Economic Forum's 2019 Global Competitiveness Report [13]. Such recognition validates Taiwan's strength internationally in terms of manufacturing and innovation. Thus, with the diversity of industries, especially combining its innovative manufacturing capability with its expertise in recycling, Taiwan is uniquely positioned to pioneer the circular economy.

In recent years, rising public awareness of ESG-related (environmental, social, and governance) risks is rapidly pushing ESG investment and regulatory change

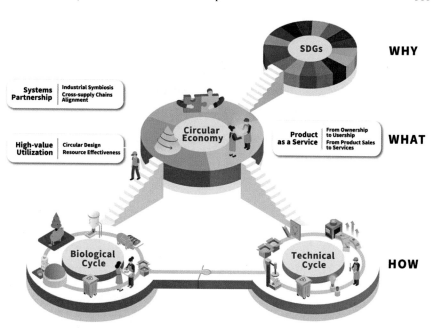

Fig. 1 Circular economy transition roadmap for enterprises

The move towards circular economy will not only make the industries more resilient but will also help Taiwan achieve many Sustainable Development Goals (SDGs) simultaneously. Together, it will contribute to a more balanced growth.

4 What Is a Circular Economy?

A circular economy is an economic and industrial system that is restorative and regenerative by design. In this roadmap, three strategic goals highlighted are "Product as a Service," "High-Value Utilization of Resources," and "Systems Collaboration."

4.1 Product as a Service

In the conventional "sell-more, sell-fast" model, companies can only generate revenue from selling products. Under this business logic, no matter how well companies perform in environmental measures, they will eventually deplete more resources and create more waste. However, globalization and convenient transportation has enabled rapid movement of population, together with the ever-evolving mobile technologies, young people are more used to adopt services then to own a product. There is a great opportunity for companies to "REDEFINE" the needs of the customers and clients. "Do they need bulbs or light? Do they need a dishwasher or clean dishes?"

In a circular economy, companies shift to provide professional and flexible services to meet the diverse needs of the market. In order to reduce the cost of maintenance and waste management, selling services instead of products drives them to design durable, easy-to-dismantle (even re-disassemble) and recyclable products, thus intrinsically more sustainable. There are emerging examples in Taiwan that echo this trend, such as Youbike, the city bike of Taipei city, WeMo and GoShare, the e-scooter services. Other examples including machinery, furniture, lighting, home appliance, etc. can be found in various fields.

4.2 High-Value Utilization of Resources

Every product we use and consume is the product of a complex supply chain where each player adds their value, from raw materials to manufacturing, to logistics, warehousing, marketing, etc. The accumulation of these values is inherent in the final product we consume. By preserving as much of that existing value we not only minimize our impact on the environment, but also rescue its intrinsic economic value. Circular design is crucial to make it possible to preserve values along the whole life cycle of products. This can be done on two fronts: the technical cycle and the biological cycle.

In technical cycles, instead of destructing the value of products after disposal extending the life or reusing a product in its original form should take precedence followed by the thinking of how to preserve the values of components. Whereas the components cannot be preserved, they go through the recycling system, processed into recycled materials.

In biological cycles, value is extracted through "biorefinery" that contemplates extracting inherent value through processes and industries beyond those traditionally associated with the product. Using the "Biomass Value Pyramid" as guide, products could flow through a series of different industries and uses along their lifecycle including in pharmaceutics, food and feed, bulk chemicals and fuels, before finally being used as energy or fuel.

4.3 Systems Partnership

Given that products today have a high degree of complexity and cut across various industries, no single company can realistically manage these cycles on their own; making collaborative efforts along the entire value chain and across industries is fundamental. Moreover, the government, the academia, research institutes, social organizations, and media need to work together to integrate the necessary measures along with the industrial upgrade. There are two specific ways that collaboration can be achieved:

1) Industrial Symbiosis: Arranging an exchange of resources (including raw material, energy, water, co-product, equipment, logistics, expertise, etc.) in a way that is most useful in a localized setting between companies to create a competitive advantage. Following the famous Kalundborg examples from Denmark, 23 industrial parks in Taiwan are practicing such type of collaboration, leading to cost reduction by approximately 3 billion NTD and GHG reduction of 803,000 metric tons [14].

2) Cross-Supply Chain Alignment: Manufacturers can partner with recyclers to provide a complete service package to clients. This turnkey solution guarantees the product until its end-of-use, when it will be recycled. The 'knowledge flows' and 'financial flows' need to be aligned accordingly with 'material flows' in supply chains.

5 How to Put the Circular Economy into Practice? Cases of the Biological Cycle

The best way to put circular economy into practice is to take the product life cycle, and consider the opportunities along every stage, from production, distribution, use, end-of-use, and back into the cycle again. In the biological cycle, biomass such as food, biomaterials, and cleaning agents are circulated in a way that they can safely turn into "nutrients" for a new product (Fig. 2).

5.1 Examples of a Biological Cycle in Taiwan

There is an old Chinese proverb that says "Do not let nutrients fall on other's land." It means that regardless of whether it's excretion from humans or animals, they both make great fertilizers and should only be used on one's own soil. This demonstrates that circular economy has long been practiced in agriculture. However, since the development of industrial agriculture, Taiwan uses more chemical fertilizer than the international community in order to produce more feed. Luckily, there are more and more owner-owned farms that are working hard toward achieving local and organic farming, which is a great start for circular agriculture. From the perspective of waste, Taiwan produces more than 10 million tons of biomass residue every year [15], including agricultural waste, food waste, etc. If it can all be fully utilized, then it can assist with local economic opportunities, local employment opportunities, and narrow the urban-rural gap. Below are some case studies in which biological cycle has been practiced in each stage.

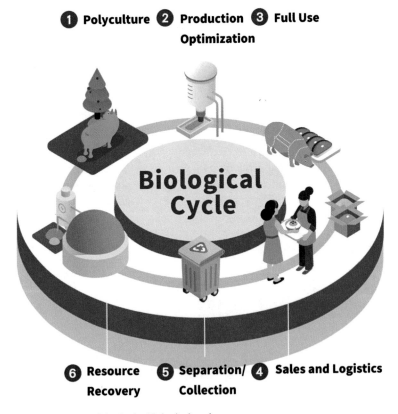

Fig. 2 Circular opportunities in the biological cycle

1. Polyculture

Unlike the conventional agricultural approaches where a single plant is cultivated, the production system is redesigned to maximize the productivity of farmlands through rotational cropping/grazing, mixed cropping/culture, or mixed culture of agriculture, forestry, fishery, and husbandry. Polyculture ensures better nutrient utilization as nutrients not used by one crop will be beneficial to another crop during rotation. It also ensures better soil utilization as the land will be able to be used year round.

2. Production Optimization

Redesigning the process in order to make the best use of the land, water, nutrients, energy, or resources required in production or processing.

3. Full Use

Taking the market needs and values into consideration, each part of biomass is put to use in the manufacturing of medicine, specialty chemicals, foods and feed, bulk materials and fertilizers, or fuels and electricity to achieve toward zero waste.

Example

AGRI-Dragon Biotech Co., Ltd. developed a biomimicry farming system that delivers water, air, and nutrient fluid directly into the soil through an underground irrigation module. This allows the soil to remain moist underground but dry on the surface, leading to a reduction in water usage by up to 70–80% and extremely low energy consumption. The pumped air and water create an environment similar to the rainforest in which the microorganisms produce metabolite to keep the soil productive without tillage. The shapes of the irrigation pipes mimic human arteries and capillaries to increase the contact area with the soil and then deliver water and nutrient fluid efficiently. In order to prevent the use of insecticide, AGRI-DRAGON began implementing polyculture techniques and mixed their main crop Danshen with scallions. Since they were able to grow Danshen without chemicals, the entire Danshen can be used in cuisine or made into spirits, fermented beverages, and even skincare products through the collaboration with several partners to achieve zero waste [11, 16].

4. Sales and Logistics

A set of process and practice with an enterprise, or in collaboration between companies, to minimize food waste during the distribution channels to table process.

Example

1919 Food Bank is a non-profit organization in Taiwan that created a platform to tackle the issue of food waste and helping people in need to get food. Through a Redistribute, Recreate, and Recovery method, it has created an effective system to ensure the full use of food resources. The food bank begins by accepting food donated by companies or markets, as well as ingredients that are close to expiring or ugly in appearance.

1919 Food Bank then creates food packets from packaged food, and redistributes them to poor families or children's accompaniment classes. In addition, donated food ingredients are recreated into frozen seasoning packets, which would prolong the timeframe in which the ingredients can be used, and sent to children's accompaniment classes in rural villages as well. Recovery takes place when leftover food is used to feed black soldier flies, who can eat 2–3 kg of food waste in their lifetime. The insects are then made into chicken or fish feed as they are rich in protein. From April 2011 to July 2019, 1919 Food Bank has distributed over 5,766 different items, worth a total retail value of 450 million NTD [11, 17].

5. Separation and Collection

The sorting and collecting and of biomass residue into different categories for the ease of resource recovery that follow.

6. Resource Recovery

Food Waste is processed into feed, compost, and/or energy as a new resource for another material or product.

Example

Taiwan Sugar Corporation (TSC) is one of Taiwan's biggest pig producers. In their latest project, Donghaifeng Agricultural Circulation Park, TSC combined negative pressure water curtain pigpens, biogas power generation, and solar photovoltaic systems into a green energy demonstration park, with scenic spots and local characteristics to promote environmental education and tourism. Anaerobic digestion is utilized to retrieve biogas for energy generation, while the slurry and residues are used to produce organic fertilizers, helping to promote the local economy and increase employment opportunities. In addition, TSC promotes a "pig house sharing" business model by leasing their modern pigpens to young farmers, which encourages them to return to their hometowns without the pressure to purchase new land and facilities. The biogas power generation produces approximately 1.2 million kWh and 5,000 tons of organic fertilizer every year [11, 18], which may lead to 2,000 tons of CO_2 eq reduction.

6 How to Put the Circular Economy into Practice? Cases of the Technical Cycle

Instead of destroying a product after disposal, the value of the original product can be retained through life cycle extension approaches such as reuse, repair, refurbish and/or remanufactured before recycling. This allows the company to maximize the value of the resources and energy invested in making the product (Fig. 3).

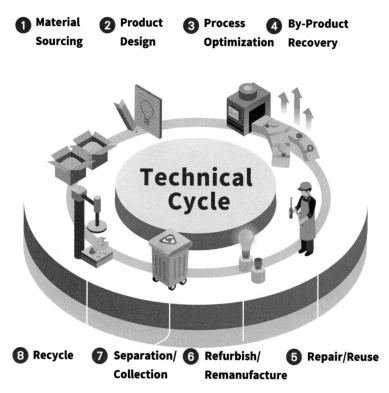

Fig. 3 Circular opportunities in the technical cycle

The linkage between biological cycle and technical cycle describes a synergy between these two cycles. For example, wood can be a raw material for home furniture, and through the technical cycle, be reused, repaired, and manufactured. At the end of its life cycle, it can enter the biological cycle via composting or anaerobic digestion, and become a newborn resource.

5.1 Examples of a Technical Cycle in the Textile Industry in Taiwan

The apparel industry is one of the biggest polluters on the planet. To make clothes, textile mills generate one-fifth of the world's industrial water pollution and use 20,000 chemicals, many of them carcinogenic. Taiwan's textile industry supplies 70% of the world's functional fabrics for outdoor wear and 40% of the fabrics for sportswear of major garment brands. As functional fabric is made from petroleum-based fibers such as polyester and nylon, Taiwan developed the technology to recycle PET bottles into polyester over 10 years ago in order to reduce the environmental impact. This practice has gained such wide acceptance that 16 out of the 32 countries participating in the 2018 FIFA World Cup chose to have their uniforms made from recycled yarn from Taiwan. In the future, the next step is to close the loop by using post-consumer clothes as raw materials, and Taiwan's textile industry is already working hard to develop this technology. Here are some examples of how the textile industry in Taiwan applies various circular approaches along the technical cycle.

1. Material Sourcing

To develop or employ safe and toxic-free materials that can be recycled and are durable; to develop or employ renewable materials including bio-based, biodegradable, or local materials; and to use recycled materials.

2. Product Design

The product is designed in a modular way, allowing for easy repair, dismantle, recycling, even upgrade. Throughout its life cycle, energy, water, and resource input are minimized.

Example

DA. AI Technology Co., Ltd. is a social enterprise dedicated to sustainability Working with the Tzu Chi Foundation, a charity with 9,000 collection points and over 80,000 volunteers, they have been producing eco-blankets and clothing made from recycled PET bottles since 2008. In addition, to solve the problem of discarded and leftover clothing, DA. AI Technology has developed the technology to recycle fiber back into PET chip form, which can then be used to create new products. In order to simplify the recycling process and improve the quality of recycled products they reduce the diversity of colors in all their clothing and developed eco-zippers and eco-buttons made of recycled PET. A physical process is selected to avoid the use of water resources and chemicals. In addition, it has developed a new composite material made of discarded fabric and recycled plastic that is then transformed into a durable DA. AI Tech Wood [11, 19].

3. Process Optimization

The water, energy, and resource utilization across stages in the manufacturing processes are maximized while the impacts on the environment are minimized. The use of equipment or chemicals in the production can be transitioned to the leasing model instead.

4. By-product Recovery

The by-product in a manufacturing process is introduced to another process or put out as a product for sale. The circulation can widen beyond the first stage as it can be distributed within the corporation, within the industry, and outside the industry.

Example

EVEREST Textile Co., Ltd., a textile company in Taiwan with vertical integration is a key fabric provider of Nike, Adidas, Patagonia, the North Face, Columbia and Spyder. To achieve their vision of zero waste and zero-emission, EVEREST optimized their production by replacing air conditioners with a composite cooling system and heat-exchanger that reduced 92% of the factory's energy consumption. The factory is also able to recover 86% of the wastewater produced by the water-jet loom. The by-product wet sludge from wastewater is used as auxiliary fuel after drying. The cinders from the incinerator are made into Eco-bricks which are verified as low-carbon construction materials [11, 20].

As the fast fashion industry is raging worldwide, Taiwanese textile manufacturers produce 21 million tons of leftover fabrics per year. To solve this problem, Tainan Enterprises Co., Ltd., collaborated with partners like EVEREST and Industrial Technology Research Institute to create a Fabric Bank. It aims to provide a marketplace for hundreds of stock fabrics for independent designers and design schools. Besides a physical site, QR code and Augmented Reality (AR) technology are introduced to ensure the disclosure of quality, thus fostering the utilization rate [11, 21].

5. Repair and Reuse

Repair is to restore a product to its original functionality via troubleshooting or parts replacement. Reuse is to extend the product life of a discarded item by having it be used multiple times by different users or for different purposes.

Example

In Taiwan, repairing clothes is not a new concept, but because of fast fashion, there are fewer repair shops nowadays. A group of youth formed a repair platform called Nothing Is Garbage [22], where they compiled the information of all the repair shops and created a city repair google map [23] to encourage the practice of maintenance.

Second-hand clothes are often reused through donation, exchanges, or second-hand shops. There are lots of used clothes donation boxes throughout Taiwan. Thus second-hand brand shops or free exchange activities are also becoming more common in Taiwan.

6. Refurbishand Remanufacture

Refurbishment is to conduct a relatively complete check-up and parts replacement in order to restore a product to a condition close to that of an almost new product while extending the life cycle of the product. Remanufacture is to ensure a used product has similar or better performance and warranty through reproduction.

Example

If a product can be used multiple times through the "Reuse, Repair, Renew, and Remanufacture" strategies to continue to provide service, not only will it lower the raw material risk, but it will produce the least amount of environmental impact. Especially during remanufacturing, if the product can maintain the same quality as a new item but is lower in cost to produce, then it is highly competitive on the market. As mentioned in the "From Manufactured in Taiwan to Remanufactured in Taiwan" report [24], Taiwan's machinery, semiconductor operations, motor, and medical equipment industry all have potential to sustain this strategy.

7.Separation and Collection

The sorting and collecting of end-of-life products into different categories for the ease of resource processing and recycling that follow.

8. Recycle

The process through which end-of-use materials or products are transformed into components or materials that can be used again.

Example

SHINKONG Synthetic Fiber Corporation developed a strategic alliance with Thread International PBC Inc. to turn marine debris into textile materials. They hire people in less developed countries like Haiti to collect abandoned plastic bottles at the beach. The plastics are then brought to SHINKONG factories and then go through a series of processes: cleaning, label removal, crushing, dehydration, then turned into PET flakes. These flakes then go through a filtration process for removal of different colors and impurities before turning into polyester fibers. The fibers are then shipped back to Thread and then weaved into the fabrics of textile products such as clothes, backpacks, and caps. The recycled PET process can reduce 51% of energy consumption and 59% of CO_2 emission [11, 25].

Service-oriented Business Model

All the above-mentioned approaches ultimately leads to a new way of thinking when it comes to the business model. When offering professional "services" which is accompanied by a particular product and generating revenue by "performance" instead of one off-trade, enterprises hence gain a full control of the flow and condition of the products and parts. It is then more feasible to arrange the supply chain and the reverse logistics that allows those value optimization processes.

Example

Clothes rental services are becoming more popular worldwide—Rent the Runway in US provides designer items for rent, while LENA Fashion Library in the Netherlands uses sustainability, texture, and design as their selection criteria for clothes, and it partners with many young designers and enduring sustainability brands. In Taiwan Amaze offers fashionable women's clothing that fits every occasion for rent. After the clients have selected their choices online, the clothes are sent directly to their home. After the clients are done wearing them, they can return it without washing the rental item. Amaze takes on the responsibility of cleaning the clothes before offering them up for rent again to control the quality and style of their clothes. Customers are able to enjoy a convenient rental experience as a result [26].

7 Case study: Circular Construction

Construction is an industry that uses more resources than others, and as such, it is the best place to implement circular economy. MINIWIZ Co., Ltd created EcoARK Pavilion, a nine-story building made from Bolli-Brick™, a revolutionary building material made from 100% recycled PET bottles. The modular 3D honeycomb self-interlocking structure makes it extremely strong without any chemical adhesives while weighting only one-fifth of traditional wall systems. The EcoARK used 300 thousand Polli-Bricks made of 1.5 million bottles. Another example is the Holland Pavilion created by The Netherlands Trade and Investment Office for a short-term exhibition during the 2018 Taiwan World Flower Expo which welcomed 3.5 million of visitors. The 3,477 kilos of wood was pieced together with dowels instead of nails that enable future reuse, with the aim of being able to be disassembled and put together again in the future as the goal. The 19 tons of heavy steel structure and 12 tons of light steel were bolted instead of welded for easier reuse.

Taisugar Circular Village: to Provide a "Residential Service" Taisugar Circular Village (TCV) is the first circular housing in Taiwan to offer a "residential service" [27]. TCV is the creation of a combined effort between Taiwan Sugar Corporation (TSC) and the Taiwanese government to promote green energy and circular economy. TSC will be renting out the apartments and provide all necessary maintenance to the residents ranging from building all the way to furniture and electrical appliances. Its design, new construction, and operation all incorporate the ideals of circular economy.

In selecting its material, TCV used old wood from a former Taisugar factory and recycled steel railroad tracks as the fence around the facility, reducing the need to use 1300 m of new fence. The design of the building is also geared toward a modular design, maintaining the flexibility to be disassembled and remanufactured into another product at its end-of-use. In addition, the life cycle and character of each material was carefully considered, and a materials passport was created so that each item can be accurately accounted for and tracked. It is estimated to save over 3000 tons of CO_2 compared to the traditional reinforced concrete structure. Solar panels were installed to collect green energy and gray water recovery systems were also installed to increase its self-sufficiency.

TCV also considered the cycle of natural resources and created an urban farm so that its inhabitants can experience the agricultural lifestyle. Through the use of fertilizer from food waste, gray water recovery systems, compost, and other circular mechanisms, it elevated the agriculture industry to a new level with a greater sense of ecosystem protection, food safety, and community preservation.

TCV rejected the purchase-once and throw away model on all its utilities, and set out to rent all its elevators, air conditioners, solar panels, bathroom equipment, lightings, etc., through businesses that offer a lifetime guarantee service model in hopes of creating a new business model within Taiwan. Unlike traditional house sales models, Taisugar not only leases these new houses, but it provides all the repair services that its customers need. Other services that it offers include shared kitchen,

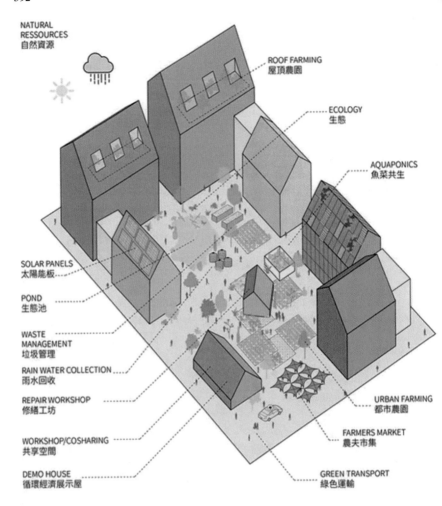

Fig. 4 Taisugar circular village

shared living rooms, shared transportation, etc. Future residents will enjoy the right
to a comfortable, healthy living environment, instead of rights to a piece of land
(Fig. 4).

8 Conclusion

Taiwan has a small landmass, few natural resources, and a high population density,
but these characteristics make the circular economy a natural goal for Taiwan to

pursue. Now that Taiwan has developed upstream and downstream capabilities, let's work together to achieve a circular Taiwan!

Questions

1. What is the "Four-In-Recycling" Program and what makes it successful?
2. Why is Taiwan uniquely positioned to transition to a circular economy?
3. What role did the Taiwan government play in promoting circular economy? What additional initiatives do you think it should consider?
4. What does the Circular Economy Roadmap for Enterprises address?
5. What is the difference between a technical cycle and a biological cycle? Please list an example for both.
6. What is your favorite Taiwan CE case study? Please explain why.

Appendix

Circular Taiwan Network (CTN), founded in 2015, is a leadership initiative in Taiwan bringing together governments, businesses, and NGOs to facilitate the country's transition to a circular economy. CTN's endeavors include C-A-N.

Communication: Until 2019, CTN has been invited to give over 170 public speeches which have reached over 14,000 audience members, including many important politicians and business leaders. CTN published the book, "Circular Economy," in 2017 and organized the first "Asia Pacific Circular Economy Roundtable" in 2019.

Advocacy: CTN attempts to bring the relevance of CE to the local context and facilitate its implementation. CTN is often solicited for its expertise and experience and has consulted for many governmental programs and projects, including the "5 + 2 innovation plan," circular economy targets and strategies, circular parks, and many others.

Network: In 2019, the "CoPartners" project was launched as a platform to accelerate the transition of the economy to a circular economy. It aims to address the bottlenecks in the transition to CE with more effective advocacy and to promote cross-sector cooperation among various working groups and clusters.

Website: https://www.eng.circular-taiwan.org/.

References

1. National statistics republic of China human development index(HDI). *Statistical Bureau, Republic of China (Taiwan)*, 15 Jan, 2020. Retrieved March 12, 2020, from https://eng.stat.gov.tw/ct.asp?xItem=25280&ctNode=6032&mp=5.
2. 2018–2020 Resource recycling promotion project. *Environmental Protection Administration, Republic of China (Taiwan)*, Jan 2018. Retrieved March 12, 2020, from https://waste1.epa.gov.tw/Ier_Web/Public/News_Item.aspx?id=916.

3. Economic development, ROC (Taiwan). *National Development Council, Republic of China*, December 2018. Retrieved March 12, 2020, from https://ws.ndc.gov.tw/Download.ashx? u=LzAwMS9hZG1pbmlzdHJhdG9yLzEwL3JlbGGZpbGUvNTYwNy8zMTk5My84Mzhh OWY3Mi0yMTIxLTQwOTMtOWM0OC1hZmMtY3MGIzOWEyYmIucGRm&n=RWNvbm 9taWMgRGV2ZWxvcG1lbnQQsIFIuTy5DLiAoVGFpd2FuKSAgMjAxOC5wZGY%3d& icon=..pdf.
4. Structure and operations. *Environmental Protection Administration Recycling Fund Management Board*, 20 Feb 2020. Retrieved March 12, 2020, from https://recycle.epa.gov.tw/epa/Show wPage2.aspx?key=1&sno=1&subsno=11.
5. Chen, K. (17 May, 2016). Taiwan: The world's geniuses of garbage disposal. *The Wall Street Journal*. Retrieved March 12, 2020, from https://www.wsj.com/articles/taiwan-the-worlds-gen iuses-of-garbage-disposal-1463519134.
6. Reference manual of Industrial waste Reuse management. *Environmental protection administration*, Jan 2020. Retrieved March 12, 2020, from https://rms.epa.gov.tw/RMS/Public/Dow nload.aspx?t=2.
7. 2018 Industrial waste statistical report. *Environmental protection administration*, May 2019 Retrieved March 12, 2020, from https://waste.epa.gov.tw/RWD/Statistics/?page=Year1.
8. Huang, C. (2017). *Circular economy*. Commonwealth Magazine Group.
9. 2018 Renewable resource recycling annual report. *Environmental protection administration*. Retrieved March 16, 2020, from https://waste1.epa.gov.tw/Ier_Web/Public/Annual.aspx.
10. Circular Economy Promotion Plan (30 Jan, 2019). *Department of information services*. Executive Yuan, Republic of China (Taiwan). Retrieved March 16, 2020, from https://english.ey gov.tw/News3/9E5540D592A5FECD/8053c7c8-e0a9-4cdd-b53a-992d6330f499.
11. Taiwan Circular Economy Network. *Towards a circular Taiwan*. October 2019. https://www circular-taiwan.org/towardsacirculartaiwan.
12. 2016 Global manufacturing competitiveness index. *Deloitte Touche Tohmatsu Limited* Retrieved March 16, 2020, from https://www2.deloitte.com/content/dam/Deloitte/global/Doc uments/Manufacturing/gx-global-mfg-competitiveness-index-2016.pdf.
13. Schwab, K. (Ed.). (2019). The global competitiveness report 2019." *World Economic Forum* Retrieved March 16, 2020, from http://www3.weforum.org/docs/WEF_TheGlobalCompetit venessReport2019.pdf.
14. The case studies of regional energy and resources integration. *Industrial Development Bureau Ministry of Economic Affairs*, 2020. Retrieved March 27, 2020, from https://eris.utrust.com tw/eris/dispPageBox/RECT.aspx?ddsPageID=NEWSE&dbid=4357926090.
15. "*2017農業資源循環利用策略研析計畫研究報告*" Council of Agriculture, Executive Yuan Republic of China (Taiwan).
16. AgriDragon biomimicry farming system heals rather than intervenes. *Viachi*, 2017. Retrieved March 16, 2020, from https://viachi.tw/portfolio-item/agridragon/?lang=en.
17. Food Bank. Retrieved March 16, 2020, from https://www.ccra.org.tw/SviAriticlePage.aspx SVSID=3.
18. Donghaifeng Agricultural Circulation Park of Taiwan Sugar Corporation. Retrieved March 16 2020, from https://www.taisugar.com.tw/Circular/cp2.aspx?n=11431.
19. DA. AI. Retrieved March 16, 2020, from http://www.daait.com/index.php/en/.
20. EVEREST Textile Co., Ltd. Retrieved March 16, 2020, from http://www.everest.com.tw index_us.aspx.
21. Tainan Enterprises Co., Ltd. Retrieved March 16, 2020, from http://www.tai-nan.com/en/news aspx.
22. 城市修理站. Retrieved March 16, 2020, from https://nothingisgarbage.com/.
23. 城市修理地圖. Retrieved March 16, 2020, from https://pse.is/EBW3G.
24. Taiwan Circular Economy Network. From manufactured in Taiwan to remanufacture in Taiwan report. *Taiwan Circular Economy Network*, 2019. Retrieved March 16, 2020, from https://www eng.circular-taiwan.org/remanreport.
25. SHINKONG Synthetic Fiber Corporation. Retrieved March 16, 2020, from http://www.sh nkong.com.tw/.

26. *AMAZE*. Retrieved March 16, 2020, from https://www.amazefashion.com.tw/.
27. *Taisugar's Circular Village*. Retrieved Marchr 16, 2020, from https://www.taisugarcircularvi
llage.com/.

Hui-Ling Chen (Shadow) currently serves as the CEO of Circular Taiwan Network. She sits in various governmental committees including 'Resource Recycling and Reuse Promotion Committee' and 'Greenhouse Gas Management Fund'. She is also the chief editor of several publications including 'Circular Economy' (in Mandarian) and 'Towards a Circular Taiwan-66 stories'. Prior to this, she was a permaculture educator, industrial designer and travel writer.

She holds a double Master of Science degree in Industrial Ecology from Leiden University in the Netherlands, and University of Graz in Austria, she also holds a Bachelor's degree in Industrial Design from National Cheng Kung University in Taiwan. She received Hans Roth Umweltpreis 2017 in Austria for her thesis "Circular Design".

Ya-hsuan Tsai currently serves as the campaigner of Circular Taiwan Network, sharing international and Taiwan's circular economy insights to Taiwan's governments, companies, and public. He is also the managing editor of the 'Circular Economy - A Model to Transform Marine Debris to Resources' and the topic introduction advisor of the book, 'Towards A Circular Taiwan - 66 Circular Stories'.

He has over 7 years of working experience in the environment and circular economy. He assisted Taiwan's Environmental Protection Administration in implementing waste recycling management projects in an environmental consulting company. He also got his Bachelor's degree in Chemistry from National University of Kaohsiung.

Chiao-ling Lyu is an Associate Researcher at the Green Energy and Environment Research Laboratories of Industrial Technology Research Institute. She is the managing editor of the book, Towards A Circular Taiwan - 66 Circular Stories, in which over 360 practitioners and partners are involved and the book was published in 2019 Asia Pacific Circular Economy Roundtable. When she worked with Circular Taiwan Network, she was a researcher responsible for food-agricultural subjects and attended the 2018 Circular Economy Hotspot in Scotland. She has over 4 years of experience promoting the circular economy in Taiwan. She took the University of Amsterdam Summer Program: the Circular City: Towards a Sustainable Urban Ecosystem in 2018. She got her Master's degree in Chemistry from National Cheng Kung University.

Yu-Lan Duggan has over 18 years of experience promoting the welfare of the environment and human rights. Prior to working with Circular Taiwan Network, she was a Senior Advisor at the Center for Child Rights and Corporate Social Responsibility (CCRCSR) and the Director of Civil Society Organizations Engagement at the Fair Labor Association (FLA). In addition, she has served as a corporate social responsibility consultant for numerous multinational corporations. She received her Master of Public Administration from the School of International and Public Affairs at Columbia University.

Youth Leadership in a Circular Economy: Education Enabled by STEAM Platform

Arslan Siddique, Panitsara Nakseemok, and Lerwen Liu

Abstract Youth leadership driven by education is inevitably important for making significant realization of a Circular Economy (CE). The success of the circular economy global movement is deeply rooted in harnessing the potential of the youth in Asian region especially. Asia is known to be a major global manufacturing hub and waste generator with over 62% of global youth population. This chapter outlines the educational practices in CE across the globe with particular focus on CE educational programs, courses/modules offered in countries including Australia, Belgium, Canada, Finland, Germany, Italy, Thailand, UK, US, and others. Furthermore, it highlights the criticality of Asian region toward CE education; how a general education module offered by the STEAM Platform helped transforming youth, and how STEAM Platform builds on existing CE practices fostering youth leadership. Of note, the role of different stakeholders in CE education such as educational institutes, government organizations, industry, and Non-Governmental Organizations (NGOs) is equally essential for working in synchronicity to develop a holistic education program in effectively developing human resources and in particular leaders in driving the CE transition. We propose that knowledge convergence (STEM knowledge), skills & mindset (strategic communication, peer-to-peer learning, life cycle, and critical thinking), and entrepreneurial practices are complimentary for the transformation of youth leading toward practical implementation of CE.

Keywords Education · Youth leadership · Sustainability · Circular economy · Industry 4.0 · STEAM platform

A. Siddique (✉)
School of Chemical Engineering, University of New South Wales, Sydney, NSW 2052, Australia
e-mail: a.siddique@unsw.edu.au

P. Nakseemok · L. Liu
STEAM Platform, King Mongkut's University of Technology Thonburi, 126 Prachauthit Road, Bangkok 10140, Thailand
e-mail: Panitsara.care@gmail.com

L. Liu
e-mail: lerwen67@gmail.com

© Springer Nature Singapore Pte Ltd. 2021
L. Liu and S. Ramakrishna (eds.), *An Introduction to Circular Economy*,
https://doi.org/10.1007/978-981-15-8510-4_29

Learning Objectives

- The global importance of practical realization of circular economy stimulates different approaches particularly CE education in fostering skilled workforce
- The role of youth leadership is indispensable in accelerating circular economy efforts across the world, especially the Asian region
- The global education providers inculcate CE education modules in degree programs to effectively promote CE and empower youth
- STEAM platform's dedicated undergraduate module on CE (GEN352) distinctively describes success factors for CE education through youth leadership.

In recent years, the circular economy movement is gaining significant momentum as a transformational economy emphasizing closed-loop systems based on the principle of "everything is input to everything else". It is particularly defined as a concept constituting reduction in consumption & production, eco-designing of products, refurbish & remanufacturing, recycling from waste to resource, etc. [5, 19]. Circular Economy (CE) is foreseen to profoundly impact different sectors of life, and it has attracted notable attention from governmental organizations, academic institutions and Non-Governmental Organizations (NGOs) as well as industries. In the wake of call for action, different countries, including China, EU, Finland, Germany, India, Netherlands, Norway, Scotland, Sweden, Taiwan, Thailand, and others, have formulated policies for implementation of circular economy contributing toward a sustainable world [27]. Different scholars approach CE differently, and more than 114 definitions of CE are used by practitioners and scholars of CE, however recover/recycle/remanufacture/refurbish/reuse/repair/reducematerials/products is fundamentally considered by all with addition of some strategies to achieve CE [17, 12]. The concept of CE has evolved overtime, and different CE terms substantially reinvigorate the concept of sustainable development as shown in Fig. 1.

The role of governments, NGOs, private sectors, and scholars is highly critical in communicating and practical realization of CE for sustainability (Environment, Economic, Social) [6]. However, the role of scholars in CE is underrepresented who build the foundation of raising awareness about CE to the world and educating the youth among the general public. Education for sustainable development received enormous traction in the past with inclusion of modules on sustainability in the curricula of high school, undergraduate/graduate students [8, 18]. In fact, the history of educating students about sustainability is dated back to 1977 when Imperial College London initiated a graduate program, M.Sc. in Environmental Technology, in 2017, the Utrecht University in Netherlands began an undergraduate course on Introductory CE with interactive and problem-based learning [16].

Similarly, Ellen Macarthur Foundation's circular economy education program is targeted on three key areas including (1) formal education through the international network of schools and universities to co-develop CE solutions and integration in existing learning modules, (2) informal education for individuals seeking self-taught CE information, and (3) business learning involving corporate partners to layout

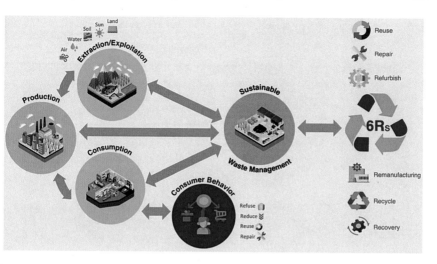

Fig. 1 Circular economy framework from Mushtaq Memon (UNEP) and Lerwen LIU (STEAM Platform)

practicable CE framework and capacity building [9]. European Circular Economy Stakeholder Platform is also greatly contributing toward the education sector by promoting CE-related events, competitions, hackathons, and other learning activities. CE education surged in Europe in recent years with some key events including "Circular Economy Competence. Making the Case for Lifelong Learning", "Circular Economy hackathon in Estonia", "Walki Circular Classroom: Co-created learning module simulates circular thinking at school" (European Circular Economy Stakeholder Platform 2019). Fixfest 2019 in Germany gathered over 200 volunteers and other stakeholders (policy-makers, companies, thinkers) to deeply discuss taking responsible ownership of things and making better products [11] as described in Table 1.

According to a World Bank's report, urbanization is expected to contribute toward 70% of global population by 2050, and waste generation would escalate twice as compared to their rural counterparts, and the amount of global waste would increase to 2.2 billion tons per year by 2025 [26]. Similarly, the United Nations (UN) reported in 2012 that the global population increase from 1 billion in 1804 to 7.4 billion in 2014 is anticipated to rise to 9.6 billion by 2050. Such an unprecedented rise in population puts a huge burden on the demand of ever depleting resources and urges CE education to the public about the new circular economic model and to live a sustainable life in the future [3]. To impart CE education, various online courses are also being offered by prominent universities regardless of regional boundaries including

a) Circular Economy French Mooc Montreal, Canada;
b) Global Leadership Program on CE 2019 Adelaide, Australia;
c) Circular Economy: Sustainable Materials Management, Lund, Sweden;

Table 1 Comparative analysis of global CE education platforms and key attributes

CE Education Platforms	Key attributes						
	LCA mindset	SDGs	Advanced technologies	Biz model for sustainability	Peer-to-peer learning /Youth Leadership	CE trainings, workshops, hackathon	CE general education
United Nations (UN), The World Bank group, ADB, WEF	✗	✓	✗	✗	✗	✓	✓
Ellen MacArthur Foundation	✗	✓	✗	✓	✗	✓	✓
CE stakeholder platform, EU	✗	✓	✗	✓	✗	✓	✓
Utrecht University, Vrijie University, TU Delft, Netherlands	✗	✗	✓	✗	✗	✓	✗
Walki circular classroom, Finland	✗	✗	✗	✗	✗	✗	✓
Fixfest, Germany	✗	✗	✗	✓	✓	✓	✗
Global Leadership Program on the CE, Australia	✗	✗	✗	✓	✓	✗	✓
UCL, University of Bradford, Cranfield University, UK	✗	✗	✓	✗	✗	✓	✗
Circular Economy Asia	✗	✗	✗	✗	✗	✗	✓
STEAM Platform, Thailand	✓	✓	✓	✓	✓	✓	✓

(d) Engineering Design in Circular Economy by TU Delft;

(e) Innovation, Enterprise, and Circular Economy by University of Bradford;

(f) And other similar modules by Wageningen University, Cranfield University, Leiden University, University of Torino and Vrijie Universiteit Amsterdam's Doctoral programs in Circular Economy [7];

(g) Turku University of Applied Sciences, Finland's program offering Circular Business with Companies, Sustainable Development, and Business development.

Particularly, Finland that is worldly renowned for its excellent school curriculum is taking exceptional lead in educating young students about Circular Economy and its enormous impact in the future. Their Circular Classroom is a very well-established and comprehensive Finnish curriculum that emphasizes sustainability and empowers students through an active learning toolkit imparting experiential learning. This learning toolkit offers an opportunity to secondary and upper-secondary school students to learn in a fun way about how to design eco-efficient products and circular economic models for building a sustainable world [1]. Such an interactive and experiential learning helps students in mastering critical thinking, life cycle thinking, design thinking, and sustainability. Also, Sitra in Finland is contributing toward rapid societal and economic transformation by providing the latest guidelines, trends, and opinions to the people for sustainability [24]. In addition, Matthew Murray published an article entitled "Teaching Circular Economics" in The Ecologists, and emphasized the role of education in shaping the future [20]. He described how critical it is to teach young students about new economic systems and more dynamic teachings expanding beyond traditional teachings about recycling integrating creative and collaborative learning. In such a holistic learning manner, the young generation feels concerned about the future when they are deeply taught about the current economic practices making them feel empowered to take a likely collaborative action for a sustainable future. (Detailed description of global CE educational modules is given in Table 2.)

The youth-led global climate movement has already begun and needs to be strengthened by promoting CE education worldwide, especially Asia where the majority of global youth resides. According to the UN's report released on International youth day (August 12, 2019), 1.2 billion people are aged between 15 and 24 years and this number is expected to increase by 7% to 1.3 billion in 2030 (deadline to achieve SDGs 2030). In 2019, Central and Southern Asia had the largest share of youth (361 million), second highest in Eastern and South-Eastern Asia (307 million) followed by Sub-Saharan Africa (211 million). Furthermore, the youth population is projected to rise by 62% in the next three decades increasing from 207 million in 2019 to 336 million by 2025, and this surge is expected in the least developed 47 countries of the UN [21]. Moreover, it is stated that increase in working-age population with substantial decline in fertility rate will offer a promising opportunity for reaping the demographic dividends which are anticipated to emerge with increase in investment and per capita economic growth leading towards achievement of SDGs [2]. These

Table 2 Detailed Overview of Global CE Education Modules

Global CE education institutes	Course/Module content	Course offering	Weblinks
University of Technology Sydney (Australia)	Responsible Production and Consumption, Business Models for Resource Efficiency	Postgraduate Research	https://www.uts.edu.au/research-and-teaching/our-research/institute-sustainable-futures/our-research/sdgs-mapping-our-0
Green Industries, South Australia (Australia)	Renewable Energy, Resource Recovery and Recycling, Water Management, Community Education and Innovation	Leadership Program	https://www.greenindustries.sa.gov.au/leadership-program
The University of Queensland (Australia)	Value from end-of-life Products, Circular Use of Metals in Australian Economy	Postgraduate Research	https://smi.uq.edu.au/
University of Sao Paulo (Brazil)	Life Cycle Management, Engineering Innovation, CE New Business Processes and Models	Undergraduate/Postgraduate	https://www.ellenmacarthurfoundation.org/news/the-university-of-sao-paulo-usp-becomes-a-pioneer-university
Ghent University (Belgium)	Sustainable and Innovative Natural Resource Management, CE Value chains, Entrepreneurship, Sustainable Development, Professional Research Projects	Postgraduate	https://studiekiezer.ugent.be/international-master-of-science-in-sustainable-and-innovative-natural-resource-management-en
University of Montreal (Canada)	Introduction to CE model, Scarcity of Resources, Circularity Strategies, Biochemical Extraction, Case Studies on Organizational Deployment, Eco-Designing	Students/Practitioners	https://cerium.umontreal.ca/en/programs-of-study/ecoles-dete-2016/economie-circulaire/

(continued)

Table 2 (continued)

Global CE education institutes	Course/Module content	Course offering	Weblinks
Turku University of Applied Sciences (Finland)	Circular Business with Companies, Sustainable Development, Business development, CE 2.0	Postgraduate Research	https://www.tuas.fi/en/research-and-development/research-groups/Circular_business_models/
Technical University Berlin (Germany)	System thinking, Built Environment, Textile & Fashion, Zero Waste, Food & Biomass, Circular Urban Systems	Summer School	https://www.tu-berlin.de/menue/summer_university/summer_university_term_4/circular_economy/
Afeka Institute of Circular Engineering and Economy (Israel)	CE Business Models, Energy, Waste Management and Logistics	Undergraduate/Postgraduate	https://www.aicee.afeka.ac.il/
University of Pavia (Italy)	Better Valorisation of Natural resources, Global Value Chains and CE, Business Model for Circular Enterprise	Postgraduate	https://mibe.unipv.it/
Delft University of Technology (Netherlands)	Introduction to Circular Economy, Business Value in CE, Longer Lasting Products, Thinking in Systems, Remanufacturing, Waste = Food	Online MOOC	https://ocw.tudelft.nl/courses/circular-economy/
Utrecht University (Netherlands)	Global Sustainable Science, Sustainable Development, Innovation Sciences, Water Science and Management	Undergraduate/Postgraduate	https://www.uu.nl/en/research/copernicus-institute-of-sustainable-development/teaching
KTH Royal Institute of Technology (Sweden)	Production Engineering, Product Realization and Industrial Engineering, Life Cycle Assessment, Environmental Management, Waste Management, Industrial Ecology, Sustainable Development	Undergraduate/Postgraduate	https://www.kth.se/ce/education/education-activities-1.812149

(continued)

Table 2 (continued)

Global CE education institutes	Course/Module content	Course offering	Weblinks
The University of Sheffield (UK)	Energy Sustainability, Resource Efficiency and Circular Economy	Postgraduate Research	https://www.she ffield.ac.uk/ene rgy/research-pil lars/energy-sus tainability-res ource-effici ency-and-cir cular-economy
University College London (UK)	Eco-Innovation, Business and Market Development, Urban Sustainability, Industrial Symbiosis, Environmental System Engineering, Environmental Design Engineering, Sustainable Economics: Policy and Transition	Undergraduate/Postgraduate	https://www.ucl. ac.uk/circular-economy-lab/ teaching1
University of Strathclyde (UK)	Circular Economy and Transformation towards Sustainability, Waste-as-a-resource, Use of Energy and Material Resources, Business Models for Green Enterprise Development, Social Trends and Consumer Behavior	Postgraduate	https://www.str ath.ac.uk/cou rses/postgradu atetaught/sustai nabilityenviron mentalstudies/# coursecontent
Cranfield University (UK)	Technology Innovation and Management for a Circular Economy, Materials Innovation, Circular Manufacturing, Circular Design, Circular Value Chains, Circular Business Models, Disruptive Innovation	Postgraduate	https://www.cra nfield.ac.uk/cou rses/taught/tec hnology-innova tion-and-man agement-for-a-circular-eco nomy
University of Arts London (UK)	Business of Fashion, Textiles and Technology, Circular Synthetics, Future Manufacturing, Circular Designing	Postgraduate Research	https://www. arts.ac.uk/res earch/research-centres/centre-for-circular-design

(continued)

Table 2 (continued)

Global CE education institutes	Course/Module content	Course offering	Weblinks
University of Bradford (UK)	Innovation, Enterprise and Circular Economy, Circular Economy Core Principles, Leadership for Transformational Change, Business Models for Circular Economy	Postgraduate	https://www.bradford.ac.uk/courses/pg/innovation-enterprise-and-circular-economy/
University of Exeter (UK)	Entrepreneurial Circular Economy, Entrepreneurial Leadership and Circular Economy, Research Projects	Postgraduate	https://business-school.exeter.ac.uk/research/centres/circular/
King Mongkut's University of Technology Thonburi (Thailand)	Life Cycle Assessment, Advanced Robotics, Internet of Things, Digital Factories, Waste Management, Sustainable Financing, Supply Chain Management, CE Value Chain Analysis, Nanotechnology, Biotechnology, CE Business Models, Social Entrepreneurship, SDGs 2030, Peer-to-peer Learning, Strategic Communication, Youth Leadership	Undergraduate	https://www.steamplatform.org/skill-mindset
University of California Davis (USA)	Industrial Ecology, Life Cycle Assessment for Sustainable Engineering, Urban Systems and Sustainability, Technology Management, Zero-Net Energy, Waste Resource Management, Green Building Design and Materials	Graduate Certificate	https://ie.ucdavis.edu/
Worcester Polytechnic Institute (USA)	Sustainable Operations and Supply Chains, Sustainability within Businesses, Circular Economy Practices and Regulatory Policy	Postgraduate	https://www.wpi.edu/news/announcements/learn-about-sustainable-circular-economy-prof-sarkis-upcoming-spring-course

(continued)

Table 2 (continued)

Global CE education institutes	Course/Module content	Course offering	Weblinks
Virginia Tech (USA)	CE Fundamentals, Sustainable Biomaterials, Resource Scarcity, Climate Change, Waste, Economic Disparity	Undergraduate	https://advising. vt.edu/advising-resources/cou rse-announcem ents/sbio-1984. html
Arizona State University (USA)	Workshop on Mapping Linear economy and Adoption of CE, Workshops on Ethical CE	Professional Certification	https://sustainab ility.asu.edu/sus tainabilitysolu tions/ece-cert/
Loyola University Chicago (USA)	Environmental Sustainability Management, Microeconomics and Marketing	Undergraduate	https://www.luc. edu/quinlan/und ergraduate/min ors/sustainable-business/

alarming statistics clearly represent harnessing the potential of youth through education to achieve a circular economy, and Asian region is particularly highlighted in this regard [15].

It is evident that European Union (EU) adopted CE as a guiding principle for achieving their zero-waste program along with the EU action plan for CE [25]. In this regard, China also incorporated CE in their "National Economic and Social Development Plan" adopting top-down approach to ensure reducing consumption of raw materials, recycling, and reusing [27, 13]. Considering the fruitful impacts of CE in Europe and other continents of the world, CE is gaining paramount importance in Asian region [15] that generates a significant proportion of the world's waste. Also many South Asian and South-East Asian countries are suffering enormously from the problem of municipal waste and imported waste from developed nations [14] The unprecedented approach is needed to ban imported waste and process municipal waste employing innovative technologies and customized local solutions. Additionally, Asian region is the manufacturing hub of most products due to inexpensive labor and poorly enforced regulations. In the value chain of products from extraction of raw materials through manufacturing, processing, production to waste management majority of production, and consumption are done in Asia leading to uncontrollable growth in waste generation [21]. The poor enforcement of environmental regulations for manufacturing has led to devastating pollution in air, water, and land in the developing world and affecting the rest of the world.

The involvement of governments, NGOs, and academia are inevitable to take collaborative action by formulating implementable policies, practices, and education programs on CE for achieving a sustainable world. CE is rooted in different disciplines, and it requires knowledge dissemination by educational institutions and

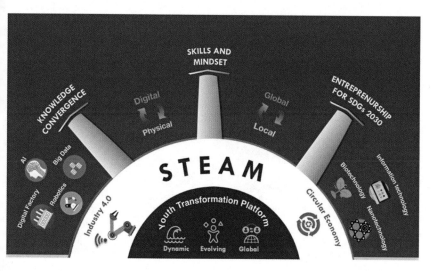

Fig. 2 STEAM platform framework

foundations such as Ellen MacArthur Foundation in Europe and Circular Economy Asia (Circular Economy Asia). Over 62% of the world's youth population resides in Asia–Pacific indicating the criticality of the region's role in circular economy transformation. Innovative solutions can be developed through the adoption of next-generation technologies led by the youth. Technical problem-based and experiential learning module at a university level for educating Asian youth is essential to train youth workforce for truly transforming the region. A Science, Technology, Engineering, Art, and Mathematics (STEAM) Platform, established in May 2017 supported by King Mongkut's University of Technology Thonburi (KMUTT), Thailand, has become a youth leadership platform for accelerating the transition of Circular Economy enabled by convergence of cutting edge technologies in the STEM field as shown in Fig. 2. The STEAM Platform aims to empower the youth to use STEM knowledge in developing effective solutions for solving problems faced in our humanity today in a sustainable way. In particular, it trains the youth to learn and adopt STEM knowledge with the "soul" as shown in Fig. 3.

To scale the STEAM Platform, allowing more students to access to STEAM training, we piloted in 2019 a general education module titled "Technology Innovation and Entrepreneurship for Sustainability – STEAM Platform Enabling Youth Leadership for Smart Circular Economy Transformation" (GEN352) for all undergraduate students [24]. This GEN352 module aims to provide an introduction to sustainability, Sustainable Development Goals 2030, circular economy and industry 4.0 and their enabling technologies and innovative solutions as shown in Fig. 4.

The module covers topics such as Life Cycle Assessment, Sustainable Development Goals, Advanced Robotics, Digital Factory, Smart Manufacturing, Big Data and Analytics, Circular Economy, Innovations in Waste Management in Thailand,

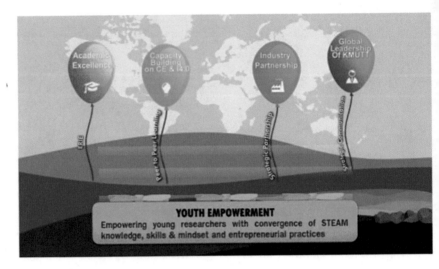

Fig. 3 Fundamental goals of STEAM platform fueled by youth empowerment

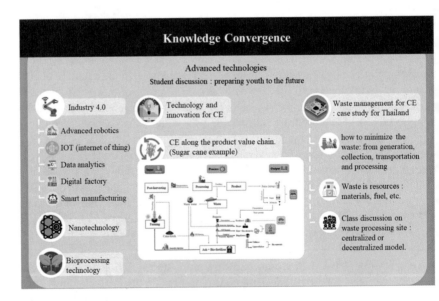

Fig. 4 Elements of knowledge convergence in STEAM module

Sustainable Financing, Supply Chain Management, Business Model for Sustain
ability, and more. We partnered with the university technology transfer team who
provided 18 patented technologies for students to choose for their project in commu
nication and business model design for CE and SDGs2030. This intends to motivate

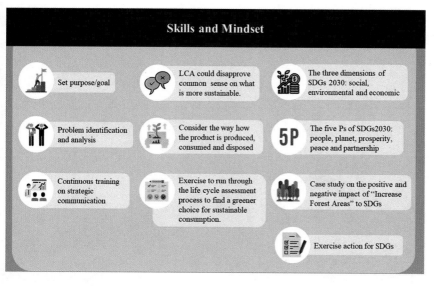

Fig. 5 Skills and mindset for creating youth workforce driving CE

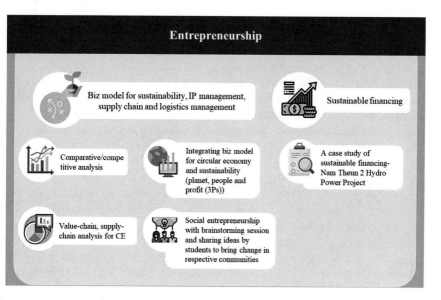

Fig. 6 Entrepreneurship practices including business models for SDGs 2030 and CE

students in continuing their future study and research in our university focusing on the right purpose: circular economy and SDGs2030 as shown in Figs. 5 and 6.

The module involved several faculty members who are experts in the topics mentioned above as well as guest speakers from UNDP, investment banking sector,

Fig. 7 Unique attributes of the STEAM undergraduate module (GEN352)

and social enterprise. The unique features of this module include learning cutting edge technologies and global trends, youth leadership for circular economy, integration of I4.0 with CE, and Entrepreneurship for SDGs 2030 and CE as represented in Fig. 7. STEAM Platform executive team, Arslan Siddique and Panitsara Nakseemok (graduate students) co-taught the module with Dr. Lerwen LIU to conduct classes and mentor students during the entire semester. Students were specifically trained on strategic communication on technology innovations, and business models design for SDGs 2030 for the selected patented technologies from KMUTT. Students received one-on-one mentoring for their final presentations and posters, and their performance was assessed by the KMUTT technology transfer team and external invited industry experts as well as the STEAM teaching team. KMUTT technology transfer team were particularly impressed by our students' outstanding performance and requested us to continue this effort and train the tech transfer team as well as using the same methodology. The STEAM team selected some of the posters to showcase at international events where the global audience could witness the transformation of young undergraduate students trained by STEAM teaching team for GEN352 in addition to learning the innovative technologies from KMUTT as presented in Fig. 8.

Industry 4.0 (I4.0), the fourth industrial revolution represents data exchange in cyber-physical systems, which follows previous 3 industrial revolutions that were attributed to discovery of steam power, assembly lines powered by electricity, and automation, respectively. The enabling technologies of I4.0 include Internet of Things (IoT), Advanced Robotics, Smart Manufacturing, Digital Factories, Data Analytics etc.

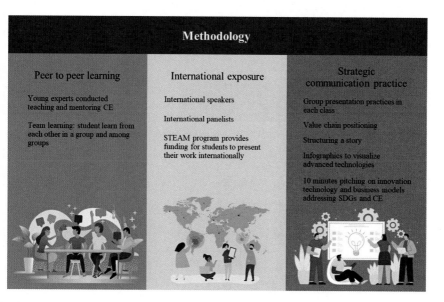

Fig. 8 STEAM methodology to train young undergraduate students particularly in GEN352

The fascinating results of these modules indicate that it truly transformed the lives of students who not only started thinking strategically but also proposing to adopt technological and business solutions for pressing issues toward CE in the Asian region. Majority of the participants belonged to the age group of 19–22 years as shown in Fig. 9 which is quite an early age to get involved with circular economy concept practices. The STEAM platform has had a huge impact on young students in terms of knowledge dissemination on advanced emerging technologies such as advanced robotics, data analytics, waste to energy, and waste to materials. The survey reveals that students were equally interested in all aforementioned areas with 60% of youth appreciating learnings in waste to energy. Our module was specifically tailored to the needs of the students and demanding technologies as shown in Fig. 10 of the future. We helped students learn and analyze some of the patented technologies of our university KMUTT which can be adopted for circular economy and addressing SDGs 2030. Asia has enormous potential to convert waste into energy owing to the fact that Asian region generates a large amount of global waste while many of its developing countries suffer from energy crisis. The students particularly find this area interesting and learnt about innovative technologies for conversion of agricultural waste into energy, because this module was run in an agricultural country like Thailand where students related these technologies in their ecosystem for potential benefit of the society.

Furthermore, it was surprising to witness significant behavioral change in students regarding consumption and reduction in waste production. Our survey indicates that 90% of the module participants changed their behavior after attending our intensive training as shown in Fig. 11. It manifests the unique teaching/training methodology

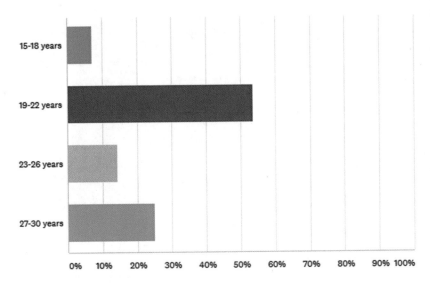

Fig. 9 Age group of the respondents

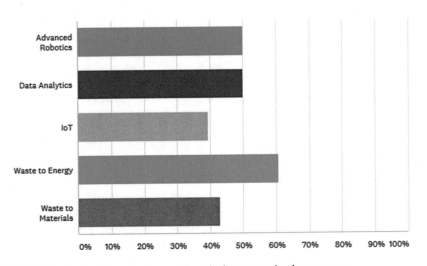

Fig. 10 Technologies learned in the module and relevance to circular economy

of STEAM platform in transforming the mindset of youth and alarming them about the detrimental impacts of the unnecessary consumption of food, waste generation, climate change, etc., with exemplary evidence.

To further explore our unique teaching strategy, we asked students about choosing one or more of the listed strategies as shown in Fig. 12. Interestingly, we found out that the majority of the students liked our workshop sessions in addition to other practices. We believe that our teaching workshops were comprehensively defined and delivered

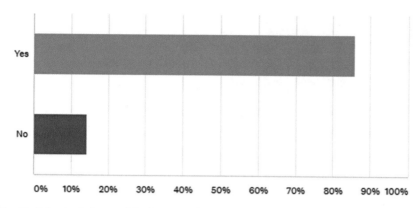

Fig. 11 Behavioral change of food consumption and waste generation

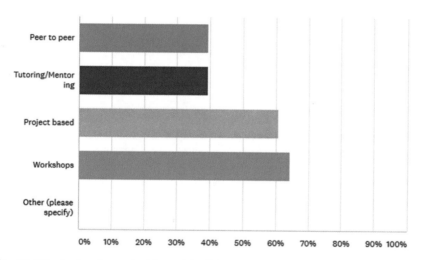

Fig. 12 Effective learning methodology of the GEN352 module

y topical experts in circular economy. Also, our workshops were purely based on xperiential learning, critical thinking, and problem-solving, which gave students an pportunity to dive deeper into the delivered technological knowledge and propose ractical and implementable solutions. Such an intensive youth-centered workshop elivered by experts hugely impacts the youth to become changemakers of smart ircular economy.

Moreover, transforming the mindset of youth in addition to providing skills was nother key pillar of the STEAM platform. Our module helped students in life ycle thinking, critical thinking, thinking out of the box, systematic and analyt- al thinking, sustainable development and global perspective. After attending our rcular economy driven education module, students experienced significant change

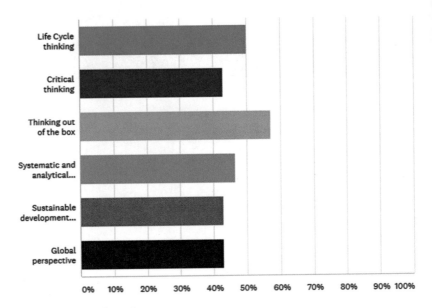

Fig. 13 Mindset transformation

in mindset as shown in Fig. 13, and they became problem solvers with great socio-economic and environmental impact. We stated earlier that systematic and critical thinking is at the forefront of innovative technological development, and certainly youth equipped with such an innovative and transformative mindset makes it happen.

In order to analyze the key soft skills gained by our module participants, we asked them to choose what they developed after attending our module. Surprisingly, the responses reveal that "teamwork" skills as shown in Fig. 14 was one of the key skills gained by individuals among others including strategic communication, leadership and problem-solving. This finding is truly aligned with our core motto of "Together Stronger" that was stressed at various instances in our workshops and training. It is exciting to know that it rhymes with the youth, and they understand the importance of teamwork and how critical it is to develop holistic solutions for a circular economy. We emphasized group work in assigned projects to help students build interpersonal and teamwork skills that are inevitably important to create leadership in circular economy.

To gain further insight into our teaching/training practices, we asked respondents to highlight the effective learning methodology for creating more impact, and unsurprisingly students predominantly chose workshops as the effective learning tool as shown in Fig. 15. This finding is in conjunction with our previous finding on our most liked training methodology. Evidently, the intensive workshops containing peer-to-peer and experiential learning along with strategic communication and entrepreneurship are proposed to have a likely impact on youth, and it is proposed to be a productive approach towards transforming the mindset of youth to help build a sustainable circular economy leadership.

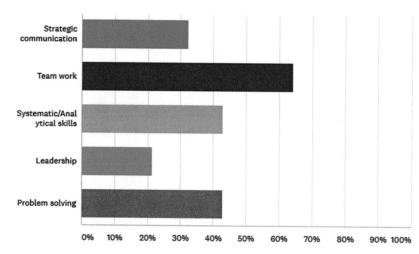

Fig. 14 Impact on personal and professional life

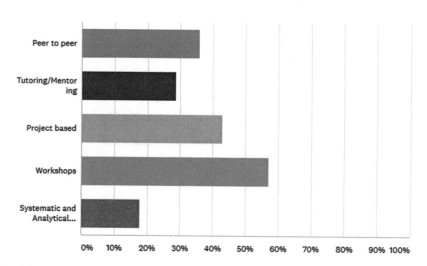

Fig. 15 Recommendation for improving effective learning

1 Suggested Improvements in GEN352:

1. The content of the GEN module could be divided into two semesters; first semester intensively focused on CE mindset and advanced technologies development and application, and second semester on eco-product design, supply chain & business model for CE, and sustainable financing for improved comprehensive learning experience.

2. Systematically synchronize successive lectures on advanced technologies and its relevance to sustainability and CE for effective outcome.

3. Create an interactive experiential learning environment for students to conduct experiments and visualize the advanced technological concept phenomena. Particularly, guide students to develop technologies to enable and accelerate CE transition.
4. Extend educating horizons by inviting more industry players to conduct in-class activities and showcase industrial practices in technologies for CE to enhance capacity building of students.
5. Guide young students to create circular business models enabled by innovative technologies.
6. Exchange students with partnering universities to accelerate the global CE movement led by youth.

To provide continuous support for the youth to further drive the CE transition, STEAM Platform will provide training/mentoring to those who are willing to venture into entrepreneurship. Together with other industry and financing partners, STEAM Platform will foster the new generation of CE entrepreneurs and leaders.

2 Conclusions

Education plays a critical role in the realization of circular economy through youth leadership harnessing the potential of technologically equipped youth and their agile mindset. STEAM platform is an Asian youth-driven leadership platform towards smart circular economy built on three key pillars including knowledge convergence, skills & mindset, and entrepreneurship. STEAM Platform piloted a revolutionary module (GEN352) on smart circular economy, which is believed to be the first of its kind module containing range of different units including life cycle thinking, sustainable development goals, advanced robotics, smart manufacturing, digital factory, waste to materials/energy, Internet of Things (IoT), data analytics, sustainable financing, supply chain management, business model for sustainability, and other futuristic technologies enabling the smart circular economy. Particularly, the module was youth-focused in both teaching and learning. Considering the fact that 62% of global youth reside in Asia–Pacific, it is highly critical to empower and equip youth of this region to be the changemakers of the circular economy. In addition to provision of highly technological knowledge in story-telling manner, we equipped students with necessary soft skills including interpersonal skills, critical/analytical thinking, and strategic communication, which we believe are equally essential for developing holistic solutions. Furthermore, entrepreneurial skills including supply chain management & business model design, technology and solution comparative/competitive analysis, and business relevance to sustainability and circular economy concepts were taught. The findings of our investigation reveal that the content and methodology has been extremely helpful for students to learn about advanced technologies and gain skills through our unique workshops-style teaching. Interestingly, we adopted peer-to-peer learning and enforced experiential earning in our module and has proven effective. This GEN352 module, with a few suggested improvements, has demonstrated a holistic methodology to create youth leadership for transforming the current economy towards a smart circular economy.

3   Future Outlook

Circular Economy (CE) education and youth leadership is foreseen as the key pillars of CE success in the next decade or so. Sustainable Development Goals (SDGs) 2030 including "SDG10- Industry, Innovation and Infrastructure", "SDG11-Sustainable Cities and Communities", "SDG12-Responsible Consumption and Production", "SDG13-Climate Action" will gain significant attention among other SDGs, and education will be the primary determining factor for achieving SDGs 2030. Furthermore, CE education is expected to truly transform current recycling practices with increased recycling rate, and consumption/dumping behavior will greatly change with rising involvement of global youth. In order to increase recycling efforts (e.g., The EU is committed to increase recycling rate of waste from 28 to 75% by 2025 and 80% by 2030), the role of intensified education and youth empowerment will become critical. Undoubtedly, youth as one of the primary stakeholders will strengthen the CE efforts enabled by advanced technologies such as advanced robotics, artificial intelligence, biotechnology, nanotechnology, and Internet of Things (IoT). Rapid advancement in the next-generation technologies in emerging Asian economies with rising youth population is expected to take a prominent lead in driving CE. Moreover, the current value chains and supply chains will be dramatically disrupted by the youth equipped with CE knowledge, skills and entrepreneurial skills without compromising the future needs for building a sustainable world.

Questions

. Why is education so important for a circular economy?
. How youth leadership can help accelerate circular economy efforts?
. Why is active involvement of Asian youth needed to excel in CE?
. Why is the STEAM platform so critical for successful circular economy transformation led by youth?

5   Answers

1. Education facilitates the dissemination of subject knowledge, and it is at the core of every success factor. It triggers ideation and critical thinking for innovation. Circular economy needs continuous and accelerated innovation for further growth employing core knowledge of subject matter. Although different universities/organizations are actively contributing toward CE education for students and the general public, it requires development of a holistic strategy for better communication of CE.

2. According to a 2019 youth report by the United Nations (UN), approximately 1.2 billion people are aged 15–24 years globally, and it is projected to significantly grow over the period of next decade or so. The increased number of young people equipped with right knowledge and skills are expected to be the asset of the future world. As we progress in a circular economy, the role of youth will become critically defined to drive the world toward sustainability and CE.

3. Asia is considered to be the manufacturing hub of the world due to inexpensive labor. It significantly contributes toward global waste production leading to health and environmental complications and poor living standards, and Asia also accounts for 62% of global youth population. The greater number of youth, and massive investments toward waste utilization are expected to make Asian economies rapidly emerge in future, and youth equipped with technologically advanced skills will be required in Asia. Hence, involvement of Asian youth is particularly inevitable for excelling the region in CE.

4. STEAM platform's CE education module significantly contributed toward dissemination of knowledge related to CE with particular emphasis on advanced technologies, experiential learning, strategic communication, critical thinking, peer-to-peer learning, etc. Apart from creating a holistic CE education module, STEAM greatly impacted the global youth through its youth-centered program, which not only helped youth in behavioral change toward CE but also empowered them to be youth leaders for taking collective action.

References

1. Acaroglu, L. (2018). Retrieved February 11, 2020, from https://medium.com/disruptive-design/the-circular-classroom-a-free-toolkit-for-activating-the-circular-economy-through-experiential-64ffe1274b9c.
2. Anderberg, E., Nordén, B., & Hansson, B. (2009). Global learning for sustainable development in higher education: Recent trends and a critique. *International Journal of Sustainability in Higher Education.*
3. Andrews, D. (2015). The circular economy, design thinking and education for sustainability. *Local Economy, 30*(3), 305–315.
4. Asia, C.E. Retrieved March 11, 2020, from https://www.circulareconomyasia.org/.
5. Bakker, C., Wever, R., Teoh, C., & De Clercq, S. (2010). Designing cradle-to-cradle products: A reality check. *International Journal of Sustainable Engineering, 3*(1), 2–8.

6. Bocken, N. M., De Pauw, I., Bakker, C., & van der Grinten, B. (2016). Product design and business model strategies for a circular economy. *Journal of Industrial and Production Engineering, 33*(5), 308–320.
7. Club, C.E. (2019). Retrieved February 11, 2020, from https://www.circulareconomyclub.com/listings/education/.
8. D'Amato, D., Droste, N., Allen, B., Kettunen, M., Lähtinen, K., Korhonen, J., et al. (2017). Green, circular, bio economy: A comparative analysis of sustainability avenues. *Journal of Cleaner Production, 168,* 716–734.
9. Ellen Macarthur Foundation. (2019). Retrieved February 11, 2020, from https://www.ellenmacarthurfoundation.org/our-work/approach/learning.
10. Erkman, S. (1997). Industrial ecology: An historical view. *Journal of Cleaner Production, 5*(1–2), 1–10.
11. Fixfest. (2019). Retrieved February 11, 2020, from https://www.fixxfest.com/.
12. Garcia-Muiña, F. E., González-Sánchez, R., Ferrari, A. M., & Settembre-Blundo, D. (2018). The paradigms of Industry 4.0 and circular economy as enabling drivers for the competitiveness of businesses and territories: The case of an Italian ceramic tiles manufacturing company. *Social Sciences, 7*(12), 255.
13. Ghisellini, P., Cialani, C., & Ulgiati, S. (2016). A review on circular economy: The expected transition to a balanced interplay of environmental and economic systems. *Journal of Cleaner Production, 114,* 11–32.
14. Group, T.W.B. (2012). Retrieved February 11, 2020, from https://siteresources.worldbank.org/INTURBANDEVELOPMENT/Resources/336387-1334852610766/Chap3.pdf.
15. Huiyao, W. (2019). In 2020, Asian economies will become larger than the rest of the world combined—here's how, from https://www.weforum.org/agenda/2019/07/the-dawn-of-the-asian-century/.
16. Kirchherr, J., & Piscicelli, L. (2019). Towards an education for the circular economy (ECE): Five teaching principles and a case study. *Resources, Conservation and Recycling, 150,* 104406.
17. Kirchherr, J., Reike, D., & Hekkert, M. (2017). Conceptualizing the circular economy: An analysis of 114 definitions. *Resources, Conservation and Recycling, 127,* 221–232.
18. Kopnina, H. (2018). Circular economy and cradle to cradle in educational practice. *Journal of Integrative Environmental Sciences, 15*(1), 119–134.
19. Merli, R., Preziosi, M., & Acampora, A. (2018). How do scholars approach the circular economy? A systematic literature review. *Journal of Cleaner Production, 178,* 703–722.
20. Murray, M. (2019). Teaching circular economics. Retrieved February 11, 2020, from https://theecologist.org/2019/sep/27/teaching-circular-economics.
21. Nations, U. (2019).
22. Pham, P. (2017). How Asia has become the world's manufacturing hub. Retrieved February 11, 2020, from https://www.forbes.com/sites/peterpham/2017/11/13/how-asia-has-become-the-worlds-manufacturing-hub/#6c3f82ea22cf.
23. Platform, E.C.E.S. (2019). Retrieved February 11, 2020, from https://circulareconomy.europa.eu/platform/en/sector/education.
24. Platform, S. (2019). Retrieved March 11, 2020, from https://www.steamplatform.org/.
25. Sitra. (2019). Retrieved February 11, 2020, from https://www.sitra.fi/en/projects/circular-economy-teaching-levels-education/.
26. Union, E. (2019). Retrieved February 11, 2020, from https://ec.europa.eu/environment/circular-economy/.
27. Yuan, Z., Bi, J., & Moriguchi, Y. (2006). The circular economy: A new development strategy in China. *Journal of Industrial Ecology, 10*(1–2), 4–8.

Mr. Arslan Siddique is a Ph.D. candidate in Chemical Engineering at University of New South Wales (UNSW) Sydney, Australia. He is also a Casual Academic at UNSW and a Guest Researcher at Australian Nuclear Science and Technology Organization (ANSTO) Sydney. Before joining UNSW, he was the project manager and technology analyst at STEAM (Science, Technology, Engineering, Arts, Mathematics) platform King Mongkut's University of Technology Thonburi (KMUTT). He completed his master's degree by research majoring in Biotechnology from KMUTT's affiliated Excellent Center of Waste Utilization (ECoWaste). He is an enthusiastic researcher, technology analyst and technopreneur. He has acquired multidisciplinary research expertise ranging from molecular surface interactions, biogas, nano-biotechnology and microfluidics technology to aerobiology and thunderstorm asthma. He strongly advocates circular economy, technology startups, greener energy production, and youth leadership to assist in building a sustainable world. Moreover, he has been an active participant of Kecti program of Malmer Knowles Family Foundation (MKFF) USA, where he was declared excellent youth fellow. He trains youth on technology analysis, entrepreneurship, and youth leadership. His presence at various international conferences/events has inspired several young people across the globe to take collaborative action for potential contribution towards sustainability and circular economy employing advanced technologies.

Panitsara Nakseemok is passionate about sustainability (environmental, social and economic) and strongly motivated in acquiring new skills and knowledge to change the world. She holds an executive position at STEAM (Science, Technology Engineering, Arts, Mathematics)-an Asian youth driven leadership platform for smart circular economy, King Mongkut's University of Technology Thonburi (KMUTT), working as a program manager. She has acquired knowledge in advanced materials and manufacturing as well as smart sensors. She is also talented in design and trained in entrepreneurship a King Mongut's University of Technology Thonburi (KMUTT) Panitsara served as the financial secretary of The Federation o the Engineering Students of Thailand during her undergraduate years. She initiated the science student leadership club during her high school studies. In 2014, she joined the volunteer program to help the Pga K'nyau Tribal in the city of Maeramae in Tak province in developing farming projects to sustain basic survival and to improve the quality of life by providing education and power systems utilizing solar energy. She joined the volunteers for cleaning plastic waste at Bangsean beach damaged by irresponsible tourists. Panitsara holds a Bachelor o Engineering in Advanced Materials and Nanotechnology from Silpakorn University in Thailand. She is currently a 2nd year Master Degree student in Nanoscience and Nanotechnology Program in KMUTT faculty of science. Panitsara Nakseemo is passionate about sustainability (environmental, social and economic) and strongly motivated in acquiring new skill and knowledge to change the world. She holds an executive position at STEAM (Science, Technology, Engineering, Arts, Mathematics)-an Asian youth driven leadership platform for smart circular economy, King Mongkut's University of Technology Thonburi (KMUTT), working as a program manager. She has acquired knowledge in advanced materials and manufacturing as well as smart sensors. She is also talented in design and trained in entrepreneurship at King Mongut's University of Technology Thonburi (KMUTT). Panitsara served as the financial secretary of The Federation of the Engineering Students of Thailand during her undergraduate years. She initiated the science student leadership club during her high school studies. In 2014, she joined the volunteers program to help the Pga K'nyau Tribal in the city of Maeramad in Tak province in developing farming projects to sustain basic survival and to improve the quality of life by providing education and power systems utilizing solar energy. She joined the volunteers for cleaning plastic waste at Bangsean beach damaged by irresponsible tourists. Panitsara holds a Bachelor of Engineering in Advanced Materials and Nanotechnology from Silpakorn University in Thailand. She is currently a 2nd year Master Degree student in Nanoscience and Nanotechnology Program in KMUTT faculty of science.

Dr. Lerwen Liu specializes in business development (including strategic communication, partnership & marketing; fund raising; and stakeholder & project management) with technical expertise in fields of nanomaterials, additive manufacturing, solar cells, nanosatellites, AI and other emerging technologies with applications in functional/green materials & manufacturing, space, agribusiness, health care, environment and social empowerment. She is known to be one of the most connected people with networks in government, universities, industry/business sectors and NGOs in the Asia Pacific, Europe, USA and beyond.

Since 2014, Dr. Liu became a strong advocate of sustainability through innovation & entrepreneurship education and training focusing in Asian region. In May 2018, she co-founded the STEAM (Science, Technology, Engineering, Arts and Mathematics) Platform (www.steamplatform.org) which empowers the youth in Asia to accelerate the transformation of the circular economy and reach the SDGs2030 through convergence of knowledge and entrepreneurship.

Lerwen is a senior advisor in KMUTT (Bangkok) playing a leadership role in (1) streamlining strategy of education, research & innovation and social impact towards sustainable development goals, industry 4.0 and circular economy; (2) building strategic partnerships with the Asia Development Bank (ADB), United Nations, leading universities, government funding agencies and industries worldwide in developing solutions for reaching SDGs and circular economy; and (3) conducting strategic training to KMUTT students and staff preparing the change agents/leaders for the SDGs and circular economy.

She has taught graduate modules (Management of Technology Program) in NUS on Corporate Entrepreneurship and High-tech Business Model where she incorporates Sustainability and SDGs into business model design and goals. She is currently offering a novel general education module, GEN352 in KMUTT Bangkok, where students learn the introduction of Industry 4.0 (I4.0) and Circular Economy (CE); Life Cycle Assessment (LCA) & SDGs2030; and Business Model design for sustainable development.

Dr. Liu also specializes in policy, technical and market/impact assessment in emerging technologies including nanotechnology and AI as well as technologies enabling the circular economy. She is co-editing with Prof. Seeram Ramakrishna the 1st Textbook "An Introduction to Circular Economy" published by Springer Nature in 2020.

She has been an evaluation panelist of the Proof-of-Concept (POC) grant for the National Research Foundation (NRF) of Singapore. She has also been in the iCAN Technology Innovation Contest and Global iCAN (G-iCAN) committee and evaluation panel of iCAN international competition. She is a member of National Advisory Panel (NAP) of 1st Sustainable Business Awards Thailand.

Future Outlook

Lerwen Liu and Seeram Ramakrishna

Abstract Given the crisis humanity is facing today, this chapter urges all stakeholders to take coherent action today and tomorrow for the transition to a circular economy. To help readers visualize a circular economy of the future, it provides a scenario of a circular economy in a community where both biological and technical cycles are closed; renewable energy drives transportation, production, and consumption. The community take care of the health of themselves and their environment and practice 6 Rs (re-use, repair, refurbish, remanufacture, recycle and recover) along the life cycle of a product at personal and professional levels. Circular supply chain is mapped. The chapter further summarizes circular economy transition enabling factors such as life cycle thinking, materials passports, and ubiquitous digitization to become integral of industries and services. In addition, it addresses the challenges ahead and concludes on the importance of education for providing circular economy workforce.

Keywords Renewable energy · Circular economy · 6Rs · Stakeholders · Action · Circular economy · Transition · Biological cycle · Technical cycle

At the time of writing this book, the world population is close to eight billion people (https://www.worldometers.info/world-population/). Assuming a growth rate of one percent per year, about eighty million people per year are added to the total population. In other words, an equivalent of Germany's population is added to the world every year. As per the current projections, the world population will reach ten billion people in the year 2057. The World Health Organization (WHO) recorded that the urban population in 2014 accounted for 54% of the total global population, and the urbanization trend is projected to grow further in the coming decades (https://www.who.int/gho/urban_health/situation_tre

L. Liu (✉)
STEAM Platform, Circular Economy Accelerator, KX Innovation Center, 110/1 Krung Thonburi Road, Banglamphulang, Khlongsan, Bangkok 10600, Thailand
e-mail: lerwen67@gmail.com

S. Ramakrishna
National University of Singapore, Singapore, Singapore

© Springer Nature Singapore Pte Ltd. 2021
L. Liu and S. Ramakrishna (eds.), *An Introduction to Circular Economy*,
https://doi.org/10.1007/978-981-15-8510-4_30

nds/urban_population_growth_text/en/). The world economic forum forecasts that over the coming decades the rural to urban migration will continue to shoot up to six billion by 2050 (https://www.weforum.org/agenda/2019/09/mapped-the-dramatic-global-rise-of-urbanization-1950-2020/). The urbanized population will continue to shape the global economy via increased spending and consumption of products and services as they seek better comforts and happiness. In other words, the Earth needs to provide for more resources and accumulate waste and pollution, if the humanity to continue the current path of modern-society. The gravity of this situation is grasped by the Earth Overshoot Day calculated by an international research organization called the Global Footprint Network (https://www.overshootday.org/about-earth-overshoot-day/). Earth Overshoot Day is defined as the date when humanity's demand for ecological resources and services in a given year exceeds what Earth can regenerate in that year. In 2019, the Earth Overshoot Day was on July 29. In 1987, the Earth Overshoot Day was on October 23. In other words, with growing consumption the Earth Overshoot Day is getting shorter and shorter. Aforementioned facts clearly indicate that the humanity has no choice but to transition from the modern-society to new-modern society. Desired characteristics of the new-modern society encompass the visions of circular economy and sustainability development and growth. Simply put it is an economic system aimed at eliminating waste and the continual use of resources (https://en.wikipedia.org/wiki/Circular_economy). If we adopt the Ellen MacArthur Foundation definition of circular economy, it is a systemic approach to economic development designed to benefit businesses, society, and the environment (https://www.ellenmacarthurfoundation.org/explore/the-circular-economy-in-detail). A circular economy practitioner, Metabolic company describes it as a new economic model for addressing human needs and fairly distributing resources without undermining the functioning of the biosphere or crossing any planetary boundaries (https://www.metabolic.nl/about/our-mission/). In other words, in the new-modern society resources and consumption are not constraints for growth and higher standards of living, thus suited for the ever-growing world population.

Given the aforementioned background, what would be the future outlook? A circular economy transformation requires all stakeholders to act coherently toward the same direction. Figure 1 represents stakeholders involved in the circular economy. Table 1 summarizes the action of different stakeholders today and tomorrow for a circular economy transformation.

A circular economy transition is undergoing three streams:

1. Continue to **Improve clean production through reduction in resource extraction, energy consumption, wastage, and emission**.
2. **Practice 4Rs (Refuse, Reduce, Reuse, and Repair) at the consumer site and 6Rs (Reuse, Repair, Refurbish, Remanufacture, Recycle, and Recover) at the production site**.
3. **Redesign** product, provide function to **Serve** the needs of consumer.

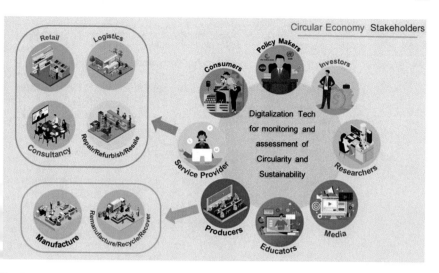

Fig. 1 Circular economy stakeholders (STEAM Platform)

The economic transition provides business opportunities in all sectors along the life cycle of a product. It also demands R&D innovations and drives business model innovations. To accelerate the transition, it needs regulations and monitoring and assessment. Most importantly, it requires life cycle thinking and circularity/sustainability mindset that drives the power of consumers as well as the stakeholders' action.

Products (from materials, component, module to system) and manufacturing processes are designed to reduce/eliminate waste. They are to retain the highest value for the longest time. Components and materials are reused, repurposed, remanufactured, and recycled. Maximum usage of energy, water, and materials are from renewable sources. Producers offer services and establish close relationships with consumers to build a business that is environmentally, socially, and economically sustainable. Consumers, driven by life cycle thinking and sustainability mindset, practice sustainable consumption achieving zero waste and choosing sustainable products. Digital technologies are adopted to provide continuous monitoring and assessment and guide both consumers and producers to ensure circularity and sustainability at local and global levels.

To help readers visualizing circularity of the future, Fig. 2 presents a scenario of a circular community where energy sources are renewable including solar, wind, and hydro managed by a community micro-grid; mining and production sites are practicing circularity with remanufacturing, recycle and recovery processes built in; consumers are practicing sustainable consumption with zero waste through composting organic food and packaging materials to fertilize the community gardens/farms; consumers receive communication, energy, and transport services from providers who will practise reuse, repair, refurbish, and resale. See Fig. 3 for the circular supply chain representation mapped in Fig. 2.

Table 1 Summary of circular economy stakeholders action today and tomorrow

Type		Today working action	Tomorrow action
Policy-makers		1. CE policy framework 2. Regulations-EPR 3. Promote best practices	1. Priortize funding programs to support PPP for circularity solutions and sustainable financing 2. CE entrepreneurship ecosystem building 3. Adopting digital technology for circularity/sustainabilit monitoring and assessment 4. iEPR implementation 5. Regional and global CE coordination
Investors		Sustainable/ESG financing	Scaling up circular and sustainable financing
Researchers		1. Open innovation platform 2. Green Chemistry for processing, recycling, and recovery 3. Bioeconomy focusing on bio-base materials 4. Energy efficiency technology and solutions	1. Prioritize R&D areas and industry partnership along the value chain focus on scalable solutions for circularity and sustainability 2. Circular product design 3. Applying digital tech and AI
Educators		CE experiential modules, programs both online and offline	1. LCA thinking is imbedded in all education program 2. Training CE educators 3. Entrepreneurship training for circularity and sustainability
Producers	Large and small corp	Industry symbiosis, environmental compliance and EPR, remanufacture, recycle, recover	1. Product service provider with zero waste by design, remanufacturing, recycling, and recovery 2. Workforce upskills 3. Partnership with the circularity ecosystem 4. Sustainability/ESG compliance in every part of the value chain

(continued

Table 1 (continued)

Type		Today working action	Tomorrow action
Service provider	Logistics	Green packaging, decentralization	1. Zero emission transportation 2. Adoption of digital technology to achieve transparency and efficiency 3. Minimizing logistics through onsite production and consumption
	Retail	Resale	Digital "farmer's market" platform bringing producer and consumers together
	End of product life	Repair, repurpose, refurbish	No end of product life
	Waste management	Sustainable waste management	1. Waste minimization included in product design and processing 2. Waste composting onsite of community garden 3. Repair and refurbish of used products
	Consultancy	1. CE advisory on strategy and implementation 2. Developing circularity indicators for monitoring and assessment	Deploy digital technology such as IoT, big data analytics, and digital twins for monitoring and assessment
Consumers		1. Refuse, reduce, reuse (3R) 2. Green Choice, Minimize waste	Making sustainability personal: 3R +Repair, practising gardening and community farm, become a sustainable and zero waste consumer
Media		1. Reporting on all circularity practices at mainstream and social media 2. Build relationships with stakeholders to promote circularity	1. Focus on sustainability broadcasting 2. Promote harmony between nature and human society 3. Featuring sustainable business practices

Fig. 2 Circular community with zero waste and zero emission (STEAM Platform)

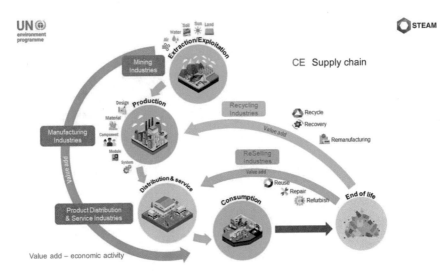

Fig. 3 Circular supply chain representation conceptualized by Dr. Mushtaq Memon (UNEP) and Dr. Lerwen LIU (STEAM Platform)

Moving ahead, life cycle thinking, materials passports, and ubiquitous digitization are integral of industries and services. More pervasive use of life cycle assessment, life cycle costing, and symbiosis methodologies to deeply understand the resources flow and trade-offs to implement policy and leadership interventions that would drive the new-modern society transition. Bottlenecks for the transition include lack of quality information and data, which are often proprietary and business confidential; transparency in regulations and incentives implemention; maturity of governing international and national standards and frameworks; and risk attitude toward new business models. Such challenges will be overcome as there is growing awareness from the public, pressure from the NGOs and more importantly humanity is capable of innovation in the face of necessity for sustainable survival.

Last, not the least, fostering the next generation of circular economy workforce is most critical for driving humanity toward a sustainable future. We hope this book inspires youth leadership with knowledge and entrepreneurship to transform our society and economy toward sustainability.

Lerwen Liu is Founding Director of STEAM Platform (www.steamplatform.org) and

Director of Circular Economy Accelerator, KX Innovation Centre, King Mongkut's University of Technology Thonburi (KMUTT), Thailand.

Dr. Lerwen Liu specializes business development and education in the emerging technologies, circular economy and sustainability. She focuses on strategic development, assessment, and support of emerging technologies including nanotechnology and Artificial Intelligence with applications in all sectors. She has 20 years of practices in global business development in strategic partnership & communication and marketing. She has worked in both the developed and developing world focusing on youth leadership and entrepreneurship development toward sustainability.

She was a founding secretary of Asia Nano Forum supporting strategic development of Nanotechnology in 17 countries in the Asia Pacific region. She founded the STEAM Platform in 2018 focusing on youth empowerment with convergent of STEM knowledge, strategic communication skills, and entrepreneurship for SDGs2030, Circular Economy and Industry 4.0. She has been an invited expert in nanotechnology, innovation, entrepreneurship, and circular economy by different agencies in the United Nations, Islamic Development Bank, Asia Development Bank (ADB), and other government bodies. She has co-founded a number of emerging technology start-ups and mentored hundreds of youth leaders in Asia.

She has developed and taught innovative modules at National University of Singapore and King Mongkut's University of Technology Thonburi on topics of nanotechnology, innovation, entrepreneurship, and sustainability. She is a co-editor of the 1st Textbook "An Introduction to Circular Economy" published by Springer Nature in June 2020. She is the editor of "Emerging Nanotechnology Power: Nanotechnology R& D and Business Trends In The Asia Pacific Rim published by World Scientific in 2009.

Lerwen holds a Ph.D. in Physics from UNSW Australia. She is an Australian citizen, currently based in Bangkok.

Professor Seeram Ramakrishna *FREng, Everest Chair*
(https://www.eng.nus.edu.sg/me/staff/ramakrishna-seeram/), is
among the top three impactful authors at the National University of Singapore, NUS (https://academic.microsoft.com/ins
titution/165932596). NUS is ranked among the top five best
global universities for engineering in the world (https://www.
usnews.com/education/best-global-universities/engineering). He
is the Chair of Circular Economy Taskforce. He is a member of
Enterprise Singapore's and ISO's Committees on ISO/TC323
Circular Economy and WG3 on Circularity. He also the Chair
of Sustainable Manufacturing TC at the Institution of Engineers
Singapore and a member of standards committee of Singapore
Manufacturing Federation (http://www.smfederation.org.sg).
He is an advisor to the Ministry of Sustainability & Environment—National Environmental Agency's CESS events, (https://
www.cleanenvirosummit.sg/programme/speakers/professor-
seeram-ramakrishna; https://bit.ly/catalyst2019video; https://
youtube.com/watch?v=ptSh_1Bgl1g). European Commission
Director-General for Environment, Excellency *Daniel Calleja*
Crespo, said, "Professor Seeram Ramakrishna should be
praised for his personal engagement leading the reflections
on how to develop a more sustainable future for all", in his
foreword for the Springer Nature book on Circular Economy
(ISBN: 978-981-15-8509-8). He is a member of UNESCO's
Global Independent Expert Group on Universities and the 2030
Agenda (EGU2030). He is the Editor-in-Chief of the Springer
NATURE Journal Materials Circular Economy—Sustainability
(https://www.springer.com/journal/42824). He is an Associate
Editor of eScience journal (http://www.keaipublishing.com/en.
journals/escience/editorial-board/). He is an opinion contributo
to the Springer Nature Sustainability Community (https://su
tainabilitycommunity.springernature.com/users/98825-seeram-
ramakrishna/posts/looking-through-covid-19-lens-for-a-sustai
nable-new-modern-society). He teaches ME6501 Materials and
Sustainability course (https://www.europeanbusinessreview
com/circular-economy-sustainability-and-business-opportuni
ties/). He also mentors Integrated Sustainable Design ISD510
project students. Microsoft Academic ranked him among
the top 25 authors out of three million materials researcher
worldwide based on H-index (https://academic.microsoft.com
authors/192562407). He is named among the World's Mos
Influential Minds (Thomson Reuters) and World's Highly
Cited Researchers (Clarivate Analytics). Listed among the top
three scientists of the world as per the Stanford University
researcher study on career-long impact of researchers or c-score
(https://drive.google.com/file/d/1bUJrvurVVBbxSl9eFZRSHFi
f7tt30-5U/view). He is an Impact Speaker at the University
of Toronto, Canada Low Carbon Renewable Materials Cente
(https://www.lcrmc.com/). He is a judge for the Mohamme
Bin Rashid Initiative for the Global Prosperity (https://www
facebook.com/Make4Prosperity/videos/innovation-inclusive-
trade/479503539339143/). He advises technology companie
with sustainability vision such as TRIA (www.triabio24.com)

Ceentek (https://ceentek.com/), Green Li-Ion (www.Greenli-ion.com) and InfraPrime (https://www.infra-prime.com/vis ion-leadership). He is a Vice-President of Asian Polymer Association (https://www.asianpolymer.org/committee.html). He is a Founding Member of Plastics Recycling Association of Singapore (PRAS). His senior academic leadership roles include University Vice-President (Research Strategy), Dean of Faculty of Engineering; Director of NUS Enterprise and Founding Chairman of Solar Energy Institute of Singapore (http://www.seris.nus.edu.sg/). He is an elected Fellow of UK Royal Academy of Engineering (FREng), Singapore Academy of Engineering and Indian National Academy of Engineering. He received PhD from the University of Cambridge, UK, and The TGMP from the Harvard University, USA.

Correction to: Circular Economy Business Models and Practices

Anna Itkin

Correction to:
Chapter "Circular Economy Business Models
and Practices" in: L. Liu and S. Ramakrishna (eds.),
An Introduction to Circular Economy,
https://doi.org/10.1007/978-981-15-8510-4_22

The original version of the book was inadvertently published with incorrect figure captions and without figure numbers in Chapter 22. Figure captions 1, 4 and 5 have been replaced with the correct figure captions and numbers in the updated version. Also, author bio and photo has included in the updated version.

The updated version of this chapter can be found at
https://doi.org/10.1007/978-981-15-8510-4_22

Printed in the United States
by Baker & Taylor Publisher Services